Studies in Advanced Mathematics

CR Manifolds and the Tangential Cauchy–Riemann Complex

Studies in Advanced Mathematics

Volumes in the Series

Real Analysis and Foundations, *Steven G. Krantz*

CR Manifolds and the Tangential Cauchy–Riemann Complex, *Albert Boggess*

Elementary Introduction to the Theory of Pseudodifferential Operators, *Xavier Saint Raymond*

Fast Fourier Transforms, *James S. Walker*

Measure Theory and Fine Properties of Functions, *L. Craig Evans and Ronald Gariepy*

ALBERT BOGGESS
Texas A & M University

CR Manifolds and the Tangential Cauchy–Riemann Complex

CRC PRESS
Boca Raton Ann Arbor Boston London

Library of Congress Cataloging-in-Publication Data

Catalog record is available from the Library of Congress

This book was formatted with LaTeX by Archetype Publishing Inc., P.O. Box 6567, Champaign, IL 61821.

Direct all inquiries to CRC Press, Inc., 2000 Corporate Blvd., N.W., Boca Raton, Florida, 33431.

International Standard Book Number 0-8493-7152-X

Printed in the United States of America 1 2 3 4 5 6 7 8 9 0

This book is dedicated to my parents

Albert and Nancy

who have always set an example worth emulating

Contents

	Contents	*vii*
	Introduction	*xiii*
	PART I: PRELIMINARIES	*1*
1	*Analysis on Euclidean Space*	2
1.1	Functions	2
1.2	Vectors and vector fields	7
1.3	Forms	9
1.4	The exterior derivative	12
1.5	Contractions	15
2	*Analysis on Manifolds*	*17*
2.1	Manifolds	17
2.2	Submanifolds	19
2.3	Vectors on manifolds	23
2.4	Forms on manifolds	26
2.5	Integration on manifolds	29
3	*Complexified Vectors and Forms*	*39*
3.1	Complexification of a real vector space	39

3.2	Complex structures	41
3.3	Higher degree complexified forms	45
4	*The Frobenius Theorem*	*51*
4.1	The real Frobenius theorem	51
4.2	The analytic Frobenius theorem	56
4.3	Almost complex structures	58
5	*Distribution Theory*	*61*
5.1	The spaces $\mathcal{D}'$ and $\mathcal{E}'$	61
5.2	Operations with distributions	65
5.3	Whitney's extension theorem	71
5.4	Fundamental solutions for partial differential equations	74
6	*Currents*	*79*
6.1	Definitions	79
6.2	Operations with currents	84
	PART II: CR MANIFOLDS	*95*
7	*CR Manifolds*	*97*
7.1	Imbedded CR manifolds	97
7.2	A normal form for a generic CR submanifold	103
7.3	Quadric submanifolds	111
7.4	Abstract CR manifolds	120
8	*The Tangential Cauchy–Riemann Complex*	*122*
8.1	Extrinsic approach	122
8.2	Intrinsic approach to $\bar{\partial}_M$	130
8.3	The equivalence of the extrinsic and intrinsic tangential Cauchy–Riemann complexes	134
9	*CR Functions and Maps*	*140*
9.1	CR functions	140
9.2	CR maps	149
10	*The Levi Form*	*156*
10.1	Definitions	156
10.2	The Levi form for an imbedded CR manifold	159

10.3	The Levi form of a real hypersurface	163
11	*The Imbeddability of CR Manifolds*	*169*
11.1	The real analytic imbedding theorem	169
11.2	Nirenberg's nonimbeddable example	172
12	*Further Results*	*179*
12.1	Bloom–Graham normal form	179
12.2	Rigid and semirigid submanifolds	183
12.3	More on the Levi form	185
12.4	Kuranishi's imbedding theorem	187
12.5	Nongeneric and non-CR manifolds	187
	PART III: THE HOLOMORPHIC EXTENSION OF CR FUNCTIONS	*189*
13	*An Approximation Theorem*	*191*
14	*The Statement of the CR Extension Theorem*	*198*
14.1	Lewy's CR extension theorem for hypersurfaces	198
14.2	The CR extension theorem for higher codimension	200
14.3	Examples	202
15	*The Analytic Disc Technique*	*206*
15.1	Reduction to analytic discs	207
15.2	Analytic discs for hypersurfaces	208
15.3	Analytic discs for quadric submanifolds	210
15.4	Bishop's equation	214
15.5	The proof of the analytic disc theorem for the general case	221
16	*The Fourier Transform Technique*	*229*
16.1	A Fourier inversion formula	230
16.2	The hypoanalytic wave front set	237
16.3	The hypoanalytic wave front set and the Levi form	244
17	*Further Results*	*251*
17.1	The Fourier integral approach in the nonrigid case	251
17.2	The holomorphic extension of CR distributions	254
17.3	CR extension near points of higher type	257

17.4 Analytic hypoellipticity 260

PART IV: SOLVABILITY OF THE TANGENTIAL CAUCHY–RIEMANN COMPLEX *263*

18 *Kernel Calculus* *265*
18.1 Definitions 265
18.2 A homotopy formula 272

19 *Fundamental Solutions for the Exterior Derivative and Cauchy–Riemann Operators* *277*
19.1 Fundamental solutions for d on $\mathbb{R}^N$ 278
19.2 Fundamental solutions for $\bar{\partial}$ on $\mathbb{C}^n$ 281
19.3 Bochner's global CR extension theorem 291

20 *The Kernels of Henkin* *294*
20.1 A general class of kernels 294
20.2 A formal identity 297
20.3 The solution to the Cauchy–Riemann equations on a convex domain 299
20.4 Boundary value results for Henkin's kernels 303

21 *Fundamental Solutions for the Tangential Cauchy–Riemann Complex on a Convex Hypersurface* *312*
21.1 The first fundamental solution for the tangential Cauchy–Riemann complex 312
21.2 A second fundamental solution to the tangential Cauchy–Riemann complex 317

22 *A Local Solution to the Tangential Cauchy–Riemann Equations* *327*

23 *Local Nonsolvability of the Tangential Cauchy–Riemann Complex* *334*
23.1 Hans Lewy's nonsolvability example 334
23.2 Henkin's criterion for local solvability at the top degree 337

24 *Further Results* 342
24.1 More on the Bochner–Martinelli kernel 342
24.2 Kernels for strictly pseudoconvex boundaries 345
24.3 Further estimates on the solution to $\overline{\partial}_M$ 348
24.4 Weakly convex boundaries 348
24.5 Solvability of the tangential Cauchy–Riemann complex in other geometries 349

Bibliography 354

Notation 359

Index 361

Introduction

The theory of complex manifolds dates back many decades so that its origins are considered classical even by the standards of mathematicians. Consequently, there are many fine references on this subject. By contrast, the origins of the theory of CR manifolds are much more recent even though this class of manifolds contains very natural objects of mathematical study (for example, real hypersurfaces in complex Euclidean space). The first formal definition of the tangential Cauchy–Riemann complex did not appear until the mid 1960s with the work of Kohn and Rossi [KR]. Since then, CR manifolds and the tangential Cauchy–Riemann complex have been extensively studied both for their intrinsic interest and because of their application to other fields of study such as partial differential equations and mathematical physics. The purpose of this book is to define CR manifolds and the associated tangential Cauchy–Riemann complex and to discuss some of their basic properties. In addition, we shall sample some of the important recent developments in the field (up to the early 1980s).

In the last two decades, research on the subject of CR manifolds has branched into many areas. Two of these areas that are of interest to us are (1) the holomorphic extension of CR functions (solutions to the homogeneous tangential Cauchy–Riemann equations) and (2) the local solvability or nonsolvability of the tangential Cauchy–Riemann complex. The first area started in the 1950s when Hans Lewy [L1] showed that under certain convexity assumptions on a real hypersurface in $\mathbb{C}^n$, CR functions locally extend to holomorphic functions. Over the years, many refinements have been made to this CR extension theorem so that it now includes manifolds of higher codimension with weaker convexity assumptions. The second area started in the 1960s with the work of Kohn. He used a Hilbert space ($\mathcal{L}^2$) approach to construct solutions to the tangential Cauchy–Riemann complex on the boundary of a strictly pseudoconvex domain (except at top degree). Later, Henkin developed integral kernels to represent solutions to the tangential Cauchy–Riemann equations.

A closely related topic is the nonsolvability of certain systems of partial differential equations. In the 1950s, Hans Lewy constructed an example of a partial differential equation with smooth coefficients that has no locally defined smooth solution. In particular, he showed that "C^∞" cannot replace "real analytic" in the statement of the Cauchy–Kowalevsky theorem. Lewy's example is closely related to the tangential Cauchy–Riemann equations on the Heisenberg group in $\mathbb{C}^2$. His example illustrates that the tangential Cauchy–Riemann complex on a strictly pseudoconvex boundary is not always solvable at the top degree. Later, Henkin developed a criterion for solvability of the tangential Cauchy–Riemann complex at the top degree.

The first half of this book contains general information on the subject of CR manifolds (Part II) and the prerequisites from real and complex analysis (Part I). In Parts III and IV, we develop the subjects of CR extension and the solvability of the tangential Cauchy–Riemann complex.

This book is not a treatise. We do not discuss the $\mathcal{L}^2$-approach to the global solvability of the tangential Cauchy–Riemann equations. This material is contained in Folland and Kohn's book [FK] and our work certainly could not offer any improvements. Instead, the integral kernel approach of Henkin is presented. The local theory dealing with points of higher type (points where the Levi form vanishes) is not presented in detail. The theory of points of higher type is too immature or too complicated for inclusion in a book at this time. Instead, the end of each part contains a chapter entitled *Further Results* where some of the recent literature on function theory near points of higher type and other topics are surveyed with few proofs given.

Each writer has his own peculiar style and tastes and this author is no different. The reader will notice that I favor the concrete over the abstract. This may offend some of the purists in the audience but I offer no apologies. I firmly believe that a student learns much more by getting his or her hands dirty with some analysis rather than by merely manipulating abstract symbols. In this book, abstract concepts are introduced after some motivation with the concrete situation.

It is hoped that the audience for this book will include researchers in several complex variables and partial differential equations along with graduate students who are beyond their first year or two of graduate study. The reader should be familiar with advanced calculus, real and complex analysis, and a little functional analysis (at least enough so that he or she does not faint at the sight of a Banach or Fréchet space). Although this book cannot reinvent the wheel, many of the prerequisites for reading Parts II through IV are given in Part I. We start with a discussion of vectors and forms, both in the Euclidean and manifold setting. A proof of Stokes' theorem is given since it is such a basic tool used throughout the book. Proofs of the smooth and real analytic versions of the Frobenius theorem are given since these theorems are used in the imbedding of CR manifolds. At the end of Part I, a rapid course in the theory of distributions and currents is given. This material will be essential in Part IV. There are other elementary topics that are not included in Part I. These include the existence and uniqueness theorem for ordinary differential equations and the Cauchy–Kowalevsky theorem for partial differential equations. Even though these topics are no more advanced than other topics covered in Part I (for example, Stokes' theorem), they are not as frequently used in this book and therefore we only give references. Surprisingly little theory from several complex variables is used. For most of the book, the reader needs only to be familiar with the definition and basic properties of holomorphic functions of several complex variables. However, the book will certainly be more meaningful to someone who has a further background in several complex variables. The Newlander–Nirenberg theorem is the most advanced topic from several complex variables that is used in the book. It is only used in the discussion of Levi flat CR manifolds. A proof of the imbedded version of the Newlander–Nirenberg theorem is provided for the reader who wishes to restrict his or her attention to imbedded Levi flat CR manifolds.

Part II covers general information about CR manifolds and the associated tangential Cauchy–Riemann complex. We start with the definitions of imbedded and abstract CR manifolds. In addition, we present a normal form that gives a convenient description of an imbedded CR manifold in local coordinates. Next, we introduce the tangential Cauchy–Riemann complex. For an imbedded CR manifold, an extrinsic approach is given that makes use of the ambient complex structure on $\mathbb{C}^n$. For an abstract CR manifold, an intrinsic approach is given that makes no use of any ambient complex structure (since none exists). In the case of an imbedded CR manifold, these two approaches are technically different, but we show they are isomorphic. Our approach to the tangential Cauchy–Riemann complex makes use of a Hermitian metric. We also mention a more invariant definition of the tangential Cauchy–Riemann complex that does make use of a metric, but this approach is not emphasized because calculations usually require a choice of a metric. CR functions and CR maps are then introduced. We prove Tomassini's theorem [Tom], which states that a real analytic CR function on an imbedded real analytic CR manifold is locally the restriction of an ambiently defined holomorphic function. This theorem does not hold for the class of smooth CR functions. However, we do show that a smooth CR function is always the restriction of an ambiently defined function that satisfies the Cauchy–Riemann equations to infinite order on the given CR submanifold. Next, we introduce the Levi form, which is the complex analysis version of the second fundamental form from differential geometry. An extensive analysis of the Levi form for the case of a hypersurface is given. In particular, we show the relationship between the Levi form and the second fundamental form. We then show that any real analytic CR manifold can be locally imbedded as a CR submanifold of $\mathbb{C}^n$. The smooth version of this theorem is false in view of Nirenberg's counter example [Nir], which is also given. In the chapter entitled *Further Results* we discuss some related results such as the Bloom–Graham normal form for CR submanifolds of higher type, rigid and semi-rigid CR structures, and Kuranishi's imbedding theorem. Most of these results are presented without proof.

Part III discusses the local holomorphic extension of CR functions from an imbedded CR manifold. We start with an approximation theorem of Baouendi and Treves [BT1], which states that CR functions can be locally approximated by entire functions. Their theorem is more general but we restrict our focus to CR functions to simplify the proof. Next, we state the CR extension theorem, which is a generalization of Hans Lewy's hypersurface theorem alluded to above. In addition, the convexity assumptions of this theorem are discussed and examples are given. We present two techniques for the proof of this theorem. Both of these techniques are used in today's research problems and thus these techniques are as important as the CR extension theorem. The first technique involves the use of analytic discs and it was originally developed by Lewy [L1] and Bishop [Bi]. This technique together with the approximation theorem yields an easy proof of Hans Lewy's CR extension theorem for hypersurfaces. An explicit proof is also given in the case of a quadric submanifold of higher codimension. The proof for the general case requires an analysis of the solution of a nonlinear integral equation involving the Hilbert Transform (Bishop's equation). The second, more recent, technique involves a modified Fourier transform approach due to Baouendi and Treves. The idea here is to obtain the holomorphic extension of a given CR function from the Fourier inversion formula — suitably modified for CR manifolds. This technique is applicable to CR distributions and points of higher type. However, to avoid technical complications, we give the details of this technique only for the case of smooth CR functions on a type two CR submanifold. Some of the extensions of this technique to CR distributions are discussed at the end of Part III in the chapter entitled *Further Results*.

Part IV deals with the solvability and nonsolvability of the tangential Cauchy–Riemann complex on a strictly pseudoconvex hypersurface in $\mathbb{C}^n$. The approach taken here involves Henkin's integral kernels although we use the notation and kernel calculus set down by Harvey and Polking [HP]. We give two fundamental solutions for the Cauchy–Riemann complex on $\mathbb{C}^n$ — the Bochner–Martinelli kernel [Boc] and the Cauchy kernel on a slice. These kernels together with the kernels of Henkin yield a solution to the Cauchy–Riemann equations on a strictly convex domain in $\mathbb{C}^n$. Furthermore, these kernels provide an easy proof of Bochner's theorem, which states that a CR function on the boundary of a bounded domain with smooth boundary globally extends to the inside as a holomorphic function. Next, a global integral kernel solution for the tangential Cauchy–Riemann complex is given for a strictly convex hypersurface. These kernels are then modified to yield Henkin's [He3] local solution to the tangential Cauchy–Riemann equations. We then present Henkin's criteria for local solvability of the tangential Cauchy–Riemann complex at the top degree. Results on the solvability of the tangential Cauchy–Riemann complex on hypersurfaces under other geometric hypotheses are given in the chapter entitled *Further Results*.

My point of view in mathematics has been influenced by a number of people whom I have the pleasure to thank. First, I owe a lot to my thesis advisor, John Polking. He along with Reese Harvey has shaped my mathematical development since my early graduate school years at Rice University. Even though I have never met him, G. M. Henkin has provided a lot of inspiration for much of my work. Other mathematicians who have influenced my mathematical point of view include Salah Baouendi, Al Taylor, Dan Burns, and Alex Nagel. The reviewers did an excellent job of finding errors and making helpful suggestions. I also wish to thank Steve Krantz (who initially encouraged me to write this book) and the rest of the editorial staff at CRC Press for having the confidence in me to complete this project.

I wish to thank Texas A&M for support during the preparation of most of this project. I did the final editing while visiting Colorado College and I want to thank their mathematics department for their hospitality during my visit.

In addition, I wish to thank Robin Campbell, who typed portions of this manuscript and answered many of my questions concerning TEX. I also wish to thank my son for putting up with me during the preparation of this manuscript. Finally, I wish to thank Steve Daniel and the rest of the Aggieland Paddle club for convincing me that from time to time, I need a break from the book writing to go whitewater kayaking.

Al Boggess, October 1990
Colorado Springs, CO

Part I

Preliminaries

In this first part, we provide most of the prerequisites for reading the rest of the book. We start with a review of certain aspects of function theory, vectors, vector fields, and differential forms on Euclidean space. These concepts are then defined in the context of manifolds. Proofs are given for Stokes' theorem and its corollaries — Green's formula and the divergence theorem. A proof of the Frobenius theorem is then given. The real analytic version of this theorem is also discussed since it will be used for the imbedding theorem for real analytic CR manifolds in Part II. We discuss distribution theory as applied to partial differential equations. Fundamental solutions for the Laplacian on $\mathbb{R}^N$ and the Cauchy–Riemann equations in one complex variable are given. They are used in Part IV, where we discuss fundamental solutions to the Cauchy–Riemann equations on $\mathbb{C}^n$ and their analogue on a real hypersurface of $\mathbb{C}^n$ — the tangential Cauchy–Riemann equations. These systems of partial differential equations act on differential forms. Therefore we shall need a distribution theory for differential forms, i.e., the theory of currents. This and related topics are reviewed at the end of Part I.

Excellent references are available for all of the topics in Part I. These include Spivak's volumes on differential geometry [Sp], Krantz's book [Kr] or Hörmander's [Ho] for several complex variables, Yosida's book [Y] for functional analysis and distribution theory (see also Schwartz [Sch]), Federer's book [Fe] for geometric measure theory, and John's [Jo] or Folland's book [Fo] for partial differential equations.

1

Analysis on Euclidean Space

Here, we discuss some function theory and define the notions of vectors and forms on Euclidean space.

1.1 Functions

There are several classes of functions we shall use. For an open set Ω in $\mathbb{R}^N$, let

$$\begin{aligned}
\mathcal{C}^k(\Omega) &= \text{the space of } k\text{-times continuously differentiable real- or complex-valued functions on } \Omega, \\
\mathcal{E}(\Omega) &= \text{the space of infinitely differentiable real- or complex-valued functions on } \Omega, \\
\mathcal{D}(\Omega) &= \text{the space of elements of } \mathcal{E}(\Omega) \text{ with compact support.}
\end{aligned}$$

We shall make use of a special class of mollifier functions $\{\chi_\epsilon; \epsilon > 0\} \subset \mathcal{D}(\Omega)$. This class is defined as follows. Let

$$\chi(x) = \begin{cases} \exp\left(\frac{-1}{1-|x|^2}\right) & \text{for } x \in \mathbb{R}^N \text{ with } |x| < 1 \\ 0 & \text{for } x \in \mathbb{R}^N \text{ with } |x| \geq 1 \end{cases}$$

and let

$$\chi_\epsilon(x) = \frac{\chi(x/\epsilon)}{\|\chi\|_{\mathcal{L}^1(\mathbb{R}^N)}} \epsilon^{-N}.$$

Each χ_ϵ is smooth. Here, $\|\cdot\|_{\mathcal{L}^1(\mathbb{R}^N)}$ denotes the usual $\mathcal{L}^1$-norm of a function.

The following properties can easily be shown:

(i) $\chi_\epsilon(x) = 0$ if $|x| \geq \epsilon$

(ii) $\int_{x \in \mathbb{R}^N} \chi_\epsilon(x)dx = 1$.

These two properties allow the construction of cutoff functions as described in the following lemma.

LEMMA 1

Given a compact subset K of an open set $\Omega \subset \mathbb{R}^N$, there is function ϕ belonging to $\mathcal{D}(\Omega)$ with $\phi \equiv 1$ on a neighborhood of K.

PROOF We first choose a compact set $K_1 \subset \Omega$ so that K is contained in the interior of K_1. Let

$$I_{K_1}(x) = \begin{cases} 1 & \text{if } x \in K_1 \\ 0 & \text{if } x \notin K_1 \end{cases}.$$

Choose $\epsilon > 0$ small enough so that 2ϵ is less than the smaller of the distance between K and $\Omega - K_1$ and the distance between K_1 and $\mathbb{R}^N - \Omega$.

Let

$$\phi = \chi_\epsilon * I_{K_1}.$$

Here, $*$ is the usual convolution operator in $\mathbb{R}^N$, so

$$\phi(x) = \int_{y \in \mathbb{R}^N} \chi_\epsilon(x - y) I_{K_1}(y) dy.$$

Since χ_ϵ is smooth, clearly ϕ is smooth. Property (i) for χ_ϵ shows that ϕ vanishes outside an ϵ-neighborhood of K_1. So ϕ has compact support in Ω by our choice of ϵ. Property (ii) for χ_ϵ and our choice of ϵ imply that $\phi \equiv 1$ on an ϵ-neighborhood of K. Therefore, ϕ is our desired cutoff function. ∎

As an immediate corollary, we can construct partitions of unity, as described in the following lemma.

LEMMA 2

Suppose U_j is an open subset of $\mathbb{R}^N$ for $j = 1, \ldots, m$. Suppose K is a compact set in $\mathbb{R}^N$ with $K \subset U_1 \cup \cdots \cup U_m$. Then there is a collection of functions $\phi_1, \ldots, \phi_m$ such that

(i) $\phi_j \in \mathcal{D}(U_j), \qquad 1 \leq j \leq m$

(ii) $0 \leq \phi_j \leq 1, \qquad 1 \leq j \leq m$

(iii) $\sum_{j=1}^{m} \phi_j \equiv 1$ *on a neighborhood of K.*

Often, the collection $\{\phi_j\}$ is called a partition of unity subordinate to the cover $\{U_j\}$.

PROOF First, we find open subsets $U_1', \ldots, U_m'$ with $\overline{U}_j' \subset U_j$ so that $K \subset U_1' \cup \cdots \cup U_m'$. Next, we choose cutoff functions $\psi_j \in \mathcal{D}(U_j)$ with $\psi_j \equiv 1$ on a neighborhood of $\overline{U}_j'$. Then let

$$\phi_j(x) = \left(\frac{\psi_j(x)}{\sum_{k=1}^{m} \psi_k(x)} \right) \cdot \phi(x) \qquad 1 \leq j \leq m$$

where $\phi \in \mathcal{D}(\cup_{j=1}^m U_j')$ with $\phi \equiv 1$ on a neighborhood of K. The reader can easily show that the set $\{\phi_j\}$ satisfies the conclusions of the lemma. ∎

The key idea in the proof of Lemma 1 is that the characteristic function for a compact set K (I_K) can be approximated by the sequence of smooth functions $\chi_\epsilon * I_K$. The approximation works because χ_ϵ satisfies properties (i) and (ii) listed just before Lemma 1. Convolving with χ_ϵ can be used to approximate other types of functions as well. For example, if f is a continuous function on $\mathbb{R}^N$, then properties (i) and (ii) can be used to show that the smooth sequence $\{\chi_\epsilon * f\}$ converges to f uniformly on compact sets as $\epsilon \to 0$.

Another important class of functions is

$$\mathcal{A}(\Omega) = \text{the space of real analytic functions on } \Omega \text{ (either real or complex valued).}$$

A function f is *real analytic* on an open set Ω if in a neighborhood of each point in Ω, f can be represented as a convergent power series. It is a standard fact that $\mathcal{A}(\Omega)$ is a subset of $\mathcal{E}(\Omega)$, i.e., real analytic functions are smooth.

A real analytic version of the χ_ϵ is given by the following. Let

$$e(x) = \pi^{-N/2} e^{-|x|^2} \quad \text{for} \quad x \in \mathbb{R}^N.$$

The power series for $e(x)$ about the origin converges for all x, so e belongs to $\mathcal{A}(\mathbb{R}^N)$. For $\epsilon > 0$, let

$$e_\epsilon(x) = e\left(\frac{x}{\epsilon}\right) \epsilon^{-N}.$$

The set of functions $\{e_\epsilon\}$ satisfies the following properties which are analogous to (i) and (ii) for χ_ϵ:

(i) Given $\delta > 0$, $\int_{\substack{y \in \mathbb{R}^N \\ |y| \geq \delta}} e_\epsilon(y) dy \to 0$ as $\epsilon \to 0$.

(ii) $\int_{\mathbb{R}^N} e_\epsilon(y) dy = 1$, for each $\epsilon > 0$.

These properties follow from a standard polar coordinate calculation after the change of variable $t = y/\epsilon$.

Using the e_ϵ, we now prove the classical Weierstrass theorem, which states that $\mathcal{A}(\Omega)$ is a dense subset of $\mathcal{E}(\Omega)$ in the following topology for $\mathcal{E}(\Omega)$: a sequence $f_n \in \mathcal{E}(\Omega)$ is said to *converge* to f in $\mathcal{E}(\Omega)$ as $n \to \infty$ if for each compact subset K of Ω and for each multiindex $\alpha = (\alpha_1, \ldots, \alpha_N)$ (α_j a non-negative integer)

$$D^\alpha f_n \to D^\alpha f \quad \text{uniformly on } K \text{ as } n \to \infty.$$

Here,

$$D^\alpha = \frac{\partial^{|\alpha|}}{\partial x_1^{\alpha_1} \ldots \partial x_N^{\alpha_N}} \quad \text{and} \quad |\alpha| = \alpha_1 + \cdots + \alpha_N.$$

THEOREM 1 WEIERSTRASS
Suppose f belongs to $\mathcal{E}(\Omega)$. Then there is a sequence of polynomials $P_1, P_2, \ldots$ that converges to f in the topology of $\mathcal{E}(\Omega)$.

PROOF Let K be an arbitrary compact subset of Ω and let $\phi \in \mathcal{D}(\Omega)$ be a cutoff function that is identically 1 on a neighborhood of K. If f belongs to $\mathcal{E}(\Omega)$ then ϕf belongs to $\mathcal{D}(\Omega)$.

For $\epsilon > 0$, define

$$\begin{aligned} F_\epsilon(x) &= (e_\epsilon * (\phi f))(x) \\ &= \int\limits_{y \in \mathbb{R}^N} e_\epsilon(x-y)\phi(y)f(y)dy \\ &= \int\limits_{y \in \mathbb{R}^N} e_\epsilon(y)\phi(x-y)f(x-y)dy \quad \text{(by a translation).} \end{aligned}$$

Note that

$$(D^\alpha F_\epsilon)(x) = \int\limits_{y \in \mathbb{R}^N} e_\epsilon(y)(D^\alpha\{\phi f\})(x-y)dy.$$

In view of property (ii) for e_ϵ

$$(D^\alpha(\phi f))(x) = \int\limits_{y \in \mathbb{R}^N} e_\epsilon(y) D^\alpha(\phi f)(x)dy.$$

Therefore, for $x \in K$ (so $\phi(x) = 1$), we have

$$D^\alpha F_\epsilon(x) - D^\alpha f(x) = \int\limits_{y \in \mathbb{R}^N} e_\epsilon(y)((D^\alpha\{\phi f\})(x-y) - (D^\alpha\{\phi f\})(x))dy.$$

For $\delta > 0$, we split the integral on the right into the sum of an integral over $\{|y| \le \delta\}$ and an integral over $\{|y| \ge \delta\}$. The first of these integrals can be made small by choosing δ small using the uniform continuity of $D^\alpha(\phi f)$. With this choice of δ, the second of these integrals converges to zero as $\epsilon \to 0$ in view of property (i) for e_ϵ. Therefore, $D^\alpha F_\epsilon$ converges to $D^\alpha f$ uniformly on K as $\epsilon \to 0$.

The power series for each F_ϵ about the origin converges for all $x \in \mathbb{R}^N$ because F_ϵ can be written

$$F_\epsilon(x) = \epsilon^{-N} \int\limits_{y\in\mathbb{R}^N} e\left(\frac{x-y}{\epsilon}\right)\phi(y)f(y)dy$$

and because the power series for $e(\cdot)$ can be integrated term by term. By truncating the power series of F_ϵ about the origin, we obtain a sequence of polynomials $P_n(x)$, $n = 1, 2, \dots$ such that for each multiindex α

$$\sup_{x\in K} |D^\alpha P_n(x) - D^\alpha f(x)| \to 0 \quad \text{as} \quad n \to \infty.$$

Now let $K_1, K_2, \dots$ be an increasing sequence of compact sets with $\cup_n K_n = \Omega$. Let P_n be a polynomial with

$$\sup_{\substack{x\in K_n \\ |\alpha|\le n}} |D^\alpha P_n(x) - D^\alpha f(x)| \le \frac{1}{n}.$$

Clearly, $P_n \to f$ in $\mathcal{E}(\Omega)$ as $n \to \infty$, and the proof of the theorem is complete. ∎

The final class of functions to consider is the class of holomorphic functions. If Ω is an open set in $\mathbb{C}^n$, then let

$$\mathcal{O}(\Omega) = \text{ the space of holomorphic functions on } \Omega.$$

A function is *holomorphic* on Ω if it satisfies the Cauchy–Riemann equations

$$\frac{\partial f}{\partial \bar{z}_j} \equiv 0 \quad \text{on} \quad \Omega, \qquad 1 \le j \le n$$

where

$$\frac{\partial}{\partial \bar{z}_j} = \frac{1}{2}\left(\frac{\partial}{\partial x_j} + i\frac{\partial}{\partial y_j}\right).$$

Here, we have labeled the coordinates for $\mathbb{C}^n$ as $(z_1, \dots, z_n)$ with $z_j = x_j + iy_j$ $(i = \sqrt{-1})$.

We assume the reader knows some basic complex analysis. If f is holomorphic in a neighborhood of a point $p = (p_1, \dots, p_n)$, then the reader should know that f can be expressed as a convergent series in powers of $(z_1 - p_1), \dots, (z_n - p_n)$. This and a connectedness argument imply the identity theorem for holomorphic functions: if f is holomorphic on a connected open set Ω and if f vanishes on an open subset of Ω, then f vanishes everywhere on Ω. This is often expressed by saying that an open set is a uniqueness set for holomorphic functions. Other types of sets are also uniqueness sets for holomorphic functions. For example, if f is holomorphic on $\mathbb{C}^n$ and if f vanishes on the copy of $\mathbb{R}^n$ given by $\{(x_1 + i0, \ldots, x_n + i0); (x_1, \ldots, x_n) \in \mathbb{R}^n\}$, then f vanishes everywhere. This is because all the x-derivatives of f vanish on this copy of $\mathbb{R}^n$. The Cauchy–Riemann equations can then be used to inductively show that all x, y derivatives of f vanish on this copy of $\mathbb{R}^n$. In particular, a power series expansion of f about the origin must vanish identically and hence $f \equiv 0$.

A function $f = (f_1, \ldots, f_m)$: $\mathbb{C}^n \to \mathbb{C}^m$ is called a *holomorphic map* if each f_j: $\mathbb{C}^n \to \mathbb{C}$ is holomorphic, $1 \leq j \leq m$. This definition can be recast in terms of the real derivative of f (denoted Df) as a map from $\mathbb{R}^{2n}$ to $\mathbb{R}^{2m}$. Define the complex structure map

$$J\colon \mathbb{R}^{2n} \to \mathbb{R}^{2n}$$

by

$$J(x_1, y_1, \ldots, x_n, y_n) = (-y_1, x_1, \ldots, -y_n, x_n).$$

By the Cauchy–Riemann equations, a C^1 map f: $\mathbb{C}^n \to \mathbb{C}^m$ is holomorphic in Ω if and only if for each point $p \in \Omega$

$$Df(p) \circ J = J \circ Df(p)$$

as linear maps from $\mathbb{R}^{2n}$ to $\mathbb{R}^{2m}$. The J on the left is the complex structure map for $\mathbb{R}^{2n}$, whereas the J on the right is the complex structure map for $\mathbb{R}^{2m}$.

1.2 Vectors and vector fields

Now, we turn to the topic of vectors and vector fields on $\mathbb{R}^N$. These concepts will be generalized to the manifold setting in Chapter 2.

A *vector* at a point $p \in \mathbb{R}^N$ is an operator of the form

$$L = \sum_{j=1}^{N} v_j \frac{\partial}{\partial x_j}, \qquad v_j \in \mathbb{R}.$$

The set of vectors at p is denoted $T_p(\mathbb{R}^N)$ and is called the (real) tangent space of $\mathbb{R}^N$ at p.

An element L in $T_p(\mathbb{R}^N)$ can be viewed as a linear map L: $\mathcal{E}(\mathbb{R}^N) \to \mathbb{R}$ by defining

$$L\{f\} = \sum_{j=1}^{N} v_j \frac{\partial f}{\partial x_j}(p).$$

A *vector field* on an open set $\Omega \subset \mathbb{R}^N$ is a smooth map that assigns to each point $p \in \Omega$ a vector in $T_p(\mathbb{R}^N)$. Here, the concept of smooth means that each vector field L can be written

$$L = \sum_{j=1}^{N} v_j \frac{\partial}{\partial x_j}$$

where each v_j is an element of $\mathcal{E}(\Omega)$. We let L_p be the value of the vector field L at p, i.e.,

$$L_p = \sum_{j=1}^{N} v_j(p) \frac{\partial}{\partial x_j}.$$

We also write

$$(Lf)(p) = L_p\{f\} = \sum_{j=1}^{N} v_j(p) \frac{\partial f}{\partial x_j}(p) \qquad (p \in \Omega),$$

for $f \in \mathcal{E}(\Omega)$. Thus, Lf is also an element of $\mathcal{E}(\Omega)$.

Now suppose F: $\mathbb{R}^N \to \mathbb{R}^{N'}$ is a smooth (C^1) map near a point $p \in \mathbb{R}^N$. F induces a map F_*: $T_p(\mathbb{R}^N) \to T_{F(p)}(\mathbb{R}^{N'})$ called the *push forward* of F at p. It is defined by

$$F_*(L)\{g\} = L\{g \circ F\}$$

for $L \in T_p(\mathbb{R}^N)$ and $g \in \mathcal{E}(\mathbb{R}^{N'})$. If L is a vector field on Ω, then the above equation reads

$$[F_*(L)]_{F(p)}\{g\} = L_p\{g \circ F\}$$

for $g \in \mathcal{E}(\mathbb{R}^{N'})$ and $p \in \Omega$. If $F : \Omega \mapsto \Omega'$ is a diffeomorphism, then $F_*(L)$ is a vector field defined on Ω'.

Let us compute $F_*(\partial/\partial x_j)$, $1 \leq j \leq N$. The $(\partial/\partial x_k)$-coefficient of a vector is obtained by computing its action on the coordinate function x_k. We have

$$\begin{aligned} F_*\left(\frac{\partial}{\partial x_j}\right)\{x_k\} &= \frac{\partial}{\partial x_j}\{x_k \circ F\} \\ &= \frac{\partial F_k}{\partial x_j} \end{aligned}$$

where we have written $F = (F_1, \ldots, F_{N'})$. Therefore

$$\left[F_*\left(\frac{\partial}{\partial x_j}\right)\right]_{F(p)} = \sum_{k=1}^{N'} \frac{\partial F_k}{\partial x_j}(p) \frac{\partial}{\partial x_k}.$$

If $L = \sum_{j=1}^{N} v_j \partial/\partial x_j$ belongs to $T_p(\mathbb{R}^N)$, then

$$\begin{aligned}[F_*(L)]_{F(p)} &= \sum_{j=1}^{N} v_j \left[F_* \left(\frac{\partial}{\partial x_j} \right) \right]_p \\ &= \sum_{k=1}^{N'} \left(\sum_{j=1}^{N} \frac{\partial F_k}{\partial x_j}(p) v_j \right) \frac{\partial}{\partial x_k}.\end{aligned}$$

Note that the coefficients of $[F_*(L)]_{F(p)}$ are obtained by multiplying the derivative matrix $DF(p)$ with the vector $(v_1, \ldots, v_N)$.

1.3 Forms

The dual space of $T_p(\mathbb{R}^N)$ is denoted $T_p^*(\mathbb{R}^N)$ and this is called the space of *1-forms on* $\mathbb{R}^N$ *at* p. By definition, this is the space of all linear functionals on $T_p(\mathbb{R}^N)$. If ϕ is a 1-form and L is a vector at p then we denote the value of ϕ on L by

$$\langle \phi, L \rangle_p.$$

Let $\{dx_j;\ 1 \leq j \leq N\}$ be the dual basis for $\{\partial/\partial x_j;\ 1 \leq j \leq N\}$. This means

$$\left\langle dx_j, \frac{\partial}{\partial x_k} \right\rangle_p = \delta_{jk} = \begin{cases} 1 & \text{if } j = k \\ 0 & \text{if } j \neq k . \end{cases}$$

If $\alpha = \sum_{j=1}^{N} \alpha_j dx_j$ is a 1-form and $L = \sum_{k=1}^{N} v_k (\partial/\partial x_k)$ is a vector, then

$$\langle \alpha, L \rangle_p = \sum_{j=1}^{N} \alpha_j v_j.$$

Note that $\langle \alpha, L \rangle$ is the same as the Euclidean inner product between $(\alpha_1, \ldots, \alpha_N)$ and $(v_1, \ldots, v_N)$ as elements of $\mathbb{R}^N$.

A *differential 1-form* on an open set $\Omega \subset \mathbb{R}^N$ is an object of the form

$$\alpha = \sum_{j=1}^{N} \alpha_j dx_j$$

where each α_j is an element of $\mathcal{E}(\Omega)$. An important example of differential 1-form is the *exterior derivative* of a function f, denoted df. This is defined as follows. For $f \in \mathcal{E}(\mathbb{R}^N)$, $p \in \mathbb{R}^N$, and $L \in T_p(\mathbb{R}^N)$, let

$$\langle df(p), L \rangle_p = L\{f\}.$$

Letting f be a coordinate function x_j and L be the vector $\partial/\partial x_k$, we see that the exterior derivative of x_j is the same as the dual basis definition of dx_j given above. By applying this definition of df to the vector $\partial/\partial x_k$, we have

$$df(p) = \sum_{k=1}^{N} \frac{\partial f}{\partial x_k}(p)dx_k.$$

If V is a vector space, then we let $\Lambda^r(V)$ be the rth *exterior power* of V. A basis of $\Lambda^r(V)$ is given by

$$\{v_{i_1} \wedge \ldots \wedge v_{i_r};\ 1 \leq i_1 < \cdots < i_r \leq N\}$$

where $\{v_1, \ldots, v_N\}$ is a basis for V. We often use the notation $v^I = v_{i_1} \wedge \ldots \wedge v_{i_r}$ where I is the increasing multiindex $\{i_1, \ldots, i_r\}$. Here, $\wedge$ denotes the wedge product which is bilinear, i.e.,

$$\begin{aligned} v \wedge (au + bw) &= a(v \wedge u) + b(v \wedge w) \qquad && a, b \in \mathbb{R} \\ (au + bw) \wedge v &= a(u \wedge v) + b(w \wedge v) && v, u, w \in V \end{aligned}$$

and anti-symmetric, i.e.,

$$v \wedge w = -w \wedge v \qquad v, w \in V.$$

In particular, $v \wedge v = 0$.

In the context of vectors and forms, we have the space of r-vectors $\Lambda^r(T_p(\mathbb{R}^N))$ and the space of r-forms $\Lambda^r(T_p^*(\mathbb{R}^N))$. A basis for each is given by

$$\left\{\frac{\partial}{\partial x^I} = \frac{\partial}{\partial x_{i_1}} \wedge \ldots \wedge \frac{\partial}{\partial x_{i_r}};\ 1 \leq i_1 < \cdots < i_r \leq N\right\}$$

and

$$\{dx^I = dx_{i_1} \wedge \ldots \wedge dx_{i_r};\ 1 \leq i_1 < \cdots < i_r \leq N\}.$$

The space $\Lambda^r(T_p^*(\mathbb{R}^N))$ can be viewed as the space of linear functionals on the space $\Lambda^r(T_p(\mathbb{R}^N))$ by defining

$$\langle \phi_1 \wedge \ldots \wedge \phi_r, L_1 \wedge \ldots \wedge L_r \rangle_p = \sum_{\sigma} \operatorname{sgn}(\sigma) \langle \phi_1, L_{\sigma(1)} \rangle_p \ldots \langle \phi_r, L_{\sigma(r)} \rangle_p$$

for $\phi_i \in T_p^*(\mathbb{R}^N)$ and $L_i \in T_p(\mathbb{R}^N)$ and where the sum runs over the set of all permutations σ of the set $\{1, \ldots, r\}$. Here, $\operatorname{sgn}(\sigma)$ is the factor $+1$ if the permutation σ is even and -1 if the permutation is odd. The collection $\{dx^I\}$ is a dual basis for $\{\partial/\partial x^J\}$, i.e.,

$$\left\langle dx^I, \frac{\partial}{\partial x^J} \right\rangle_p = \begin{cases} 1 & \text{if } I = J \\ 0 & \text{otherwise.} \end{cases}$$

If $F\colon \mathbb{R}^N \to \mathbb{R}^{N'}$ is a C^1 map then the push forward map $F_*\colon T_p(\mathbb{R}^N) \to T_{F(p)}(\mathbb{R}^{N'})$ can be extended to a linear map $F_*\colon \Lambda^r(T_p(\mathbb{R}^N)) \to \Lambda^r(T_{F(p)}(\mathbb{R}^{N'}))$

by defining

$$F_*(L_1 \wedge \ldots \wedge L_r) = F_*(L_1) \wedge \ldots \wedge F_*(L_r).$$

We can dualize this map to obtain the *pull back* map $F^*\colon \Lambda^r(T^*_{F(p)}(\mathbb{R}^{N'})) \to \Lambda^r(T^*_p(\mathbb{R}^N))$. For $r = 1$, the pull back is defined by

$$\langle F^*(\phi), L\rangle_p = \langle \phi, F_*(L)\rangle_{F(p)} \quad \text{for} \quad \phi \in T^*_{F(p)}(\mathbb{R}^{N'}), L \in T_p(\mathbb{R}^N).$$

For $r \geq 1$, define

$$F^*(\phi_1 \wedge \ldots \wedge \phi_r) = F^*(\phi_1) \wedge \ldots \wedge F^*(\phi_r), \quad \text{for} \quad \phi_i \in T^*_p(\mathbb{R}^{N'}).$$

The equation

$$\langle F^*(\phi_1)\wedge\ldots\wedge F^*(\phi_r), L_1\wedge\ldots\wedge L_r\rangle_p = \langle \phi_1\wedge\ldots\wedge\phi_r, F_*(L_1)\wedge\ldots\wedge F_*(L_r)\rangle_{F(p)}$$

follows easily from the above definitions.

It is useful to know how to compute the pull back of a form using coordinates. We start with the 1-form dx_j. From the definitions, we have

$$\begin{aligned}\left\langle F^*(dx_j), \frac{\partial}{\partial x_k}\right\rangle_p &= \left\langle dx_j, F_*\left(\frac{\partial}{\partial x_k}\right)\right\rangle_{F(p)} \\ &= \left\langle dx_j, \sum_{\ell=1}^{N'} \frac{\partial F_\ell}{\partial x_k}\frac{\partial}{\partial x_\ell}\right\rangle_{F(p)} \\ &= \frac{\partial F_j}{\partial x_k}(p).\end{aligned}$$

So at any point $p \in \mathbb{R}^N$, we have

$$\begin{aligned}(F^* dx_j)(p) &= \sum_{k=1}^{N} \frac{\partial F_j(p)}{\partial x_k} dx_k \\ &= dF_j(p).\end{aligned}$$

If $\phi = \alpha dx^I = \alpha dx_{i_1} \wedge \ldots \wedge dx_{i_r} (\alpha \in \mathbb{R})$, then

$$\begin{aligned}F^*(\phi) &= \alpha dF_{i_1} \wedge \ldots \wedge dF_{i_r} \\ &= \alpha dF^I\end{aligned}$$

by the definitions and the above computations. If α is a smooth function, then on the right, α must be composed with F, because the pairing on the right side of the definition $\langle F^*\phi, X\rangle_p = \langle \phi, F_*(X)\rangle_{F(p)}$ is occurring at the point $F(p)$. We have

$$\begin{aligned}F^*\phi &= (\alpha \circ F) \cdot dF_{i_1} \wedge \ldots \wedge dF_{i_r} \\ &= (\alpha \circ F) \cdot dF^I \quad \text{for} \quad \alpha \in \mathcal{E}(\mathbb{R}^{N'}).\end{aligned}$$

The pull back of the volume form $dx = dx_1 \wedge \ldots \wedge dx_N$ should be singled out as a special case. Suppose $N = N'$. We have

$$F^* dx = dF_1 \wedge \ldots \wedge dF_N.$$

By expanding each dF_j and using the multilinearity of the wedge product, we obtain

$$F^* dx = (\det DF) dx.$$

1.4 The exterior derivative

We have already defined the exterior derivative of a smooth function and we have shown

$$df = \sum_{j=1}^{N} \frac{\partial f}{\partial x_j} dx_j \quad \text{for} \quad f \in \mathcal{E}(\mathbb{R}^N).$$

In this section, we extend the definition of the exterior derivative to higher degree forms and then discuss some of its properties.

First, we require a definition. A *smooth differential form of degree* r, or more simply, a *smooth* r*-form* on an open set $\Omega \subset \mathbb{R}^N$ is an object of the form

$$\phi = \sum_{|I|=r} f_I dx^I$$

where each f_I is an element of $\mathcal{E}(\Omega)$. The sum is over all increasing indices I of length $= |I| = r$. The space of all smooth r-forms on Ω is denoted $\mathcal{E}^r(\Omega)$. We denote by $\mathcal{D}^r(\Omega)$ the space of smooth r-forms with compact support.

DEFINITION 1 *The exterior derivative* d: $\mathcal{E}^r(\Omega) \to \mathcal{E}^{r+1}(\Omega)$ *is defined by*

$$d\left\{\sum_{|I|=r} f_I dx^I\right\} = \sum_{|I|=r} df_I \wedge dx^I.$$

Since df_I is a smooth 1-form, the right side is a smooth $(r+1)$-form.

LEMMA 1

(i) *(product rule)* $d(\phi_1 \wedge \phi_2) = d\phi_1 \wedge \phi_2 + (-1)^r \phi_1 \wedge d\phi_2$, *for* $\phi_1 \in \mathcal{E}^r(\Omega)$ *and* $\phi_2 \in \mathcal{E}^s(\Omega)$.

(ii) $d^2 = 0$, *i.e., if* $\phi \in \mathcal{E}^r(\Omega)$, *then* $d(d\phi) = 0$.

PROOF Property (i) follows from the product rule for differentiation. From the definition of the exterior derivative, it suffices to show property (ii) for functions

(0-forms). This follows by an easy computation using the fact that the mixed partial derivatives of a smooth function are independent of the order in which they are taken. ∎

Property (ii) is often described by stating that $\{d\colon \mathcal{E}^r(\Omega) \to \mathcal{E}^{r+1}(\Omega);\ 0 \le r \le N\}$ forms a *complex.*

The exterior derivative commutes with pull back, as the following lemma shows.

LEMMA 2
Suppose $\Omega \subset \mathbb{R}^N$ and $\Omega' \subset \mathbb{R}^{N'}$ and suppose $F\colon \Omega \to \Omega'$ is a C^1 map. Then $F^* \circ d = d \circ F^*$ as maps from $\mathcal{E}^r(\Omega')$ to $\mathcal{E}^{r+1}(\Omega)$, for $0 \le r \le N'$.

PROOF Suppose $\phi = \alpha dx^I$ with $\alpha \in \mathcal{E}(\Omega')$ and $|I| = r$. From the formula for $F^*\phi$ at the end of the previous section, we have

$$F^*\phi = (\alpha \circ F)dF^I.$$

Since $d^2 = 0$, clearly

$$d(F^*\phi) = d(\alpha \circ F) \wedge dF^I.$$

On the other hand,

$$\begin{aligned} F^* d\phi &= F^*(d\alpha \wedge dx^I) \\ &= F^*(d\alpha) \wedge dF^I. \end{aligned}$$

Comparing these two expressions, we see that the lemma will follow provided we show

$$F^*(d\alpha) = d(\alpha \circ F) \quad \text{for} \quad \alpha \in \mathcal{E}(\Omega').$$

Using the notation $F^*(\alpha) = \alpha \circ F$ for a function α, this equation can be viewed as the statement of the lemma for the case $r = 0$. This can be proved by using the chain rule. ∎

In the previous section, we defined the exterior derivative of a smooth function f by its action on a vector field X, i.e.,

$$\langle df, X\rangle_p = X_p\{f\}.$$

There is a corresponding formula that relates the exterior derivative of a higher degree form in terms of its action on a wedge product of vector fields. The formula is rather messy. Moreover, we shall only need this relationship for the exterior derivative of 1-forms. We present this case in the following lemma.

LEMMA 3

Suppose ϕ is a smooth 1-form and L^1, L^2 are smooth vector fields. Then

$$\begin{aligned}\langle d\phi, L^1 \wedge L^2\rangle = L^1\{\langle\phi, L^2\rangle\} - L^2\{\langle\phi, L^1\rangle\}\\ -\langle\phi, [L^1, L^2]\rangle.\end{aligned}$$

Some explanation is in order. First, ϕ is a 1-form; so the pairings $\langle\phi, L^1\rangle$ and $\langle\phi, L^2\rangle$ are smooth functions. Therefore, $L^2\{\langle\phi, L^1\rangle\}$ and $L^1\{\langle\phi, L^2\rangle\}$ are the actions of the vector fields L^2 and L^1 on the smooth functions $\langle\phi, L^1\rangle$ and $\langle\phi, L^2\rangle$. The notation $[L^1, L^2]$ in the third term on the right is the *Lie bracket* of the vector fields L^1 and L^2. At a point $p \in \mathbb{R}^N$, the Lie bracket is defined by

$$[L^1, L^2]_p\{f\} = L^1_p\{L^2 f\} - L^2_p\{L^1 f\}, \text{ for } f \in \mathcal{E}(\mathbb{R}^N).$$

An easy computation using the coordinate functions $f = x_j$, $1 \leq j \leq N$, shows that if

$$L^1 = \sum_{j=1}^N a_j \frac{\partial}{\partial x_j}, \qquad L^2 = \sum_{j=1}^N b_j \frac{\partial}{\partial x_j} \qquad a_j, b_j \in \mathcal{E}(\mathbb{R}^N)$$

then

$$[L^1, L^2] = \sum_{j=1}^N (L^1\{b_j\} - L^2\{a_j\}) \frac{\partial}{\partial x_j}.$$

PROOF We first note that both sides of the equation of the lemma are multilinear (over $\mathcal{E}(\mathbb{R}^N)$) as functions of the vector fields L^1 and L^2. This is true for the left side since the wedge product is multilinear. For the right side, this follows from the computation

$$\begin{aligned}[a_1L^1, a_2L^2] = a_1(L^1\{a_2\}) \cdot L^2 - a_2(L^2\{a_1\}) \cdot L^1\\ +a_1a_2[L^1, L^2], \text{ for } a_1, a_2 \in \mathcal{E}(\mathbb{R}^N).\end{aligned}$$

Since both sides of the equation of the lemma are multilinear (over $\mathcal{E}(\mathbb{R}^N)$) as functions of L^1 and L^2, then it suffices to verify the lemma when $L^1 = \partial/\partial x_j$ and $L^2 = \partial/\partial x_k$ for $1 \leq j, k \leq N$. In this case, $[L^1, L^2] \equiv 0$ and the equation follows from an easy computation. ∎

1.5 Contractions

The *contraction* operator $\lrcorner$ is defined as the dual to wedge product.

DEFINITION 1 *Suppose L is a vector in $T_p(\mathbb{R}^N)$ and ϕ is an r-form at $p \in \mathbb{R}^N$. The $(r-1)$-form $L \lrcorner \phi$ is defined by*

$$\langle L \lrcorner \phi, L_1 \wedge \ldots \wedge L_{r-1} \rangle_p = \langle \phi, L \wedge L_1 \wedge \ldots \wedge L_r \rangle_p .$$

$L \lrcorner \phi$ is called the *contraction* of ϕ by L.

For example, we have

$$\frac{\partial}{\partial x_j} \lrcorner (dx_{i_1} \wedge \ldots \wedge dx_{i_r}) = \begin{cases} 0 & \text{if } j \notin \{i_1, \ldots, i_r\} \\ (-1)^{k-1} dx_{i_1} \wedge \ldots \wedge \widehat{dx_{i_k}} & \\ \qquad\qquad \wedge \ldots \wedge dx_{i_r} & \text{if } j = i_k . \end{cases}$$

The notation $\widehat{dx_{i_k}}$ indicates that dx_{i_k} is removed.

The contraction operator also satisfies a product rule.

LEMMA 1

Suppose L is a vector in $T_p^*(\mathbb{R}^N)$ and suppose ϕ_1 belongs to $\Lambda^r T_p^*(\mathbb{R}^N)$ and ϕ_2 belongs to $\Lambda^s T_p^*(\mathbb{R}^N)$. Then

$$L \lrcorner (\phi_1 \wedge \phi_2) = (L \lrcorner \phi_1) \wedge \phi_2 + (-1)^r \phi_1 \wedge (L \lrcorner \phi_2).$$

This lemma is established by first proving it for basis elements, i.e., $L = \partial/\partial x_j$, $\phi_1 = dx^I$, $|I| = r$ and $\phi_2 = dx^J$, $|J| = s$. The general case then follows since both sides of the equation are multilinear as functions of L, ϕ_1 and ϕ_2.

The contraction operator makes it easy to compute the $\mathcal{L}^2$-adjoint of the exterior derivative. The Euclidean inner product $(\cdot)$ on $\mathbb{R}^N$ naturally extends to $\Lambda^r T^* \mathbb{R}^N$ by declaring that the collection $\{dx^I; |I| = r \text{ and } I \text{ increasing}\}$ is an orthonormal basis. Note that

$$dx^I \cdot dx^J = \left\langle dx^I, \frac{\partial}{\partial x^J} \right\rangle_p \qquad |I| = |J| = r, \qquad p \in \mathbb{R}^N$$

where the pairing on the right is the pairing between r-forms and r-vectors discussed in Section 1.3. Let L be a vector and let η be a 1-form that is dual to L (i.e., $\eta \lrcorner L = 1$ and $\eta \lrcorner L' = 0$ for all vectors L' which are orthogonal to L). From the definition of contraction, we have

$$(L \lrcorner \phi) \cdot \psi = \phi \cdot (\eta \wedge \psi)$$

for all $\phi \in \Lambda^r T^*(\mathbb{R}^N)$ and $\psi \in \Lambda^{r-1}(\mathbb{R}^N)$.

We define an $\mathcal{L}^2$-inner product on $\mathcal{D}^r(\mathbb{R}^N)$, by setting

$$(\phi, \psi)_{\mathcal{L}^2} = \int_{x\in\mathbb{R}^N} (\phi(x) \cdot \psi(x))dx \quad \text{for} \quad \phi, \psi \in \mathcal{D}^r(\mathbb{R}^N).$$

The $\mathcal{L}^2$-*adjoint* of d (denoted d^*: $\mathcal{D}^r(\mathbb{R}^N) \to \mathcal{D}^{r-1}(\mathbb{R}^N)$) is defined by

$$(d^*\phi, \psi)_{\mathcal{L}^2} = (\phi, d\psi)_{\mathcal{L}^2}, \quad \phi \in \mathcal{D}^r(\mathbb{R}^N), \psi \in \mathcal{D}^{r-1}(\mathbb{R}^N).$$

LEMMA 2
For $\phi = g dx^I \in \mathcal{D}^r(\mathcal{R}^N)$

$$d^*\phi = -\sum_{j=1}^{N} \frac{\partial g}{\partial x_j}\left(\frac{\partial}{\partial x_j} \lrcorner dx^I\right).$$

PROOF To establish the lemma, we take the inner product of $d^*\phi$ with an arbitrary $\psi = h dx^J \in \mathcal{D}^{r-1}(\mathbb{R}^N)$. From the definition of d^* and d and integration by parts, we have

$$\begin{aligned}(d^*\phi, \psi)_{\mathcal{L}^2} &= (\phi, d\psi)_{\mathcal{L}^2} \\ &= \int_{\mathbb{R}^N} \sum_{j=1}^{N} g(x)\frac{\partial h(x)}{\partial x_j}(dx^I \cdot (dx_j \wedge dx^J))dx \\ &= -\int_{\mathbb{R}^N} \sum_{j=1}^{N} \frac{\partial g}{\partial x_j}(x)h(x)(dx^I \cdot (dx_j \wedge dx^J))dx.\end{aligned}$$

Since $dx^I \cdot (dx_j \wedge dx^J) = \langle dx^I, (\partial/\partial x_j) \wedge (\partial/\partial x^J)\rangle = \langle (\partial/\partial x_j) \lrcorner dx^I, (\partial/\partial x^J)\rangle$, we can write

$$\begin{aligned}(d^*\phi, \psi)_{\mathcal{L}^2} &= -\int_{\mathbb{R}^N} \sum_{j=1}^{N} \left(\frac{\partial g}{\partial x_j}(x)\left(\frac{\partial}{\partial x_j} \lrcorner dx^I\right) \cdot h(x)dx^J\right) dx \\ &= \left(-\sum_{j=1}^{N} \frac{\partial g}{\partial x_j}\frac{\partial}{\partial x_j} \lrcorner dx^I, \psi\right)_{\mathcal{L}^2}\end{aligned}$$

as desired. ∎

2

Analysis on Manifolds

After one is comfortable with analysis on Euclidean space, the next step is to study analysis on spaces that only locally look like Euclidean space. This leads us to the concept of a manifold. After some definitions, we generalize the notions of vectors and forms from Chapter 1 to the manifold setting. Integration of forms on orientable manifolds is discussed and the chapter ends with the proofs of Stokes' theorem and some of its corollaries.

2.1 Manifolds

DEFINITION 1 *A real N-dimensional smooth manifold is a Hausdorff topological space X together with an open covering $\{U_\alpha\}$ of X (where α runs over some index set) and open maps χ_α: $U_\alpha \to \mathbb{R}^N$ with the following properties:*

(i) *χ_α is a homeomorphism with its image in $\mathbb{R}^N$.*

(ii) *Whenever $U_\alpha \cap U_\beta \neq \emptyset$, then $\chi_\alpha \circ \chi_\beta^{-1}$ is a diffeomorphism from $\chi_\beta\{U_\alpha \cap U_\beta\} \subset \mathbb{R}^N$ to the set $\chi_\alpha\{U_\alpha \cap U_\beta\} \subset \mathbb{R}^N$.*

A smooth manifold locally "looks like" $\mathbb{R}^N$ because about any point there is an open set U_α (often called a *coordinate patch*) and a map χ_α: $U_\alpha \to \mathbb{R}^N$ (often called a *coordinate chart*) that takes U_α homeomorphically to an open subset of $\mathbb{R}^N$. As we shall see, familiar concepts from analysis on Euclidean space such as differentiation, vectors, and forms can be defined on a manifold via these coordinate charts.

In requirement (ii) of the definition, it makes sense to talk about the differentiability of $\chi_\alpha \circ \chi_\beta^{-1}$ because this map is defined on an open subset of $\mathbb{R}^N$ (namely $\chi_\beta\{U_\alpha \cap U_\beta\}$). This requirement also distinguishes different categories of manifolds. For example, if we require $\chi_\alpha \circ \chi_\beta^{-1}$ to be a real analytic diffeomorphism, then X becomes a *real analytic manifold.* If, instead of $\mathbb{R}^N$, we have $\mathbb{C}^n$ and we require $\chi_\alpha \circ \chi_\beta^{-1}$ to be a biholomorphism (i.e., holomorphic with a holomorphic inverse), then by definition, X is a *complex manifold.*

Note that any open subset of a manifold is again a manifold.

The easiest example of a smooth manifold is Euclidean space itself. Here, there is only one coordinate patch (i.e., all of $\mathbb{R}^N$) and there is only one coordinate chart (the identity map). In the same way, $\mathbb{C}^n$ is a complex manifold.

The next easiest example of a manifold is projective space $\mathbb{P}^N$, which is the collection of equivalence classes in $\mathbb{R}^{N+1}$ under the equivalence relation

$$(x_0, \dots, x_N) \sim (y_0, \dots, y_N)$$

if there exists a nonzero real number λ with

$$(x_0, \dots, x_N) = \lambda(y_0, \dots, y_N).$$

The equivalence class of $(x_0, \dots, x_N)$ is denoted $[x_0, \dots, x_N]$ and it is the line in $\mathbb{R}^{N+1}$ that passes through the origin and the point $(x_0, \dots, x_N)$. The coordinate patches are

$$U_i = \{[x_0, \dots, x_N]; x_i \neq 0\} \qquad 0 \leq i \leq N.$$

The corresponding coordinate charts are the maps χ_i: $U_i \to \mathbb{R}^N$ given by

$$\chi_i([x_0, \dots, x_N]) = \left(\frac{x_0}{x_i}, \dots, \frac{x_{i-1}}{x_i}, \frac{x_{i+1}}{x_i}, \dots, \frac{x_N}{x_i}\right).$$

Complex projective space is defined analogously by replacing $\mathbb{R}^{N+1}$ by $\mathbb{C}^{N+1}$ and by letting λ be a nonzero complex number in the above equivalence relation. More examples of manifolds will follow the definition of a submanifold.

The concept of a differentiable function on Euclidean space generalizes to the manifold setting.

DEFINITION 2 *Suppose X is a smooth manifold. A function f: $X \to \mathbb{C}$ is said to be of class C^k $(k \geq 0)$ on X if for each coordinate patch U_α, the function $f \circ \chi_\alpha^{-1}$: $\chi_\alpha\{U_\alpha\} \to \mathbb{C}$ is of class C^k in the Euclidean sense.*

The point is that $f \circ \chi_\alpha^{-1}$ is defined on an open subset of $\mathbb{R}^N$ and therefore it makes sense to talk about the differentiability of $f \circ \chi_\alpha^{-1}$ in the usual Euclidean sense. Requirement (ii) in the definition of a manifold ensures that if $f \circ \chi_\alpha^{-1}$ is differentiable then so is $f \circ \chi_\beta^{-1}$ on $\chi_\beta\{U_\alpha \cap U_\beta\}$ for any other coordinate chart χ_β: $U_\beta \to \mathbb{R}^N$. This follows from $f \circ \chi_\beta^{-1} = (f \circ \chi_\alpha^{-1}) \circ (\chi_\alpha \circ \chi_\beta^{-1})$ and noting that $\chi_\alpha \circ \chi_\beta^{-1}$ is smooth.

The above definition can be generalized to define the concept of a smooth map between two manifolds.

DEFINITION 3 *Suppose X is a smooth manifold with coordinate charts χ_α: $U_\alpha \to \mathbb{R}^N$ and suppose Y is a smooth manifold with coordinate charts $\mathcal{Y}_\beta$: $V_\beta \to$*

$\mathbb{R}^{N'}$. *We say that* $f\colon X \to Y$ *is a* C^k *map if for every* α *and* β, $\mathcal{Y}_\beta \circ f \circ \chi_\alpha^{-1}$ *is a* C^k *map from* $\chi_\alpha\{U_\alpha\} \subset \mathbb{R}^N$ *to* $\chi_\beta\{V_\beta\} \subset \mathbb{R}^{N'}$, *provided* $\mathcal{Y}_\beta \circ f \circ \chi_\alpha^{-1}$ *is defined. The map* f *is called a diffeomorphism if* $f\colon X \to Y$ *is* C^1 *and if* f *has a* C^1 *inverse* $f^{-1}\colon Y \to X$.

These definitions can be modified for other categories of manifolds. For example, if X is a real analytic manifold, then $f\colon X \to \mathbb{C}$ is called *real analytic* if for each coordinate chart χ_α, $f \circ \chi_\alpha^{-1}$ is real analytic in the Euclidean sense. If X is a complex manifold, then $f\colon X \to \mathbb{C}$ is called *holomorphic* if for each coordinate chart χ_α, $f \circ \chi_\alpha^{-1}$ is holomorphic in the Euclidean sense.

It is important to realize that different manifold structures can be put on the same topological space by changing the coordinate charts. For example, $\mathbb{C}^n$ is a complex manifold whose usual manifold structure is given by taking the identity map for its coordinate chart. However, suppose we are given a smooth map $\chi = (\chi^1, \ldots, \chi^n)\colon \mathbb{C}^n \to \mathbb{C}^n$ whose real derivative (as a map from $\mathbb{R}^{2n}$ to $\mathbb{R}^{2n}$) at each point in some open set $\Omega \subset \mathbb{C}^n$ is nonsingular. We can define a new complex manifold structure on Ω by declaring χ to be the coordinate chart. If χ is not holomorphic in the usual sense, then this new complex manifold structure on Ω is different than the standard Euclidean complex manifold structure. A function $f\colon \Omega \to \mathbb{C}$ is holomorphic with respect to this new structure provided $f \circ \chi^{-1}$ is holomorphic in the usual sense, or equivalently, if there is a locally defined holomorphic function F in the usual sense with

$$f = F \circ \chi.$$

This idea will be used in Section 9.1, where we shall use a nonstandard complex manifold structure for $\mathbb{C}^n$.

2.2 Submanifolds

Next, we define the concept of a submanifold of another manifold.

DEFINITION 1 *Suppose* X *is a smooth* N*-dimensional real manifold and let* M *be a subset of* X. *We say that* M *is a smooth (imbedded) submanifold of* X *of real dimension* ℓ *if for each point* $p_0 \in M$, *there is a neighborhood* U *of* p_0 *in* X *and a smooth map* $\chi\colon U \to \mathbb{R}^N$ *such that*

(i) $\chi\colon U \to \chi(U) \subset \mathbb{R}^N$ *is a diffeomorphism.*

(ii) $\chi(p_0)$ *is the origin in* $\mathbb{R}^N$.

(iii) $\chi\{U \cap M\}$ *is an open subset of the origin in the copy of* $\mathbb{R}^\ell$ *given by* $\{(t_1, \ldots, t_\ell, 0, \ldots, 0) \in \mathbb{R}^N; (t_1, \ldots, t_\ell) \in \mathbb{R}^\ell\}$.

The collection of all such $\chi|_{M\cap U}$: $M \cap U \to \mathbb{R}^\ell$ serves as a collection of coordinate charts for M and thus M is a smooth ℓ-dimensional manifold in its own right.

Our main concern will be with submanifolds of Euclidean space. In this case, the map $\chi = (\chi^1, \ldots, \chi^N)$ is defined on an open subset U of $\mathbb{R}^N$. From (iii), the set $M \cap U$ can be viewed as the common zero set of the functions $\chi^{\ell+1}, \ldots, \chi^N$. Since χ is a diffeomorphism, the derivative $D\{(\chi^{\ell+1}, \ldots, \chi^N)\}$ has maximal rank at each point in U. This is equivalent to the requirement

$$d\chi^{\ell+1} \wedge \ldots \wedge d\chi^N \neq 0 \quad \text{on} \quad U.$$

A collection of functions $\{\chi^{\ell+1}, \ldots, \chi^N\}$ with this property will be referred to as a *local defining system for* M. So about each point of a submanifold, there is a local defining system. The converse also holds.

LEMMA 1

Suppose $M \subset \mathbb{R}^N$ is a subset with the property that for each point p_0 in M there is a neighborhood U in $\mathbb{R}^N$ and smooth functions $\rho_1, \ldots, \rho_{N-\ell}$: $U \to \mathbb{R}$ so that $M \cap U = \{x \in U; \rho_1(x) = \cdots = \rho_{N-\ell}(x) = 0\}$ and with $d\rho_1 \wedge \ldots \wedge d\rho_{N-\ell} \neq 0$ at p_0. Then there is a diffeomorphism $\chi = (\chi^1, \ldots, \chi^N)$ defined near p_0, with $\chi^{\ell+1} = \rho_1, \ldots, \chi^N = \rho_{N-\ell}$. In particular, M is an ℓ-dimensional smooth submanifold of $\mathbb{R}^N$.

PROOF Suppose $p_0 \in M$ and $\rho_1, \ldots, \rho_{N-\ell}$ are given. Since $d\rho_1 \wedge \ldots \wedge d\rho_{N-\ell} \neq 0$ at p_0, we can choose coordinates (x, y) for $\mathbb{R}^N$ with $x \in \mathbb{R}^\ell$ and $y \in \mathbb{R}^{N-\ell}$ so that $D_y\{(\rho_1, \ldots, \rho_{N-\ell})\}$ at p_0 is a nonsingular $(N-\ell) \times (N-\ell)$ matrix. Then we define χ: $\mathbb{R}^N \to \mathbb{R}^N$ by

$$\chi(x, y) = (x, \rho(x, y)) \quad \text{for} \quad (x, y) \in \mathbb{R}^\ell \times \mathbb{R}^{N-\ell}$$

where $\rho = (\rho_1, \ldots, \rho_{N-\ell})$. By the choice of coordinates, $D\chi$ at p_0 is nonsingular and so χ is a local diffeomorphism by the inverse function theorem. Therefore, χ has all the properties required for the lemma. ∎

The above lemma allows us to construct many examples of submanifolds of $\mathbb{R}^N$. For example, the unit sphere $\{x \in \mathbb{R}^N; |x|^2 - 1 = 0\}$ is an $(N-1)$-dimensional submanifold or *hypersurface* in $\mathbb{R}^N$. Another important example is the graph of a smooth function. Give $\mathbb{R}^N$ the coordinates (x, y) with $x \in \mathbb{R}^\ell$ and $y \in \mathbb{R}^{N-\ell}$. Let h: $\mathbb{R}^\ell \to \mathbb{R}^{N-\ell}$ be a smooth function. Define

$$M = \{(x, y) \in \mathbb{R}^N; y = h(x)\}.$$

M is called the *graph of* h. Let $\rho_j(x, y) = y_j - h_j(x)$ for $1 \leq j \leq N - \ell$ where we have written $h = (h_1, \ldots, h_{N-\ell})$. M is the common zero set of $\rho_1, \ldots, \rho_{N-\ell}$. Moreover, $d\rho_1 \wedge \ldots \wedge d\rho_{N-\ell}$ is nonzero (everywhere) and so M is an ℓ-dimensional submanifold by the lemma. In this case, χ is given by

$\chi(x,y) = (x, y - h(x))$. The converse also holds locally. That is, a submanifold can always be written locally as the graph of a smooth function.

LEMMA 2

Suppose M is a smooth ℓ-dimensional submanifold of $\mathbb{R}^N$ and p_0 is a point on M. There is an affine linear map L: $\mathbb{R}^N \to \mathbb{R}^N$ and a neighborhood U of the origin and a smooth function h: $\mathbb{R}^\ell \to \mathbb{R}^{N-\ell}$ such that

(i) *$L(p_0)$ is the origin.*

(ii) *$L\{M\} \cap U = \{(x,y) \in \mathbb{R}^\ell \times \mathbb{R}^{N-\ell}; y = h(x)\}$.*

(iii) *$h(0) = 0$ and $Dh(0) = 0$.*

PROOF First, we translate coordinates so that p_0 is the origin. Give $\mathbb{R}^N$ the coordinates (x,y) with $x \in \mathbb{R}^\ell$ and $y \in \mathbb{R}^{N-\ell}$. We let $\rho_1, \ldots, \rho_{N-\ell}$ be a local defining system for M near the origin. The gradient vectors $\nabla\rho_1, \ldots, \nabla\rho_{N-\ell}$ are linearly independent at the origin. Therefore, we can find a nonsingular linear map L: $\mathbb{R}^N \to \mathbb{R}^N$ which sends $\nabla\rho_j(0)$ to $\partial/\partial y_j$ for $1 \leq j \leq N - \ell$. The manifold $L\{M\}$ has a local defining system $\{\tilde{\rho}_1, \ldots, \tilde{\rho}_{N-\ell}\}$ with $\tilde{\rho}_j = \rho_j \circ L^{-1}$ and

$$\nabla\tilde{\rho}_j(0) = \frac{\partial}{\partial y_j} \qquad 1 \leq j \leq N - \ell.$$

The matrix $(D_y\tilde{\rho})(0)$ with $\tilde{\rho} = (\tilde{\rho}_1, \ldots, \tilde{\rho}_{N-\ell})$ is the identity matrix. By the implicit function theorem, there is a smooth function h: $\mathbb{R}^\ell \to \mathbb{R}^{N-\ell}$ with $h(0) = 0$ and

$$\tilde{\rho}(x, h(x)) = 0, \text{ for } x \text{ near } 0.$$

Near the origin, $L\{M\}$ is the graph of h. By differentiating this equation and then using $\nabla\tilde{\rho}_j(0) = \partial/\partial y_j$ for $1 \leq j \leq N - \ell$, we obtain $Dh(0) = 0$, as desired. ▮

The previous two lemmas describe the way submanifolds are most often presented when doing local analysis. When a submanifold is presented by a local defining system, the functions in that defining system locally generate all functions that vanish on M, as the following lemma shows.

LEMMA 3

Suppose M is an ℓ-dimensional smooth submanifold of $\mathbb{R}^N$ given by $M = \{\rho_1 = \cdots = \rho_{N-\ell} = 0\}$ with $d\rho_1 \wedge \ldots \wedge d\rho_{N-\ell} \neq 0$ on M. Suppose f: $\mathbb{R}^N \to \mathbb{R}$ is a smooth function that vanishes on M. Then there are smooth functions $\alpha_1, \ldots, \alpha_{N-\ell}$ defined near M so that

$$f = \sum_{j=1}^{N-\ell} \alpha_j \rho_j \quad \textit{near} \quad M.$$

PROOF By a partition of unity argument, we need only to establish the existence of the α's in a neighborhood of each point in M. Give $\mathbb{R}^N$ the coordinates (x, y) with $x \in \mathbb{R}^\ell$, $y \in \mathbb{R}^{N-\ell}$. Near a given point in M, we use Lemma 1 to find a local diffeomorphism χ that takes a neighborhood of M to the set $\{y = 0\}$. Our given function f satisfies $f \circ \chi^{-1} = 0$ on $\{y = 0\}$. We shall show

$$f \circ \chi^{-1} = \sum_{j=1}^{N-\ell} \alpha_j y_j$$

for some choice of smooth functions $\alpha_1, \dots, \alpha_{N-\ell}$. The lemma will then follow by composing both sides with χ and noting that $y_j \circ \chi = \rho_j$ by Lemma 1. Let $\tilde{f} = f \circ \chi^{-1}$. To find the α_j, we write

$$\begin{aligned}
\tilde{f}(x, y) &= \tilde{f}(x, y) - \tilde{f}(x, 0) \\
&= \sum_{j=1}^{N-\ell} \tilde{f}(x, y_1, \dots, y_j, 0 \dots 0) - \tilde{f}(x, y_1, \dots, y_{j-1}, 0 \dots 0) \\
&= \sum_{j=1}^{N-\ell} \int_0^1 \frac{d}{dt}\{\tilde{f}(x, y_1, \dots, y_{j-1}, ty_j, 0 \dots 0)\} dt \\
&= \sum_{j=1}^{N-\ell} \left[\int_0^1 \frac{\partial \tilde{f}}{\partial y_j}(x, y_1, \dots, y_{j-1}, ty_j, 0 \dots 0) dt \right] \cdot y_j.
\end{aligned}$$

The lemma now follows by letting α_j be the expression in the brackets on the right. ∎

The concept of a submanifold can be defined for other categories of functions. Thus, we have the concept of a real analytic submanifold of a real analytic manifold and the concept of a complex submanifold of a complex manifold. The above three lemmas also locally hold in the real analytic and complex analytic categories. We leave the minor modifications of the above proofs to the reader.

We now discuss the concept of a submanifold with boundary. From Definition 1, an ℓ-dimensional submanifold is a set that can be locally straightened to an open subset of $\mathbb{R}^\ell$. Roughly speaking, an ℓ-dimensional submanifold with boundary is a set that can be locally straightened to a half space in $\mathbb{R}^\ell$. More precisely, we have the following definition.

DEFINITION 2 *A subset S of an N-dimensional manifold X is called an ℓ-dimensional submanifold with boundary if for each point $p \in S$, there is a neighborhood U of p in X and a diffeomorphism χ: $U \to \chi\{U\} \subset \mathbb{R}^N$ with*

$\chi(p) = 0$ *such that one of the following holds:*

(i) $\chi\{U \cap S\}$ *is an open neighborhood of the origin in* $\{t = (t_1, \ldots, t_N) \in \mathbb{R}^N;\, t_{\ell+1} = \cdots = t_N = 0\}$.

(ii) $\chi\{U \cap S\}$ *is an open neighborhood of the origin in* $\{t = (t_1, \ldots, t_N) \in \mathbb{R}^N; t_\ell \geq 0$ *and* $t_{\ell+1} = \cdots = t_N = 0\}$.

Both conditions cannot hold simultaneously for a given point p. A point in S that satisfies condition (i) is called a *manifold point.* A point in S that satisfies condition (ii) is called a *boundary point.* The set of boundary points in S is denoted ∂S. Note that ∂S is an $(\ell - 1)$-dimensional smooth submanifold of X. For example, suppose $S = \{x \in \mathbb{R}^N, \rho(x) \leq 0\}$ where ρ: $\mathbb{R}^N \to \mathbb{R}$ is smooth with $d\rho \neq 0$ on S. From Lemma 1, S is an N-dimensional submanifold with boundary in $\mathbb{R}^N$. Its boundary is the set $\partial S = \{x \in \mathbb{R}^N; \rho(x) = 0\}$.

2.3 Vectors on manifolds

We now define the concept of vectors in the manifold setting. Suppose X is a smooth N-dimensional manifold and let p_0 be a point on X. Let χ: $U \to \mathbb{R}^N$ be a coordinate chart defined on the coordinate patch U containing p_0. Here and for the rest of this chapter, we change our notation a bit. We denote the component functions for our coordinate chart χ by $(x_1, \ldots, x_N)$ instead of $\chi_1, \ldots, \chi_N$. We want to think of $x_1, \ldots, x_N$ as coordinates on X, rather than just component functions. To distinguish these coordinates from the coordinates in $\mathbb{R}^N$, we denote the latter by $t = (t_1, \ldots, t_N)$. For $p \in U$, define the vector $(\partial/\partial x_j)_p$ by the equation

$$\left(\frac{\partial}{\partial x_j}\right)_p \{f\} = \left(\frac{\partial}{\partial t_j}\right)_{\chi(p)} \{f \circ \chi^{-1}\}, \text{ for } f \in \mathcal{E}(U).$$

Since $f \circ \chi^{-1}$ is smoothly defined on the open subset $\chi\{U\}$ of $\mathbb{R}^N$, the Euclidean derivative on the right is well defined. Note that since $x_k \circ \chi^{-1}(t) = t_k$ for $t = (t_1, \ldots, t_N) \in \mathbb{R}^N$, we have

$$\frac{\partial}{\partial x_j}\{x_k\} = \begin{cases} 1 & \text{if } j = k \\ 0 & \text{otherwise.} \end{cases}$$

A *vector* at a point $p \in X$ is any operator of the form

$$\sum_{j=1}^{N} a_j \left(\frac{\partial}{\partial x_j}\right)_p \quad \text{where each} \quad a_j \in \mathbb{R}.$$

The set of all vectors on X at p is called the (real) tangent space of X at p and is denoted $T_p(X)$. So each vector in $T_p(X)$ is a linear functional on the space $\mathcal{E}(U)$.

It is particularly important to be able to identify the tangent space of a submanifold. We have the following lemma.

LEMMA 1

Suppose M is an ℓ-dimensional submanifold of $\mathbb{R}^N$ defined by $M = \{t \in \mathbb{R}^N; \rho_1(t) = \cdots = \rho_{N-\ell}(t) = 0\}$ with $d\rho_1 \wedge \ldots \wedge d\rho_{N-\ell} \neq 0$ on M. For a point $p \in M$

$$T_p(M) = \{L \in T_p(\mathbb{R}^N);\ L\{\rho_j\} = 0 \text{ at } p \text{ for } 1 \leq j \leq N - \ell\}.$$

PROOF This lemma has an intuitive geometric proof. From advanced calculus, each $\nabla\rho_j$ is orthogonal to M. $T_p(M)$ consists of all vectors in $\mathbb{R}^N$ that are orthogonal to $\{\nabla\rho_j(p);\ 1 \leq j \leq N - \ell\}$. Since $L\{\rho_j\} = \nabla\rho_j \cdot L$ (where $(\cdot)$ is the Euclidean inner product), the lemma follows.

This lemma also follows from Lemma 1 in Section 2.2. Near $p \in M$ we can find a local diffeomorphism $\chi = (x_1, \ldots, x_N)$ with $x_{\ell+1} = \rho_1, \ldots, x_N = \rho_{N-\ell}$. As mentioned in Section 2.2, the map $t \mapsto (x_1(t), \ldots, x_\ell(t)) \in \mathbb{R}^\ell$ serves as a coordinate chart for M. By definition, $T_p(M)$ is spanned by $(\partial/\partial x_1)_p, \ldots, (\partial/\partial x_\ell)_p$. Since $(\partial/\partial x_j)(x_k) = 0$ for $j \neq k$, clearly $(\partial/\partial x_j)(\rho_k) = 0$ for $1 \leq j \leq \ell$ and $1 \leq k \leq N - \ell$, and the lemma follows. ∎

Let us return to the general case of an N-dimensional manifold X. The collection of all $\{T_p(X)\}$ for $p \in X$ is called the *tangent bundle* of X and is denoted $T(X)$. If X is a complex manifold, then we will refer to $T(X)$ as the *real* tangent bundle in order to distinguish it from the complexified tangent bundle that is defined in Chapter 3.

A smooth section of the tangent bundle of X is called a *vector field* on X. More precisely, a vector field L on X is a smooth function that assigns to each point $p \in X$ a vector $L_p \in T_p(X)$. Here, smooth means that in any coordinate patch $U \subset X$ with coordinate chart χ: $U \to \mathbb{R}^N$, L can be expressed as

$$L_p = \sum_{j=1}^{N} a_j(p) \left(\frac{\partial}{\partial x_j}\right)_p \quad \text{for} \quad p \in U$$

where each a_j: $U \to \mathbb{R}$ is a smooth function. Note that if f is a smooth function on X and L is a vector field on X, then $L\{f\}$ is a smooth function on X.

We call $\mathbb{L} \subset T(M)$ an m-dimensional *subbundle* of $T(X)$ if $\mathbb{L}$ assigns to each point $p \in X$ an m-dimensional vector space $\mathbb{L}_p \subset T_p(M)$. These vector spaces are required to fit together smoothly in the sense that near each fixed point $p_0 \in X$, there are smooth vector fields $L_1, \ldots, L_m$ so that $\{(L_1)_p, \ldots, (L_m)_p\}$ forms a basis for $\mathbb{L}_p$. As an example, let X be an open subset of $\mathbb{R}^N$ and suppose

$\rho_1, \ldots, \rho_{N-\ell}$: $X \to \mathbb{R}$ are smooth functions with $d\rho_1 \wedge \ldots \wedge d\rho_{N-\ell} \neq 0$ on X. We let $\mathbb{L}_p$ be the set of vectors L_p with $L_p\{\rho_j\} = 0$ for $1 \leq j \leq N - \ell$. We leave it to the reader to show that $\mathbb{L} = \cup_{p \in X} \mathbb{L}_p$ is a subbundle of $T(X)$. By Lemma 1, each $\mathbb{L}_p$ is the tangent space at p for the submanifold given by the unique level set of $\rho = (\rho_1, \ldots, \rho_{N-\ell})$ that passes through p.

The tangent bundle is an example of a vector bundle over a smooth manifold. We shall not need the definition of a vector bundle in its most abstract form. Instead, we refer the reader to [W2].

Often, different coordinate patches are used near the same point $p_0 \in X$ and so it is useful to know how the description of a given vector in $T_{p_0}(X)$ changes as the coordinate chart changes. Let $\mathcal{Y} = (y_1, \ldots, y_N)$: $U' \to \mathbb{R}^N$ be another coordinate chart. For $p \in U \cap U'$, we describe the vector $L_p \in T_p(X)$ in terms of $\partial/\partial x_1, \ldots, \partial/\partial x_N$ and $\partial/\partial y_1, \ldots, \partial/\partial y_N$

$$L_p = \sum_{j=1}^{N} a_j \left(\partial/\partial x_j\right)_p$$

$$L_p = \sum_{j=1}^{N} b_j \left(\frac{\partial}{\partial y_j}\right)_p .$$

Since $\frac{\partial}{\partial y_j}(y_k) = \delta_{jk}$, we have

$$\begin{aligned} b_j &= L_p\{y_j\} \\ &= \sum_{k=1}^{N} a_k \left(\frac{\partial}{\partial x_k}\right)_p \{y_j\}. \end{aligned}$$

This gives us a relation between the a's and the b's. Let $L = \partial/\partial x_k$ (so $a_k = 1$ and all other $a_j = 0$); we have

$$b_j = \left(\frac{\partial}{\partial x_k}\right)_p \{y_j\}$$

and so

$$\left(\frac{\partial}{\partial x_k}\right)_p = \sum_{j=1}^{N} \left(\frac{\partial}{\partial x_k}\right)_p \{y_j\} \left(\frac{\partial}{\partial y_j}\right)_p .$$

We now describe the push forward of a vector under a smooth map between manifolds. This is given in Chapter 1 in the Euclidean setting and the definitions easily generalize to the manifold setting. Suppose $F: X \to Y$ is a smooth map between the smooth manifolds X and Y. This induces the *push forward* map $F_*(p)$: $T_p(X) \to T_{F(p)}(Y)$, which is defined by

$$F_*(p)(L_p)\{g\} = L_p\{g \circ F\}, \text{ for } L_p \in T_p(X).$$

Here, g is any smooth function defined in a neighborhood of $F(p)$. If L is a vector field on X then the above equation reads

$$(F_* L)_{F(p)}\{g\} = L_p\{g \circ F\}.$$

for each $p \in X$. Note that $(F \circ G)_* = F_* \circ G_*$ for two smooth maps F and G. If F has a smooth inverse, then $(F^{-1})_* = (F_*)^{-1}$.

From the definition of $\partial/\partial x_j$ given at the beginning of this section, $\partial/\partial x_j$ is the push forward of $\partial/\partial t_j$ (from $\mathbb{R}^N$) via the coordinate chart χ^{-1}, i.e., $\partial/\partial x_j = \chi_*^{-1}(\partial/\partial t_j)$.

In Chapter 1, we gave a formula for the push forward of a vector in terms of the derivative of the map. The analogous formula holds for maps between manifolds. Let F: $X \to Y$ be a smooth map. Suppose p is a point in X. Let $\chi = (x_1, \ldots, x_N)$: $U \to \mathbb{R}^N$ be a coordinate chart for X with $p \in U$ and let $\mathcal{Y} = (y_1, \ldots, y_{N'})$: $V \mapsto \mathbb{R}^{N'}$ be a coordinate chart for Y with $F(p) \in V$. Here $N' = \dim_{\mathbb{R}} Y$. We let $F_j = y_j \circ F$: $U \to \mathbb{R}$, $1 \leq j \leq N'$. The coefficient of $\partial/\partial y_j$ for the vector $[F_*(\partial/\partial x_k)]_{F(p)}$ is

$$\left[F_*\left(\frac{\partial}{\partial x_k}\right)\right]_{F(p)} \{y_j\}$$

which in view of the definition of push forward equals

$$\left(\frac{\partial}{\partial x_k}\right)_p \{F_j\}.$$

So

$$\left[F_*\left(\frac{\partial}{\partial x_k}\right)\right]_{F(p)} = \sum_{j=1}^{N'} \left(\frac{\partial}{\partial x_k}\right)_p \{F_j\} \cdot \left(\frac{\partial}{\partial y_j}\right)_{F(p)}, \qquad 1 \leq k \leq N.$$

If F is the identity map from X to itself, then we recover the change of coordinate formula given above which relates the $(\partial/\partial x)$-vectors to the $(\partial/\partial y)$-vectors.

2.4 Forms on manifolds

As with Euclidean space, the space of 1-forms at a point $p \in X$, denoted $T_p^*(X)$, is the space of all linear functionals on $T_p(X)$. The pairing between forms and vectors at p is denoted $\langle , \rangle_p$. For a coordinate chart $\chi = (x_1, \ldots, x_N)$: $U \to \mathbb{R}^N$, we let $dx_1, \ldots, dx_N$ be the dual basis to $\partial/\partial x_1, \ldots, \partial/\partial x_N$. The space of r-forms at p is the rth exterior power $\Lambda^r(T_p^*(X))$. This can be considered as the space of all linear functionals on $\Lambda^r(T_p(X))$, as in the Euclidean case. A basis (over $\mathbb{R}$) for $\Lambda^r(T_p^*(X))$ is given by

$\{dx^I = dx_{i_1} \wedge \ldots \wedge dx_{i_r};\ I = (i_1, \ldots, i_r)$ is an increasing index of length $r\}$.

By definition, the bundle of r forms $\Lambda^r(T^*(X))$ is the union $\cup_{p \in X} \Lambda^r(T_p^*(X))$. A smooth section of $\Lambda^r(T^*(X))$ is called a *differential form of degree* r, or more simply, a *smooth r-form* on X. By smooth, we mean that in each coordinate patch $U \subset X$ with coordinate chart $\chi := (x_1, \ldots, x_N)$: $U \to \mathbb{R}^N$, we can express the form as

$$\phi(p) = \sum_{|I|=r} \phi_I(p) dx^I$$

where each ϕ_I is a smooth function on U. The space of smooth r-forms on X will be denoted $\mathcal{E}^r(X)$. By pairing each side with $\partial/\partial x^J = \partial/\partial x_{j_1} \wedge \ldots \wedge \partial/\partial x_{j_r}$, we can express the coefficients as

$$\phi_J(p) = \left\langle \phi(p), \frac{\partial}{\partial x_J} \right\rangle_p .$$

So the coordinate description of an r-form depends on the coordinate chart used, as one would expect.

A smooth map F: $X \to Y$ between manifolds induces the pull back operator F^*: $\Lambda^r(T^*_{F(p)}(Y)) \to \Lambda^r(T_p^*(X))$. This is the dual of the operation of push forward of vectors, i.e.,

$$\langle F^*\phi, L\rangle_p = \langle \phi, F_*L\rangle_{F(p)} \text{ for } \phi \in T^*_{F(p)}(Y),\ L \in T_p(X).$$

For a coordinate chart $\chi = (x_1, \ldots, x_N)$, we have $dx_j = \chi^* dt_j$, $1 \le j \le N$, where $t = (t_1, \ldots, t_N)$ are the coordinates for $\mathbb{R}^N$. For higher degree forms we define

$$F^*(\phi_1 \wedge \ldots \wedge \phi_r) = (F^*\phi_1) \wedge \ldots \wedge (F^*\phi_r),\ \phi_i \in T^*(Y).$$

Note that $(F \circ G)^* = G^* \circ F^*$. In particular, if F is invertible then $(F^{-1})^* = (F^*)^{-1}$.

Now we define the exterior derivative d: $\mathcal{E}^r(X) \to \mathcal{E}^{r+1}(X)$. The definition given for the exterior derivative of an r-form on Euclidean space is a candidate for the definition in the manifold setting using a coordinate chart $\chi = (x_1, \ldots, x_N)$. However, we must show that the resulting coordinate definition is independent of the coordinate chart. The key lemma is the following.

LEMMA 1

Suppose χ: $U \to \mathbb{R}^N$ and $\mathcal{Y}$: $V \to \mathbb{R}^N$ are two coordinate charts for X. For $\phi \in \mathcal{E}^r(X)$,

$$\chi^*\{d_{\mathbb{R}^N}(\chi^{-1^*}\phi)\} = \mathcal{Y}^*\{d_{\mathbb{R}^N}(\mathcal{Y}^{-1^*}\phi)\} \quad \text{on} \quad U \cap V.$$

PROOF The notation $d_{\mathbb{R}^N}$ refers to the Euclidean exterior derivative. The assertion of the lemma is equivalent to

$$\mathcal{Y}^{-1^*} \circ \chi^* \{ d_{\mathbb{R}^N} (\chi^{-1^*} \phi) \} = d_{\mathbb{R}^N} \{ \mathcal{Y}^{-1^*} \phi \}.$$

Recall, $\mathcal{Y}^{-1^*} \circ \chi^* = (\chi \circ \mathcal{Y}^{-1})^*$. Also $\chi \circ \mathcal{Y}^{-1}$ is a smooth map between open subsets of $\mathbb{R}^N$. Lemma 2 in Section 1.4 states that the exterior derivative on $\mathbb{R}^N$ commutes with pull back. Therefore

$$\begin{aligned}(\chi \circ \mathcal{Y}^{-1})^* d_{\mathbb{R}^N} \{ \chi^{-1^*} \phi \} &= d_{\mathbb{R}^N} \{ (\chi \circ \mathcal{Y}^{-1})^* \chi^{-1^*} \phi \} \\ &= d_{\mathbb{R}^N} \{ (\chi^{-1} \circ \chi \circ \mathcal{Y}^{-1})^* \phi \} \\ &= d_{\mathbb{R}^N} \{ \mathcal{Y}^{-1^*} \phi \}\end{aligned}$$

as desired. ∎

The above lemma allows us to define the exterior derivative unambiguously.

DEFINITION 1 *The exterior derivative on* X d_X: $\mathcal{E}^r(X) \to \mathcal{E}^{r+1}(X)$ *is defined in a coordinate patch* $U \subset \mathbb{R}^N$ *by*

$$d_X \phi = \chi^* (d_{\mathbb{R}^N} \{ \chi^{-1^*} \phi \})$$

where χ: $U \to \mathbb{R}^N$ *is the corresponding coordinate chart.*

If the manifold X is clear from the context, then the "X" often will be omitted from the notation of the exterior derivative.

If $\chi = (x_1, \ldots, x_N)$ is a coordinate chart then $\chi_*(\partial/\partial x_j) = \partial/\partial t_j$ and $\chi^*(dt_j) = dx_j$ for $1 \leq j \leq N$. Thus, from the above definition

$$d_X \{ \phi_I(p) dx^I \} = \sum_{j=1}^{N} \frac{\partial \phi_I}{\partial x_j}(p) dx_j \wedge dx^I$$

where we have written $(\partial \phi_I / \partial x_j)(p) = (\partial/\partial x_j)_p \{ \phi_I \}$.

The fact that the exterior derivative on Euclidean space commutes with pull backs generalizes to manifolds.

LEMMA 2

Suppose F: $X \to Y$ *is a smooth map between manifolds. Then* $d_X \circ F^* = F^* \circ d_Y$.

PROOF Let χ: $U \to \mathbb{R}^N$ and $\mathcal{Y}$: $V \to \mathbb{R}^{N'}$ be coordinate charts for X and Y respectively with $F\{U\} \cap V \neq \emptyset$. Using the definition of the exterior derivative, the statement of the lemma is equivalent to

$$\chi^* \circ d_{\mathbb{R}^N} \circ \chi^{-1^*} \circ F^* = F^* \circ \mathcal{Y}^* \circ d_{\mathbb{R}^{N'}} \circ \mathcal{Y}^{-1^*}$$

which in turn is equivalent to

$$d_{\mathbb{R}^N} \circ (F \circ \chi^{-1})^* = (\mathcal{Y} \circ F \circ \chi^{-1})^* \circ d_{\mathbb{R}^{N'}} \circ \mathcal{Y}^{-1^*}.$$

As in the proof of the previous lemma, this equation follows by commuting the exterior derivative on Euclidean space with the pull back of the map $\mathcal{Y} \circ F \circ \chi^{-1}$: $\mathbb{R}^N \to \mathbb{R}^{N'}$. ∎

As an example, suppose $X = \{x \in \mathbb{R}^N; \rho(x) = 0\}$ where ρ: $\mathbb{R}^N \to \mathbb{R}$ is smooth with $d\rho \neq 0$ on M. Let j: $X \to \mathbb{R}^N$ be the inclusion map. Then $j^* \circ d_{\mathbb{R}^N} = d_X \circ j^*$. In particular, $j^* d_{\mathbb{R}^N}(\rho) = 0$ because $j^*\rho = \rho \circ j = 0$ on X.

2.5 Integration on manifolds

We first discuss orientation. The form $dt_1 \wedge \ldots \wedge dt_N$ on $\mathbb{R}^N$ induces the standard orientation on $\mathbb{R}^N$. A basis of 1-forms $\phi_1, \ldots, \phi_N$ for $T_p^*(\mathbb{R}^N)$ is said to be *oriented* if $\phi_1 \wedge \ldots \wedge \phi_N = \alpha dt_1 \wedge \ldots \wedge dt_N$ with $\alpha > 0$. A linear map L: $\mathbb{R}^N \to \mathbb{R}^N$ is said to be *orientation preserving* if $L^*\phi_1, \ldots, L^*\phi_N$ is oriented whenever $\phi_1, \ldots, \phi_N$ is oriented. This is equivalent to the condition that the determinant of the matrix that represents L is positive. A smooth map F: $\mathbb{R}^N \to \mathbb{R}^N$ is said to be *orientation preserving* if $F^*(p)$: $T_{F(p)}^*(\mathbb{R}^N) \to T_p^*(\mathbb{R}^N)$ is orientation preserving for each point p in its domain of definition. This is equivalent to the condition that $\det DF(p) > 0$ for each point $p \in \mathbb{R}^N$ in the domain of definition of F.

Let X be a smooth N-dimensional manifold. A collection C of coordinate charts $\{\chi$: $U \to \mathbb{R}^N\}$ for X is said to be orientation preserving if $\chi \circ \mathcal{Y}^{-1}$ is orientation preserving as a map from $\mathbb{R}^N$ to $\mathbb{R}^N$ for each $\chi, \mathcal{Y}$ belonging to C. The manifold X is said to be *orientable* if there is a collection of orientation-preserving coordinate charts such that the corresponding coordinate patches cover X. If X is a submanifold with boundary, then we define X to be orientable in the same way. It is an easy exercise to show that if a manifold with boundary is orientable then its boundary is also orientable.

Every complex manifold is orientable. For if F is a holomorphic map between any two open sets in $\mathbb{C}^n$, then $\det DF = |\det(\partial F/\partial z)|^2$. Here, DF is the real derivative of F as a map from $\mathbb{R}^{2n}$ to $\mathbb{R}^{2n}$ whereas $\partial F/\partial z$ is the complex derivative of F as a map from $\mathbb{C}^n$ to $\mathbb{C}^n$. Therefore, $\chi \circ \mathcal{Y}^{-1}$ is orientation preserving for any holomorphic coordinate charts χ and $\mathcal{Y}$ on a complex manifold.

If the manifold X is orientable, then its collection C of orientation-preserving coordinate charts determines an orientation for $T_p^*(X)$ as follows: a set of one forms $\{\phi_1, \ldots, \phi_N\} \in T_p^*(X)$ is said to be oriented if $\phi_1 \wedge \ldots \wedge \phi_N$ at p is a positive multiple of $dx_1 \wedge \ldots \wedge dx_N$ where the chart $\chi = (x_1, \ldots, x_N)$: $U \to \mathbb{R}^N$ belongs to C. Since $\chi \circ \mathcal{Y}^{-1}$ is orientation preserving for $\chi, \mathcal{Y} \in C$, this definition is independent of the chart $\chi \in C$. With such an orientation for $T_p^*(X)$ for each $p \in X$, we call X an *oriented manifold* and we call the chart $\chi \in C$ an *oriented coordinate chart*.

If M is an ℓ-dimensional oriented submanifold with boundary contained in an N-dimensional manifold X, then we define the induced orientation on ∂M. For $p \in \partial M$, let $\chi = (x_1, \ldots, x_N)$: $U \subset X \to \mathbb{R}^N$ be an oriented boundary coordinate chart, i.e., a coordinate chart $\chi \in C$ with $\chi(p) = 0$ and $\chi\{U \cap M\} = \{(t_1, \ldots, t_N) \in \mathbb{R}^N; t_\ell \geq 0 \text{ and } t_{\ell+1} = \cdots = t_N = 0\}$. Let $n = -dx_\ell = -\chi^* dt_\ell$. This 1-form is called the outward pointing co-normal for ∂M. If $\mathcal{Y}$: $V \subset X \to \mathbb{R}^N$ (with $p \in V$) is another oriented boundary coordinate chart on M, then dy_ℓ at p is a positive multiple of dx_ℓ at p. Hence, we can make the following unambiguous definition: a collection of one forms $\{\phi_1, \ldots, \phi_{\ell-1}\} \subset T_p^*(\partial M)$ is said to have the *induced orientation* on ∂M provided $\{n, \phi_1, \ldots, \phi_{\ell-1}\}$ has the orientation on M, i.e., if $n \wedge \phi_1 \wedge \ldots \wedge \phi_{\ell-1}$ is a positive multiple of $dx_1 \wedge \ldots \wedge dx_\ell$. This orientation differs from the orientation on ∂M obtained by restricting the oriented coordinate charts for M to ∂M. A collection of forms $\{\phi_1, \ldots, \phi_{\ell-1}\}$ has this orientation if $\phi_1 \wedge \ldots \wedge \phi_{\ell-1}$ is a positive multiple of $dx_1 \wedge \ldots \wedge dx_{\ell-1}$ or equivalently, if $\phi_1 \wedge \ldots \wedge \phi_{\ell-1} \wedge n$ is a positive multiple of $dx_1 \wedge \ldots \wedge dx_{\ell-1} \wedge n$. Since

$$n \wedge dx_1 \wedge \ldots \wedge dx_{\ell-1} = (-1)^\ell dx_1 \wedge \ldots \wedge dx_\ell,$$

the induced boundary orientation on ∂M differs by a factor of $(-1)^\ell$ from the orientation obtained by restricting the oriented coordinate charts from M to ∂M. As an example, let $M = \{(x_1, x_2, x_3); x_3 \geq 0\}$. Then $\partial M = \{(x_1, x_2, 0)\}$ and $n = -dx_3$. The orientation on ∂M as a subspace of $\mathbb{R}^3$ is determined by $dx_1 \wedge dx_2$. However $dx_1 \wedge dx_2$ has the opposite orientation from the induced orientation on ∂M as the boundary of M.

Now we define the integral of a differential form. For $\alpha \in \mathcal{D}(\mathbb{R}^N)$, we define

$$\int_{\mathbb{R}^N} \alpha dx_1 \wedge \ldots \wedge dx_N = \int_{\mathbb{R}^N} \alpha dx_1 \ldots dx_N$$

where the right side is computed in the usual way.

Suppose U and V are open sets in $\mathbb{R}^N$ and let F: $U \to V$ be a diffeomorphism. From Section 1.3, $F^*(\alpha dx) = (\alpha \circ F)(\det DF) dx$ where $dx = dx_1 \wedge \ldots \wedge dx_N$. Using the language of differential forms, the change of variables formula for integration is

$$\int_V \alpha dx = \begin{cases} \int_U F^*(\alpha dx) & \text{if Det } DF > 0 \\ -\int_U F^*(\alpha dx) & \text{if Det } DF < 0. \end{cases}$$

Now suppose X is an oriented N-dimensional manifold and let ϕ be a smooth

N-form with support contained in a coordinate patch U of X. If χ: $U \to \mathbb{R}^N$ is the corresponding oriented coordinate chart, then define

$$\int_X \phi = \int_{\mathbb{R}^N} \chi^{-1^*} \phi.$$

We formally define the integral of an r-form on an N-dimensional manifold to be zero if $r \neq N$.

The first thing to check is whether or not the above definition is independent of the oriented coordinate chart. If $\mathcal{Y}$: $V \to \mathbb{R}^N$ is another such chart with supp $\phi \subset V$, then we must show

$$\int_{\mathbb{R}^N} \mathcal{Y}^{-1^*} \phi = \int_{\mathbb{R}^N} \chi^{-1^*} \phi.$$

Since $\mathcal{Y} \circ \chi^{-1}$ is a smooth, orientation-preserving map from $\mathbb{R}^N \to \mathbb{R}^N$, $\det D(\mathcal{Y} \circ \chi^{-1}) > 0$. Using the above change of variables formula for integration, we have

$$\begin{aligned} \int_{\mathbb{R}^N} \mathcal{Y}^{-1^*} \phi &= \int_{\mathbb{R}^N} (\mathcal{Y} \circ \chi^{-1})^* (\mathcal{Y}^{-1^*} \phi) \\ &= \int_{\mathbb{R}^N} \chi^{-1^*} \mathcal{Y}^* \mathcal{Y}^{-1^*} \phi \\ &= \int_{\mathbb{R}^N} \chi^{-1^*} \phi \end{aligned}$$

as desired. We conclude that the integral of a form supported in a coordinate patch is unambiguously defined.

Suppose ϕ is a compactly supported N-form on X (but not necessarily supported in *one* coordinate patch). Suppose $\{U_1, \ldots, U_m\}$ is a collection of coordinate patches that cover supp ϕ and let χ^j: $U_j \to \mathbb{R}^N$ be the corresponding oriented coordinate charts. Let $\{\phi_1, \ldots, \phi_m\}$ be a partition of unity for supp ϕ with $\phi_j \in \mathcal{D}(U_j)$. Then we define

$$\int_X \phi = \sum_{j=1}^{m} \int_{\mathbb{R}^N} (\chi^j)^{-1^*} \{\phi_j \phi\}.$$

Again we check that this definition is independent of the choices made. Suppose $\{\mathcal{Y}_j$: $V_j \to \mathbb{R}^N$, $j = 1, \ldots, n\}$ is another set of oriented coordinate charts whose corresponding coordinate patches cover supp ϕ. Let $\{\psi_1, \ldots, \psi_n\}$ be a partition of unity for supp ϕ with $\psi_j \in \mathcal{D}(V_j)$. Since $\sum \psi_k \equiv 1$,

$$\sum_{j=1}^{m} \int_{\mathbb{R}^N} (\chi^j)^{-1^*} \{\phi_j \phi\} = \sum_{k=1}^{n} \sum_{j=1}^{m} \int_{\mathbb{R}^N} (\chi^j)^{-1^*} \{\phi_j \psi_k \phi\}.$$

Since $\phi_j \psi_k \phi$ has support in a single coordinate patch, we have

$$\int_{\mathbb{R}^N} (\chi^j)^{-1^*} \{\phi_j \psi_k \phi\} = \int_{\mathbb{R}^N} (\mathcal{Y}^k)^{-1^*} \{\phi_j \psi_k \phi\}.$$

Summing the right side over j and using $\sum \phi_j \equiv 1$, we obtain

$$\sum_{j=1}^{m} \int_{\mathbb{R}^N} (\chi^j)^{-1^*} \{\phi_j \phi\} = \sum_{k=1}^{n} \int_{\mathbb{R}^N} (\mathcal{Y}^k)^{-1^*} \{\psi_k \phi\}$$

and so our definition of the integral of ϕ is well defined.

If M is an ℓ-dimensional submanifold of X, then we define

$$\int_M \phi = \int_M j^* \phi \qquad \phi \in \mathcal{D}^\ell(X)$$

where j: $M \to X$ is the inclusion map. The point is that $j^*\phi$ is an intrinsically defined form on the ℓ-dimensional manifold M and so the right side can be evaluated as described above. If ϕ contains a factor of $d\rho$ where ρ: $X \to \mathbb{R}$ vanishes on M, then $\int_M \phi$ must also vanish since $j^* d\rho = dj^*\rho = 0$ on M. Also if $r \neq \ell$, then $\int_M \phi = 0$ for $\phi \in \mathcal{D}^r(X)$.

If M is a smooth ℓ-dimensional submanifold with boundary contained in an N-dimensional manifold X, then $\int_M \phi$ is defined in a similar way to the case when M is a manifold. Here, the only difference is that M has two types of coordinate charts. The first type is a coordinate chart about a manifold point and the second is a coordinate chart about a boundary point. If χ: $U \to \mathbb{R}^N$ is a coordinate chart about a manifold point, then $\chi\{U \cap M\}$ is an open subset of $\mathbb{R}^\ell = \{(t_1, \ldots, t_N); t_{\ell+1} = \cdots = t_N = 0\}$. If ϕ is an ℓ-form with support in $U \cap M$, then by definition,

$$\int_M \phi = \int_{\mathbb{R}^\ell} \chi^{-1^*}(\phi).$$

If χ: $U \to \mathbb{R}^N$ is a coordinate chart containing a boundary point, then $\chi\{M \cap U\}$ is an open subset of the half space of $\mathbb{R}^\ell$ given by $\{(t_1, \ldots, t_N); t_\ell \geq 0$ and $t_{\ell+1} = \cdots = t_N = 0\}$. If ϕ is a form of degree ℓ with support in $U \cap M$, then by definition,

$$\int_M \phi = \int_{\substack{t \in \mathbb{R}^\ell \\ t_\ell \geq 0}} (\chi^{-1^*} \phi)(t).$$

As with the case of a manifold without boundary, the above definitions are independent of the oriented coordinate chart. If ϕ is a compactly supported ℓ-form on M (but not necessarily supported in one coordinate patch), then its integral is defined in the same way as the case where M has no boundary by using a partition of unity.

Now suppose X and Y are smooth, oriented N-dimensional manifolds and suppose $F\colon X \to Y$ is a smooth map. We say that F preserves orientation if $\mathcal{Y} \circ F \circ \chi^{-1}$ preserves orientation on $\mathbb{R}^N$ for each oriented coordinate chart χ for X and $\mathcal{Y}$ for Y. The change of variables formula for $\mathbb{R}^N$ easily generalizes to manifolds.

LEMMA 1
Suppose X and Y are smooth, oriented N-dimensional manifolds. Suppose $F\colon X \to Y$ is a smooth, orientation-preserving map. If $\phi \in \mathcal{D}^N(Y)$, then

$$\int_Y \phi = \int_X F^*\phi.$$

The proof of this lemma follows from the definitions together with the change of variables formula on $\mathbb{R}^N$. Details are left to the reader.

We are now ready to state and prove Stokes' theorem, which for an oriented submanifold M with boundary, equates the integral of $d\phi$ over M with the integral of ϕ over ∂M. The simplest example of Stokes' theorem is the fundamental theorem of calculus, where M is the interval $\{a \le x \le b\}$ in $\mathbb{R}$. We shall show that the definitions of forms, orientation, and integration reduce the proof of Stokes' theorem for more general manifolds to the fundamental theorem of calculus.

THEOREM 1 STOKES
Suppose M is a smooth, oriented ℓ-dimensional submanifold with boundary that is contained in a smooth N-dimensional manifold X. Suppose $\phi \in \mathcal{D}^{\ell-1}(M)$, then

$$\int_M d_M\phi = \int_{\partial M} \phi.$$

Here, ∂M has the induced boundary orientation.

PROOF We first cover supp ϕ with a finite number of oriented coordinate charts $\chi^i\colon U_i \subset X \mapsto \mathbb{R}^N$ $i = 1, \dots, m$ that are either of manifold type (i.e., $\chi^i\{U_i \cap M\} = \{(t_1, \dots, t_N); t_{\ell+1} = \cdots = t_N = 0\}$) or boundary type (i.e., $\chi^i\{U_i \cap M\} = \{(t_1, \dots, t_N); t_\ell \ge 0, t_{\ell+1} = \cdots = t_N = 0\}$). Let $\{\phi_i \in \mathcal{D}(U_i)\}$ be a partition of unity for supp ϕ. It suffices to prove

$$\int_M d_M\{\phi_i\phi\} = \int_{\partial M} \phi_i\phi$$

for each $i = 1, \ldots, m$. Stokes' theorem then follows by summing over i and using $\sum \phi_i = 1$ on supp ϕ. So from now on, we assume that supp ϕ is contained in a coordinate patch (one of the U_i's).

We first consider the case where $\chi\colon U \cap M \to \mathbb{R}^\ell$ is a coordinate chart of manifold type. Using the definitions of the integral and the exterior derivative on M, we have

$$\int_M d_M\phi = \int_{\chi\{U\cap M\}\subset\mathbb{R}^\ell} d\{\chi^{-1^*}\phi\} = \int_{\mathbb{R}^\ell} d\{\chi^{-1^*}\phi\}.$$

Here and below, d refers to the exterior derivative on $\mathbb{R}^\ell$. Since $(\chi^{-1^*}\phi)$ is a compactly supported $(\ell-1)$-form on $\mathbb{R}^\ell$, we may write

$$\chi^{-1^*}\phi = \sum_{j=1}^{\ell} \alpha_j dt_1 \wedge \ldots \wedge \widehat{dt_j} \wedge \ldots \wedge dt_\ell$$

where each α_j belongs to $\mathcal{D}(\mathbb{R}^\ell)$. The notation $\widehat{dt_j}$ indicates that dt_j has been omitted. We have

$$d\{\chi^{-1^*}\phi\} = \left(\sum_{j=1}^{\ell}(-1)^{j-1}\frac{\partial\alpha_j}{\partial t_j}\right) dt_1 \wedge \ldots \wedge dt_\ell.$$

So

$$\begin{aligned}\int_{\mathbb{R}^\ell} d\{\chi^{-1^*}\phi\} &= \sum_{j=1}^{\ell}(-1)^{j-1}\int_{-\infty}^{\infty}\cdots\int_{-\infty}^{\infty}\frac{\partial\alpha_j}{\partial t_j}dt_1\ldots dt_\ell\\ &= 0.\end{aligned}$$

The last equality follows from the fundamental theorem of calculus in the t_j-variable, using the fact that α_j has compact support.

Therefore, we have shown that if ϕ has support in a coordinate patch, U, of manifold type, then $\int_M d_M\phi = 0$. On the other hand, we have $U \cap \partial M = \emptyset$. Therefore, $\int_{\partial M}\phi$ also vanishes and we have established Stokes' theorem: $\int_M d_M\phi = \int_{\partial M}\phi$ in this case.

The remaining case to consider is where ϕ has support in a coordinate patch U of boundary type. In this case, we have

$$\begin{aligned}\int_M d_M\phi &= \int_{\chi\{U\cap M\}} d\{\chi^{-1^*}\phi\}\\ &= \int_{\substack{t\in\mathbb{R}^\ell\\ t_\ell\geq 0}} d\{\chi^{-1^*}\phi\}.\end{aligned}$$

As above, we write

$$\chi^{-1^*}\phi = \sum_{j=1}^{\ell} \alpha_j dt_1 \wedge \ldots \wedge \widehat{dt_j} \wedge \ldots dt_\ell$$

where $\alpha_j \in \mathcal{D}(\mathbb{R}^\ell)$. We claim that for each $j = 1, \ldots, \ell$,

$$\int_{\{t_\ell \geq 0\}} d\{\alpha_j dt_1 \wedge \ldots \widehat{dt_j} \wedge \ldots \wedge dt_\ell\} = (-1)^\ell \int_{\{t_\ell = 0\}} \alpha_j dt_1 \wedge \ldots \wedge \widehat{dt_j} \wedge \ldots \wedge dt_\ell. \quad (1)$$

This is just the special case of Stokes' theorem where M is the half space $\{t_\ell \geq 0\}$ because, as mentioned earlier, the induced boundary orientation of $\{t_\ell = 0\}$ differs from its inherited orientation from $\mathbb{R}^\ell$ by the factor $(-1)^\ell$.

If $j < \ell$, then

$$\begin{aligned}\int_{\{t_\ell \geq 0\}} d\{\alpha_j dt_1 \wedge \ldots \wedge \widehat{dt_j} \wedge \ldots \wedge dt_\ell\} &= \int_{\{t_\ell \geq 0\}} (-1)^{j-1} \frac{\partial \alpha_j}{\partial t_j} dt_1 \wedge \ldots \wedge dt_\ell \\ &= 0.\end{aligned}$$

The last equation follows by the fundamental theorem of calculus in the t_j-variable, using the fact that α_j has compact support. Since $j < \ell$, there is no boundary term. If $j < \ell$, then the right side of (1) also vanishes because the pull back of dt_ℓ to $\{t_\ell = 0\}$ is zero. Thus, both sides of (1) vanish for $j < \ell$.

If $j = \ell$, then

$$\begin{aligned}\int_{\{t_\ell \geq 0\}} d\{\alpha_\ell dt_1 \wedge \ldots \wedge dt_{\ell-1}\} &= (-1)^{\ell-1} \int_{\{t_\ell \geq 0\}} \frac{\partial \alpha_\ell}{\partial t_\ell} dt_1 \wedge \ldots \wedge dt_\ell \\ &= (-1)^\ell \int_{\{t_\ell = 0\}} \alpha_\ell(t_1, \ldots, t_{\ell-1}, 0) dt_1 \wedge \ldots \wedge dt_{\ell-1}\end{aligned}$$

where the last equality follows from the fundamental theorem of calculus in the t_ℓ-variable. This proves (1) for $j = \ell$.

Summing (1) over j yields

$$\int_{\{t_\ell \geq 0\}} d\{\chi^{-1^*}\phi\} = (-1)^\ell \int_{\substack{\mathbb{R}^{\ell-1} \\ \{t_\ell = 0\}}} \chi^{-1^*}\phi.$$

Therefore, Stokes' theorem follows for the case of a coordinate chart χ: $U \cap M \to \mathbb{R}^\ell$ of boundary type. The proof of Stokes' theorem is now complete. ∎

We prove two useful corollaries: the divergence theorem and Green's formula. Let $M = \Omega$ be an open subset of $\mathbb{R}^N$ with smooth boundary. Define

ρ: $\mathbb{R}^N \to \mathbb{R}$ by

$$\rho(x) = \begin{cases} -\text{dist}(x, \mathbb{R}^N - \Omega) & \text{if } x \in \Omega \\ \text{dist}(x, \Omega) & \text{if } x \in \mathbb{R}^N - \Omega. \end{cases}$$

It is an easy exercise using the inverse function theorem to show that ρ is smoothly defined on a neighborhood of $\partial\Omega$. Define the vector field $\mathbb{N} = \nabla\rho = \sum_{j=1}^N (\partial\rho/\partial x_j)(\partial/\partial x_j)$. $\mathbb{N}$ is the unit outward normal to $\partial\Omega$. The volume form for $\partial\Omega$ is given by the $(N-1)$-form

$$\begin{aligned} d\sigma &= \mathbb{N}\lrcorner dx \quad \text{where} \quad dx = dx_1 \wedge \ldots \wedge dx_N \\ &= \sum_{j=1}^N (-1)^{j-1} \frac{\partial\rho}{\partial x_j} dx_1 \wedge \ldots \wedge \widehat{dx_j} \wedge \ldots \wedge dx_N. \end{aligned}$$

COROLLARY 1 DIVERGENCE THEOREM

Suppose Ω is a bounded open set with smooth boundary in $\mathbb{R}^N$. Let $F = \sum_{j=1}^N F_j(\partial/\partial x_j)$ be a smooth vector field on $\overline{\Omega}$. Then

$$\int_{\partial\Omega} (F \cdot \mathbb{N}) d\sigma = \int_{\Omega} (Div\ F) dx$$

where $Div\ F = \sum_{j=1}^N \partial F_j/\partial x_j$ and $F \cdot \mathbb{N} = \sum_{j=1}^N F_j(\partial\rho/\partial x_j)$.

PROOF Since $\mathbb{N}$ has unit length, $\mathbb{N}\lrcorner d\rho = 1$ on $\partial\Omega$. So if ϕ is a smooth $(N-1)$-form on $\partial\Omega$, then from the product rule for $\lrcorner$ (see Lemma 1 in Section 1.5), we have

$$\begin{aligned} \mathbb{N}\lrcorner(d\rho \wedge \phi) &= (\mathbb{N}\lrcorner d\rho)\phi - d\rho \wedge (\mathbb{N}\lrcorner\phi) \\ &= \phi - d\rho \wedge (\mathbb{N}\lrcorner\phi). \end{aligned}$$

From the definition of the integral over a submanifold, we obtain

$$\begin{aligned} \int_{\partial\Omega} d\rho \wedge (\mathbb{N}\lrcorner\phi) &= \int_{\partial\Omega} j^*(d\rho) \wedge j^*(\mathbb{N}\lrcorner\phi) \\ &= 0 \end{aligned}$$

because $j^* d\rho = d(j^*\rho) = 0$ on $\partial\Omega$. Together with the previous equation, we have

$$\int_{\partial\Omega} \phi = \int_{\partial\Omega} \mathbb{N}\lrcorner(d\rho \wedge \phi).$$

Applying this equation to the form $\phi = \sum_{j=1}^N (-1)^{j-1} F_j dx_1 \wedge \ldots \wedge \widehat{dx_j} \wedge \ldots \wedge dx_N$

yields

$$\sum_{j=1}^{N} \int_{\partial\Omega} (-1)^{j-1} F_j dx_1 \wedge \ldots \wedge \widehat{dx_j} \wedge \ldots \wedge dx_N$$

$$= \sum_{j=1}^{N} \int_{\partial\Omega} F_j \frac{\partial \rho}{\partial x_j} \mathbb{N} \lrcorner (dx_1 \wedge \ldots \wedge dx_N)$$

$$= \int_{\partial\Omega} (F \cdot \mathbb{N}) d\sigma.$$

On thc other hand, by Stokes' theorem

$$\int_{\partial\Omega} \sum_{j=1}^{N} (-1)^{j-1} F_j dx_1 \wedge \ldots \wedge \widehat{dx_j} \wedge \ldots \wedge dx_N$$

$$= \int_{\Omega} \sum_{j=1}^{N} d((-1)^{j-1} F_j dx_1 \wedge \ldots \wedge \widehat{dx_j} \wedge \ldots \wedge dx_N)$$

$$= \int_{\Omega} \left(\sum_{j=1}^{N} \frac{\partial F_j}{\partial x_j} \right) dx_1 \wedge \ldots \wedge dx_N$$

$$= \int_{\Omega} (\text{Div } F) dx.$$

The proof of the divergence theorem is now complete. ∎

COROLLARY 2 GREEN'S FORMULA
Suppose $\Omega \subset \mathbb{R}^N$ is a bounded domain with smooth boundary and suppose u and v are smooth functions on $\overline{\Omega}$. Then

$$\int_{\partial\Omega} (u\mathbb{N}v - v\mathbb{N}u) d\sigma = \int_{\Omega} (u\Delta v - v\Delta u) dx$$

where $\mathbb{N}$ is the unit outward normal to $\partial\Omega$.

PROOF We apply the divergence theorem to the vector field

$$F = v(\nabla u) = v \sum_{j=1}^{N} \frac{\partial u}{\partial x_j} \frac{\partial}{\partial x_j}$$

to obtain

$$\int_{\partial\Omega} v(\mathbb{N}\cdot\nabla u)d\sigma = \int_{\Omega} v(\Delta u)dx + \int_{\Omega}\left(\sum_{j=1}^{N}\frac{\partial v}{\partial x_j}\frac{\partial u}{\partial x_j}\right)dx.$$

Next, we apply the divergence theorem to the vector field

$$G = u(\nabla v) = u\sum_{j=1}^{N}\frac{\partial v}{\partial x_j}\frac{\partial}{\partial x_j}$$

to obtain

$$\int_{\partial\Omega} u(\mathbb{N}\cdot\nabla v)d\sigma = \int_{\Omega} u(\Delta v)dx + \int_{\Omega}\left(\sum_{j=1}^{N}\frac{\partial u}{\partial x_j}\frac{\partial v}{\partial x_j}\right)dx.$$

Green's formula follows by subtracting these two integral equations. ∎

3

Complexified Vectors and Forms

Objects such as

$$\frac{\partial}{\partial z_j} = \frac{1}{2}\left(\frac{\partial}{\partial x_j} - i\frac{\partial}{\partial y_j}\right)$$

or

$$dz_j = dx_j + idy_j$$

are often encountered in complex analysis. The former is a complexified vector and the latter is a complexified form (due to the presence of $i = \sqrt{-1}$). In this chapter, we make these notions precise and generalize some of the concepts in Chapters 1 and 2 to the complexified setting.

3.1 Complexification of a real vector space

Suppose V is a real vector space. The *complexification* of V is the tensor product $V \otimes_{\mathbb{R}} \mathbb{C}$ (or for brevity, $V \otimes \mathbb{C}$). As a vector space over the reals, $V \otimes \mathbb{C}$ is generated by $v \otimes 1$ and $v \otimes i$ for $v \in V$. $V \otimes \mathbb{C}$ can be made into a complex vector space by defining

$$\alpha(v \otimes \beta) = v \otimes \alpha\beta \qquad \text{for } \alpha, \beta \in \mathbb{C} \text{ and } v \in V.$$

As a complex vector space, $V \otimes \mathbb{C}$ is generated over $\mathbb{C}$ by $v \otimes 1$ for $v \in V$. If V is a vector space of real dimension N, then $\dim_{\mathbb{R}} V \otimes \mathbb{C} = 2N$ and $\dim_{\mathbb{C}} V \otimes \mathbb{C} = N$. For shorthand, we write $v\alpha$ or αv for $v \otimes \alpha$. We identify v with $1v = v \otimes 1$. In this way, V is naturally imbedded into $V \otimes \mathbb{C}$ by identifying V with $V \otimes 1$. There is also a natural conjugation operator for $V \otimes \mathbb{C}$. This is defined by

$$\overline{\alpha v} = \bar{\alpha} v = v \otimes \bar{\alpha}.$$

As an example, let M be a smooth manifold of real dimension N. For $p \in M$, $T_p(M) \otimes \mathbb{C}$ is called the complexified tangent space and $T_p^*(M) \otimes \mathbb{C}$ is called the complexified cotangent space. The space $T_p^*(M) \otimes \mathbb{C}$ can be viewed as the complex dual space of $T_p(M) \otimes \mathbb{C}$ by defining the pairing

$$\langle \phi \otimes \alpha, L \otimes \beta \rangle_p = \langle \phi, L \rangle_p \alpha\beta$$

for $\phi \in T_p^*(M)$, $L \in T_p(M)$, and $\alpha, \beta \in \mathbb{C}$.

If $F : M \to N$ is a smooth map between smooth manifolds, then the push forward and pull back operators extend to the complexified setting in a complex linear fashion. We define $F_*(L \otimes \alpha) = F_*(L) \otimes \alpha$ and $F^*(\phi \otimes \alpha) = F^*\phi \otimes \alpha$ for $L \in T_p(M)$, $\phi \in T^*_{F(p)}(N)$, and $\alpha \in \mathbb{C}$. Note that these operators commute with the conjugation operator.

The *complexified tangent bundle* $T^{\mathbb{C}}(M)$ is defined analogously to the real tangent bundle. We let $T^{\mathbb{C}}(M) = \cup_{p \in M} T_p(M) \otimes \mathbb{C}$. Similarly, the complexified cotangent bundle is defined by $T^{*^{\mathbb{C}}}(M) = \cup_{p \in M} T_p^*(M) \otimes \mathbb{C}$. A *complexified vector field* L on M is a smooth section of $T^{\mathbb{C}}(M)$. This means that L assigns to each $p \in M$ a vector L_p belonging to $T_p(M) \otimes \mathbb{C}$. In any smooth coordinate system $\chi = (x_1, \ldots, x_N) : U \subset M \to \mathbb{R}^N$, we can express L by

$$L_p = \sum_{j=1}^{N} a_j(p) \frac{\partial}{\partial x_j}$$

where each a_j is a smooth, complex-valued function defined on U. A *subbundle* $\mathbb{L}$ of $T^{\mathbb{C}}(M)$ of complex dimension m assigns a subspace $\mathbb{L}_p$ of $T_p(M) \otimes \mathbb{C}$ of complex dimension m to each point p in M. These subspaces are required to fit together smoothly in the sense that near each point $p_0 \in M$ there are m linearly independent smooth complex vector fields $L^1, \ldots, L^m$ such that $\mathbb{L}_p$ is generated (over $\mathbb{C}$) by $L_p^1, \ldots, L_p^m$.

If M is a complex manifold of complex dimension n, then it is important to distinguish between the real tangent bundle and the complexified tangent bundle. The former, denoted $T(M)$, is defined in Chapter 2, where we think of M as a smooth manifold of real dimension $2n$. Its fiber, $T_p(M)$, has real dimension $2n$. The fiber of the complexified tangent bundle is $T_p(M) \otimes \mathbb{C}$ and this has complex dimension $2n$.

The rth exterior power of $T_p(M) \otimes \mathbb{C}$ and $T_p^*(M) \otimes \mathbb{C}$ are called the spaces of complexified r-vectors and r-forms at p, respectively. The latter can be viewed as the space of complex linear functionals on the former by using the same pairing formula given in Section 2. Letting the point $p \in M$ vary, we obtain the bundle of r-vectors $\Lambda^r T^{\mathbb{C}}(M)$ and the bundle of r-forms $\Lambda^r T^{*^{\mathbb{C}}}(M)$, respectively. The space of smooth sections of $\Lambda^r T^{*^{\mathbb{C}}}(M)$ is called the space of complex differential forms of degree r (or, more simply, the space of smooth r-forms) and is denoted by $\mathcal{E}^r(M)$. An element ϕ of $\mathcal{E}^r(M)$ can be expressed in local coordinates $\chi = (x_1, \ldots, x_N) : U \subset M \to \mathbb{R}^N$ by

$$\phi = \sum_{|I|=r} \phi_I dx^I$$

where each ϕ_I is a smooth, complex-valued function on U.

The exterior derivative and the integral easily generalize to the complex setting. In local coordinates, we define

$$d\phi = \sum_{|I|=r} d\phi_I \wedge dx^I \in \mathcal{E}^{r+1}(M)$$

where

$$d\phi_I = \sum_{j=1}^{N} \frac{\partial \phi_I}{\partial x_j} dx_j.$$

The proof that the exterior derivative is independent of the choice of local coordinates is unchanged from the proof for the real case given in the previous chapter. The integral of a complex differential form over a manifold is defined the same way as in the real case. Of course, the result of such an integral is generally a complex number rather than a real number.

3.2 Complex structures

DEFINITION 1 *Suppose V is a real vector space. A linear map $J : V \to V$ is called a complex structure map if $J \circ J = -I$ where $I : V \to V$ is the identity map.*

A complex structure map can only be defined on an even-dimensional real vector space, because $(\det J)^2 = (-1)^N$ where $N = \dim_{\mathbb{R}} V$.

As an example, let $V = T_p(\mathbb{R}^{2n}) \simeq T_p(\mathbb{C}^n)$. We give $\mathbb{R}^{2n}$ the coordinates $(x_1, y_1, \ldots, x_n, y_n)$. The standard complex structure for $T_p(\mathbb{R}^{2n})$ is defined by setting

$$J\left(\frac{\partial}{\partial x_j}\right) = \frac{\partial}{\partial y_j}$$
$$J\left(\frac{\partial}{\partial y_j}\right) = -\frac{\partial}{\partial x_j} \qquad 1 \le j \le n,$$

and then by extending J to all of $T_p(\mathbb{R}^{2n})$ by real linearity. This complex structure map is designed to simulate multiplication by $i = \sqrt{-1}$.

The standard complex structure map for $\mathbb{R}^{2n}$ is an isometry with respect to the Euclidean metric $(\cdot)$ on $\mathbb{R}^{2n}$, as the following lemma shows.

LEMMA 1
For $v, w \in T_p(\mathbb{R}^{2n})$, $Jv \cdot w = -v \cdot Jw$. In particular $Jv \cdot Jw = v \cdot w$.

This lemma is established by showing that it holds when v and w are basis vectors (i.e., $\partial/\partial x_j$ or $\partial/\partial y_j$).

The standard complex structure J^* for the cotangent space $T_p^*(\mathbb{R}^{2n})$ is defined as the dual of J. For $p \in \mathbb{R}^{2n}$, we have

$$\langle J^*\phi, L\rangle_p = \langle \phi, JL\rangle_p \quad \text{for } \phi \in T_p^*(\mathbb{R}^{2n}) \text{ and } L \in T_p(\mathbb{R}^{2n}).$$

Setting $\phi = dx_j$ or dy_j and $L = \partial/\partial x_j$ or $\partial/\partial y_j$, we obtain

$$\begin{aligned} J^* dx_j &= -dy_j \\ J^* dy_j &= dx_j \qquad 1 \le j \le n. \end{aligned}$$

A complex structure can be defined on the real tangent bundle of a complex manifold M by pushing forward the complex structure from $\mathbb{C}^n$ up to M via a coordinate chart. For $p \in M$ and holomorphic coordinate chart $Z : U \subset M \to \mathbb{C}^n$ (with $p \in U$), we define the complex structure map $J_p : T_p(M) \to T_p(M)$ by

$$J_p(L) = Z_*^{-1}(Z(p))\{JZ_*(p)(L)\}.$$

Here, the J on the right is the complex structure map on $\mathbb{C}^n \simeq \mathbb{R}^{2n}$. It follows from this definition that if $Z = (z_1, \ldots, z_n)$ with $z_j = x_j + iy_j$, then $J_p(\partial/\partial x_j) = \partial/\partial y_j$ and $J_p(\partial/\partial y_j) = -(\partial/\partial x_j)$. This description of J_p is independent of the choice of holomorphic coordinates. For if $W : U' \to \mathbb{C}^n$ is another holomorphic coordinate chart, then $Z \circ W^{-1}$ is a holomorphic map from $\mathbb{C}^n$ to itself and therefore its push forward map commutes with J (the complex structure on $\mathbb{C}^n$). Thus, $J_p : T_p(M) \to T_p(M)$ is well defined for each $p \in M$. In a similar manner, a complex structure J_p^* can be defined on $T_p^*(M)$.

We now return to the general case. If $J : V \to V$ is a complex structure for the real vector space V, then J can be extended as a complex linear map on the complexification of V by setting

$$J(\alpha v) = \alpha J(v) \quad \text{for } v \in V \text{ and } \alpha \in \mathbb{C}.$$

Note that

$$J\bar{v} = \overline{Jv} \quad \text{for} \quad v \in V \otimes \mathbb{C}$$

for if $v_1 \in V$ and $\alpha \in \mathbb{C}$, then

$$\begin{aligned} J(\overline{\alpha v_1}) &= J(\bar{\alpha} v_1) && \text{by the definition of conjugation} \\ &= \bar{\alpha} J(v_1) && \text{by the definition of } J \text{ on } V \otimes \mathbb{C} \\ &= \overline{\alpha J(v_1)} \\ &= \overline{J(\alpha v_1)}. \end{aligned}$$

Since $J \circ J = -I$ on V, the same holds true on $V \otimes \mathbb{C}$. Therefore, $J : V \otimes \mathbb{C} \to V \otimes \mathbb{C}$ has eigenvalues $+i$ and $-i$ with corresponding eigenspaces denoted by $V^{1,0}$ and $V^{0,1}$. We have

$$V \otimes \mathbb{C} = V^{1,0} \oplus V^{0,1}$$

from elementary linear algebra. Since $J\bar{v} = \overline{Jv}$ for $v \in V \otimes \mathbb{C}$, we have

$$\overline{V^{1,0}} = V^{0,1}.$$

A basis for $V^{1,0}$ and for $V^{0,1}$ can be easily constructed. First note that for $v \in V$, the vectors v and Jv are linearly independent over $\mathbb{R}$ because J has no real eigenvalues. We can inductively build a basis for V (over $\mathbb{R}$) of the form

$$v_1, Jv_1, \ldots, v_n, Jv_n$$

where $2n = \dim_{\mathbb{R}} V$. We claim the set

$$\{v_1 - iJv_1, \ldots, v_n - iJv_n\}$$

is a basis for the complex-n-dimensional vector space $V^{1,0}$. Each vector $v_j - iJv_j$ belongs to $V^{1,0}$ because

$$\begin{aligned} J(v_j - iJv_j) &= -iJ^2 v_j + Jv_j \\ &= i(v_j - iJv_j). \end{aligned}$$

Moreover, this set of vectors is linearly independent over $\mathbb{C}$.

Likewise, the set

$$\{v_1 + iJv_1, \ldots, v_n + iJv_n\}$$

is a basis for $V^{0,1}$.

We return to the example $T_p(M)$ where M is an n-dimensional complex manifold. Let $(z_1, \ldots, z_n)$ with $z_j = x_j + iy_j$ be a set of local holomorphic coordinates for M. As mentioned earlier, the complex structure map for $T_p(M)$ satisfies $J_p(\partial/\partial x_j) = \partial/\partial y_j$ and $J_p(\partial/\partial y_j) = -(\partial/\partial x_j)$. Define the vector fields

$$\frac{\partial}{\partial z_j} = \frac{1}{2}\left(\frac{\partial}{\partial x_j} - i\frac{\partial}{\partial y_j}\right) \text{ and } \frac{\partial}{\partial \bar{z}_j} = \frac{1}{2}\left(\frac{\partial}{\partial x_j} + i\frac{\partial}{\partial y_j}\right), \quad 1 \le j \le n.$$

In view of the above discussion, a basis for $T_p^{1,0}(M)$ is given by $\{\partial/\partial z_1, \ldots, \partial/\partial z_n\}$ and a basis for $T_p^{0,1}(M)$ is given by $\{\partial/\partial \bar{z}_1, \ldots, \partial/\partial \bar{z}_n\}$.

In a similar manner, define the forms

$$dz_j = dx_j + i dy_j \text{ and } d\bar{z}_j = dx_j - i dy_j, \quad 1 \le j \le n.$$

A basis for $T_p^{*1,0}(M)$ is given by $\{dz_1, \ldots, dz_n\}$ and a basis for $T_p^{*0,1}(M)$ is given by $\{d\bar{z}_1, \ldots, d\bar{z}_n\}$.

The constants are arranged so that

$$\left\langle dz_j, \frac{\partial}{\partial z_k} \right\rangle_p = \begin{cases} 1 & \text{if } j = k \\ 0 & \text{if } j \neq k \end{cases} \qquad \left\langle dz_j, \frac{\partial}{\partial \bar{z}_k} \right\rangle_p = 0$$

$$\left\langle d\bar{z}_j, \frac{\partial}{\partial z_k} \right\rangle_p = 0 \qquad \left\langle d\bar{z}_j, \frac{\partial}{\partial \bar{z}_k} \right\rangle_p = \begin{cases} 1 & \text{if } j = k \\ 0 & \text{if } j \neq k. \end{cases}$$

In particular, $\langle \phi, L\rangle_p = 0$ for $\phi \in T_p^{*^{1,0}}(M)$ and $L \in T_p^{0,1}(M)$ and $\langle \psi, X\rangle_p = 0$ for $\psi \in T_p^{*^{0,1}}(M)$ and $X \in T_p^{1,0}(M)$.

The Hermitian inner product $(\cdot)$ on $T_p(\mathbb{C}^n) \otimes \mathbb{C}$ is defined by declaring that

$$\left\{ \frac{\partial}{\partial z_1}, \ldots, \frac{\partial}{\partial z_n}, \frac{\partial}{\partial \bar{z}_1}, \ldots, \frac{\partial}{\partial \bar{z}_n} \right\}$$

is an orthonormal basis. By the definition of a Hermitian inner product, we have

$$(\alpha U) \cdot V = \alpha (U \cdot V)$$

and

$$U \cdot (\alpha V) = \overline{\alpha} (U \cdot V)$$

for $\alpha \in \mathbb{C}$ and U, $V \in T_p(\mathbb{C}^n) \otimes \mathbb{C}$. Note that $T_p^{1,0}(\mathbb{C}^n)$ and $T_p^{0,1}(\mathbb{C}^n)$ are orthogonal under this metric. We can identify $\mathbb{C}^n$ with $T_p^{1,0}(\mathbb{C}^n)$ via the map

$$(u_1, \ldots, u_n) \mapsto \sum_{j=1}^{n} u_j \frac{\partial}{\partial z_j}$$

for $U = (u_1, \ldots, u_n) \in \mathbb{C}^n$. With this identification, we have $U \cdot V = \sum u_j \overline{v}_j$ for $U = (u_1, \ldots, u_n)$, $V = (v_1, \ldots, v_n) \in \mathbb{C}^n$.

We have also used $(\cdot)$ to denote the Euclidean inner product on $\mathbb{R}^N$. It will be clear from the context which inner product $(\cdot)$ refers to. On the (rare) occasion in which both inner products are discussed simultaneously, we will denote the Hermitian inner product of U and $V \in \mathbb{C}^n$ by $H(U, V)$ and the Euclidean inner product of U and V by $S(U, V)$, where in the latter case, we view U and V as vectors in $\mathbb{R}^{2n}$. We leave the proof of the following lemma to the reader.

LEMMA 2
For U, $V \in \mathbb{C}^n$, $H(U, V) = S(U, V) + iS(U, JV)$.

In an analogous fashion, the Hermitian inner product on $T_p^*(\mathbb{C}^n) \otimes \mathbb{C}$ (also denoted by $(\cdot)$) is defined by declaring that

$$\{dz_1, \ldots, dz_n, d\bar{z}_1, \ldots, d\bar{z}_n\}$$

is an orthonormal basis.

Let us return to the case of a general even-dimensional real vector space V. The construction of $V^{1,0}$ and $V^{0,1}$ from a given complex structure map can be reversed as the next lemma shows.

LEMMA 3
Suppose V is an even-dimensional real vector space V. Suppose $\mathbb{L}$ is a complex subspace of $V \otimes \mathbb{C}$ with the following properties

(i) $\mathbb{L} \cap \overline{\mathbb{L}} = \{0\}$

(ii) $\mathbb{L} \oplus \overline{\mathbb{L}} = V \otimes \mathbb{C}$.

Then there is a unique complex structure map J on V so that $\mathbb{L}$ and $\overline{\mathbb{L}}$ are the $+i$ and $-i$ eigenspaces of the extension of J to $V \otimes \mathbb{C}$.

PROOF We first define $J^{\mathbb{C}} : V \otimes \mathbb{C} \to V \otimes \mathbb{C}$ by

$$J^{\mathbb{C}}(L) = iL \quad \text{for} \quad L \in \mathbb{L}$$
$$J^{\mathbb{C}}(L) = -iL \quad \text{for} \quad L \in \overline{\mathbb{L}}.$$

Property (i) for $\mathbb{L}$ shows that $J^{\mathbb{C}}$ is well defined. Property (ii) for $\mathbb{L}$ shows that $J^{\mathbb{C}}$ can be extended to a complex linear map defined on all of $V \otimes \mathbb{C}$. Clearly, $\mathbb{L}$ and $\overline{\mathbb{L}}$ are the $+i$ and $-i$ eigenspaces for $J^{\mathbb{C}}$. It remains to show that $J^{\mathbb{C}}$ maps $V(= V \otimes 1)$ to itself, for then we can define our desired J to be the restriction of $J^{\mathbb{C}}$ to V. Note that the vectors

$$X = L + \overline{L} \qquad L \in \mathbb{L}$$
$$Y = i(L - \overline{L}) \qquad L \in \mathbb{L}$$

are real, i.e., $\overline{X} = X$ and $\overline{Y} = Y$. Hence, X and Y belong to $V \otimes 1$. Since $\mathbb{L} \oplus \overline{\mathbb{L}} = V \otimes \mathbb{C}$, V is generated by vectors of the form X and Y given above. Since $J^{\mathbb{C}}X = Y$ and $J^{\mathbb{C}}Y = -X$, which both belong to V, clearly $J^{\mathbb{C}}$ maps V to itself, as desired. ∎

Sometimes in the literature, attention is focused on the $-i$ eigenspace of J and therefore it is denoted by $\mathbb{L}$. We have chosen to denote the $-i$ eigenspace by $\overline{\mathbb{L}}$ since it seems more natural to think of vectors of type $(0,1)$ as conjugated rather than unconjugated. This is consistent with the fact that the set $\{\partial/\partial \overline{z}_j,\ 1 \leq j \leq n\}$ is a basis for $T_p^{0,1}(\mathbb{C}^n)$.

3.3 Higher degree complexified forms

Let M be a complex manifold of dimension n. For $0 \leq r \leq 2n$, we have defined $\Lambda^r T^{*^{\mathbb{C}}}(M)$ to be the bundle of complexified r-forms and $\mathcal{E}^r(M)$ to be the space of r-forms on M whose coefficients are smooth, complex-valued functions.

For $0 \leq p, q \leq n$ and a point $z \in M$, define the space

$$\Lambda_z^{p,q} T^*(M) = \Lambda^p \{T_z^{*1,0}(M)\} \widehat{\otimes} \Lambda^q \{T_z^{*0,1}(M)\}.$$

Here, $\widehat{\otimes}$ denotes the antisymmetric tensor product. Let $(z_1, \ldots, z_n)$ be a set of local holomorphic coordinates for M. In more sensible terms, $\Lambda_z^{p,q} T^*(M)$ is the vector space spanned over $\mathbb{C}$ by the set

$$\{dz^I \wedge d\bar{z}^J = dz_{i_1} \wedge \ldots \wedge dz_{i_p} \wedge d\bar{z}_{j_1} \wedge \ldots \wedge d\bar{z}_{i_q}\}$$

where $I = \{i_1, \ldots, i_p\}$ and $J = \{j_1, \ldots, j_q\}$ run over the set of all increasing multiindices of length p and q, respectively.

In the same way, we define the space

$$\Lambda_z^{p,q} T(M) = \Lambda^p T_z^{1,0}(M) \widehat{\otimes} \Lambda^q T_z^{0,1}(M)$$

which is spanned (over $\mathbb{C}$) by

$$\left\{ \frac{\partial}{\partial z^I} \wedge \frac{\partial}{\partial \bar{z}^J};\ |I| = p, |J| = q \right\}.$$

By letting the point $z \in M$ vary, we obtain the bundles $\Lambda^{p,q}(T^*(M))$ and $\Lambda^{p,q} T(M)$ in the usual way. The space of smooth sections of $\Lambda^{p,q}(T^*(M))$ is denoted $\mathcal{E}^{p,q}(M)$ and is called the space of differential forms on M of bidegree (p, q). An element ϕ of $\mathcal{E}^{p,q}(M)$ can be expressed in local coordinates as

$$\phi = \sum_{\substack{|I|=p \\ |J|=q}} \phi_{IJ} dz^I \wedge d\bar{z}^J$$

where each ϕ_{IJ} is a smooth, complex-valued function.

In the local coordinates $z_j = x_j + iy_j$, we write $dx_j = (1/2)(dz_j + d\bar{z}_j)$ and $dy_j = -(i/2)(dz_j - d\bar{z}_j)$ for $1 \leq j \leq n$. In this way, any element of $\Lambda^r(T^{*\mathbb{C}}(M))$ can be written uniquely as a sum of forms of various bidegrees, i.e.,

$$\Lambda^r(T^{*^{\mathbb{C}}}(M)) = \Lambda^{r,0}(T^*(M)) \oplus \cdots \oplus \Lambda^{0,r}(T^*(M)).$$

Therefore

$$\mathcal{E}^r(M) = \mathcal{E}^{r,0}(M) \oplus \cdots \oplus \mathcal{E}^{0,r}(M).$$

For $0 \leq p, q \leq n$ with $p + q = r$, let

$$\pi^{p,q} : \Lambda^r(T^{*^{\mathbb{C}}}(M)) \to \Lambda^{p,q}(T^*(M))$$

be the natural projection.

DEFINITION 1 *The Cauchy–Riemann operator* $\bar{\partial} : \mathcal{E}^{p,q}(M) \to \mathcal{E}^{p,q+1}(M)$ *and the operator* $\partial : \mathcal{E}^{p,q}(M) \to \mathcal{E}^{p+1,q}(M)$ *are defined by*

$$\bar{\partial} = \pi^{p,q+1} \circ d$$

$$\partial = \pi^{p+1,q} \circ d.$$

For a smooth function $f : M \to \mathbb{C}$, we have

$$df = \sum_1^n \frac{\partial f}{\partial x_j} dx_j + \frac{\partial f}{\partial y_j} dy_j$$

which can be rewritten

$$df = \sum_{j=1}^n \frac{\partial f}{\partial z_j} dz_j + \sum_{j=1}^n \frac{\partial f}{\partial \bar{z}_j} d\bar{z}_j.$$

The first term on the right is an element of $\mathcal{E}^{1,0}(M)$ and the second term is an element of $\mathcal{E}^{0,1}(M)$. Therefore

$$\partial f = \sum_{j=1}^n \frac{\partial f}{\partial z_j} dz_j$$

$$\bar{\partial} f = \sum_{j=1}^n \frac{\partial f}{\partial \bar{z}_j} d\bar{z}_j.$$

Note that $df = \partial f + \bar{\partial} f$. Also note that a C^1 function $f : M \mapsto \mathbb{C}$ is holomorphic if and only if $\bar{\partial} f = 0$.

For higher degree forms, we have

$$\partial\{f dz^I \wedge d\bar{z}^J\} = \partial f \wedge dz^I \wedge d\bar{z}^J \quad \text{for} \quad f \in \mathcal{E}(M)$$

$$\bar{\partial}\{f dz^I \wedge d\bar{z}^J\} = \bar{\partial} f \wedge dz^I \wedge d\bar{z}^J$$

so

$$df = \partial f + \bar{\partial} f.$$

Another important observation is that $\partial^2 \phi = 0$, $\bar{\partial}^2 \phi = 0$, and $\partial\bar{\partial}\phi = -\bar{\partial}\partial\phi$ for $\phi \in \mathcal{E}^{p,q}(M)$. This follows from $0 = d^2\phi = \partial^2\phi + (\partial\bar{\partial} + \bar{\partial}\partial)\phi + \bar{\partial}^2\phi$ and from noting that $\partial^2\phi$, $(\partial\bar{\partial} + \bar{\partial}\partial)\phi$ and $\bar{\partial}^2\phi$ have different bidegrees ($(p+2, q)$, $(p+1, q+1)$, and $(p, q+2)$, respectively).

We summarize the above discussion in the following lemma.

LEMMA 1

(i) $d = \partial + \bar{\partial}$

(ii) $\partial^2 = 0, \partial\bar{\partial} = -\bar{\partial}\partial$ *and* $\bar{\partial}^2 = 0$

The operators ∂ and $\bar{\partial}$ satisfy a product rule whose proof follows from the product rule for the exterior derivative (Lemma 1 in Section 1.4).

LEMMA 2
If $f \in \mathcal{E}^{p,q}(M)$ and $g \in \mathcal{E}^{r,s}(M)$, then

$$\bar{\partial}(f \wedge g) = \bar{\partial} f \wedge g + (-1)^{p+q} f \wedge \bar{\partial} g$$
$$\partial(f \wedge g) = \partial f \wedge g + (-1)^{p+q} f \wedge \partial g.$$

The pull back of a smooth map between smooth manifolds M and N preserves degree (see Section 1.3). If M and N are complex manifolds and the map is holomorphic, then its pull back also preserves bidegree, as the next lemma shows.

LEMMA 3
Suppose M and N are complex manifolds and $F : M \to N$ is a holomorphic map. If ϕ is an element of $\Lambda^{p,q}_{F(z)} T^*(N)$, then $F^*\phi$ is an element of $\Lambda^{p,q}_z T^*(M)$.

PROOF If $w_1, \ldots, w_n$ is a set of local coordinates for N and if $F_j = w_j \circ F : M \to \mathbb{C}$, $1 \leq j \leq n$, then

$$F^* dw_j = dF_j \quad \text{and} \quad F^* d\bar{w}_j = d\overline{F}_j.$$

Since F is holomorphic, $\bar{\partial} F_j = 0$. Conjugating this gives $\partial \overline{F}_j = 0$. So from the equation $d = \partial + \bar{\partial}$, we obtain

$$F^* dw_j = \partial F_j \in \Lambda^{1,0} T^*(M)$$

and

$$F^* d\bar{w}_j = \bar{\partial} \overline{F}_j \in \Lambda^{0,1} T^*(M).$$

If $|I| = p, |J| = q$, then $F^*(dw^I \wedge d\bar{w}^J) \in \Lambda^{p,q} T^*(M)$. Since $\{dw^I \wedge d\bar{w}^J; |I| = p, |J| = q\}$ is a local basis for $\Lambda^{p,q} T^*(N)$, the proof of the lemma is now complete. ∎

In Lemma 2 of Section 1.4 we showed that the pull back operator commutes with the exterior derivative. If the map is a holomorphic map between complex manifolds, then the pull back operator also commutes with ∂ and $\bar{\partial}$, as the next lemma shows.

LEMMA 4
Suppose M and N are complex manifolds and $F : M \to N$ is a holomorphic map. Then $F^* \circ \bar{\partial} = \bar{\partial} \circ F^*$ and $F^* \circ \partial = \partial \circ F^*$.

The $\bar{\partial}$ on the left side of the equation $F^* \circ \bar{\partial} = \bar{\partial} \circ F^*$ is the $\bar{\partial}$-operator for N and the $\bar{\partial}$ on the right is the $\bar{\partial}$-operator on M. Since F^* commutes with the exterior derivative and since F^* preserves bidegree, the proof of the lemma follows easily.

For functions, there is a useful relationship between the exterior derivative, ∂, $\bar{\partial}$, and J^* — the complex structure map on the space of 1-forms.

LEMMA 5

If f is a smooth, complex-valued function defined on a complex manifold M, then

$$\partial f = \frac{1}{2}(df - iJ^*df)$$
$$\bar{\partial} f = \frac{1}{2}(df + iJ^*df).$$

The operator $J^* \circ d$ is often denoted d^c in the literature. Thus, $(1/2)d^c = (1/2)J^* \circ d$ is the imaginary part of the $\bar{\partial}$ operator.

PROOF We start with the equation $d = \partial + \bar{\partial}$. For a function f, the one form ∂f belongs to $T^{*^{1,0}}(M)$, which is the $+i$ eigenspace of J^*. Likewise, $\bar{\partial} f$ belongs to $T^{*^{0,1}}(M)$, which is the $-i$ eigenspace of J^*. By applying J^* to the equation $df = \partial f + \bar{\partial} f$, we obtain

$$J^*df = i\partial f - i\bar{\partial} f.$$

Adding this equation to the equation $idf = i\partial f + i\bar{\partial} f$ and then dividing the result by $2i$ yields the first assertion of the lemma. The second assertion is derived similarly. ▌

We end this chapter with the computation of the $\mathcal{L}^2$-adjoint of $\bar{\partial}$ on $\mathbb{C}^n$. This is analogous to the computation of the $\mathcal{L}^2$-adjoint of the exterior derivative on $\mathbb{R}^N$ given in Section 1.5. The Hermitian inner product for $T^{*^{\mathbb{C}}}(\mathbb{C}^n)$ can be extended to an inner product on $\Lambda^{p,q}T^*(\mathbb{C}^n)$ by declaring that the set

$$\{dz^I \wedge d\bar{z}^J; |I| = p, |J| = p, \ I, \ J \text{ increasing}\}$$

is an orthonormal basis for $\Lambda^{p,q}T^*(\mathbb{C}^n)$. Let $\mathcal{D}^{p,q}(\mathbb{C}^n)$ be the space of compactly supported elements of $\mathcal{E}^{p,q}(\mathbb{C}^n)$. We endow $\mathcal{D}^{p,q}(\mathbb{C}^n)$ with the following $\mathcal{L}^2$-inner product

$$(\phi, \psi)_{\mathcal{L}^2} = \int\limits_{z \in \mathbb{C}^n} (\phi(z) \cdot \psi(z))dv \qquad \phi, \psi \in \mathcal{D}^{p,q}(\mathbb{C}^n)$$

where $dv = dx_1 \wedge dy_1 \wedge \ldots \wedge dx_n \wedge dy_n$ is the usual volume form for $\mathbb{C}^n$. The adjoint of $\bar{\partial}$ with respect to this inner product is denoted $\bar{\partial}^* : \mathcal{D}^{p,q}(\mathbb{C}^n) \to \mathcal{D}^{p,q-1}(\mathbb{C}^n)$. It is defined by $(\bar{\partial}^* f, g)_{\mathcal{L}^2} = (f, \bar{\partial} g)_{\mathcal{L}^2}$ for $f \in \mathcal{D}^{p,q}(\mathbb{C}^n)$ and $g \in \mathcal{D}^{p,q-1}(\mathbb{C}^n)$. The coordinate formula for $\bar{\partial}^*$ involves the contraction operator $\lrcorner$ whose definition in Section 1.5 easily generalizes to the complex setting. Note that for $|I| = p, |J| = q$,

$$\frac{\partial}{\partial \bar{z}_j} \lrcorner (dz^I \wedge d\bar{z}^J) = \begin{cases} 0 & \text{if } j \notin J \\ (-1)^{p+k-1} dz^I \wedge d\bar{z}^{J'} & \text{if } j \in J. \end{cases}$$

Here, we assume $J = \{j_1, \ldots, j_q\}$ and that if j belongs to J then $j = j_k$. In this case, J' is defined to be the index of length $q - 1$ given by $\{j_1, \ldots, \hat{j}_k, \ldots, j_q\}$ where $\hat{j}_k$ indicates that j_k is removed. We now state the formula for $\bar{\partial}^*$ whose derivation is similar to the proof of Lemma 2 in Section 1.5 and is left to the reader.

LEMMA 6
Let $\phi = f dz^I \wedge d\bar{z}^J$ where f is a smooth, compactly supported function on $\mathbb{C}^n$. Then

$$\bar{\partial}^* \phi = -\sum_{j=1}^{n} \frac{\partial f}{\partial z_j} \frac{\partial}{\partial \bar{z}_j} \lrcorner (dz^I \wedge d\bar{z}^J).$$

4

The Frobenius Theorem

In this section, we discuss the Frobenius theorem. We also discuss the complex analytic version of this theorem, which will be used in Part II when we discuss imbeddings of real analytic CR manifolds.

4.1 The real Frobenius theorem

Let $\mathbb{L}$ be an m-dimensional subbundle of the real tangent bundle to $\mathbb{R}^N$. The Frobenius theorem gives conditions on the subbundle $\mathbb{L}$ in a neighborhood U of a given point p_0 which guarantee the existence of locally defined smooth functions $\rho_1, \ldots, \rho_{N-m} : U \mapsto \mathbb{R}$ with $\rho_j(p_0) = 0,\ 1 \le j \le N-m$ and $d\rho_1 \wedge \ldots \wedge d\rho_{N-m} \neq 0$ on U such that

$$L\{\rho_j\} = 0 \quad \text{on} \quad U \quad \text{for} \quad 1 \le j \le N-m, L \in \mathbb{L}. \tag{1}$$

Let us assume $\rho_1, \ldots, \rho_{N-m}$ exist and satisfy (1). For each $c = (c_1, \ldots, c_{N-m}) \in \mathbb{R}^{N-m}$ near the origin, the set

$$M_c = \{x \in U \subset \mathbb{R}^N; \rho_1(x) = c_1, \ldots, \rho_{N-m}(x) = c_{N-m}\}$$

is an m-dimensional submanifold of U by Lemma 1 in Section 2.2. By Lemma 1 in Section 2.3, the tangent bundle of each M_c is generated by those vector fields on $\mathbb{R}^N$ which satisfy condition (1) on M_c. An easy argument using the inverse function theorem or Lemma 1 in Section 2.2 shows that the family of submanifolds $\{M_c; c \in \mathbb{R}^{N-m}\}$ fills out a possibly smaller open set U' in $\mathbb{R}^N$ which contains p_0. It follows that the existence of $\rho_1, \ldots, \rho_{N-m}$ leads to a foliation of an open set in $\mathbb{R}^N$ containing p_0 by submanifolds of dimension m whose tangent bundle can be identified with $\mathbb{L}$.

If L_1 and L_2 belong to $\mathbb{L}$, then $L_1\{\rho_j\} = L_2\{\rho_j\} = 0$ on U for $1 \leq j \leq N-m$. Therefore

$$[L_1, L_2]\{\rho_j\} = L_1\{L_2\rho_j\} - L_2\{L_1\rho_j\} = 0 \quad \text{on} \quad U.$$

The vector field $[L_1, L_2]$ satisfies condition (1) and so $[L_1, L_2]$ also belongs to $\mathbb{L}$. A subbundle $\mathbb{L}$ is said to be *involutive* if $[L_1, L_2]$ belongs to $\mathbb{L}$ whenever L_1 and L_2 belong to $\mathbb{L}$. So the existence of $\rho_1, \ldots, \rho_{N-m}$ satisfying (1) implies that $\mathbb{L}$ is involutive.

The Frobenius theorem states that the converse holds. For simplicity, we state this theorem for Euclidean space. However, since it is local in nature, it can easily be generalized to the manifold setting.

THEOREM 1 FROBENIUS

Let $\mathbb{L}$ be an m-dimensional subbundle of the real tangent bundle of $\mathbb{R}^N$. Suppose $\mathbb{L}$ is involutive, i.e., $[L_1, L_2]$ belongs to $\mathbb{L}$ whenever L_1 and L_2 belong to $\mathbb{L}$. Then given a point $p_0 \in \mathbb{R}^N$ there is a neighborhood U of p_0 and a diffeomorphism $\chi = (\chi^1, \ldots, \chi^N) : U \to \chi\{U\} \subset \mathbb{R}^N$ such that

$$L\{\chi^j\} = 0 \quad \textit{on} \quad U \quad \textit{for} \quad m+1 \leq j \leq N \quad \textit{and} \quad L \in \mathbb{L}.$$

PROOF We assume the given point p_0 is the origin. Near the origin, there is a set of linearly independent vector fields $\tilde{L}^1, \ldots, \tilde{L}^m$ that span the subbundle $\mathbb{L}$ over $\mathcal{E}(\mathbb{R}^N)$. Give $\mathbb{R}^N$ the coordinates (y, x) with $y \in \mathbb{R}^m$ and $x \in \mathbb{R}^{N-m}$. We write

$$\tilde{L}^j = \sum_{k=1}^{m} \mu_{jk} \frac{\partial}{\partial y_k} + \sum_{k=1}^{N-m} \gamma_{jk} \frac{\partial}{\partial x_k} \qquad 1 \leq j \leq m$$

where the μ_{jk} and γ_{jk} are smooth functions defined near the origin. Since $\tilde{L}^1, \ldots, \tilde{L}^m$ are linearly independent near the origin, we may reorder the coordinates if necessary so that the $m \times m$ matrix $\mu = (\mu_{jk})_{j,k=1}^m$ is nonsingular near the origin. Multiplying through by μ^{-1} yields another locally defined basis for $\mathbb{L}$ of the form $L^1, \ldots, L^m$ where

$$L^j = \frac{\partial}{\partial y_j} + \sum_{k=1}^{N-m} \lambda_{jk} \frac{\partial}{\partial x_k} \qquad 1 \leq j \leq m.$$

Each λ_{jk} is a smooth function of (y, x) and equals the (j, k)th entry of the matrix product $\mu^{-1} \cdot \gamma$, where γ is the $m \times (N-m)$ matrix with entries γ_{jk}.

By explicitly computing $[L^j, L^k]$, it is clear that the $(\partial/\partial y_\ell)$-coefficient of $[L^j, L^k]$ vanishes. On the other hand, by hypothesis, $[L^j, L^k]$ is a linear combination of $\{L^1, \ldots, L^m\}$. Any nontrivial linear combination of $\{L^1, \ldots, L^m\}$ must involve a nontrivial linear combination of $\{\partial/\partial y_1, \ldots, \partial/\partial y_m\}$.

Therefore, we conclude

$$[L^j, L^k] = 0 \quad \text{near } 0, \quad 1 \le j, k \le m.$$

The proof of the Frobenius theorem will be complete after we prove the following key lemma, which is important in its own right.

LEMMA 1

Let $1 \le m \le N$ and give $\mathbb{R}^N$ the coordinates (y, x) with $y \in \mathbb{R}^m$ and $x \in \mathbb{R}^{N-m}$. Suppose

$$L^j = \frac{\partial}{\partial y_j} + \sum_{k=1}^{N-m} \lambda_{jk}(y, x) \frac{\partial}{\partial x_k} \qquad 1 \le j \le m$$

where each λ_{jk} is a smooth function defined near the origin. In addition, suppose $[L^j, L^k] = 0$ for $1 \le j, k \le m$. Then there is a neighborhood U of the origin and a smooth diffeomorphism $\chi = (\chi_1, \ldots, \chi_N) : U \to \chi\{U\} \subset \mathbb{R}^N$ such that

$$L^j\{\chi_k\} = 0 \quad \text{on} \quad U, \quad 1 \le j \le m, \quad m+1 \le k \le N$$

and

$$\chi(0, x) = (0, x) \quad \text{for} \quad x \in \mathbb{R}^{N-m} \text{ with } (0, x) \in U.$$

PROOF The proof proceeds by induction on m, $1 \le m \le N$. We first consider the case $m = 1$. Of course, for any vector field L, the condition $[L, L] = 0$ always holds, and so this condition provides no new information.

We shall construct a smooth map $\chi = (\chi_1, \ldots, \chi_N)$ such that $L = \partial/\partial\chi_1$ where the vector field $\partial/\partial\chi_1$ was defined in Section 2.3 as the push forward of $\partial/\partial t_1$ under the map χ^{-1}. Since $(\partial/\partial\chi_1)\{\chi_j\} = 0$ for $2 \le j \le N$, this will complete the proof of the lemma for the case $m = 1$.

Let (t, x) be coordinates for $\mathbb{R}^N$ with $t \in \mathbb{R}$ and $x \in \mathbb{R}^{N-1}$. We shall find a local diffeomorphism $F = (F_1, \ldots, F_N) : \mathbb{R} \times \mathbb{R}^{N-1} \to \mathbb{R}^N$ with

$$\left[F_*\left(\frac{\partial}{\partial t}\right)\right]_{F(t,x)} = L_{F(t,x)} \quad \text{for } (t, x) \text{ near the origin} \tag{2}$$

and

$$F(0, x) = (0, x).$$

Then the map $\chi = F^{-1}$ will satisfy the initial condition $\chi(0, x) = (0, x)$. Moreover,

$$\frac{\partial}{\partial \chi_1} = \chi_*^{-1}\left(\frac{\partial}{\partial t}\right) = F_*\left(\frac{\partial}{\partial t}\right) = L$$

as desired.

In the coordinates $(y, x) = F(t, x)$, we have

$$\left[F_*\left(\frac{\partial}{\partial t}\right)\right]_{F(t,x)} = \frac{\partial F_1}{\partial t}(t, x)\frac{\partial}{\partial y} + \sum_{k=1}^{N-1} \frac{\partial F_{k+1}}{\partial t}(t, x)\frac{\partial}{\partial x_k}$$

(see Section 2.3). Comparing this expression with

$$L = \frac{\partial}{\partial y} + \sum_{k=1}^{N-1} \lambda_{1k}(y, x)\frac{\partial}{\partial x_k},$$

we see that (2) is equivalent to the following initial value problem

$$\begin{aligned}
\frac{\partial F_1(t,x)}{\partial t} &= 1 \\
\frac{\partial F_2(t,x)}{\partial t} &= \lambda_{1,1}(F(t,x)) \\
&\vdots \\
\frac{\partial F_N(t,x)}{\partial t} &= \lambda_{1,(N-1)}(F(t,x)) \\
F(0,x) = (0,x) &\quad \text{for} \quad x \in \mathbb{R}^{N-1}.
\end{aligned} \tag{3}$$

The existence and uniqueness theorem from ordinary differential equations guarantees a smooth solution $F : \mathbb{R} \times \mathbb{R}^{N-1} \to \mathbb{R}^N$ which is defined on $\{|t| < \epsilon, |x| < \epsilon\}$ for some sufficiently small $\epsilon > 0$. Note that the initial condition $F(0, x) = (0, x)$ implies that $F(0, 0) = 0$ and

$$\frac{\partial F_j}{\partial x_k}(0,0) = \begin{cases} 1 & \text{if } j = k \\ 0 & \text{if } j \neq k. \end{cases}$$

Together with the above differential equation at $t = 0$ and $x = 0$, we obtain

$$DF(0,0) = \begin{pmatrix} 1 & & & \\ \lambda_{11}(0) & 1 & & \bigcirc \\ \vdots & & \ddots & \\ & \bigcirc & & \\ \lambda_{1(N-1)}(0) & & & 1 \end{pmatrix}$$

In particular, $DF(0, 0)$ is nonsingular. By the inverse function theorem, F has a smooth locally defined inverse $\chi : \mathbb{R}^N \to \mathbb{R}^N$ with $\chi(0) = 0$. As mentioned above, χ is the desired map for the lemma for the case $m = 1$.

The geometric interpretation of the above analysis is the following. For fixed x, the curve $\{F(t, x);\ |t| < \epsilon\}$ is called an *integral curve* for L. The vector field $L = F_*(\partial/\partial t)$ is tangent to this curve. The solution to (2) determines a unique family of integral curves for L which satisfy the given initial condition.

Now we assume the lemma is true for $m-1$ $(2 \le m \le N)$ and we prove the lemma for m. So by assumption, there is a local diffeomorphism $\hat{\chi} = (\hat{\chi}_1, \dots, \hat{\chi}_N) : \mathbb{R}^N \to \mathbb{R}^N$ such that

$$L^j\{\hat{\chi}_k\} = 0 \quad \text{for} \quad 1 \le j \le m-1,\ m \le k \le N \tag{4}$$

and

$$\hat{\chi}(0, \dots, 0, y_m, x_1, \dots, x_{N-m}) = (0, \dots, 0, y_m, x_1, \dots, x_{N-m}).$$

In particular, $\hat{\chi}(0,x) = (0,x)$, for $x \in \mathbb{R}^{N-m}$ near the origin. Let $\widehat{F}$ be the inverse of $\hat{\chi}$. Since $\hat{\chi}(0) = 0$, we have $\widehat{F}(0) = 0$. From (4), we see that L^j is tangent to any level set of $\hat{\chi}_k$ for $m \le k \le N$. In particular, for $1 \le j \le m-1$, L^j is tangent to the hypersurface

$$M_0 = \{\widehat{F}(t', 0, x);\ t' \in \mathbb{R}^{m-1},\ x \in \mathbb{R}^{N-m}\}$$

which is the level set $\{\hat{\chi}_m = 0\}$.

Let (t', t_m, x) be the coordinates for $\mathbb{R}^N$ where $t' = (t_1, \dots, t_{m-1}) \in \mathbb{R}^{m-1}$, $t_m \in \mathbb{R}$, $x = (x_1, \dots, x_{N-m}) \in \mathbb{R}^{N-m}$. We find a local diffeomorphism $F : \mathbb{R}^{m-1} \times \mathbb{R} \times \mathbb{R}^{N-m} \to \mathbb{R}^N$ with

$$\left[F_*\left(\frac{\partial}{\partial t_m}\right)\right]_{F(t,x)} = [L^m]_{F(t,x)} \tag{5}$$

$$F(t', 0, x) = \widehat{F}(t', 0, x) \quad t' \in \mathbb{R}^{m-1},\ x \in \mathbb{R}^{N-m}.$$

This is accomplished by solving a system of ordinary differential equations that is analogous to (3). The solution to (5) determines a family of integral curves for L^m that satisfy the given initial condition.

We claim that F is a diffeomorphism on a neighborhood of the origin in $\mathbb{R}^N$. This follows from two observations. First, the map $(t', 0, x) \mapsto F(t', 0, x)$ parameterizes M_0. Second, the map $t_m \mapsto F(t', t_m, x)$ parameterizes an integral curve for L_m that is transverse to M_0 (because L_m is transverse to $L_1, \dots, L_{m-1}$ and the set $\{(0,x); x \in \mathbb{R}^{N-m}\}$).

We let χ be the local inverse of F. We claim that χ is the desired map for the lemma. We must show that $\chi(0,x) = (0,x)$ and that $L^j(\chi_k) = 0$ for $1 \le j \le m$ and $m+1 \le k \le N$. Clearly, $\chi(0,x) = (0,x)$ because χ is the inverse of F and $F(0,x) = \widehat{F}(0,x) = (0,x)$.

Next we show that $L^m(\chi_k) = 0$ for $m+1 \le k \le N$. From (5), we have $L^m = F_*(\partial/\partial t_m) = \chi_*^{-1}(\partial/\partial t_m)$. Therefore

$$\begin{aligned} L^m(\chi_k) &= \chi_*^{-1}\left(\frac{\partial}{\partial t_m}\right)\{\chi_k\} \\ &= \frac{\partial}{\partial t_m}\{\chi_k \circ \chi^{-1}\} \\ &= 0 \quad \text{if} \quad m+1 \le k \le N. \end{aligned}$$

The last equation follows because $\chi_k \circ \chi^{-1}(t', t_m, x) = x_{k-m}$ for $m+1 \le k \le N$ and because t_m and x_{k-m} are independent variables.

It remains to show $L^j\{\chi_k\} = 0$ for $1 \le j \le m-1$ and $m+1 \le k \le N$. From the initial condition in (5), we have $\chi = \hat{\chi}$ on the hypersurface M_0. As already mentioned, each vector field L^j, $1 \le j \le m-1$, is tangent to M_0. Therefore, (4) implies

$$L^j \chi_k = 0 \ \text{ on } M_0 \quad m+1 \le k \le N.$$

Since $[L^j, L^m] = 0$ and $L^m \chi_k = 0$, we obtain

$$\begin{aligned} L^m(L^j \chi_k) &= L^j(L^m \chi_k) \\ &= 0 \quad m+1 \le k \le N. \end{aligned}$$

Since L^m is transverse to the hypersurface M_0, we see that $L^j \chi_k = 0$ for $m+1 \le k \le N$, as desired. This establishes the inductive step and so the proof of the lemma is now complete.

As mentioned earlier, the lemma completes the proof of the Frobenius theorem. ∎

4.2 The analytic Frobenius theorem

The complex analytic version of the Frobenius theorem requires the following lemma.

LEMMA 1

Suppose $1 \le m \le n$ and suppose $\mathbb{C}^n$ has coordinates (ζ, z) with $\zeta \in \mathbb{C}^m$, $z \in \mathbb{C}^{n-m}$. Let

$$L^j = \frac{\partial}{\partial \zeta_j} + \sum_{k=1}^{n-m} \lambda_{jk}(\zeta, z) \frac{\partial}{\partial z_k} \qquad 1 \le j \le m$$

where each λ_{jk} is a holomorphic function defined in a neighborhood of the origin in $\mathbb{C}^n$. Then there is a neighborhood U of the origin in $\mathbb{C}^n$ and a *biholomorphic map*

$$Z = (Z_1, \ldots, Z_n) : U \to Z\{U\} \subset \mathbb{C}^n$$

with

$$L^j\{Z_k\} = 0 \quad on \quad U \quad 1 \le j \le m, \quad m+1 \le k \le n$$

and

$$Z(0, z) = (0, z) \quad for \quad z \in \mathbb{C}^{n-m} \text{ with } (0, z) \in U.$$

The proof of this lemma is similar to the proof of the corresponding lemma in Section 4.1. The only major change is that instead of solving a system of ordinary differential equations, we must solve a system of partial differential equations. For example, when $m = 1$, the following system of partial differential equations replaces (3) in Section 4.1:

$$\begin{aligned}
&\frac{\partial F_1}{\partial w}(w, z) = 1 \\
&\frac{\partial F_2}{\partial w}(w, z) = \lambda_{11}(F(w, z)) \\
&\vdots \\
&\frac{\partial F_n}{\partial w}(w, z) = \lambda_{1,n-1}(F(w, z))
\end{aligned}$$

and

$$F(0, z) = (0, z), \ z \in \mathbb{C}^{n-1}.$$

Here, the coordinates for $\mathbb{C}^n$ are (w, z) with $w \in \mathbb{C}$ and $z \in \mathbb{C}^{n-1}$. Since each λ_{jk} is holomorphic, a locally defined holomorphic solution F can be found by using the Cauchy–Kowalevsky theorem (see [Jo] or [Fo]).

The rest of the proof is the same as the proof of Lemma 1 in Section 4.1 except that the holomorphic version of the inverse function theorem must be used instead of the real version of the inverse function theorem. We leave the details to the reader.

This lemma implies the analytic version of the Frobenius theorem in the same way that the lemma in Section 4.1 implies the real Frobenius theorem. First, we make two definitions. A vector field $\sum_{k=1}^n a_k(\partial/\partial z_k)$ in $T^{1,0}(\mathbb{C}^n)$ is called a *holomorphic vector field* if each $a_j : \mathbb{C}^n \to \mathbb{C}$ is holomorphic. A *holomorphic* or *analytic subbundle* of dimension m is a subbundle of $T^{1,0}(\mathbb{C}^n)$ that is locally generated over $\mathcal{O}(\mathbb{C}^n)$ by m-holomorphic vector fields that are linearly independent.

THEOREM 1 THE ANALYTIC FROBENIUS THEOREM
Suppose $\mathbb{L}$ is an m-dimensional holomorphic, involutive subbundle of $T^{1,0}(\mathbb{C}^n)$. Then given a point $p_0 \in \mathbb{C}^n$ there is a neighborhood U of $\mathbb{C}^n$ and a biholomorphic map $Z = (Z_1, \ldots, Z_n) : U \to Z(U) \subset \mathbb{C}^n$ such that

$$L\{Z_j\} = 0 \text{ on } U \text{ for } m+1 \leq j \leq n \text{ and } L \in \mathbb{L}.$$

Note that for $p \in U$, each $L^j|_p$ is tangent to the m-dimensional complex manifold

$$M_p = \{z \in U \subset \mathbb{C}^n; Z_j(z) = Z_j(p),\ m+1 \leq j \leq n\}.$$

Thus, the above theorem foliates U into complex manifolds M_p of dimension m so that $\mathbb{L}_p = T_p^{1,0}(M_p)$, for each $p \in U$.

4.3 Almost complex structures

In the analytic version of the Frobenius theorem, the subbundle $\mathbb{L}$ is defined on an open subset of complex Euclidean space. This theorem has an abstract version, which we shall state but not prove. First, we make a definition.

DEFINITION 1 *Let M be a smooth manifold (not necessarily imbedded in $\mathbb{C}^n$). Let $\mathbb{L}$ be a smooth subbundle of $T^{\mathbb{C}}(M)$ (the complexified tangent bundle of M). We say that the pair $(M, \mathbb{L})$ is an almost complex structure if*

1. $\mathbb{L} \cap \overline{\mathbb{L}} = \{0\}$
2. $\mathbb{L} \oplus \overline{\mathbb{L}} = T^{\mathbb{C}}(M)$.

An almost complex structure can only exist if the real dimension of M is even. Note that a complex manifold M together with $\mathbb{L} = T^{1,0}(M)$ is an example of an almost complex structure.

From Lemma 3 in Section 3.2, specifying $\mathbb{L}$ in the above definition is equivalent to specifying a complex structure map $J_p : T_p(M) \to T_p(M)$ which varies smoothly with $p \in M$.

If M is a complex manifold then $\overline{\mathbb{L}} = T^{0,1}(M)$ is involutive. This follows by observing that a Lie bracket between any two vector fields that only involve $\partial/\partial\bar{z}_1, \ldots, \partial/\partial\bar{z}_n$ is again a linear combination (over $\mathcal{E}(M)$) of $\partial/\partial\bar{z}_1, \ldots$, $\partial/\partial\bar{z}_n$. Here, $(z_1, \ldots, z_n)$ is a set of local holomorphic coordinates for M. Similarly, $\mathbb{L} = T^{1,0}(M)$ is also involutive.

A natural question is to ask if the converse is true: if $(M, \mathbb{L})$ is an involutive almost complex structure, then does there exist a complex structure for M so that $\overline{\mathbb{L}} = T^{0,1}(M)$? The answer is yes, locally, and this is the content of the following theorem.

THEOREM 1 NEWLANDER–NIRENBERG [NN]

Suppose $(M, \mathbb{L})$ is an almost complex structure and suppose $\mathbb{L}$ is involutive. Then near any given point $p_0 \in M$, there is a complex structure for M which makes M a complex manifold so that $\overline{\mathbb{L}} = T^{0,1}(M)$.

The proof of this theorem involves a lot of machinery that would take us too far afield and so we only give a reference (see Section 5.7 in [Ho]). The reader who is not familiar with this theorem can take comfort in knowing that it will only be used in Section 10.1 where we discuss Levi flat CR manifolds. If the reader restricts his or her attention to imbedded Levi flat CR submanifolds of $\mathbb{C}^n$, then the reader will only need the following imbedded version of the Newlander–Nirenberg theorem whose easy proof we provide.

THEOREM 2

Let J be the usual complex structure map on $\mathbb{C}^n$. Suppose M is a smooth submanifold of $\mathbb{C}^n$ such that $J\{T_p(M)\} = T_p(M)$ for each $p \in M$. Then M is a complex submanifold of $\mathbb{C}^n$.

The proof of Theorem 2 does not follow from the complex analytic version of the Frobenius theorem in the preceding section because we do not know that $\mathbb{L} = T^{\mathbb{C}}(M) \cap T^{1,0}(\mathbb{C}^n)$ is an analytic subbundle of $T^{1,0}(\mathbb{C}^n)$ (M is only assumed to be smooth). Theorem 2 does follow from Theorem 1 because $\mathbb{L}$ is an involutive, almost complex structure. However, we give a proof that does not require the use of Theorem 1.

PROOF We shall show that near any fixed point $p_0 \in M$, we can exhibit M as the graph over its tangent space at p_0 of a holomorphic mapping. By a translation, we assume the point p_0 is the origin. Then we use a complex linear map so that the tangent space of M at 0 is the copy of $\mathbb{C}^k$ given by $\{(0, w) \in \mathbb{C}^n; w \in \mathbb{C}^k\}$ where $2k = \dim_{\mathbb{R}} M$. This is accomplished as follows. Since $T_0(M)$ is J-invariant by hypothesis, the orthogonal complement of $T_0(M)$, $T_0(M)^\perp$ (using the Euclidean metric for $\mathbb{R}^{2n}$) is also J-invariant, in view of Lemma 1 in Section 3.2. In addition, J has no real eigenvalues. Therefore, there is a real basis for $T_0(M)^\perp$ of the form

$$v_1, Jv_1, \ldots, v_{n-k}, Jv_{n-k}$$

and a real basis for $T_0(M)$ of the form

$$v_{n-k+1}, Jv_{n-k+1}, \ldots, v_n, Jv_n.$$

Give $\mathbb{R}^{2n}$ the coordinates $(x_1, y_1, \ldots, x_n, y_n)$. Recall

$$J\left(\frac{\partial}{\partial x_j}\right) = \frac{\partial}{\partial y_j} \qquad 1 \le j \le n.$$

Define the linear map $A : \mathbb{R}^{2n} \to \mathbb{R}^{2n}$ by setting

$$A(v_j) = \frac{\partial}{\partial x_j}$$

$$A(Jv_j) = \frac{\partial}{\partial y_j}.$$

Then extend A to all of $T_0(\mathbb{R}^{2n})$ by (real) linearity. Note that $A(Jv_j) = JA(v_j)$, $1 \le j \le n$. Therefore $A \circ J = J \circ A$ on $T_0(\mathbb{R}^{2n})$. Viewed as a map from $\mathbb{C}^n$ to $\mathbb{C}^n$, A is complex linear. In particular, A is holomorphic and so $A\{M\}$ also satisfies the hypothesis of Theorem 2.

Let $w_1 = x_{n-k+1} + iy_{n-k+1}, \ldots, w_k = x_n + iy_n$. The real tangent space of $A\{M\}$ at the origin can be identified with

$$\{(0, w) \in \mathbb{C}^{n-k} \times \mathbb{C}^k ; w \in \mathbb{C}^k\}.$$

Therefore, near the origin, $A\{M\}$ is the graph of a *smooth* map $h : \mathbb{C}^k \to \mathbb{C}^{n-k}$, i.e.,

$$A\{M\} = \{(h(w), w); w \in \mathbb{C}^k\}.$$

To show $A\{M\}$ is a complex submanifold of $\mathbb{C}^n$, we must show h is holomorphic. This is equivalent to showing

$$h_*(p) \circ J = J \circ h_*(p)$$

for each $p \in \mathbb{R}^{2k} \simeq \mathbb{C}^k$. Here, $h_*(p)$ is the push forward map of h at p (see Chapter 2) where we view h as a map from $\mathbb{R}^{2k} \mapsto \mathbb{R}^{2n-2k}$. The J on the left side of the above equation is the complex structure map for $\mathbb{R}^{2k}$, whereas the J on the right is the complex structure map for $\mathbb{R}^{2(n-k)}$.

Define $H : \mathbb{C}^k \to \mathbb{C}^{n-k} \times \mathbb{C}^k$ by

$$H(w) = (h(w), w) \quad \text{for} \quad w \in \mathbb{C}^k.$$

Let L be a vector in $T_0(\mathbb{R}^{2k})$. For $p \in \mathbb{R}^{2k} \simeq \mathbb{C}^k$ near 0, we have

$$\begin{aligned} J\{H_*(p)(L)\} &= J\{(h_*(p)(L), L)\} \in \mathbb{R}^{2(n-k)} \times \mathbb{R}^{2k} \\ &= (J\{h_*(p)(L)\}, JL). \end{aligned}$$

Since H is the graphing map for $A\{M\}$ over $\mathbb{R}^{2k}$, $H_*(p)(L)$ belongs to $T_{H(p)}(A\{M\})$. Since $T_{H(p)}(A\{M\})$ is J-invariant, $J\{H_*(p)(L)\}$ also belongs to $T_{H(p)}(A\{M\})$. Now any vector in $T_{H(p)}(A\{M\})$ is of the form

$$(h_*(p)(W), W)$$

for some vector $W \in T_0(\mathbb{R}^{2k})$. Equating these two expressions for $J\{H_*(p)(L)\}$, we see that $W = J(L)$ and therefore $h_*(p)(JL) = Jh_*(p)(L)$. This establishes $h_*(p) \circ J = J \circ h_*(p)$ and so h is holomorphic, as desired. ▮

5

Distribution Theory

The discussion of solutions to partial differential equations is facilitated by the language of distribution theory. In this chapter we define distributions and discuss the basic operations with distributions. We then discuss a version of Whitney's extension theorem. The chapter ends with the computations of a fundamental solution for the Cauchy–Riemann operator on $\mathbb{C}$ and a fundamental solution for the Laplacian on $\mathbb{R}^N$.

5.1 The spaces $\mathcal{D}'$ and $\mathcal{E}'$

Let Ω be an open set in $\mathbb{R}^N$. In Section 1.1, we defined $\mathcal{E}(\Omega)$ to be the space of smooth, complex-valued functions on Ω. This space is topologized by stating that a sequence $\{f_n;\ n = 1, 2, \ldots\}$ converges to f in $\mathcal{E}(\Omega)$ provided $D^\alpha f_n$ converges to $D^\alpha f$ uniformly on compact subsets of Ω, for each multiindex α. This topology on $\mathcal{E}(\Omega)$ can also be described by the following seminorms. Let $\{K_n;\ n = 1, 2, \ldots\}$ be a nested sequence of compact subsets of Ω that exhaust Ω, i.e., $\cup_n K_n = \Omega$. The seminorms $\rho_n : \mathcal{E}(\Omega) \to \mathbb{R}$ for $n = 1, 2, \ldots$ are defined by

$$\rho_n(f) = \sup_{\substack{x \in K_n \\ |\alpha| \leq n}} |D^\alpha f(x)| \quad \text{for} \quad f \in \mathcal{E}(\Omega).$$

These seminorms satisfy the following properties:

(i) $\rho_n(\alpha f) = |\alpha| \rho_n(f)$ for $\alpha \in \mathbb{C}, f \in \mathcal{E}(\Omega)$

(ii) $\rho_n(f + g) \leq \rho_n(f) + \rho_n(g)$ for $f, g \in \mathcal{E}(\Omega)$

(iii) $f \equiv 0$ if and only if $\rho_n(f) = 0$ for all $n = 1, 2, \ldots$.

We have $f_m \to f$ in $\mathcal{E}(\Omega)$ if and only if $\rho_n(f_m) - \rho_n(f) \to 0$ as $m \to \infty$ for each $n = 1, 2, \ldots$. It is a routine matter to show that $\mathcal{E}(\Omega)$ is complete under this topology. A complete topological vector space whose topology is defined by a countable collection of seminorms is called a *Fréchet space*.

In Section 1.1, we also defined $\mathcal{D}(\Omega)$ to be the space of elements in $\mathcal{E}(\Omega)$ with compact support. A sequence $\{\phi_n,\ n = 1, 2, \ldots\}$ is said to converge to ϕ in $\mathcal{D}(\Omega)$ if there exists a fixed compact set K that contains the support of each ϕ_n and such that $D^\alpha \phi_n \to D^\alpha \phi$ uniformly on K for each multiindex α. The space $\mathcal{D}(\Omega)$ is *not* a Fréchet space.

Using coordinate charts, the above definitions of the topologies of $\mathcal{E}(\Omega)$ and $\mathcal{D}(\Omega)$ carry over to the case where Ω is an open subset of a smooth manifold. Most of the time in this chapter, our discussion will be focused on $\mathbb{R}^N$. However, much of what is to follow (with the exception of the convolution operator) generalizes to the manifold setting. We leave these routine generalizations to the reader.

We now define the space of distributions.

DEFINITION 1 *For an open subset Ω of $\mathbb{R}^N$ (or a smooth manifold), the space of distributions $\mathcal{D}'(\Omega)$ is defined to be the dual space of $\mathcal{D}(\Omega)$. That is, $\mathcal{D}'(\Omega)$ is the space of all continuous linear maps from $\mathcal{D}(\Omega)$ to $\mathbb{C}$. $\mathcal{E}'(\Omega)$ is defined to be the dual space of $\mathcal{E}(\Omega)$.*

Note that $\mathcal{E}'(\Omega)$ is a subspace of $\mathcal{D}'(\Omega)$ because $\mathcal{D}(\Omega) \subset \mathcal{E}(\Omega)$ and since a sequence that converges in $\mathcal{D}(\Omega)$ also converges in $\mathcal{E}(\Omega)$. The pairing between an element T of $\mathcal{D}'(\Omega)$ and an element ϕ of $\mathcal{D}(\Omega)$ will be denoted by $(T, \phi)_\Omega \in \mathbb{C}$.

Example 1

Let T be a locally integrable function on Ω. Then T can be viewed as an element of $\mathcal{D}'(\Omega)$ by defining

$$(T, \phi)_\Omega = \int_{x \in \Omega} T(x)\phi(x)\,dx \quad \text{for} \quad \phi \in \mathcal{D}(\Omega).$$

If $\phi_n \to \phi$ in $\mathcal{D}(\Omega)$ then clearly $(T, \phi_n)_\Omega \to (T, \phi)_\Omega$. Thus, $T = (T, \cdot)_\Omega$ defines an element of $\mathcal{D}'(\Omega)$. If supp T is a compact subset of Ω, then $T = (T, \cdot)_\Omega$ is an element of $\mathcal{E}'(\Omega)$. $\square$

Example 2

For a point $p \in \mathbb{R}^N$, the delta function at p, δ_p, defines an $\mathcal{E}'(\mathbb{R}^N)$-distribution by the formula

$$(\delta_p, f)_{\mathbb{R}^N} = f(p) \quad \text{for} \quad f \in \mathcal{E}(\mathbb{R}^N). \qquad \square$$

Example 3

Suppose M is a smooth, oriented ℓ-dimensional submanifold of $\mathbb{R}^N$. Let $d\mu_M$ be Hausdorff ℓ-dimensional measure restricted to M. The measure $d\mu_M$ is constructed as follows. Let $d\sigma$ be the volume form for M, which is the unique positive ℓ-form on M of unit length. To construct $d\sigma$, we first focus on a coordinate patch $U \subset M$ corresponding to an oriented coordinate chart χ: $U \to \mathbb{R}^\ell$. On U, let

$$d\sigma = \alpha\chi^*(dt_1 \wedge \ldots \wedge dt_\ell)$$

where α is a smooth function on U chosen so that $|d\sigma| = 1$ (here $|\cdot|$ denotes the pointwise Euclidean norm as extended to forms (Section 1.5)). The volume form $d\sigma$ is then globally constructed on M by covering M with a collection of coordinate patches corresponding to oriented coordinate charts together with a partition of unity.

For a compactly supported continuous function g: $\mathbb{R}^N \to \mathbb{C}$, define

$$\Lambda(g) = \int_M g d\sigma.$$

Note that Λ is linear and $\Lambda(g) \geq 0$ for $g \geq 0$. By the Riesz representation theorem, there is a unique positive Borel measure $d\mu_M$ that represents Λ, i.e.,

$$\Lambda(g) = \int g d\mu_M.$$

As a measure, $d\mu_M$ acts on elements of $\mathcal{D}(\mathbb{R}^N)$ via integration. Hence, $d\mu_M$ is an element of $\mathcal{D}'(\mathbb{R}^N)$. If M is compact, then $d\mu_M$ defines an element of $\mathcal{E}'(\mathbb{R}^N)$. □

Sometimes, we shall emphasize the "variable of integration" on Ω by writing

$$(T(x), \phi(x))_{x\in\Omega} = (T, \phi)_\Omega \quad \text{for} \quad T \in \mathcal{D}'(\Omega), \phi \in \mathcal{D}(\Omega).$$

This notation is motivated by Example 1 above where T is an $\mathcal{L}^1_{\text{loc}}$ function and the pairing $(T, \phi)_\Omega$ is given by integration. However, we shall sometimes use this notation even when T is not a function defined on Ω in the traditional sense.

A distribution T in $\mathcal{D}'(\Omega)$ or $\mathcal{E}'(\Omega)$ is said to *vanish* on an open subset U of Ω if $(T, \phi)_\Omega = 0$ for all $\phi \in \mathcal{D}(U)$. The *support* of a distribution T (denoted supp T) is defined to be the smallest closed set in Ω outside of which T vanishes. In Example 1 above, the support of T as a distribution is the same as the usual support of T as a function defined on Ω. In Example 2, supp $\{\delta_p\}$ is the single point set $\{p\}$. In Example 3, supp $d\mu_M$ is M. In all three examples, if supp T is compact in Ω, then T defines an element of $\mathcal{E}'(\Omega)$. This is not just a coincidence, as the following lemma shows.

LEMMA 1
If T is an element of $\mathcal{E}'(\Omega)$, then there is a compact set $K \subset \Omega$, an integer $M > 0$, and a constant $C > 0$ such that

$$|(T, f)_\Omega| \leq C \sup_{\substack{|\alpha| \leq M \\ x \in K}} |D^\alpha f(x)| \quad \text{for all} \quad f \in \mathcal{E}(\Omega).$$

In particular, supp $T \subset K$. Conversely, if T is an element of $\mathcal{D}'(\Omega)$ and supp T is compact in Ω, then T defines an element of $\mathcal{E}'(\Omega)$.

PROOF It is a standard fact from topology that a continuous linear complex-valued map defined on a Fréchet space must be bounded in absolute value by a constant multiple of a finite sum of seminorms. Thus, the above inequality follows for some integer M, constant C, and compact set K. In particular, supp T must be contained in K, for if not, then there must exist an element $f \in \mathcal{D}(\Omega - K)$ with $(T, f)_\Omega \neq 0$. This contradicts the above inequality.

For the converse, suppose T belongs to $\mathcal{D}'(\Omega)$ and supp T is compact. Let $\phi \in \mathcal{D}(\Omega)$ be a smooth cutoff function (see Section 1.1) with $\phi = 1$ on a neighborhood of supp T. Define

$$(T, f)_\Omega = (T, \phi f)_\Omega \quad \text{for} \quad f \in \mathcal{E}(\Omega).$$

The right side is well defined because ϕf belongs to $\mathcal{D}(\Omega)$ and T belongs to $\mathcal{D}'(\Omega)$. Also, if $f_n \to f$ in $\mathcal{E}(\Omega)$, then $\phi f_n \to \phi f$ in $\mathcal{D}(\Omega)$. Therefore, this definition of T makes it a well-defined distribution in $\mathcal{E}'(\Omega)$. Since $1 - \phi$ vanishes on a neighborhood of supp T, clearly

$$(T, (1 - \phi)f)_\Omega = 0 \quad \text{for each} \quad f \in \mathcal{D}(\Omega).$$

Hence, the above definition of T as a linear map acting on $\mathcal{E}(\Omega)$ agrees with the original definition of T as a linear map acting on $\mathcal{D}(\Omega) \subset \mathcal{E}(\Omega)$. ∎

The spaces $\mathcal{D}'(\Omega)$ and $\mathcal{E}'(\Omega)$ are endowed with a topology called the *weak topology*. This is defined by declaring that a sequence $\{T_n,\ n = 1, 2, \ldots\}$ converges to T in $\mathcal{D}'(\Omega)$ (or $\mathcal{E}'(\Omega)$) if

$$(T_n, \phi)_\Omega \to (T, \phi)_\Omega$$

for each ϕ in $\mathcal{D}(\Omega)$ (or $\mathcal{E}(\Omega)$). This convergence is *not* required to be uniform in ϕ. As an example, if T_n is a sequence of continuous functions on Ω that converges to the function T uniformly on compact subsets of Ω, then the sequence T_n also converges to T as distributions in $\mathcal{D}'(\Omega)$. However, the topology on $\mathcal{D}'(\Omega)$ allows for much weaker types of convergence. For example, define the sequence $e_\epsilon(x) = e(x/\epsilon)\epsilon^{-N}$, for $\epsilon > 0$, where $e(y) = \pi^{-N/2} e^{-|y|^2}$. This sequence was used in Section 1.1 for the proof of the Weierstrass theorem. Using the ideas from its proof, the reader can show $e_\epsilon \to \delta_0$ in $\mathcal{D}'(\mathbb{R}^N)$ as $\epsilon \to 0$. Note that $e_\epsilon(x) \to 0$ for $x \neq 0$ (but not uniformly) and $e_\epsilon(0) \to \infty$.

5.2 Operations with distributions

All of the definitions of operations with distributions are motivated by considering the case where the distribution is a smooth function.

Multiplication of a distribution by a smooth function

If T and ψ are smooth functions on Ω, then $T \cdot \psi$ defines a $\mathcal{D}'(\Omega)$-distribution by

$$(T\psi, \phi)_\Omega = \int_{x \in \Omega} T(x)\psi(x)\phi(x)dx \quad \text{for} \quad \phi \in \mathcal{D}(\Omega).$$

Note that $(T\psi, \phi)_\Omega = (T, \psi\phi)_\Omega$. So if T is an arbitrary $\mathcal{D}'(\Omega)$-distribution and if ψ belongs to $\mathcal{E}(\Omega)$, then we *define* the distribution $T\psi$ by

$$(T\psi, \phi)_\Omega = (T, \psi\phi)_\Omega \quad \text{for} \quad \phi \in \mathcal{D}(\Omega).$$

With this definition, $T\psi$ satisfies all the required properties for a distribution. Moreover, this definition agrees with the action (as a distribution) of $T\psi$ when both T and ψ are smooth functions.

Differentiation

If T is a smooth function on Ω and $\phi \in \mathcal{D}(\Omega)$, then

$$\begin{aligned}(D^\alpha T, \phi)_\Omega &= \int_{x \in \Omega} (D^\alpha T)\phi dx = (-1)^{|\alpha|} \int_{x \in \Omega} T D^\alpha \phi dx \\ &= (-1)^{|\alpha|}(T, D^\alpha \phi)_\Omega.\end{aligned}$$

If T is an arbitrary $\mathcal{D}'(\Omega)$-distribution, then we *define* the distribution $D^\alpha T$ by

$$(D^\alpha T, \phi)_\Omega = (-1)^{|\alpha|}(T, D^\alpha \phi)_\Omega.$$

If $\phi_n \to \phi$ in $\mathcal{D}(\Omega)$ then the sequence $D^\alpha \phi_n$ also converges to $D^\alpha \phi$ in $\mathcal{D}(\Omega)$. Thus, $(D^\alpha T, \cdot)_\Omega$ is continuous and so $D^\alpha T$ is a well-defined $\mathcal{D}'(\Omega)$-distribution.

As an example, we have

$$(D^\alpha \delta_p, \phi)_{\mathbb{R}^N} = (-1)^{|\alpha|}(D^\alpha \phi)(p) \quad \text{for} \quad \phi \in \mathcal{E}(\mathbb{R}^N).$$

If a sequence of distributions $\{T_n\}$ in $\mathcal{D}'(\Omega)$ converges to the distribution T as $n \to \infty$, then $D^\alpha T_n$ also converges to $D^\alpha T$. This follows from the definition of the weak topology on $\mathcal{D}'(\Omega)$. This is in contrast to other types of topologies on function spaces where this does not hold. For example, the sequence $T_n(x) = n^{-1} \sin nx$ converges uniformly to 0 but $T_n'(x)$ is not even pointwise convergent.

Convolution

Here, we specialize to the case $\Omega = \mathbb{R}^N$. If T belongs to $\mathcal{E}(\mathbb{R}^N)$ and ψ belongs to $\mathcal{D}(\mathbb{R}^N)$, then the convolution $T * \psi$ is the smooth function defined by

$$(T * \psi)(x) = \int\limits_{y \in \mathbb{R}^N} T(y)\psi(x-y)dy.$$

As a distribution on $\mathbb{R}^N$, we have

$$\begin{aligned}(T * \psi, \phi)_{\mathbb{R}^N} &= \int\limits_{x \in \mathbb{R}^N} \left(\int\limits_{y \in \mathbb{R}^N} T(y)\psi(x-y)dy \right) \phi(x)dx, \quad \phi \in \mathcal{D}(\mathbb{R}^N) \\ &= \int\limits_{y} T(y) \left(\int\limits_{x} \psi(x-y)\phi(x)dx \right) dy.\end{aligned}$$

With the notation $\check{\psi}(t) = \psi(-t)$, this becomes

$$\begin{aligned}(T * \psi, \phi)_{\mathbb{R}^N} &= \int\limits_{y \in \mathbb{R}^N} T(y)(\check{\psi} * \phi)(y) \\ &= (T, \check{\psi} * \phi)_{\mathbb{R}^N}.\end{aligned}$$

Now $\check{\psi} * \phi$ belongs to $\mathcal{D}(\mathbb{R}^N)$ because both ψ and ϕ belong to $\mathcal{D}(\mathbb{R}^N)$. Therefore, $(T, \check{\psi} * \phi)_{\mathbb{R}^N}$ is well defined for T in $\mathcal{D}'(\mathbb{R}^N)$. If T is an arbitrary $\mathcal{D}'(\mathbb{R}^N)$-distribution and if $\psi \in \mathcal{D}(\mathbb{R}^N)$, then we define

$$(T * \psi, \phi)_{\mathbb{R}^N} = (T, \check{\psi} * \phi)_{\mathbb{R}^N} \quad \text{for} \quad \phi \in \mathcal{D}(\mathbb{R}^N).$$

If $\phi_n \to \phi$ in $\mathcal{D}(\mathbb{R}^N)$ then $\check{\psi} * \phi_n \to \check{\psi} * \phi$ in $\mathcal{D}(\mathbb{R}^N)$. Therefore, $T * \psi$ is a well-defined element of $\mathcal{D}'(\mathbb{R}^N)$. From the above discussion, this definition agrees with the action of $T * \psi$ as a distribution when T is a smooth function.

If ψ belongs to $\mathcal{E}(\mathbb{R}^N)$ and ϕ belongs to $\mathcal{D}(\mathbb{R}^N)$, then $\check{\psi} * \phi$ belongs to $\mathcal{E}(\mathbb{R}^N)$. So the same definition can be used to define the convolution of an $\mathcal{E}'(\mathbb{R}^N)$-distribution T with an element ψ of $\mathcal{E}(\mathbb{R}^N)$. The result is a $\mathcal{D}'(\mathbb{R}^N)$-distribution.

Let us examine the definition of $T * \psi$ more carefully. If T is an element of $\mathcal{D}'(\mathbb{R}^N)$ and ψ and ϕ are elements of $\mathcal{D}(\mathbb{R}^N)$, then

$$\begin{aligned}(T * \psi, \phi)_{\mathbb{R}^N} &= (T, \check{\psi} * \phi)_{\mathbb{R}^N} \\ &= \left(T(y), \int\limits_{x \in \mathbb{R}^N} \psi(x-y)\phi(x)dx \right)_{y \in \mathbb{R}^N}.\end{aligned}$$

As already mentioned, the above notation indicates that the distribution T pairs with the function $\int_x \psi(x-y)\phi(x)dx$ in the y-variable. The term $\int_x \psi(x-y)\phi(x)dx$ can be approximated by Riemann sums in x. Moreover, this approximation can be accomplished in the $\mathcal{D}(\mathbb{R}^N)$-topology (in the y-variable). Therefore, using the linearity and continuity of T, we can interchange the order of the pairing of T with the integral in x to obtain

$$\begin{aligned}(T*\psi,\phi)_{\mathbb{R}^N} &= \int\limits_{x\in\mathbb{R}^N} (T(y),\psi(x-y))_{y\in\mathbb{R}^N}\,\phi(x)dx \\ &= ((T(y),\psi(x-y))_{y\in\mathbb{R}^N}, \phi(x))_{x\in\mathbb{R}^N}.\end{aligned}$$

Therefore, the distribution $T*\psi$ is given by pairing with the following function:

$$x \longmapsto (T(y),\psi(x-y))_{y\in\mathbb{R}^N}.$$

This formula generalizes the usual convolution formula when T is a smooth function on $\mathbb{R}^N$.

If ψ is an element of $\mathcal{D}(\mathbb{R}^N)$, then a difference quotient for ψ converges in $\mathcal{D}(\mathbb{R}^N)$ to the corresponding derivative of ψ. So if $T\in\mathcal{D}'(\mathbb{R}^N)$, then

$$D^\alpha\{T*\psi\}(x) = (T(y),(D^\alpha\psi)(x-y))_{y\in\mathbb{R}^N}.$$

This shows that $T*\psi$ is a smooth function on $\mathbb{R}^N$. By the definition of the derivative of a distribution, we also have

$$D^\alpha\{T*\psi\}(x) = (D^\alpha T,\psi(x-y))_{y\in\mathbb{R}^N}.$$

So $D^\alpha\{T*\psi\}$ can be written as either $T*D^\alpha\psi$ or $D^\alpha T*\psi$. This discussion is summarized in the following lemma.

LEMMA 1

Suppose T is an element of $\mathcal{D}'(\mathbb{R}^N)$ and ψ is an element of $\mathcal{D}(\mathbb{R}^N)$. Then the distribution $T*\psi$ is given by pairing with the smooth function

$$T*\psi(x) = (T(y),\psi(x-y))_{y\in\mathbb{R}^N}.$$

Moreover, $D^\alpha\{T*\psi\} = D^\alpha T*\psi = T*D^\alpha\psi$. The operator $\psi\mapsto T*\psi$ is a continuous linear map from $\mathcal{D}(\mathbb{R}^N)$ to $\mathcal{E}(\mathbb{R}^N)$.

We can also define the convolution of a $\mathcal{D}'$-distribution with an $\mathcal{E}'$-distribution by mimicking the above definition of the convolution of an element of $\mathcal{D}'(\mathbb{R}^N)$

with an element of $\mathcal{D}(\mathbb{R}^N)$. First we define $\check{T}$ for $T \in \mathcal{D}'(\mathbb{R}^N)$ by

$$(\check{T}, \phi)_{\mathbb{R}^N} = (T, \check{\phi})_{\mathbb{R}^N} \quad \text{for} \quad \phi \in \mathcal{D}(\mathbb{R}^N).$$

This definition agrees with the formula $\check{T}(x) = T(-x)$ when T is a smooth function. For $T \in \mathcal{D}'(\mathbb{R}^N)$ and $S \in \mathcal{E}'(\mathbb{R}^N)$, the distribution $T * S$ is defined by

$$(T * S, \phi)_{\mathbb{R}^N} = (T, \check{S} * \phi)_{\mathbb{R}^N} \quad \text{for} \quad \phi \in \mathcal{D}(\mathbb{R}^N).$$

Since $\check{S} * \phi$ is a smooth function (Lemma 1) with compact support, the right side is well defined. The same definition can be used if T is an $\mathcal{E}'(\mathbb{R}^N)$-distribution and S is a $\mathcal{D}'(\mathbb{R}^N)$-distribution.

LEMMA 2
For $T \in \mathcal{D}'(\mathbb{R}^N), \delta_0 * T = T$.

PROOF We have

$$\begin{aligned}
(\delta_0 * T, \phi)_{\mathbb{R}^N} &= (\delta_0, \check{T} * \phi)_{\mathbb{R}^N} \quad \text{for} \quad \phi \in \mathcal{D}(\mathbb{R}^N) \\
&= (\check{T} * \phi)(0) \\
&= (\check{T}(y), \phi(0 - y))_{y \in \mathbb{R}^N} \quad \text{(by Lemma 1)} \\
&= (T(y), \check{\phi}(-y))_{y \in \mathbb{R}^N} \quad \text{(by Definition of } \check{T}) \\
&= (T, \phi)_{\mathbb{R}^N} .
\end{aligned}$$

Hence, $\delta_0 * T = T$ and the lemma follows. ∎

If T is a smooth function on $\mathbb{R}^N$, then Lemma 2 can be established more directly by using Lemma 1

$$(\delta_0 * T)(x) = (\delta_0(y), T(x - y))_{y \in \mathbb{R}^N} = T(x).$$

Lemma 2 shows that convolution with δ_0 represents the identity operator. This will be important in Section 5.4 when we discuss fundamental solutions for partial differential operators with constant coefficients.

THEOREM 1
For an open set $\Omega \subset \mathbb{R}^N, \mathcal{D}(\Omega)$ *is a dense subset of* $\mathcal{D}'(\Omega)$ *with the weak topology.*

PROOF Let T be an element of $\mathcal{D}'(\Omega)$. Exhaust Ω by an increasing sequence $\{K_n\}$ of compact sets in Ω. Let $\phi_n \in \mathcal{D}(\Omega)$ be a cutoff function that is 1 on a neighborhood of K_n. Clearly $\phi_n T$ converges to T in $\mathcal{D}'(\Omega)$ as $n \to \infty$. So it suffices to approximate any element T of $\mathcal{E}'(\Omega)$ by a sequence of smooth functions on Ω' where $\Omega' \subset \Omega$ is a neighborhood of supp T. Let $\{\phi_\epsilon; \epsilon > 0\}$ be the mollifier sequence defined in Section 1.1. It follows from the ideas in Section 1.1 that $\phi_\epsilon * \phi \to \phi$ in $\mathcal{E}(\Omega')$ as $\epsilon \to 0$, provided ϕ is an element of $\mathcal{E}(\Omega)$. From Lemma 1, $T * \phi_\epsilon$ is a smooth function whose support is contained in Ω, provided $\epsilon > 0$ is suitably small. From the definitions, we have

$$\begin{aligned}(T * \phi_\epsilon, \phi)_{\Omega'} &= (T, \check{\phi}_\epsilon * \phi)_{\Omega'} \quad \text{for} \quad \phi \in \mathcal{E}(\Omega)\\ &= (T, \phi_\epsilon * \phi)_{\Omega'} \quad \text{since} \quad \check{\phi}_\epsilon = \phi_\epsilon\\ &\to (T, \phi)_{\Omega'} \quad \text{since} \quad \phi_\epsilon * \phi \to \phi \text{ in } \mathcal{E}(\Omega').\end{aligned}$$

Therefore, $T * \phi_\epsilon \to T$ in $\mathcal{E}'(\Omega')$, as desired. ∎

Tensor products

Suppose T is an element of $\mathcal{D}'(\mathbb{R}^N)$ and S is an element of $\mathcal{D}'(\mathbb{R}^k)$. We wish to define the tensor product $T \otimes S \in \mathcal{D}'(\mathbb{R}^N \times \mathbb{R}^k)$ in such a way that the following holds:

$$((T \otimes S)(x,y), \phi(x)\psi(y))_{(x,y)\in\mathbb{R}^N\times\mathbb{R}^k} = (T, \phi)_{\mathbb{R}^N} \cdot (S, \psi)_{\mathbb{R}^k} \tag{1}$$

for $\phi \in \mathcal{D}(\mathbb{R}^N), \psi \in \mathcal{D}(\mathbb{R}^k)$. Let g be an element of $\mathcal{D}(\mathbb{R}^N \times \mathbb{R}^k)$. The function

$$x \longmapsto (S(y), g(x,y))_{y\in\mathbb{R}^k} \quad \text{for} \quad x \in \mathbb{R}^N \tag{2}$$

is smooth with compact support. This follows from the fact that S is linear and that difference quotients of $g(x,y)$ in the x-variable converge in $\mathcal{D}(\mathbb{R}^N \times \mathbb{R}^k)$ to the corresponding x-derivative of $g(x,y)$. Therefore the pairing

$$(T(x), (S(y), g(x,y))_{y\in\mathbb{R}^k})_{x\in\mathbb{R}^N} \tag{3}$$

is well defined for $T \in \mathcal{D}'(\mathbb{R}^N)$. If $g_n \to g$ in $\mathcal{D}(\mathbb{R}^N \times \mathbb{R}^k)$, then the corresponding sequence of functions defined in (2) with g replaced by g_n is a convergent sequence in $\mathcal{D}(\mathbb{R}^N)$. So, the corresponding sequence of complex numbers in (3) also converges. We define $(T \otimes S, g)_{\mathbb{R}^N\times\mathbb{R}^k}$ by (3). By the above discussion, $T \otimes S$ is a well-defined element in $\mathcal{D}'(\mathbb{R}^N \times \mathbb{R}^k)$. If $g(x,y) = \phi(x)\psi(y)$, then (3) reduces to (1), as desired.

Instead of (3), we could have used

$$(S(y), (T(x), g(x,y))_{x\in\mathbb{R}^N})_{y\in\mathbb{R}^k} \tag{4}$$

as the definition of $T \otimes S$. This formula also yields (1) in the case $g(x,y) = \phi(x)\psi(y)$. So the expressions in (4) and (3) agree for functions of the form $g(x,y) = \phi(x)\psi(y)$. Since the set of functions that are finite sums of terms of the form $\phi(x) \cdot \psi(y)$ for $\phi \in \mathcal{D}(\mathbb{R}^N), \psi \in \mathcal{D}(\mathbb{R}^k)$ is dense in $\mathcal{D}(\mathbb{R}^N \times \mathbb{R}^k)$ by the Weierstrass theorem (see Section 1.1), clearly (4) and (3) agree for all $g \in \mathcal{D}(\mathbb{R}^N \times \mathbb{R}^k)$.

As an example, consider the tensor product of δ_0 in $\mathcal{D}'(\mathbb{R}^N)$ with the function 1 as an element of $\mathcal{D}'(\mathbb{R}^k)$. For $g \in \mathcal{D}(\mathbb{R}^N \times \mathbb{R}^k)$, we have

$$\begin{aligned}(\delta_0 \otimes 1, g(x,y))_{(x,y)\in\mathbb{R}^N\times\mathbb{R}^k} &= (\delta_0(x), (1(y), g(x,y))_{y\in\mathbb{R}^k})_{x\in\mathbb{R}^N} \\ &= \left(\delta_0(x), \int\limits_{y\in\mathbb{R}^k} g(x,y)dy\right)_{x\in\mathbb{R}^N} \\ &= \int\limits_{y\in\mathbb{R}^k} g(0,y)dy.\end{aligned}$$

Composing a distribution with a diffeomorphism

Suppose Ω and Ω' are connected open sets in $\mathbb{R}^N$ and $F\colon \Omega \to \Omega'$ is a diffeomorphism. If T is a smooth, complex-valued function on Ω', then $T \circ F$ is a smooth function on Ω. As a distribution on Ω, we have

$$\begin{aligned}(T \circ F, \phi)_\Omega &= \int\limits_{y\in\Omega} T(F(y))\phi(y)dy \quad \text{for} \quad \phi \in \mathcal{D}(\Omega) \\ &= \int\limits_{\Omega'} T(x)\phi(F^{-1}(x))|\det DF^{-1}(x)|dx\end{aligned}$$

by the change of variables $x = F(y)$. So

$$(T \circ F, \phi)_\Omega = (T, (\phi \circ F^{-1})|\text{Det } DF^{-1}|)_{\Omega'}.$$

Since F is a diffeomorphism, Det DF^{-1} is either always positive or always negative on Ω'. So $(\phi \circ F^{-1})|\text{Det } DF^{-1}|$ is a smooth function with compact support in Ω'. We therefore define, for $T \in \mathcal{D}(\Omega')$, the distribution $T \circ F \in \mathcal{D}'(\Omega)$ by the above formula.

As an application of these last two operations, we show how to extend a distribution defined on a submanifold of $\mathbb{R}^N$ to a distribution defined on all of $\mathbb{R}^N$. Suppose $N = \ell + k$ where ℓ is the real dimension of the submanifold in $\mathbb{R}^N$. By a partition of unity, we can localize the problem. By a coordinate change and the above change of variables formula, we may assume that the submanifold is the copy of $\mathbb{R}^\ell$ given by $\{(x,0) \in \mathbb{R}^\ell \times \mathbb{R}^k; x \in \mathbb{R}^\ell\}$. A distribution on $\mathbb{R}^\ell$ can then be extended to all of $\mathbb{R}^\ell \times \mathbb{R}^k$ by tensoring it with the function 1 on $\mathbb{R}^k$.

5.3 Whitney's extension theorem

In its simplest form (due to Borel), Whitney's extension theorem states that given any infinite sequence of real numbers $\{a_0, a_1, \ldots\}$, there is a smooth function $f\colon \mathbb{R} \to \mathbb{R}$ with $f^{(n)}(0) = a_n$ for $n = 0, 1, \ldots$. Here, $f^{(n)}$ is the nth derivative of f. We wish to prove this theorem along with its natural generalization to higher dimensions.

First, we need a preliminary result from distribution theory which is interesting in its own right.

THEOREM 1
Let (x, y) be coordinates for $\mathbb{R}^N$ with $x \in \mathbb{R}^k$ and $y \in \mathbb{R}^{N-k}$. Let Ω be an open set in $\mathbb{R}^N$ and let $\Omega_0 = \{(x,0) \in \Omega; x \in \mathbb{R}^k\}$. Suppose T is an element of $\mathcal{E}'(\Omega)$ with support in Ω_0. Then there is an integer $M \geq 0$ and a collection of distributions $\{T_\alpha \in \mathcal{E}'(\Omega_0); \alpha = (\alpha_1, \ldots, \alpha_{N-k})$ with $|\alpha| \leq M\}$ such that

$$T(x,y) = \sum_{|\alpha| \leq M} \frac{1}{\alpha!} T_\alpha(x) \otimes \frac{\partial^{|\alpha|}}{\partial y^\alpha} \delta_0(y).$$

PROOF We assume that Ω is all of $\mathbb{R}^N$ and leave to the reader the minor modifications required in the case where Ω is not all of $\mathbb{R}^N$. Since supp $T \subset \{y = 0\}$, by Lemma 1 in Section 5.1 there is a compact set K in $\mathbb{R}^N$, a constant $C > 0$, and an integer $M \geq 0$ such that

$$|(T, f)_{\mathbb{R}^N}| \leq C \sup_{\substack{|\alpha| \leq M \\ (x,0) \in K}} |(D^\alpha_{x,y} f)(x, 0)| \quad \text{for} \quad f \in \mathcal{E}(\mathbb{R}^N). \tag{1}$$

Here, D^α_{xy} is a partial differential operator in both x and y of order $|\alpha| \leq M$. For $\alpha = (\alpha_1, \ldots, \alpha_{N-k})$ with $|\alpha| \leq M$, define the distribution $T_\alpha \in \mathcal{E}'(\mathbb{R}^k)$ by

$$(T_\alpha(x), \phi(x))_{x \in \mathbb{R}^k} = (-1)^{|\alpha|} (T(x,y), \phi(x) y^\alpha)_{(x,y) \in \mathbb{R}^N} \quad \text{for} \quad \phi \in \mathcal{E}(\mathbb{R}^k).$$

We claim $\{T_\alpha\}$ satisfies the requirements of the theorem.

Given $f \in \mathcal{E}(\mathbb{R}^N)$, its Taylor expansion in the y-variable is

$$f(x,y) = \sum_{|\beta| \leq M} \frac{1}{\beta!} \frac{\partial^{|\beta|} f}{\partial y^\beta}(x, 0) y^\beta + e(x,y).$$

The Taylor remainder e satisfies

$$(D^\beta_{x,y} e)(x, 0) = 0 \qquad |\beta| \leq M.$$

So in view of (1)

$$(T, f)_{\mathbb{R}^N} = \sum_{|\beta| \leq M} \frac{1}{\beta!} \left(T(x,y), \frac{\partial^{|\beta|} f}{\partial y^\beta}(x, 0) y^\beta \right)_{(x,y) \in \mathbb{R}^N}. \tag{2}$$

On the other hand, from the Taylor expansion of f, we have

$$\left(\sum_{|\alpha|\le M}\frac{1}{\alpha!}T_\alpha\otimes\frac{\partial^{|\alpha|}\delta_0}{\partial y^\alpha},f\right)_{\mathbb{R}^N}$$

$$=\sum_{|\alpha|\le M}\sum_{|\beta|\le M}\frac{1}{\alpha!\beta!}\left(T_\alpha(x),\frac{\partial^{|\beta|}f}{\partial y^\beta}(x,0)\right)_{x\in\mathbb{R}^k}\cdot\left(\frac{\partial^{|\alpha|}\delta_0}{\partial y^\alpha},y^\beta\right)_{y\in\mathbb{R}^{N-k}}.$$

Clearly

$$\left(\frac{\partial^{|\alpha|}\delta_0(y)}{\partial y^\alpha},y^\beta\right)_{y\in\mathbb{R}^{N-k}}=\begin{cases}0 & \text{if } (\alpha_1,\ldots,\alpha_{N-k})\neq(\beta_1,\ldots,\beta_{N-k})\\ (-1)^{|\alpha|}\alpha! & \text{if } (\alpha_1,\ldots,\alpha_{N-k})=(\beta_1,\ldots,\beta_{N-k}).\end{cases}$$

Therefore

$$\left(\sum_{|\alpha|\le M}\frac{1}{\alpha!}T_\alpha\otimes\frac{\partial^{|\alpha|}\delta_0(y)}{\partial y^\alpha},f\right)_{\mathbb{R}^N}=\sum_{|\alpha|\le M}\frac{(-1)^{|\alpha|}}{\alpha!}\left(T_\alpha(x),\frac{\partial^{|\alpha|}f}{\partial y^\alpha}(x,0)\right)_{x\in\mathbb{R}^k}.$$

By the definition of T_α, the right side equals

$$\left(T(x,y),\sum_{|\alpha|\le M}\frac{1}{\alpha!}\frac{\partial^{|\alpha|}f}{\partial y^\alpha}(x,0)y^\alpha\right)_{(x,y)\in\mathbb{R}^N}.$$

By (2), this equals $(T,f)_{\mathbb{R}^N}$. This completes the proof of the theorem. ∎

Now we prove Whitney's theorem.

THEOREM 2

Suppose $\mathbb{R}^N$ has the coordinates (x,y) with $x\in\mathbb{R}^k$ and $y\in\mathbb{R}^{N-k}$. Let Ω_0 be an open set in $\mathbb{R}^k$. Suppose a_α is an element of $\mathcal{E}(\Omega_0)$ for each index $\alpha=(\alpha_1,\ldots,\alpha_{N-k})$ with $\alpha_j\ge 0$, $1\le j\le N-k$. Then there is a function $f\in\mathcal{E}(\Omega_0\times\mathbb{R}^{N-k})$ such that for each α

$$\frac{\partial^{|\alpha|}f}{\partial y^\alpha}(x,0)=a_\alpha(x)\quad\textit{for}\quad x\in\Omega_0.$$

The function f is unique modulo the space of smooth functions on $\Omega_0\times\mathbb{R}^{N-k}$ which vanish to infinite order on $\Omega_0\times\{0\}$.

PROOF The uniqueness assertion is obvious. So we shall concentrate on the existence part of the theorem. As in the proof of Theorem 1, we shall assume Ω_0 is all of $\mathbb{R}^k$.

If $N=1$ and $k=0$, then the theorem reduces to showing that for a given sequence of real numbers $\{a_0,a_1,\ldots\}$, there is a smooth function $f\colon\mathbb{R}\to\mathbb{R}$

with

$$f^{(n)}(0) = a_n, \text{ for } n = 0, 1, \dots .$$

This can be done explicitly by first choosing a cutoff function $\phi \in \mathcal{D}(\mathbb{R})$ with

$$\phi(y) = \begin{cases} 1 & \text{if } |y| \le 1 \\ 0 & \text{if } |y| \ge 2. \end{cases}$$

Then we let

$$f(y) = \sum_{n=0}^{\infty} \frac{a_n y^n}{n!} \phi(\epsilon_n^{-1} y). \tag{3}$$

There exist ϵ_n (depending on $|a_n|$) so that for any integer $\alpha \ge 0$

$$\left| D_y^\alpha \left\{ \frac{a_n y^n}{n!} \phi(\epsilon_n^{-1} y) \right\} \right| \le C 2^{-n}.$$

where C is a uniform constant depending only on α. Therefore, the sum in (3) converges in the topology of $\mathcal{E}(\mathbb{R})$. Since $\phi \equiv 1$ in a neighborhood of the origin, $f^{(n)}(0) = a_n$, for $n = 0, 1, 2, \dots$, as desired.

The reader can modify this proof so that it will work in the general case. However, a slicker proof is available with the help of a little functional analysis. We consider the space

$$\prod_\alpha \mathcal{E}(\mathbb{R}^k)$$

where $\prod_\alpha$ denotes the infinite cartesian product indexed by $\alpha = (\alpha_1, \dots, \alpha_{N-k})$ where each α_i is a nonnegative integer. Elements of this space are infinite tuples (a_α) where each a_α is an element of $\mathcal{E}(\mathbb{R}^k)$. This space is given the product topology, that is, a sequence (a_α^n) for $n = 1, 2, \dots$ is said to converge to (a_α) in $\prod_\alpha \mathcal{E}(\mathbb{R}^k)$ as $n \to \infty$ if for each fixed α, $a_\alpha^n \to a_\alpha$ in $\mathcal{E}(\mathbb{R}^k)$. This makes $\prod_\alpha \mathcal{E}(\mathbb{R}^k)$ into a Fréchet space (since each $\mathcal{E}(\mathbb{R}^k)$ is a Fréchet space and since the index set $\{\alpha = (\alpha_1, \dots, \alpha_{N-k}); \alpha_i \text{ is a nonnegative integer}\}$ is countable).

Define the map π: $\mathcal{E}(\mathbb{R}^k \times \mathbb{R}^{N-k}) \to \prod_\alpha \mathcal{E}(\mathbb{R}^k)$ by

$$(\pi f)_\alpha(x) = \frac{1}{\alpha!} \frac{\partial^{|\alpha|} f}{\partial y^\alpha}(x, 0) \quad \text{for } \textit{each} \quad \alpha.$$

We want to show the map π is surjective. By looking at the subspace of polynomials on $\mathbb{R}^N$, it is clear from the definition of the topology of $\prod_\alpha \mathcal{E}(\mathbb{R}^k)$ that the image of π is dense. So it suffices to show the image of π is closed. By the closed range theorem for Fréchet spaces, it suffices to show the range of the dual to π

$$\pi' \colon \left\{ \prod_\alpha \mathcal{E}(\mathbb{R}^k) \right\}' \longrightarrow \mathcal{E}'(\mathbb{R}^N)$$

is closed. Here, π' is defined by

$$(\pi'(T), f)_{\mathbb{R}^N} = (T, \pi f)$$

for $T \in \left\{\prod_\alpha \mathcal{E}(\mathbb{R}^k)\right\}'$ and $f \in \mathcal{E}(\mathbb{R}^N)$. We have

$$\left\{\prod_\alpha \mathcal{E}(\mathbb{R}^k)\right\}' = \sum_\alpha \mathcal{E}'(\mathbb{R}^k)$$

where, by definition, the elements of $\sum_\alpha \mathcal{E}'(\mathbb{R}^k)$ are infinite tuples (T_α) with $T_\alpha \in \mathcal{E}'(\mathbb{R}^k)$ such that only a finite number of the T_α are nonzero. An element (T_α) in $\sum_\alpha \mathcal{E}'(\mathbb{R}^k)$ acts on an element $(f_\beta) \in \prod_\beta \mathcal{E}(\mathbb{R}^k)$ by

$$((T_\alpha), (f_\beta)) = \sum_\alpha (T_\alpha, f_\alpha)_{\mathbb{R}^k}.$$

The sum on the right is well defined since only a finite number of the T_α are nonzero.

We now compute π'. We have

$$\begin{aligned}(\pi'\{(T_\alpha)\}, f)_{\mathbb{R}^N} &= ((T_\alpha), \pi f) \\ &= \sum_\alpha \left(T_\alpha(x), \frac{1}{\alpha!}\frac{\partial^{|\alpha|} f}{\partial y^\alpha}(x, 0)\right)_{x \in \mathbb{R}^k} \\ &= \sum_\alpha \left(T_\alpha(x) \otimes \frac{(-1)^{|\alpha|}}{\alpha!}\frac{\partial^{|\alpha|}\delta_0(y)}{\partial y^\alpha}, f(x, y)\right)_{(x,y) \in \mathbb{R}^N}.\end{aligned}$$

Therefore

$$\pi'\{(T_\alpha)\}(x, y) = \sum_\alpha T_\alpha(x) \otimes \frac{(-1)^{|\alpha|}}{\alpha!}\frac{\partial^{|\alpha|}\delta_0(y)}{\partial y^\alpha}$$

which has its support in the set $\Omega_0 = \{(x, 0) \in \mathbb{R}^N; x \in \mathbb{R}^k\}$. This equation together with Theorem 1 implies that the range of π' is the set of all elements in $\mathcal{E}'(\mathbb{R}^k \times \mathbb{R}^{N-k})$ whose support is contained in Ω_0. This set of distributions is closed with respect to the weak topology on $\mathcal{E}'(\mathbb{R}^k \times \mathbb{R}^{N-k})$. As mentioned earlier, the closed range theorem implies that the range of π is closed. Since π also has dense range, π must be surjective and so the proof of the theorem is complete. ∎

5.4 Fundamental solutions for partial differential equations

As mentioned at the beginning of this chapter, one of the reasons for the introduction of distribution theory is that it provides a convenient language to discuss solutions to partial differential equations. We start by defining a fundamental solution for a constant coefficient partial differential operator

$$P(D) = \sum_{|\alpha| \le M} a_\alpha \frac{\partial^{|\alpha|}}{\partial x^\alpha} \qquad a_\alpha \in \mathbb{C}.$$

DEFINITION 1 *$T \in \mathcal{D}'(\mathbb{R}^N)$ is a fundamental solution for $P(D)$ if*

$$P(D)\{T\} = \delta_0.$$

The reason for the name "fundamental solution," is that a solution to the equation $P(D)\{u\} = \phi$ for $\phi \in \mathcal{D}(\mathbb{R}^N)$ can be found by convolution with T, as the following theorem shows.

THEOREM 1
Suppose $P(D)$ is a partial differential operator with constant coefficients. Suppose T is a fundamental solution for $P(D)$. If $\phi \in \mathcal{D}(\mathbb{R}^N)$ then $u = T * \phi$ is a solution to the differential equation $P(D)\{u\} = \phi$.

PROOF From Lemmas 1 and 2 in Section 5.2, we have

$$P(D)\{T * \phi\} = (P(D)\{T\}) * \phi = \delta_0 * \phi = \phi.$$

Therefore, $P(D)\{T * \phi\} = \phi$, as desired. ∎

We remark that if $P(D)$ has variable coefficients, then this theorem is not true. This is because the step $P(D)\{T * \phi\} = (P(D)\{T\}) * \phi$ is not valid; for in order to apply the derivatives to T, an integration by parts is required and expressions involving the derivatives of the coefficients of $P(D)$ will appear.

In Part IV, we shall need fundamental solutions for the Cauchy–Riemann operator on $\mathbb{C}$ and the Laplacian on $\mathbb{R}^N$.

THEOREM 2
The distribution $T(z) = 1/\pi z$ is a fundamental solution for the Cauchy–Riemann operator $\partial/\partial\bar{z} = 1/2(\partial/\partial x + i\partial/\partial y)$ on $\mathbb{C}$.

PROOF Note that $T(z) = 1/\pi z$ is a locally integrable function on $\mathbb{C}$ and so T defines an element of $\mathcal{D}'(\mathbb{C})$. To prove this theorem, we must show

$$\frac{\partial}{\partial \bar{z}} \left\{ \frac{1}{\pi z} \right\} = \delta_0$$

which is equivalent to

$$-\iint\limits_{z\in\mathbb{C}} \frac{\partial\phi(z)}{\partial\bar{z}}\frac{1}{\pi z}dxdy = \phi(0) \quad \text{for} \quad \phi\in\mathcal{D}(\mathbb{C}). \tag{1}$$

This is a generalized version of the Cauchy integral formula which is established by the following safety disc argument. For $0 < \epsilon < R < \infty$, let

$$\Delta_{R,\epsilon} = \{z\in\mathbb{C}; \epsilon < |z| < R\}.$$

Here, R is chosen so that supp $\phi \subset \{|z| < R\}$. Later we shall let $\epsilon \to 0$.

By Stokes' theorem, we have

$$\int\limits_{\partial\Delta_{\epsilon,R}} \frac{\phi(z)dz}{2\pi iz} = \iint\limits_{\Delta_{\epsilon,R}} d\left\{\frac{\phi(z)dz}{2\pi iz}\right\}.$$

The curve $\partial\Delta_{\epsilon,R}$ is oriented as the boundary of the open set $\Delta_{\epsilon,R}$, which means that $\{|z| = R\}$ is oriented in the counterclockwise direction and $\{|z| = \epsilon\}$ is oriented in the clockwise direction.

Since $d = \partial + \bar{\partial}$ and since $\phi = 0$ on $\{|z| = R\}$, we obtain

$$-\oint\limits_{\{|z|=\epsilon\}} \frac{\phi(z)dz}{2\pi iz} = \iint\limits_{\Delta_{\epsilon,R}} \frac{\partial\phi(z)}{\partial\bar{z}}\frac{d\bar{z}\wedge dz}{2\pi iz}.$$

Since $d\bar{z}\wedge dz = 2idx\wedge dy$, this becomes

$$-\oint\limits_{\{|z|=\epsilon\}} \frac{\phi(z)dz}{2\pi iz} = \iint\limits_{\Delta_{\epsilon,R}} \frac{\partial\phi(z)}{\partial\bar{z}}\frac{dx\wedge dy}{\pi z}.$$

As $\epsilon \to 0$, the right side converges to

$$\iint\limits_{\mathbb{C}} \frac{\partial\phi}{\partial\bar{z}}(z)\frac{dx\wedge dy}{\pi z}$$

because the integrand is locally integrable on $\mathbb{C}$. By parameterizing $\{|z| = \epsilon\}$ by $z = \epsilon e^{it}$ and using the continuity of ϕ at 0, the left side converges to $-\phi(0)$ as $\epsilon \to 0$. This proves (1) and completes the proof of the theorem. ∎

THEOREM 3

Let

$$T(x) = \begin{cases} \frac{1}{2\pi}\log|x| & \text{if } N = 2 \\ \frac{1}{(2-N)\omega_{N-1}}|x|^{2-N} & \text{if } N \geq 3 \end{cases}$$

where $\omega_{N-1} = (2\pi^{N/2}/\Gamma(N/2))$ *is the volume of the unit sphere in* $\mathbb{R}^N$. *Then* T *is a fundamental solution for the Laplacian* $\Delta = \sum_{j=1}^{N}(\partial^2/\partial x_j^2)$ *on* $\mathbb{R}^N$.

PROOF For $N = 2$, we have

$$\frac{\partial}{\partial z}\left\{\frac{1}{2\pi}\log|z|\right\} = \frac{1}{4\pi z}, \qquad z \in \mathbb{C}$$

in the sense of distribution theory. This follows from straightforward calculus if $z \neq 0$. Since $1/\pi z$ is locally integrable, the above equation holds as distributions in a neighborhood of the origin as well. Theorem 3 now follows from the above equation together with Theorem 2 and the fact that $\Delta = 4(\partial^2/\partial\bar{z}\partial z)$.

For $N \geq 3$, we use a safety disc argument similar to the one in the proof of Theorem 2. We must show

$$\int\limits_{x\in\mathbb{R}^N} T(x)\Delta\phi(x)dx = \phi(0) \quad \text{for} \quad \phi \in \mathcal{D}(\mathbb{R}^N). \tag{2}$$

For $0 < \epsilon < R < \infty$, define

$$\Delta_{\epsilon,R} = \{x \in \mathbb{R}^N; \epsilon < |x| < R\}$$

where R is chosen with supp $\phi \subset \{|x| < R\}$. This time we apply Green's formula (see Corollary 2 in Section 2.5) on $\Delta_{\epsilon,R}$ with $u = T$ and $v = \phi$. After noting that $\Delta u = 0$ on $\Delta_{\epsilon,R}$ (by explicit computation), we obtain

$$\int\limits_{\Delta_{\epsilon,R}} T(x)\Delta\phi(x)dx = \int\limits_{\partial\Delta_{\epsilon,R}} [T(x)(\mathbb{N}\phi)(x) - \phi(x)(\mathbb{N}T)(x)]d\sigma(x)$$

where $\mathbb{N}$ is the unit outward normal vector field to $\partial\Delta_{\epsilon,R}$. Now $\phi = 0$ on a neighborhood of $\{|x| = R\}$ by the choice of R. Therefore, the integral on the right is just an integral over $\{|x| = \epsilon\}$. The above equation reduces to

$$\int\limits_{\Delta_{\epsilon,R}} T(x)\Delta\phi(x)dx = \int\limits_{\{|x|=\epsilon\}} [T(x)(\mathbb{N}\phi)(x) - \phi(x)\mathbb{N}T(x)]d\sigma(x). \tag{3}$$

Since T is locally integrable on $\mathbb{R}^N$, the left side converges to $\int_{\mathbb{R}^N} T\Delta\phi dx$ as $\epsilon \to 0$. For the right side, note that $|T(x)| \leq C|x|^{2-N} = C\epsilon^{2-N}$ on $\{|x| = \epsilon\}$, where C is a uniform constant. Thus

$$\begin{aligned}\left|\int\limits_{\{|x|=\epsilon\}} T(x)\mathbb{N}\phi(x)d\sigma(x)\right| &\leq C\epsilon^{2-N}\cdot\epsilon^{N-1} \\ &= C\epsilon \to 0,\end{aligned} \tag{4}$$

as $\epsilon \to 0$. In polar coordinates, $\mathbb{N} = -(\partial/\partial r)$ on $\{|x| = \epsilon\}$. So

$$\mathbb{N}T(x) = -(2-N)^{-1}\omega_{N-1}^{-1}\frac{\partial}{\partial r}\{r^{2-N}\} = -\omega_{N-1}^{-1}r^{1-N}.$$

Therefore, we have

$$-\int\limits_{\{|x|=\epsilon\}} \phi(x)\mathbb{N}T(x)d\sigma(x) = \omega_{N-1}^{-1}\epsilon^{1-N}\int\limits_{\{|x|=\epsilon\}} \phi(x)d\sigma(x).$$

By the continuity of ϕ, the right side converges to $\phi(0)$ as $\epsilon \to 0$. This together with (3) and (4) show that (2) holds and the theorem follows. ∎

6

Currents

Roughly speaking, currents are to forms what distributions are to functions. Therefore, much of this chapter parallels the previous chapter on distributions. However, we leave to Part IV the discussion of fundamental solutions for various overdetermined systems of partial differential equations acting on differential forms.

6.1 Definitions

We consider two spaces of forms. For an open subset Ω of a smooth manifold, let

$\mathcal{E}^r(\Omega)$ be the space of smooth differential forms of degree r

$\mathcal{D}^r(\Omega)$ be the space of elements in $\mathcal{E}^r(\Omega)$ with compact support.

The space $\mathcal{E}^r(\Omega)$ is topologized as follows. In local coordinates, $\chi = (x_1, \ldots, x_N) : U \subset \Omega \to \mathbb{R}^N$, a sequence $f_n \in \mathcal{E}^r(\Omega)$ $n = 1, 2, \ldots$, can be written

$$f_n(x) = \sum_{|I|=r} f_n^I(x) dx^I$$

where each f_n^I is an element of $\mathcal{E}(U)$. We say that f_n converges to $f = \sum_{|I|=r} f^I dx^I$ in the coordinate patch U if the component functions f_n^I converge to f^I as $n \to \infty$ in the topology of $\mathcal{E}(U)$ (i.e., uniform convergence on compact sets of each derivative). The sequence $f_n \in \mathcal{E}^r(\Omega)$ is said to converge to the form f in $\mathcal{E}^r(\Omega)$ if f_n converges to f in each coordinate patch U of Ω. This definition of convergence is independent of the choice of coordinate charts.

The topology of $\mathcal{D}^r(\Omega)$ is defined analogously. Here, a sequence ϕ_n, $n = 1, 2, \ldots$ is said to converge to ϕ in $\mathcal{D}^r(\Omega)$ if there is a compact set $K \subset \Omega$ with supp $\phi_n \subset K$ for $n = 1, 2, \ldots$ and each derivative of the coefficient functions of ϕ_n converge to the corresponding derivative of the coefficient functions of ϕ.

We now define the space of currents as the dual of the space of forms.

DEFINITION 1 *For an open subset Ω of a smooth manifold, the dual space of $\mathcal{D}^r(\Omega)$ is denoted by $\{\mathcal{D}^r(\Omega)\}'$ and it is called the space of currents of dimension r. Likewise, the dual of the space $\mathcal{E}^r(\Omega)$ is denoted by $\{\mathcal{E}^r(\Omega)\}'$.*

The dual space of $\mathcal{D}^r(\Omega)$ is the space of all continuous complex linear functionals defined on $\mathcal{D}^r(\Omega)$. Since $\mathcal{D}^r(\Omega)$ is a subspace of $\mathcal{E}^r(\Omega)$ and since the inclusion map is continuous, clearly $\{\mathcal{E}^r(\Omega)\}'$ is a subspace of $\{\mathcal{D}^r(\Omega)\}'$.

The pairing between elements of $\{\mathcal{D}^r(\Omega)\}'$ and elements of $\mathcal{D}^r(\Omega)$ is denoted by $\langle\ ,\ \rangle_\Omega$ (as opposed to $(\ ,\ )_\Omega$ for distributions). Occasionally, we shall emphasize the variable in Ω by writing

$$\langle T(x), f(x)\rangle_{x\in\Omega} = \langle T, f\rangle_\Omega \quad \text{for} \quad T \in \{\mathcal{D}^r(\Omega)\}', f \in \mathcal{D}^r(\Omega).$$

The following examples parallel the examples of distributions given in the previous chapter.

Example 1

Let Ω be an open subset of $\mathbb{R}^N$ and let $T = T_J dx^J$ be a form of degree $N - r$ (i.e., $|J| = N - r$). Suppose T_J: $\Omega \to \mathbb{C}$ is locally integrable. T is viewed as a current in $\{\mathcal{D}^r(\Omega)\}'$ by defining

$$\langle T, f\rangle_\Omega = \int_{x\in\Omega} T(x) \wedge f(x) \quad \text{for} \quad f \in \mathcal{D}^r(\Omega).$$

$T \wedge f$ is an N-form on $\Omega \subset \mathbb{R}^N$ with integrable coefficients and so the right side is well defined.

If $f = f_I dx^I, |I| = r$, for some function $f_I \in \mathcal{D}(\Omega)$, then

$$\langle T, f\rangle_\Omega = \begin{cases} 0 & \text{if } dx^J \wedge dx^I = 0 \\ (-1)^{\epsilon_{IJ}} \int\limits_{x\in\Omega} T_J(x) f_I(x) dx & \text{if } dx^J \wedge dx^I \neq 0. \end{cases}$$

where $(-1)^{\epsilon_{IJ}}$ is defined by

$$dx^J \wedge dx^I = (-1)^{\epsilon_{IJ}} dx_1 \wedge \ldots \wedge dx_N.$$

Note that $\int_\Omega T_J f_I dx$ is the same as the pairing between the distribution $T_J \in \mathcal{D}'(\Omega)$ with the function $f_I \in \mathcal{D}(\Omega)$. In this way, currents are closely related to distributions. This relationship will be made precise in Lemma 1. $\square$

Example 2

Consider the point $p \in \mathbb{R}^N$. We view $[p]$ as a current in $\{\mathcal{E}^0(\mathbb{R}^N)\}'$ by defining

$$\langle [p], f\rangle_{\mathbb{R}^N} = f(p) \quad \text{for} \quad f \in \mathcal{E}^0(\mathbb{R}^N) = \mathcal{E}(\mathbb{R}^N).$$

This example is analogous to the delta function at p given in the last chapter. $\square$

Example 3

Suppose M is an oriented submanifold of $\mathbb{R}^N$ of dimension r. We view $[M]$ as a current in $\{\mathcal{D}^r(M)\}'$ by defining

$$\langle [M], f\rangle_{\mathbb{R}^N} = \int_M f \quad \text{for} \quad f \in \mathcal{D}^r(\mathbb{R}^N).$$

If M is compact, then we can allow f to be in $\mathcal{E}^r(\mathbb{R}^N)$ and so in this case, $[M]$ is an element of $\{\mathcal{E}^r(\mathbb{R}^N)\}'$. Note that Example 2 is a special case of Example 3. $\square$

The last two examples illustrate the reason the space $\{\mathcal{D}^r(\Omega)\}'$ is called the space of currents of dimension r. A basic example of such a current is given by integration over an r-dimensional submanifold.

Since these three classes of examples parallel the three classes of examples of distributions from the previous chapter, there should be a connection between distributions and currents.

DEFINITION 2 *Let Ω be an open subset of an oriented manifold of dimension N and let $0 \le q \le N$. We let $\mathcal{D}'^q(\Omega)$ be the space of q-forms whose coefficients are $\mathcal{D}'(\Omega)$-distributions. This space will be called the space of currents of degree q. Likewise, we let $\mathcal{E}'^q(\Omega)$ be the space of q-forms whose coefficients are $\mathcal{E}'(\Omega)$-distributions.*

LEMMA 1

Suppose Ω is an open subset of an oriented N-dimensional manifold and let $0 \le r \le N$. Then $\{\mathcal{D}^r(\Omega)\}'$ is isomorphic to $\mathcal{D}'^{N-r}(\Omega)$.

PROOF We concentrate first on the case where Ω is an open subset of $\mathbb{R}^N$. The idea of the proof comes from examining Example 1 above. For $T = T_J dx^J \in \mathcal{D}'^{N-r}(\Omega)$ and $f = f_I dx^I \in \mathcal{D}^r(\Omega)$, we define T as a current in $\{\mathcal{D}^r(\Omega)\}'$ by setting

$$\langle T, f\rangle_\Omega = \begin{cases} 0 & \text{if } dx^I \wedge dx^J = 0 \\ (-1)^{\epsilon_{IJ}}(T_J, f_I)_\Omega & \text{if } dx^I \wedge dx^J \neq 0 \end{cases}$$

where as in Example 1, $(-1)^{\epsilon_{IJ}}$ is defined by $dx^I \wedge dx^J = (-1)^{\epsilon_{IJ}} dx$. Since T_J is a $\mathcal{D}'(\Omega)$-distribution and f_I is an element of $\mathcal{D}(\Omega)$, clearly the right side is well defined. Therefore, any current of degree $N - r$ can be considered as an element of $\{\mathcal{D}^r(\Omega)\}'$.

Conversely, suppose T is an element of $\{\mathcal{D}^r(\Omega)\}'$. For any increasing multiindex J of length $N-r$, define the distribution $T_J \in \mathcal{D}'(\Omega)$ by

$$(T_J, \phi)_\Omega = \langle T, \phi dx^{J'} \rangle_\Omega \quad \text{for} \quad \phi \in \mathcal{D}(\Omega)$$

where J' is the increasing multiindex of length r which is comprised of the indices in $\{1, \ldots, N\}$ that do not belong to J. It is then an easy exercise to show

$$T = \sum_{|J|=N-r} (-1)^{\epsilon_{IJ}} T_J dx^J. \qquad \blacksquare$$

The proof of Lemma 1 for oriented manifolds is the same. In this case, the form dx is replaced by the volume form $d\sigma$ (constructed as in Section 5.1 with some suitably chosen metric replacing the Euclidean metric on $\mathbb{R}^N$).

Let us describe the above three examples from the degree point of view. The first example is already presented as a form with distribution coefficients. For Example 2, we have $[p] = \delta_p dx \in \mathcal{E}'^N(\mathbb{R}^N)$. For Example 3, we specialize to the case where M is the smooth boundary of an open set $\Omega \subset \mathbb{R}^N$ with the usual boundary orientation. Suppose $\Omega = \{x \in \mathbb{R}^N; \rho(x) < 0\}$ where ρ: $\mathbb{R}^N \to \mathbb{R}$ is smooth. By rescaling, we assume that $|\nabla \rho| = 1$ on M. Let μ_M be Hausdorff $(N-1)$-dimensional measure on M. Then we claim

$$[M] = \mu_M d\rho$$

which exhibits M as a degree 1 current. To see this, we first need a working formula for μ_M. If $\phi \in \mathcal{D}(\mathbb{R}^N)$, then from Section 5.1,

$$(\mu_M, \phi)_{\mathbb{R}^N} = \int_M \phi d\sigma$$

where $d\sigma$ is the volume form on M. Since $|\nabla \rho| = 1$ on M

$$d\sigma = \nabla \rho \lrcorner dx.$$

Let $g = \phi dx_1 \wedge \ldots \wedge \widehat{dx_j} \wedge \ldots \wedge dx_N$ with $\phi \in \mathcal{D}(\mathbb{R}^N)$. From the definitions, we have

$$\begin{aligned} \langle \mu_M d\rho, g \rangle_{\mathbb{R}^N} &= (-1)^{j-1} \left(\mu_M, \phi \frac{\partial \rho}{\partial x_j} \right)_{\mathbb{R}^N} \\ &= (-1)^{j-1} \int_M \phi \frac{\partial \rho}{\partial x_j} d\sigma. \end{aligned} \tag{1}$$

On the other hand

$$\begin{aligned} g &= (\nabla \rho \lrcorner d\rho) g \qquad (\text{since } |\nabla \rho| = 1) \\ &= \nabla \rho \lrcorner (d\rho \wedge g) + d\rho \wedge (\nabla \rho \lrcorner g) \end{aligned}$$

where the last equality follows from the product rule for $\lrcorner$ (Lemma 1 in Section 1.5). Since $j^*d\rho = 0$ on M where $j\colon M \to \mathbb{R}^N$ is inclusion, we have

$$\begin{aligned} \langle [M], g\rangle_{\mathbb{R}^N} &= \int\limits_M g \\ &= \int\limits_M \nabla\rho\lrcorner(d\rho \wedge g). \end{aligned}$$

Inserting $g = \phi dx_1 \wedge \ldots \wedge \widehat{dx_j} \wedge \ldots \wedge dx_N$ and using $d\sigma = \nabla\rho\lrcorner dx$, we obtain

$$\langle [M], g\rangle_{\mathbb{R}^N} = (-1)^{j-1} \int\limits_M \phi \frac{\partial \rho}{\partial x_j} d\sigma,$$

which by (1) above proves our claim that $[M] = \mu_M d\rho$.

In a similar manner, if $M = \{x \in \mathbb{R}^N; \rho_1(x) = \cdots = \rho_d(x) = 0\}$, then the current [M] is given by

$$\mu_M \alpha d\rho_1 \wedge \ldots \wedge d\rho_d$$

where $\alpha = |d\rho_1 \wedge \ldots \wedge d\rho_d|^{-1}$ or $-|d\rho_1 \wedge \ldots \wedge d\rho_d|^{-1}$ depending on the orientation given to M.

As a further example, let $\Delta = \{(x,x) \in \mathbb{R}^N \times \mathbb{R}^N; x \in \mathbb{R}^N\}$. ($\Delta$ is called the *diagonal* of $\mathbb{R}^N \times \mathbb{R}^N$.) The current $[\Delta]$ (integration over Δ) has dimension N and therefore degree N in $\mathbb{R}^N \times \mathbb{R}^N$. From the degree point of view, $[\Delta] = \delta_0(x-y)d(x_1 - y_1) \wedge \ldots \wedge d(x_N - y_N)$ where $\delta_0(x-y)$ is the distribution on $\mathbb{R}^N \times \mathbb{R}^N$ defined by

$$(\delta_0(x-y), \phi(x,y))_{(x,y)\in\mathbb{R}^N\times\mathbb{R}^N} = \int\limits_{x\in\mathbb{R}^N} \phi(x,x)dx.$$

We leave the verification of this to the reader.

The definition of the weak topology for $\{\mathcal{D}^r(\Omega)\}'$ is analogous to the definition of the weak topology for $\mathcal{D}'(\Omega)$. A sequence T_n, $n = 1, 2, \ldots$ converges to T in $\{\mathcal{D}^r(\Omega)\}'$ if $\langle T_n, g\rangle_\Omega \to \langle T, g\rangle_\Omega$ for each form $g \in \mathcal{D}^r(\Omega)$. This is equivalent to the following characterization of convergence using the degree point of view: a sequence of currents

$$T_n = T_n^I dx^I, \quad |I| = N - r, \quad T_n^I \in \mathcal{D}'(\Omega) \quad \text{for} \quad n = 1, 2, \ldots$$

converges to $T^I dx^I$ if and only if $T_n^I \to T^I$ in $\mathcal{D}'(\Omega)$. Since $\mathcal{D}(\Omega)$ is a dense subset of $\mathcal{D}'(\Omega)$, $\mathcal{D}^{N-r}(\Omega)$ is a dense subset of $\mathcal{D}'^{N-r}(\Omega)$.

We should say a few words about currents on a complex manifold. For an open subset Ω of a complex n-dimensional manifold, recall that $\mathcal{D}^{p,q}(\Omega)$ is the space of smooth compactly supported forms of bidegree (p,q) $(0 \le p, q \le n)$. Its dual space $\{D^{p,q}(\Omega)\}'$ is the space of currents of *bidimension* (p,q). By definition, the space of currents of *bidegree* (p,q) (denoted $\mathcal{D}'^{p,q}(\Omega)$) is the space of (p, q)-forms with coefficients belonging to $\mathcal{D}'(\Omega)$. The analogue of Lemma 1 for this case is that $\{\mathcal{D}^{p,q}(\Omega)\}'$ is isomorphic to $\mathcal{D}'^{n-p,n-q}(\Omega)$. All of these statements also hold with $\mathcal{D}$ replaced by $\mathcal{E}$.

Recall there is a splitting of an r-form on a complex manifold into its various bidegrees. We have

$$\mathcal{D}^r(\Omega) = \mathcal{D}^{r,0}(\Omega) \oplus \cdots \oplus \mathcal{D}^{0,r}(\Omega).$$

The same applies to currents, i.e.,

$$\mathcal{D}'^r(\Omega) = \mathcal{D}'^{r,0}(\Omega) \oplus \cdots \oplus \mathcal{D}'^{0,r}(\Omega).$$

Let $\pi^{p,q}$: $\mathcal{D}'^r(\Omega) \to \mathcal{D}'^{p,q}(\Omega)$ with $p + q = r$ be the natural projection. For a current T, we often write $T^{p,q}$ for $\pi^{p,q}\{T\}$. As an example, let $\Omega = \{z \in \mathbb{C}^n; \rho(z) < 0\}$ where ρ: $\mathbb{C}^n \to \mathbb{R}$ is smooth with $|\nabla\rho| = 1$ on $\partial\Omega$. We have shown that $[\partial\Omega] = \mu d\rho$ where μ is Hausdorff $(2n - 1)$-dimensional measure. So in view of the above discussion

$$[M]^{1,0} = \mu\partial\rho$$

and

$$[M]^{0,1} = \mu\bar{\partial}\rho.$$

6.2 Operations with currents

Analogous to distribution theory, the definitions of operations with currents are motivated by considering the case when the current is a smooth form.

Wedge product of a current with a smooth form

If T is a smooth form of degree q on Ω and f is a smooth form of degree r on Ω, then $T \wedge f$ is a smooth form of degree $q + r$. As a current on Ω

$$\begin{aligned}\langle T \wedge f, \phi\rangle_\Omega &= \int_\Omega (T \wedge f) \wedge \phi \quad \text{for} \quad \phi \in \mathcal{D}^{N-q-r}(\Omega) \\ &= \langle T, f \wedge \phi\rangle_\Omega.\end{aligned}$$

We *define* $T \wedge f$ for a current $T \in \mathcal{D}'^q(\Omega)$ and a smooth form $f \in \mathcal{E}^r(\Omega)$ by the formula

$$\langle T \wedge f, \phi\rangle_\Omega = \langle T, f \wedge \phi\rangle_\Omega \quad \text{for} \quad \phi \in \mathcal{D}^{N-q-r}(\Omega).$$

The result, $T \wedge f$, is a current in $\mathcal{D}'^{q+r}(\Omega)$.

Exterior derivative

Suppose T is a smooth form of degree $N-r$ and let $\phi \in \mathcal{D}^{r-1}(\Omega)$. Then dT is an element of $\mathcal{E}^{N-r+1}(\Omega)$ and

$$\langle dT, \phi \rangle_\Omega = \int_\Omega dT \wedge \phi.$$

By the product rule for the exterior derivative (Lemma 1 in Section 1.4), we have

$$\langle dT, \phi \rangle_\Omega = \int_\Omega d(T \wedge \phi) + (-1)^{N-r+1} \int_\Omega T \wedge d\phi.$$

Since $T \wedge \phi$ has compact support in Ω, the first integral on the right vanishes by Stokes' theorem. So

$$\begin{aligned} \langle dT, \phi \rangle_\Omega &= (-1)^{N-r+1} \int_\Omega T \wedge d\phi \\ &= (-1)^{N-r+1} \langle T, d\phi \rangle_\Omega. \end{aligned}$$

Therefore, we define dT for $T \in \mathcal{D}'^{N-r}(\Omega)$ by the formula

$$\langle dT, \phi \rangle_\Omega = (-1)^{N-r+1} \langle T, d\phi \rangle_\Omega \quad \text{for} \quad \phi \in \mathcal{D}^{r-1}(\Omega).$$

Note that the exterior derivative raises the degree of a current by one and therefore lowers the dimension by one.

Suppose Ω is an open subset of $\mathbb{R}^N$. If $T = T_I dx^I$, $|I| = N - r$, and $T_I \in \mathcal{D}'(\Omega)$ then an equivalent expression for dT is given by

$$dT = \sum_{j=1}^{N} \frac{\partial T_I}{\partial x_j} dx_j \wedge dx^I$$

where $\partial T_I / \partial x_j$ is the derivative of T_I in the sense of distributions. This formula generalizes the usual exterior derivative formula for smooth forms.

As with distributions, if $T_n \to T$ in $\{\mathcal{D}^r(\Omega)\}'$ then the sequence dT_n converges to dT in $\{\mathcal{D}^{r-1}(\Omega)\}'$.

As stated in Section 6.1, an important class of currents is the class of submanifolds. So it is natural to compute the exterior derivative of such a current.

THEOREM 1
Suppose M is an oriented r-dimensional submanifold with boundary contained in a smooth N-dimensional manifold X. Then $d[M] = (-1)^{N-r+1}[\partial M]$.

PROOF The proof will follow from Stokes' theorem. Suppose ϕ is an element of $\mathcal{D}^{r-1}(X)$. Then

$$\begin{aligned}\langle d[M], \phi\rangle_X &= (-1)^{N-r+1}\langle [M], d\phi\rangle_X \\ &= (-1)^{N-r+1}\int_M d\phi \\ &= (-1)^{N-r+1}\int_{\partial M}\phi \qquad \text{(by Stokes' theorem)} \\ &= (-1)^{N-r+1}\langle [\partial M], \phi\rangle_X,\end{aligned}$$

as desired. ∎

For an open subset Ω of a complex manifold, recall that $d = \partial + \bar{\partial}$ where $\partial: \mathcal{E}^{p,q}(\Omega) \to \mathcal{E}^{p+1,q}(\Omega)$ and $\bar{\partial}: \mathcal{E}^{p,q}(\Omega) \to \mathcal{E}^{p,q+1}(\Omega)$ are defined by $\partial = \pi^{p+1,q} \circ d$ and $\bar{\partial} = \pi^{p,q+1} \circ d$. These same formulas extend the definition of ∂ and $\bar{\partial}$ to currents. From the definition of the exterior derivative of a current, if $T \in \mathcal{D}'^{p,q}(\Omega)$ then

$$\begin{aligned}\langle \bar{\partial} T, \phi\rangle_\Omega &= (-1)^{p+q+1}\langle T, \bar{\partial}\phi\rangle \qquad \text{for} \quad \phi \in \mathcal{D}^{n-p,n-q-1}(\Omega) \\ \langle \partial T, \phi\rangle_\Omega &= (-1)^{p+q+1}\langle T, \partial\phi\rangle_\Omega \quad \text{for} \quad \phi \in \mathcal{D}^{n-p-1,n-q}(\Omega).\end{aligned}$$

As an example, suppose Ω is an open set given by $\Omega = \{z \in \mathbb{C}^n; \rho(z) < 0\}$ where $\rho: \mathbb{C}^n \to \mathbb{R}$ is smooth with $|\nabla\rho| = 1$ on $\partial\Omega$. The current $[\Omega]$ has degree 0 and dimension $2n$. From Theorem 1

$$d[\Omega] = -[\partial\Omega].$$

Thus

$$\partial[\Omega] = -[\partial\Omega]^{1,0} \quad \text{and} \quad \bar{\partial}[\Omega] = -[\partial\Omega]^{0,1}.$$

Note that

$$\begin{aligned}\langle [\partial\Omega]^{1,0}, \phi\rangle_{\mathbb{C}^n} &= \langle [\partial\Omega], \phi^{n-1,n}\rangle_{\mathbb{C}^n} \quad \text{for} \quad \phi \in \mathcal{D}^{2n-1}(\mathbb{C}^n) \\ &= \int_{\partial\Omega}\phi^{n-1,n}\end{aligned}$$

and

$$\langle [\partial\Omega]^{0,1}, \phi\rangle_{\mathbb{C}^n} = \langle [\partial\Omega], \phi^{n,n-1}\rangle_{\mathbb{C}^n} = \int_{\partial\Omega}\phi^{n,n-1}.$$

In words, the pairing between $[\partial\Omega]^{0,1}$ and a form ϕ of degree $2n-1$ is the same as the usual integral over $\partial\Omega$ of the piece of ϕ of bidegree $n, n-1$.

The push forward of a current under a smooth map

Suppose Ω and Ω' are open subsets of smooth manifolds, and let $F\colon \Omega \to \Omega'$ be a smooth map. The pull back operator $F^*\colon \mathcal{E}^r(\Omega') \to \mathcal{E}^r(\Omega)$ can be dualized to obtain a map defined on currents.

DEFINITION 1 *Let $F\colon \Omega \to \Omega'$ be a smooth map. For $T \in \{\mathcal{E}^r(\Omega)\}'$ the push forward of T via F, denoted F_*T, is the current in $\{\mathcal{E}^r(\Omega')\}'$ defined by*

$$\langle F_*T, \phi\rangle_{\Omega'} = \langle T, F^*\phi\rangle_{\Omega}, \text{ for } \phi \in \mathcal{E}^r(\Omega').$$

Since $F^*\phi$ is an element of $\mathcal{E}^r(\Omega)$, the right side of this definition is well defined. Moreover if $\phi_n \to \phi$ in $\mathcal{E}^r(\Omega')$ then $F^*\phi_n \to F^*\phi$ in $\mathcal{E}^r(\Omega)$ and so F_*T is a continuous linear functional on $\mathcal{E}^r(\Omega')$. Hence, F_*T is a well defined $\{\mathcal{E}^r(\Omega)\}'$-current.

The push forward preserves dimension (but not degree) because F^* preserves degree. Note that $F^*\phi$ does not necessarily have compact support even when ϕ has compact support (unless F is proper). For this reason, the push forward of an element in $\{\mathcal{D}^r(\Omega)\}'$ is not well defined (unless F is proper).

For smooth maps F and G we have $(F \circ G)^* = G^* \circ F^*$ on forms. Therefore, we have $(F \circ G)_* = F_* \circ G_*$.

We now give some examples. Suppose M and M' are smooth, oriented submanifolds of $\mathbb{R}^N$ and $\mathbb{R}^{N'}$, respectively. Let $F\colon \mathbb{R}^N \to \mathbb{R}^{N'}$ be an orientation-preserving bijective smooth map with $M' = F\{M\}$. Then $F_*[M] = [M']$ because

$$\begin{aligned}
\langle F_*[M], \phi\rangle_{\mathbb{R}^{N'}} &= \langle [M], F^*\phi\rangle_{\mathbb{R}^N} \quad \text{for} \quad \phi \in \mathcal{D}^*(\mathbb{R}^{N'}) \\
&= \int_M F^*\phi \\
&= \int_{M'} \phi \qquad \text{(since } F \text{ is orientation preserving)} \\
&= \langle [M'], \phi\rangle_{\mathbb{R}^{N'}}.
\end{aligned}$$

As another example, suppose $\pi\colon \mathbb{R}^N \times \mathbb{R}^k \to \mathbb{R}^k$ is the projection $\pi(x, y) = y$ for $x \in \mathbb{R}^N, y \in \mathbb{R}^k$. Let us compute π_* on the subspace $\mathcal{D}^{N+k-r}(\mathbb{R}^N \times \mathbb{R}^k) \subset$

$\{\mathcal{E}^r(\mathbb{R}^N \times \mathbb{R}^k)\}'$. For $T \in \mathcal{D}^{N+k-r}(\mathbb{R}^N \times \mathbb{R}^k)$ and $\phi \in \mathcal{D}^r(\mathbb{R}^k)$

$$\begin{aligned}\langle \pi_* T, \phi\rangle_{\mathbb{R}^k} &= \langle T, \pi^*\phi\rangle_{\mathbb{R}^N\times\mathbb{R}^k} \\ &= \int\limits_{\mathbb{R}^N\times\mathbb{R}^k} T \wedge \pi^*\phi \\ &= \int\limits_{y\in\mathbb{R}^k}\left(\int\limits_{x\in\mathbb{R}^N} T(x,y)\right)\wedge\phi(y).\end{aligned}$$

Now the inner x-integral is zero unless all the dx's are present. So we write

$$T(x,y) = \sum_{r=0}^{N}\sum_{|I|=r} dx^I \wedge T_I(x,y)$$

where each T_I is a form in the y-variables with coefficients that depend on x and y. With this notation, we have

$$\int\limits_{x\in\mathbb{R}^N} T(x,y) = \int\limits_{x\in\mathbb{R}^N} dx_1\wedge\ldots\wedge dx_N \wedge T_{1\ldots N}(x,y).$$

All the x's and dx's are integrated leaving a differential form in y. From the above, we have

$$\langle \pi_* T, \phi\rangle_{\mathbb{R}^k} = \left\langle \int\limits_{x\in\mathbb{R}^N} T(x,y), \phi(y)\right\rangle_{y\in\mathbb{R}^k}$$

and so

$$(\pi_* T)(y) = \int\limits_{x\in\mathbb{R}^N} T(x,y) \quad \text{for} \quad y \in \mathbb{R}^k.$$

In words, the push forward under π is the same as the operation of integrating out the fiber of π.

The above idea is illustrated in the next example, which we state as a lemma (to be used in Part IV).

LEMMA 1
Define τ: $\mathbb{R}^N \times \mathbb{R}^N \to \mathbb{R}^N$ by $\tau(y,x) = x - y$. If $T \in \mathcal{D}^{2N-r}(\mathbb{R}^N \times \mathbb{R}^N) \subset \{\mathcal{E}^r(\mathbb{R}^N \times \mathbb{R}^N)\}'$, then

$$(\tau_* T)(x) = \int\limits_{y\in\mathbb{R}^N} (s^*T)(y,x)$$

where s: $\mathbb{R}^N \times \mathbb{R}^N \to \mathbb{R}^N \times \mathbb{R}^N$ is defined by $s(y,x) = (y, x+y)$.

Sometimes, we shall use the notation $T(y, x+y)$ for $(s^*T)(y,x)$. The form $T(y, x+y)$ is the differential form obtained by replacing x by $x+y$ in each coefficient of T and by replacing each dx_j by $dx_j + dy_j$. On the right side of the equation in the lemma, the only nontrivial contribution in the integral comes from the terms with all the dy's. The y's and dy's are then integrated leaving a differential form in x.

PROOF Suppose ϕ is an element of $\mathcal{D}^r(\mathbb{R}^N)$. Then

$$\begin{aligned}\langle \tau_* T, \phi\rangle_{\mathbb{R}^N} &= \langle T, \tau^*\phi\rangle_{\mathbb{R}^N\times\mathbb{R}^N}\\ &= \int\limits_{\mathbb{R}^N\times\mathbb{R}^N} T\wedge \tau^*\phi.\end{aligned}$$

Now, $\mathrm{Det}(Ds) = 1$ and so s is an orientation-preserving linear isomorphism. By the change of variables formula for integration

$$\begin{aligned}\langle \tau_* T, \phi\rangle_{\mathbb{R}^N} &= \int\limits_{\mathbb{R}^N\times\mathbb{R}^N} (s^*T)\wedge(s^*\tau^*\phi)\\ &= \int\limits_{x\in\mathbb{R}^N}\int\limits_{y\in\mathbb{R}^N} s^*T(y,x)\wedge\phi(x)\end{aligned}$$

where the last equation follows from $s^*\circ\tau^* = (\tau\circ s)^*$ and because $(\tau\circ s)(y,x) = x$. Therefore

$$\langle \tau_* T, \phi\rangle_{\mathbb{R}^N} = \left\langle \int\limits_{y\in\mathbb{R}^n} (s^*T)(y,x), \phi(x)\right\rangle_{x\in\mathbb{R}^N}$$

and the proof of the lemma is complete. ∎

For each $x\in\mathbb{R}^N$, note that $\{s(y,x);\ y\in\mathbb{R}^N\} = \tau^{-1}(x)$. So again, we see that the operation of push forward is the same as the operation of integrating out the fiber. This idea is illustrated in a more general context in the proof of Lemma 5, below.

As the next lemma shows, the push forward operation via an orientation-preserving diffeomorphism is the same as the operation of pull back via its inverse.

LEMMA 2

Suppose Ω and Ω' are open subsets of oriented N-dimensional manifolds. Let $F\colon \Omega\to\Omega'$ be an orientation-preserving diffeomorphism. For $T\in\mathcal{D}^{N-r}(\Omega)\subset\{\mathcal{E}^r(\Omega)\}'$

$$F_*T = (F^{-1})^*T.$$

From the proof given below, it will be clear that if F is orientation reversing, then $F_*T = -(F^{-1})^*T$.

PROOF Let ϕ be an element of $\mathcal{D}^r(\Omega')$. Then

$$\begin{aligned}\langle F_*T, \phi\rangle_{\Omega'} &= \langle T, F^*\phi\rangle_\Omega \\ &= \int_\Omega T \wedge F^*\phi.\end{aligned}$$

Since F^{-1}: $\Omega' \to \Omega$ is an orientation-preserving diffeomorphism, the change of variables formula for integrals yields

$$\begin{aligned}\langle F_*T, \phi\rangle_{\Omega'} &= \int_{\Omega'} (F^{-1^*}T) \wedge \phi \\ &= \langle F^{-1^*}T, \phi\rangle_{\Omega'}\end{aligned}$$

and the proof of the lemma is complete. ∎

Since the pull back operator on differential forms commutes with the exterior derivative, the push forward operator commutes with the exterior derivative up to a sign factor.

LEMMA 3
Suppose Ω and Ω' are open subsets of oriented manifolds X and Y with real dimensions N and N', respectively. Let F: $\Omega \to \Omega'$ be a smooth map. Then

$$F_* \circ d_X = (-1)^{N+N'} d_Y \circ F_*$$

as operators from $\{\mathcal{E}^r(\Omega)\}'$ to $\{\mathcal{E}^{r-1}(\Omega')\}'$.

PROOF Suppose that T is an element of $\{\mathcal{E}^r(\Omega)\}' \simeq \mathcal{E}'^{N-r}(\Omega)$, and let $\phi \in \mathcal{E}^{r-1}(\Omega')$. Then

$$\begin{aligned}\langle F_*d_XT, \phi\rangle_{\Omega'} &= \langle d_XT, F^*\phi\rangle_\Omega \\ &= (-1)^{N-r+1}\langle T, d_X(F^*\phi)\rangle_\Omega \\ &= (-1)^{N-r+1}\langle T, F^*(d_Y\phi)\rangle_\Omega \\ &= (-1)^{N-r+1}\langle F_*T, d_Y\phi\rangle_{\Omega'} \\ &= (-1)^{N+N'}\langle d_YF_*T, \phi\rangle_{\Omega'}.\end{aligned}$$

To see that the minus sign is correct in the last equality, recall that F_* preserves dimension and so F_*T has dimension r, or equivalently, degree $N' - r$ as a current on the N'-dimensional set Ω'. Thus $\langle d_YF_*T, \phi\rangle_{\Omega'} = (-1)^{N'-r+1}$ $\langle F_*T, d_Y\phi\rangle_{\Omega'}$ from the definition of the exterior derivative of a current. This completes the proof of the lemma. ∎

Now suppose Ω and Ω' are open subsets of complex manifolds and let F: $\Omega \to \Omega'$ be a holomorphic map. From Lemma 3 in Section 3.3, F^* preserves bidegree. The dual of this statement is that F_* preserves the bidimension of currents. This together with Lemma 3 imply that F_* commutes with the Cauchy–Riemann operator.

LEMMA 4
Suppose Ω and Ω' are open subsets of complex manifolds of complex dimensions n and n' respectively. Let F: $\Omega \to \Omega'$ be a holomorphic map. Then

$$F_* \circ \bar{\partial} = \bar{\partial} \circ F_*$$

as operators from $\{\mathcal{E}^{p,q}(\Omega)\}'$ to $\{\mathcal{E}^{p,q-1}(\Omega')\}'$.

Since the real dimension of a complex manifold is even, there is no sign factor in this lemma.

The pull back of a current via a smooth map

Defining the pull back of a current is perhaps a little nonstandard. However, since the space of differential forms make up a large subclass of currents and since the pull back of a differential form is well defined, we find it convenient to extend (where possible) the pull back operator to currents. Simply put, the pull back operator is defined to be the dual of the push forward operator. To make this precise, we must show that the push forward operator sends smooth forms to smooth forms in a continuous manner. This is clear for the maps π and τ as shown earlier. We shall now prove that this holds more generally.

LEMMA 5
Suppose Ω and Ω' are open subsets of oriented manifolds of dimensions N and N', respectively with $N \geq N'$. Let F: $\Omega \to \Omega'$ be a smooth surjective map such that $F_*(p)$: $T_p(\Omega) \to T_{F(p)}(\Omega')$ has rank N' (i.e., maximal rank) at each point $p \in \Omega$. Then F_* is a continuous map from $\mathcal{D}^r(\Omega)$ to $\mathcal{D}^{r+N'-N}(\Omega')$.

PROOF By a partition of unity argument and by the use of local coordinates, we may assume that Ω and Ω' are open subsets of $\mathbb{R}^N$ and $\mathbb{R}^{N'}$, respectively. Let p_0 be an arbitrary point in Ω. Since $DF(p_0)$ has maximal rank $= N'$, we may arrange coordinates (x, y) for $\mathbb{R}^N$ so that $x \in \mathbb{R}^{N-N'}$ and $y \in \mathbb{R}^{N'}$ and so that $(D_yF)(p_0)$ is a nonsingular $N' \times N'$ matrix. Let $p_0 = (x_0, y_0)$. From the inverse function theorem, there is a neighborhood of $(x_0, F(p_0))$ in $\mathbb{R}^{N-N'} \times \mathbb{R}^{N'}$ of the form $U' \times V'$ with $V' \subset \Omega'$ and a diffeomorphism G: $U' \times V' \to G\{U' \times V'\} \subset \Omega$ such that $F(G(x, y)) = y$ for $y \in V'$. That is, $F \circ G$: $U' \times V' \to V'$ is the projection π: $U' \times V' \to V', \pi(x, y) = y$.

Let ϕ be an element of $\mathcal{D}^r(G\{U' \times V'\})$. Then

$$F_*\phi = (F \circ G)_* \circ (G_*^{-1}\phi).$$

Since $G_*^{-1}\phi = G^*\phi$ or $-G^*\phi$ (by Lemma 2), $G_*^{-1}\phi$ is a smooth form with compact support in $U' \times V'$. Since G is a diffeomorphism, the map $\phi \longmapsto G_*^{-1}\phi$ is a continuous linear map from $\mathcal{D}^r(G\{U' \times V'\})$ to $\mathcal{D}^r(U' \times V')$. In addition, since $F \circ G = \pi$, the computation of π_* preceding Lemma 1 yields

$$(F_*\phi)(y) = (\pi_* G_*^{-1}\phi)(y) = \int\limits_{x \in \mathbb{R}^{N-N'}} (G_*^{-1}\phi)(x, y) \tag{1}$$

where the integral on the right involves the variable x and all the dx's, leaving a differential form in y. From this expression, it follows that $F_*\phi$ is a smooth form with compact support in $V' \subset \Omega'$ and that the map $\phi \longmapsto F_*\phi$ is a continuous linear map from $\mathcal{D}^r(G\{U \times V'\})$ to $\mathcal{D}^{N'-N+r}(V')$. The dimension of $F_*\phi$ (as a current) is $N - r$ since F_* preserves dimension. Therefore, the degree of $F_*\phi$ in $\Omega' \subset \mathbb{R}^{N'}$ is $N' - N + r$.

The general case for $\phi \in \mathcal{D}^r(\Omega)$ now follows by a partition of unity argument subordinate to an open cover of Ω by open sets of the form $G\{U' \times V'\}$ as above. The proof of the lemma is complete. ∎

Since $G_*^{-1}\phi = G^*\phi$ or $-G^*\phi$, (1) shows that the operation of push forward is the same as the operation of integrating out the fiber (for each fixed $y \in V'$, the fiber $F^{-1}(y)$ is the set $\{G(x, y);\ x \in \mathbb{R}^{N-N'}\}$).

We now define the pull back of a current.

DEFINITION 2 *Suppose Ω and Ω' are open subsets of the oriented manifolds X and Y with dimensions N and N' ($N \geq N'$) respectively. Suppose F: $\Omega \to \Omega'$ is a smooth surjective map so that $F_*(p)$: $T_p(\Omega) \to T_{F(p)}(\Omega')$ has maximal rank $= N'$ at each point $p \in \Omega$. For a current $T \in \mathcal{D}'^r(\Omega')$, define the pull back $F^*T \in \mathcal{D}'^r(\Omega)$ by*

$$\langle F^*T, \phi \rangle_\Omega = \langle T, F_*\phi \rangle_{\Omega'} \quad \textit{for} \quad \phi \in \mathcal{D}^{N-r}(\Omega).$$

Lemma 5 ensures that F^*T is a well-defined element of $\mathcal{D}'^r(\Omega)$. Note that since the push forward operator preserves dimension, the pull back operator preserves the degree of a current. In the case where the current T is given by a smooth form, then F^*T as defined above is the same as the usual pull back defined in Section 1.3 or 2.4. This is a tautology because the push forward has already been defined as the dual to the usual pull back.

Since $\mathcal{D}^r(\Omega')$ is dense in $\mathcal{D}'^r(\Omega')$, many currents can be pulled back in the same way that differential forms are pulled back. For example, suppose

$$T = \sum_{|I|=r} T_I dx^I$$

where each T_I is a locally integrable function on Ω'. Then

$$F^*T = \sum_{|I|=r} (T_I \circ F)(dF^I).$$

Another example of pull back is given in the following lemma, which we will need for Part IV.

LEMMA 6
Let τ: $\mathbb{R}^N \times \mathbb{R}^N \to \mathbb{R}^N$ be given by $\tau(y,x) = x - y$. Let $\Delta = \{(x,x) \in \mathbb{R}^N \times \mathbb{R}^N; x \in \mathbb{R}^N\}$ (the diagonal of $\mathbb{R}^N \times \mathbb{R}^N$). Then $\tau^*[0] = [\Delta]$.

PROOF An appealing formal proof can be given by writing [0] and $[\Delta]$ as forms with distribution coefficients. From the examples early in this chapter, we have

$$\begin{aligned} [0] &= \delta_0(x)dx_1 \wedge \ldots \wedge dx_N \\ [\Delta] &= \delta_0(x-y)d(x_1-y_1) \wedge \ldots \wedge d(x_N - y_N). \end{aligned}$$

The lemma is then formally proved by replacing x by $\tau(y,x) = x - y$ and dx_j by $d\tau_j(y,x) = d(x_j - y_j)$.

The above argument can be made rigorous by approximating [0] by a sequence of smooth forms. However, here is another approach using Definition 2. Let $\phi \in \mathcal{D}^N(\mathbb{R}^N \times \mathbb{R}^N)$. We write

$$\phi(y,x) = \sum_{|I|+|J|=N} \phi_{IJ}(y,x)dy^I \wedge dx^J$$

where each ϕ_{IJ} is an element of $\mathcal{D}(\mathbb{R}^N \times \mathbb{R}^N)$. Then

$$\begin{aligned} \langle \tau^*[0], \phi\rangle_{\mathbb{R}^N\times\mathbb{R}^N} &= \langle [0], \tau_*\phi\rangle_{\mathbb{R}^N} \quad \text{(by Definition 2)} \\ &= \langle [0], \int_{y\in\mathbb{R}^N} \phi(y, x+y)\rangle_{x\in\mathbb{R}^N} \qquad \text{(by Lemma 1)} \end{aligned}$$

where

$$\phi(y, x+y) = \sum_{|I|+|J|=N} \phi_{IJ}(y, x+y)dy^I \wedge d(x+y)^J.$$

Since only the piece of $\phi(y, x+y)$ of degree N in dy contributes to the above integral, we obtain

$$\int\limits_{y\in\mathbb{R}^N} \phi(y, x+y) = \sum_{|I|+|J|=N} \int\limits_{y\in\mathbb{R}^N} \phi_{IJ}(y, x+y)dy^I \wedge dy^J.$$

Therefore

$$\begin{aligned}\langle\tau^*[0], \phi\rangle_{\mathbb{R}^N\times\mathbb{R}^N} &= \left\langle [0], \sum_{|I|+|J|=N} \int\limits_{y\in\mathbb{R}^N} \phi_{IJ}(y, x+y)dy^I \wedge dy^J \right\rangle_{x\in\mathbb{R}^N} \\ &= \sum_{|I|+|J|=N} \int\limits_{y\in\mathbb{R}^N} \phi_{IJ}(y, y)dy^I \wedge dy^J \\ &= \int\limits_{\Delta} \phi \\ &= \langle[\Delta], \phi\rangle_{\mathbb{R}^N\times\mathbb{R}^N}\end{aligned}$$

as desired. ▮

Finally, we mention the analogues of Lemmas 3 and 4 for pull backs. We leave the easy proof to the reader.

LEMMA 7

Suppose Ω and Ω' are open subsets of oriented manifolds of dimensions N and N' $(N \geq N')$, respectively. Suppose F: $\Omega \to \Omega'$ is a smooth surjective map such that $F_*(p)$: $T_p(\Omega) \to T_p(\Omega')$ has maximal rank $= N'$ for each $p \in \Omega$. Then $F^* \circ d = d \circ F^*$ as operators from $\mathcal{D}'^r(\Omega')$ to $\mathcal{D}'^{r+1}(\Omega)$. If in addition Ω and Ω' are complex manifolds and F is holomorphic, then $F^* \circ \bar\partial = \bar\partial \circ F^*$ as operators from $\mathcal{D}'^{p,q}(\Omega')$ to $\mathcal{D}'^{p,q+1}(\Omega)$.

Part II

CR Manifolds

In the first part of this book, we defined two types of manifolds: the smooth manifold and the complex manifold. Real analytic manifolds were also defined, but aside from the real analyticity of the coordinate functions, the structure of this class of manifolds differs little from the structure of the class of smooth manifolds. Complex manifolds are fundamentally different because of their additional structure. A complex manifold gives rise to a complex structure map (J) and a Cauchy–Riemann operator $(\bar{\partial})$ which are not part of the structure of a smooth or real analytic manifold.

In Part II of this book, a third class of manifolds is introduced — the CR manifolds. The CR stands for Cauchy–Riemann and as the name suggests, a CR manifold retains some of the additional complex structure of a complex manifold. Indeed, any complex manifold is a CR manifold. However, the class of CR manifolds is much more general. It includes the class of all real hypersurfaces in $\mathbb{C}^n$. In fact, "most" submanifolds of $\mathbb{C}^n$ are CR manifolds.

The basic definition of a CR manifold is given in Chapter 7. Both abstract and imbedded CR manifolds are discussed. In Chapter 8, the tangential Cauchy–Riemann complex is introduced. This is analogous to the $\bar{\partial}$-complex for a complex manifold. Here, two points of view are discussed. For a CR submanifold of $\mathbb{C}^n$, an extrinsic point of view is given which relates the tangential Cauchy–Riemann complex to the $\bar{\partial}$-complex on the ambient $\mathbb{C}^n$. For an abstract CR manifold, an intrinsic point of view is presented which requires no ambiently defined Cauchy–Riemann complex. These two points of view are then shown to be isomorphic in the case of a CR submanifold of $\mathbb{C}^n$. In Chapter 9, the concept of a CR function is introduced. A CR function on a CR manifold is analogous to a holomorphic function on a complex manifold. However, the behavior of a CR function can be much different. For example, a CR function is not always smooth or even continuous. Holomorphic functions on $\mathbb{C}^n$ always restrict to CR functions on a CR submanifold. However, a CR function on a CR submanifold does not always extend to a holomorphic function on some open subset of $\mathbb{C}^n$. The question of the holomorphic extension of CR functions will be taken up in detail in Part III of this book.

In Chapter 10, we introduce the Levi form, which is analogous to the second fundamental form in differential geometry. The Levi form is the key differential geometric object which determines many function theoretic properties of a CR manifold (for example, the holomorphic extendability of CR functions and the local solvability of the tangential Cauchy–Riemann complex).

In Chapter 11, we discuss the imbeddability of CR manifolds. We show that a real analytic CR manifold can always be imbedded as a CR submanifold of $\mathbb{C}^n$. The C^∞ version of this theorem does not hold. Nirenberg's example of a three-dimensional C^∞ strictly pseudoconvex manifold that cannot be imbedded in any $\mathbb{C}^n$ is presented at the end of Chapter 11.

7

CR Manifolds

We start this chapter with the definition of an imbedded CR manifold, which is the simplest class of CR manifolds. Later, we will present the definition of an abstract CR manifold.

7.1 Imbedded CR manifolds

Here, we will concentrate on the case of a CR manifold imbedded in $\mathbb{C}^n$. We could easily replace $\mathbb{C}^n$ by a general complex manifold but this would unnecessarily complicate matters and add little to the understanding of the subject of CR manifolds.

For a smooth submanifold M of $\mathbb{C}^n$, recall that $T_p(M)$ is the real tangent space of M at a point $p \in M$. In general, $T_p(M)$ is not invariant under the complex structure map J for $T_p(\mathbb{C}^n)$. Therefore, we give special designation to the largest J-invariant subspace of $T_p(M)$.

DEFINITION 1 *For a point $p \in M$, the complex tangent space of M at p is the vector space*

$$H_p(M) = T_p(M) \cap J\{T_p(M)\}.$$

The space $H_p(M)$ is sometimes called the *holomorphic tangent space*. This must be an even-dimensional real vector space because

$$J \circ J|_{H^p(M)} = -I$$

and therefore

$$[\det J|_{H_p(M)}]^2 = (-1)^m$$

where $m = \dim_{\mathbb{R}} H_p(M)$. Note that if $A : \mathbb{R}^{2n} \mapsto \mathbb{R}^{2n}$ is a complex linear map (i.e., $J \circ A = A \circ J$), then $A\{H_p(M)\} \subset H_{A(p)}(A\{M\})$.

We also give special designation for the "other directions" in $T_p(M)$ which do not lie in $H_p(M)$.

DEFINITION 2 *The totally real part of the tangent space of M is the quotient space*

$$X_p(M) = T_p(M)/H_p(M).$$

Using the Euclidean inner product on $T_p(\mathbb{R}^{2n})$, we can identify $X_p(M)$ with the orthogonal complement of $H_p(M)$ (denoted $H_p(M)^{\perp}$) in $T_p(M)$. With this identification, note that $J\{X_p(M)\} \cap X_p(M) = \{0\}$, because $H_p(M)$ is the largest J-invariant subspace of $T_p(M)$. We have $T_p(M) = H_p(M) \oplus X_p(M)$. From Lemma 1 in Section 3.2, $J\{X_p(M)\}$ is orthogonal to $H_p(M)$. Therefore, $J\{X_p(M)\}$ is transverse to $T_p(M)$.

The dimensions of $H_p(M)$ and $X_p(M)$ are of crucial importance.

LEMMA 1
Suppose M is a real submanifold of $\mathbb{C}^n$ of real dimension $2n - d$. Then

$$2n - 2d \leq \dim_{\mathbb{R}} H_p(M) \leq 2n - d$$

and

$$0 \leq \dim_{\mathbb{R}} X_p(M) \leq d.$$

PROOF First note that $H_p(M) \subset T_p(M)$ and so

$$\dim_{\mathbb{R}} H_p(M) \leq \dim_{\mathbb{R}} T_p(M) = 2n - d.$$

To establish the other inequality, note

$$T_p(\mathbb{R}^{2n}) \supset T_p(M) + J\{T_p(M)\}$$

and so

$$\dim T_p(\mathbb{R}^{2n}) \geq \dim_{\mathbb{R}} T_p(M) + \dim_{\mathbb{R}} J\{T_p(M)\} - \dim_{\mathbb{R}}\{H_p(M)\}$$

from elementary linear algebra. Since J is an isometry, $\dim_{\mathbb{R}} J\{T_p(M)\} = \dim_{\mathbb{R}} T_p(M) = 2n - d$, from which $\dim_{\mathbb{R}} H_p(M) \geq 2n - 2d$ follows. Since

$$\dim_{\mathbb{R}} X_p(M) = \dim_{\mathbb{R}} T_p(M) - \dim_{\mathbb{R}} H_p(M),$$

the statement of the dimension of $X_p(M)$ in the lemma also follows. ∎

The real dimension of $X_p(M)$ is called the *CR codimension* of M.

The lemma states that $\dim_{\mathbb{R}} H_p(M)$ is an even number between $2n - 2d$ and $2n - d$. If M is a real hypersurface, then $d = 1$ and so the only possibility is $\dim_{\mathbb{R}} H_p(M) = 2n - 2$. In particular, the dimension of $H_p(M)$ never changes. If $d > 1$, then there are more possibilities. Consider the following example.

Example 1
Let $M = \{z \in \mathbb{C}^n; |z| = 1 \text{ and } \operatorname{Im} z_1 = 0\}$. M is just the equator of the unit sphere in $\mathbb{C}^n$. Here, $d = 2$ and so $2n - 4 \leq \dim_{\mathbb{R}} H_p(M) \leq 2n - 2$, for $p \in M$. At the point $p_1 = (z_1 = 0, z_2 = 1, z_3 = 0, \ldots, z_n = 0) \in M$, $T_{p_1}(M)$ is spanned over $\mathbb{R}$ by $\{\partial/\partial x_1, \partial/\partial y_2, \partial/\partial x_3, \partial/\partial y_3, \ldots, \partial/\partial x_n, \partial/\partial y_n\}$. The vectors $J(\partial/\partial x_1) = \partial/\partial y_1$ and $J(\partial/\partial y_2) = -(\partial/\partial x_2)$ are orthogonal to $T_{p_1}(M)$ and therefore $\partial/\partial x_1, \partial/\partial y_2$ span $X_{p_1}(M)$. The vectors $\{\partial/\partial x_3, \partial/\partial y_3, \ldots, \partial/\partial x_n, \partial/\partial y_n\}$ span the J-invariant subspace $H_{p_1}(M)$. So in this case, $\dim_{\mathbb{R}} H_{p_1}(M) = 2n - 4$ and $\dim_{\mathbb{R}} X_{p_1}(M) = 2$.

Now consider the point $p_2 = (z_1 = 1, z_2 = 0, \ldots, z_n = 0) \in M$. Here, $T_{p_2}(M)$ is spanned (over $\mathbb{R}$) by $\{\partial/\partial x_2, \partial/\partial y_2, \ldots, \partial/\partial x_n, \partial/\partial y_n\}$ which is J-invariant. Therefore, $H_{p_2}(M) = T_{p_2}(M)$ and $X_{p_2}(M) = \{0\}$. In this case, $\dim_{\mathbb{R}} H_{p_2}(M) = 2n - 2$ and $\dim X_{p_2}(M) = 0$. □

In the above example, the dimension of $H_p(M)$ varies with p. The basic requirement of a CR manifold is that $\dim_{\mathbb{R}} H_p(M)$ is independent of $p \in M$.

DEFINITION 3 *A submanifold M of $\mathbb{C}^n$ is called an imbedded CR manifold or a CR submanifold of $\mathbb{C}^n$ if* $\dim_{\mathbb{R}} H_p(M)$ *is independent of $p \in M$.*

Any real hypersurface in $\mathbb{C}^n$ is a CR submanifold of $\mathbb{C}^n$. Another class of CR submanifolds is the class of complex submanifolds of $\mathbb{C}^n$. For a complex submanifold M, the real tangent space is already J-invariant and so $T_p(M) = H_p(M)$.

Another example of a CR submanifold is a totally real submanifold, which is on the opposite end of the spectrum from a complex manifold.

DEFINITION 4 *A submanifold M in $\mathbb{C}^n$ is said to be totally real if $H_p(M) = \{0\}$, for each $p \in M$.*

An equivalent definition of a totally real submanifold is that $X_p(M) = T_p(M)$, for $p \in M$. From Lemma 1, the real dimension of a totally real submanifold is at most n. An example of a totally real submanifold is the copy of $\mathbb{R}^n$ given by $\{(x + iy) \in \mathbb{C}^n; y = 0\}$. Any smooth graph over this copy of $\mathbb{R}^n$ is also a totally real submanifold.

The complexifications of $T_p(M)$, $H_p(M)$, and $X_p(M)$ are denoted by $T_p(M) \otimes \mathbb{C}$, $H_p(M) \otimes \mathbb{C}$, and $X_p(M) \otimes \mathbb{C}$, respectively. The complex structure map J on $T_p(\mathbb{R}^{2n}) \otimes \mathbb{C}$ restricts to a complex structure map on $H_p(M) \otimes \mathbb{C}$ because $H_p(M)$ is J-invariant. From Section 3.2, $H_p(M) \otimes \mathbb{C}$ is the direct sum of the $+i$ and $-i$ eigenspaces of J which are denoted by $H_p^{1,0}(M)$ and $H_p^{0,1}(M)$, respectively. We have

$$H_p^{1,0}(M) = T_p^{1,0}(\mathbb{C}^n) \cap \{T_p(M) \otimes \mathbb{C}\}$$
$$H_p^{0,1}(M) = T_p^{0,1}(\mathbb{C}^n) \cap \{T_p(M) \otimes \mathbb{C}\}$$
$$H_p^{0,1}(M) = \overline{H_p^{1,0}(M)}.$$

It will be useful to have a way of identifying the above spaces in terms of a local defining system for M. We have the following lemma.

LEMMA 2
Let M be a smooth submanifold of $\mathbb{C}^n$ defined near a point $p \in M$ by $M = \{z \in \mathbb{C}^n; \rho_1(z) = \cdots = \rho_d(z) = 0\}$, where $\rho_1, \ldots, \rho_d$ are smooth real-valued functions with $d\rho_1 \wedge \ldots \wedge d\rho_d \neq 0$ near p.

(a) *A vector $W = \sum_{j=1}^n w_j \frac{\partial}{\partial z_j} \in T_p^{1,0}(\mathbb{C}^n)$ belongs to $H_p^{1,0}(M)$ if and only if*

$$W\{\rho_k\}(p) = \langle \partial\rho_k, W\rangle_p = \sum_{j=1}^n \frac{\partial \rho_k}{\partial z_j}(p) w_j = 0, \qquad 1 \leq k \leq d.$$

(b) *A vector $W = \sum_{j=1}^n w_j \frac{\partial}{\partial \bar{z}_j} \in T_p^{0,1}(\mathbb{C}^n)$ belongs to $H_p^{0,1}(M)$ if and only if*

$$W\{\rho_k\}(p) = \langle \bar{\partial}\rho_k, W\rangle_p = \sum_{j=1}^n \frac{\partial \rho_k}{\partial \bar{z}_j}(p) w_j = 0, \qquad 1 \leq k \leq d.$$

Recall that $W\{\rho_k\}$ denotes the action of a vector W on a function ρ_k, and that $\langle\ ,\ \rangle_p$ denotes the pairing between forms and vectors.

PROOF We have $H_p^{1,0}(M) = T_p^{1,0}(\mathbb{C}^n) \cap \{T_p(M) \otimes \mathbb{C}\}$ and

$$T_p(M) \otimes \mathbb{C} = \{W \in T_p(\mathbb{C}^n) \otimes \mathbb{C}; \langle d\rho_k, W\rangle_p = 0 \quad \text{for} \quad 1 \leq k \leq d\}.$$

Clearly, $\langle \bar{\partial}\rho_k, W\rangle_p = 0$ for $W \in T_p^{1,0}(\mathbb{C}^n)$ because $\bar{\partial}\rho_k$ is a form of bidegree $(0,1)$. In addition, $d\rho_k = \partial\rho_k + \bar{\partial}\rho_k$. Therefore, $W \in H_p^{1,0}(M)$ if and only if $\langle \partial\rho_k, W\rangle_p = 0$. Part (b) is proved the same way. ∎

If M is a CR submanifold of $\mathbb{C}^n$, then the dimensions of $H_p^{1,0}(M)$, $H_p^{0,1}(M)$, and $H_p(M) \otimes \mathbb{C}$ are independent of the point $p \in M$. We define the following subsets of $T^{\mathbb{C}}(M)$:

$$H^{\mathbb{C}}(M) = \bigcup_{p \in M} H_p(M) \otimes \mathbb{C}$$

$$H^{1,0}(M) = \bigcup_{p \in M} H_p^{1,0}(M)$$

$$H^{0,1}(M) = \bigcup_{p \in M} H_p^{0,1}(M).$$

As mentioned in Part I, a subbundle of $T^{\mathbb{C}}(M)$ is an object that assigns to each point $p \in M$ a subspace of $T_p(M) \otimes \mathbb{C}$ whose dimension is independent of p. In addition, these subspaces are required to fit together smoothly in the sense that they are locally generated by a basis of smooth vector fields. This latter requirement is easily seen to be satisfied by the spaces $H^{1,0}(M)$, $H^{0,1}(M)$, and $H^{\mathbb{C}}(M)$. For if M is locally defined by $\{\rho_1, \dots, \rho_d\}$, then near a point $p_0 \in M$, we can choose from $\{\partial\rho_1, \dots, \partial\rho_d\}$ a collection of k-elements $\partial\rho_{i_1}, \dots, \partial\rho_{i_k}$ $(1 \leq k \leq d)$ that are linearly independent. The number k is the CR codimension of M. From elementary linear algebra, there are smooth linearly independent vector fields $L_1, \dots, L_{n-k}$ that are annihilated by $\partial\rho_{i_1}, \dots, \partial\rho_{i_k}$. These vector fields locally generate $H^{1,0}(M)$ by Lemma 2, and so $H^{1,0}(M)$ is a subbundle of $T^{\mathbb{C}}(M)$. Similarly, $H^{0,1}(M)$ and $H^{\mathbb{C}}(M)$ are also subbundles of $T^{\mathbb{C}}(M)$. In the next section, we will construct local bases for $H^{1,0}(M)$ and $H^{0,1}(M)$ in a canonical way after locally graphing M over its tangent space.

LEMMA 3
Suppose M is a CR submanifold of $\mathbb{C}^n$. Then

(a) $H_p^{0,1}(M) \cap H_p^{1,0}(M) = \{0\}$ *for each* $p \in M$.
(b) *The subbundles $H^{0,1}(M)$ and $H^{1,0}(M)$ are involutive.*

Recall that a subbundle is *involutive* if it is closed under the Lie bracket.

PROOF The proof of part (a) follows from the fact that the intersection of eigenspaces of any linear map corresponding to different eigenvalues is always trivial. For part (b), we first note

$$H^{1,0}(M) = T^{\mathbb{C}}(M) \cap \{T^{1,0}(\mathbb{C}^n)|_M\}.$$

The bundle $T^{1,0}(\mathbb{C}^n)$ is involutive because the Lie bracket of any two vector fields spanned by $\partial/\partial z_1, \dots, \partial/\partial z_n$ is again spanned by $\{\partial/\partial z_1, \dots, \partial/\partial z_n\}$. In addition, $T^{\mathbb{C}}(M)$ is involutive because the tangent bundle of *any* manifold is involutive. So, $H^{1,0}(M)$ is involutive, as desired. Since $H^{0,1}(M) = \overline{H^{1,0}(M)}$, $H^{0,1}(M)$ is also involutive. ∎

The above lemma is important because properties (a) and (b) are the defining properties of an abstract CR manifold (see Section 7.4). The lemma does *not* imply that $H^{\mathbb{C}}(M) = H^{1,0}(M) \oplus H^{0,1}(M)$ is involutive. In general, this is not true. In fact the Levi form, discussed in Chapter 10, measures the degree to which $H^{\mathbb{C}}(M)$ fails to be involutive.

Lemma 2 implies that $\dim_{\mathbb{C}} H_p^{1,0}(M) = n - k$ where k is the number of linearly independent elements of $\{\partial\rho_1(p), \ldots, \partial\rho_d(p)\}$. If $\partial\rho_1 \wedge \ldots \wedge \partial\rho_d \neq 0$ then $\dim_{\mathbb{C}} H_p^{1,0}(M) = n - d = \dim_{\mathbb{C}} H_p^{0,1}(M)$ and so $\dim_{\mathbb{R}} H_p(M) = 2n - 2d$. According to Lemma 1, this is the minimum value of the dimension for $H_p(M)$. The condition $\partial\rho_1 \wedge \ldots \wedge \partial\rho_d \neq 0$ is an open condition. It is also generic in the sense that a random collection of defining functions will, with high probability, satisfy this condition.

DEFINITION 5 *A CR submanifold M is called generic if* $\dim_{\mathbb{R}} H_p(M)$ *is minimal.*

A real hypersurface in $\mathbb{C}^n$ is always generic. By Lemma 1, a generic CR submanifold of $\mathbb{C}^n$ whose real codimension is at least n must be totally real. Any complex submanifold of $\mathbb{C}^n$ that is not an open subset of $\mathbb{C}^n$ is an example of a nongeneric CR submanifold.

The following lemma follows easily from the definitions and Lemma 2.

LEMMA 4
Suppose M is a CR submanifold of $\mathbb{C}^n$ with $\dim_{\mathbb{R}} M = 2n - d$, $0 \leq d \leq n$. *The following are equivalent.*

(a) *M is generic.*

(b) $\dim_{\mathbb{R}} H_p(M) = 2n - 2d$, *for* $p \in M$.

(c) *The CR codimension of M equals d.*

(d) $\partial\rho_1 \wedge \ldots \wedge \partial\rho_d \neq 0$ *on M for each local defining system* $\{\rho_1, \ldots, \rho_d\}$ *for* M.

(e) $T_p(\mathbb{C}^n) = T_p(M) \oplus J\{X_p(M)\}$.

Let us return to the above example of the equator in the unit sphere, $M = \{z \in \mathbb{C}^n; |z| = 1 \text{ and } \operatorname{Im} z_1 = 0\}$. In this case, M is the common zero set of the defining functions $\rho_1(z) = |z|^2 - 1$ and $\rho_2(z) = (1/2i)(z_1 - \bar{z}_1)$. We have

$$\partial\rho_1(z) = \bar{z} \cdot dz = \sum_{j=1}^{n} \bar{z}_j dz_j$$

$$\partial\rho_2(z) = \frac{1}{2i} dz_1.$$

Clearly, $\partial\rho_1 \wedge \partial\rho_2 = 0$ only at the points $z = (z_1, \ldots, z_n) \in M$ with $z_2 = \cdots =$

$z_n = 0, z_1 = \pm 1$. So, M is not a CR submanifold. However, the set

$$\widetilde{M} = \{z = (z_1, \ldots, z_n) \in M; z_1 \neq \pm 1\}$$

is a generic (noncompact) CR submanifold of $\mathbb{C}^n$.

If M is a real hypersurface, then $X_p(M)$ is a one (real) dimensional subspace of $T_p(M)$. By Lemma 1 in Section 3.2, $J\{X_p(M)\}$ is orthogonal to both $X_p(M)$ and $H_p(M)$. Therefore, $J\{X_p(M)\}$ can be identified with the orthogonal complement of $T_p(M)$. For higher codimension, both $X_p(M)$ and $J\{X_p(M)\}$ are orthogonal to $H_p(M)$ (by Lemma 1 in Section 3.2). However, in general, $J\{X_p(M)\}$ is not orthogonal to $X_p(M)$ and therefore $J\{X_p(M)\}$ cannot be identified with the orthogonal complement of $T_p(M)$ in any natural way. Consider the following example: let

$$M = \{(z_1, z_2) \in \mathbb{C}^2;\ \mathrm{Im} z_1 = \mathrm{Re} z_2,\ \mathrm{Im} z_2 = 0\}.$$

M is totally real and a basis for $T_0(M)$ is given by

$$\left\{ \frac{\partial}{\partial x_1}, \frac{\partial}{\partial y_1} + \frac{\partial}{\partial x_2} \right\}$$

Note that $J(\partial/\partial x_1) = \partial/\partial y_1$ is not orthogonal to $\partial/\partial y_1 + \partial/\partial x_2$.

7.2 A normal form for a generic CR submanifold

In this section, we present a convenient coordinate description of a generic CR submanifold. As a start, we locally graph the CR submanifold over its real tangent space. Then, we show that the graphing function can be chosen so that certain "pure terms" in its Taylor expansion vanish. The real analytic case is handled first. From this, a C^k version will easily follow.

We start with an easy warm-up.

LEMMA 1

Suppose M is a smooth generic CR submanifold of $\mathbb{C}^n$ with $\dim_{\mathbb{R}} M = 2n - d$, $1 \leq d \leq n$. Suppose p_0 is a point in M. There is a nonsingular, complex affine linear map A: $\mathbb{C}^n \to \mathbb{C}^n$; an open neighborhood U of p_0; and a smooth function $h : \mathbb{R}^d \times \mathbb{C}^{n-d} \mapsto \mathbb{R}^d$ with $h(0) = 0$ and $Dh(0) = 0$ such that

$$A\{M \cap U\} = \{(x + iy, w) \in \mathbb{C}^d \times \mathbb{C}^{n-d}; y = h(x, w)\}.$$

PROOF First, we translate so that p_0 is the origin. It suffices to find a nonsingular complex linear map A: $\mathbb{C}^n \to \mathbb{C}^n$ that takes $T_0(M)$ to the space $\{y = 0\}$ in $\mathbb{R}^{2n}$, where $y = \mathrm{Im}\, z \in \mathbb{R}^d$. For then the graphing function, h, for $A\{M\}$ will satisfy the required properties.

To find the desired map A, let $\hat{v}_1, \ldots, \hat{v}_d$ be an orthonormal basis for $X_0(M)$. The J-invariant space $H_0(M)$ is orthogonal to both $X_0(M)$ and $J\{X_0(M)\}$. In addition, J has no real eigenvalues and $Jv \cdot w = -v \cdot Jw$ for $v, w \in T_0(\mathbb{R}^{2n})$. Therefore, there is an orthonormal basis for $H_0(M)$ of the form $\hat{v}_{d+1}, J\hat{v}_{d+1}, \ldots, \hat{v}_n, J\hat{v}_n$. Since M is generic, the set

$$\{\hat{v}_1, J\hat{v}_1, \ldots, \hat{v}_n, J\hat{v}_n\}.$$

is a basis for $T_0(\mathbb{R}^{2n})$, by Lemma 4.

Now let $z = x + iy \in \mathbb{C}^d$ and $w = u + iv \in \mathbb{C}^{n-d}$. Here, x and y belong to $\mathbb{R}^d$ and u, v belong to $\mathbb{R}^{n-d}$. We define the real linear map A: $\mathbb{R}^{2n} \to \mathbb{R}^{2n}$ by its action on basis vectors

$$A(\hat{v}_j) = \frac{\partial}{\partial x_j}, \qquad A(J\hat{v}_j) = \frac{\partial}{\partial y_j} \qquad 1 \le j \le d$$

$$A(\hat{v}_j) = \frac{\partial}{\partial u_{j-d}}, \qquad A(J\hat{v}_j) = \frac{\partial}{\partial v_{j-d}} \qquad d+1 \le j \le n.$$

Since $J(\partial/\partial x_j) = \partial/\partial y_j$ and $J(\partial/\partial u_k) = \partial/\partial v_k$, clearly $A \circ J = J \circ A$ on the basis for $T_0(\mathbb{R}^{2n})$ and therefore $A \circ J = J \circ A$ on all of $T_0(\mathbb{R}^{2n})$. Viewed as a map from $\mathbb{C}^n$ to $\mathbb{C}^n$, A is complex linear. Clearly $A\{T_0(M)\} = \{(x, 0, u, v); x \in \mathbb{R}^d$ and $u, v \in \mathbb{R}^{n-d}\} = \{y = 0\}$, as desired. This completes the proof of the lemma. ▮

REMARK Since A is complex linear, we have $A\{H_p(M)\} = H_0(A\{M\})$ and so the extension of A to $T_p(\mathbb{C}^n) \otimes \mathbb{C}$ satisfies $A\{H_p^{1,0}(M)\} = H_0^{1,0}(A\{M\})$ and $A\{H_p^{0,1}(M)\} = H_0^{0,1}(A\{M\})$. In addition, A sends an orthonormal basis for $T_p(M)$ to an orthonormal basis for $T_0(A\{M\})$. In particular, $A\{X_p(M)\} = X_0(A\{M\})$. In the new coordinates — relabeling $A\{M\}$ by M — we have

$$T_0(M) = \{(x, 0, u, v); x \in \mathbb{R}^d \quad \text{and} \quad u, v \in \mathbb{R}^{n-d}\}$$

$$H_0(M) = \{(0, 0, u, v); u, v \in \mathbb{R}^{n-d}\}$$

$$X_0(M) = \{(x, 0, 0, 0); x \in \mathbb{R}^d\}.$$

$H_0^{1,0}(M)$ can be identified with $\{(0, w); w \in \mathbb{C}^{n-d}\}$ in a complex linear fashion, where $w = u + iv$.

Most of our concern in this book will be with generic CR submanifolds. However, we should point out that a different version of Lemma 1 holds in the nongeneric case. Let k be the CR codimension of M and let d be the real codimension of M. For the nongeneric case, we have $0 \le k < d$ and so $n \ge \dim_{\mathbb{C}} H^{1,0}(M) > n - d$ by Lemma 1 in Section 7.1. Define the integer j

by $\dim H^{1,0}_{\mathbb{C}}(M) = n - d + j$. We have

$$\begin{aligned} 2n - d &= \dim_{\mathbb{C}} H^{1,0}(M) + \dim_{\mathbb{C}} H^{0,1}(M) + \dim_{\mathbb{R}} X(M) \\ &= 2n - 2d + 2j + k. \end{aligned}$$

Therefore, $2j + k = d$. Since $J\{X_p(M)\}$ is a k-dimensional subspace that is transverse to $T_p(M)$, there is a J-invariant subspace of $T_p(\mathbb{C}^n)^\perp$ of real dimension $2j$ that is transverse to $T_p(M) \oplus J\{X_p(M)\}$. The proof of Lemma 1 can be modified so that after a complex linear change of coordinates, the defining equations for M become

$$\begin{aligned} z_1 &= H_1(x_{j+1}, \dots, x_{d-j}, w_1, \dots, w_{n-d+j}) \\ &\vdots \\ z_j &= H_j(x_{j+1}, \dots, x_{d-j}, w_1, \dots, w_{n-d+j}) \\ y_{j+1} &= h_{j+1}(x_{j+1}, \dots, x_{d-j}, w_1, \dots, w_{n-d+j}) \\ &\vdots \\ y_{d-j} &= h_{d-j}(x_{j+1}, \dots, x_{d-j}, w_1, \dots, w_{n-d+j}) \end{aligned}$$

where $H_1, \dots, H_j$ are smooth *complex*-valued functions with $H_\ell(0) = 0$ and $DH_\ell(0) = 0$, $1 \le \ell \le j$, and where $h_{j+1}, \dots, h_{d-j}$ are smooth *real*-valued functions with $h_\ell(0) = 0$ and $Dh_\ell(0) = 0$ for $j + 1 \le \ell \le d - j$.

Now we present a normal form for a real analytic, CR, generic submanifold. This is the first step toward a more complicated normal form of Bloom and Graham [BG], which we will discuss in Chapter 12.

THEOREM 1

Suppose M is a real analytic, generic CR submanifold of $\mathbb{C}^n$ with $\dim_{\mathbb{R}} M = 2n - d$ $(1 \le d \le n)$. Suppose that p_0 is a point in M. There is a neighborhood U of p_0 in $\mathbb{C}^n$; a biholomorphism Φ: $U \to \Phi\{U\} \subset \mathbb{C}^n$; and a real analytic h: $\mathbb{R}^d \times \mathbb{C}^{n-d} \to \mathbb{R}^d$ with $h(0) = 0$ and $Dh(0) = 0$ such that

$$\Phi\{M \cap U\} = \{(x + iy, w) \in \Phi\{U\} \subset \mathbb{C}^d \times \mathbb{C}^{n-d}; y = h(x, w)\}.$$

Furthermore

$$\begin{aligned} \frac{\partial^{|\alpha|+|\beta|} h}{\partial x^\alpha \partial w^\beta}(0) = 0 \\ \frac{\partial^{|\alpha|+|\beta|} h}{\partial x^\alpha \partial \bar{w}^\beta}(0) = 0 \end{aligned}$$

for all multiindices α and β.

Another way to describe the graphing function h is to look at its Taylor expansion about the origin

$$h(x,w) = \sum_{\alpha,\beta,\gamma} a_{\alpha,\beta,\gamma} x^\alpha w^\beta \bar{w}^\gamma$$

where

$$a_{\alpha,\beta,\gamma} = \frac{1}{\alpha!\beta!\gamma!} \frac{\partial^{|\alpha|+|\beta|+|\gamma|} h}{\partial x^\alpha \partial w^\beta \partial \bar{w}^\gamma}(0).$$

The terms in the Taylor expansion with either $\beta = 0$ or $\gamma = 0$ are called *pure terms.* The content of the theorem is that with the proper choice of holomorphic coordinates, the graphing function for M has no pure terms in its Taylor expansion.

PROOF From Lemma 1, we may assume that the given point p_0 is the origin and

$$M = \{(x+iy,w) \in \mathbb{C}^d \times \mathbb{C}^{n-d}; y = h(x,w)\}$$

where $h(0) = 0$ and $Dh(0) = 0$. Since M is real analytic, its graphing function h: $\mathbb{R}^d \times \mathbb{C}^{n-d} \to \mathbb{R}^d$ is real analytic. The Taylor expansion of h is given by

$$h(x,w,\bar{w}) = \sum a_{\alpha\beta\gamma} x^\alpha w^\beta \bar{w}^\gamma.$$

Note we have emphasized that h is *not* holomorphic in w by the notation $h(x,w,\bar{w})$, which illustrates the dependence of h on $\bar{w}$. We replace $\bar{w}$ by an independent coordinate $\eta \in \mathbb{C}^{n-d}$ and x by $z \in \mathbb{C}^d$ and we define the holomorphic map $\tilde{h}$: $\mathbb{C}^d \times \mathbb{C}^{n-d} \times \mathbb{C}^{n-d} \to \mathbb{C}^d$ by

$$\tilde{h}(z,w,\eta) = \sum a_{\alpha\beta\gamma} z^\alpha w^\beta \eta^\gamma.$$

This series converges for $\{|z|,|w|,|\eta| < \delta\}$ for some $\delta > 0$.

Since $h(0) = 0$ and $Dh(0) = 0$, we have $\tilde{h}(0) = 0$ and $D\tilde{h}(0) = 0$. By the implicit function theorem for holomorphic maps, there is a unique holomorphic map ϕ: $\mathbb{C}^d \times \mathbb{C}^{n-d} \to \mathbb{C}^d$ defined near the origin such that

$$\phi(z + i\tilde{h}(z,w,0), w) = z \quad \text{for} \quad |z|,|w| < \delta \tag{1}$$

for some, possibly smaller $\delta > 0$.

Define the holomorphic change of variables $(\hat{z},\hat{w}) = \Phi(z,w)$ where

$$\begin{aligned} \hat{z} &= z - i\tilde{h}(\phi(z,w),w,0) \in \mathbb{C}^d \\ \hat{w} &= w \in \mathbb{C}^{n-d}. \end{aligned} \tag{2}$$

Since $D\tilde{h}(0,0,0) = 0$, $D\Phi(0,0)$ is the identity. So for an appropriate $\delta > 0$, the map Φ is a biholomorphism from $\{|z|,|w| < \delta\}$ to a neighborhood of the origin in $\mathbb{C}^n$.

Let $\widehat{M}$ be the image of $M \cap \{|z|, |w| < \delta\}$ under Φ. We wish to find a defining function for $\widehat{M}$ in the $\hat{z} = \hat{x} + i\hat{y}$, $\hat{w}$ coordinates of the form

$$\hat{y} = \hat{h}(\hat{x}, \hat{w})$$

so that $\hat{h}$ satisfies the required properties stated in the theorem.

From (2),

$$\hat{y} = y - \text{Re}\{\tilde{h}(\phi(z, w), w, 0)\}.$$

If $(\hat{z}, \hat{w})$ belongs to $\widehat{M}$, then (z, w) belongs to M and so

$$y = \tilde{h}(x, w, \bar{w}) \in \mathbb{R}^d.$$

Substituting this equation into the previous one, we have

$$\hat{y} = \text{Re}\{\tilde{h}(x, w, \bar{w}) - \tilde{h}(\phi(x + i\tilde{h}(x, w, \bar{w}), w), w, 0)\} \tag{3}$$

for $(\hat{z}, \hat{w}) = \Phi(z, w) \in \widehat{M}$.

To obtain a defining equation for $\widehat{M}$ in the $(\hat{z}, \hat{w})$ coordinates, we must transform the right side of (3) into a function of $\hat{x}$ and $\hat{w}$. Now since Φ is a local biholomorphism, clearly $(z, w) = \Phi^{-1}(\hat{z}, \hat{w})$ and we write

$$z = z(\hat{z}, \hat{w})$$
$$w = \hat{w}$$

where z: $\mathbb{C}^d \times \mathbb{C}^{n-d} \to \mathbb{C}^d$ is holomorphic near the origin. Therefore $x =$ Re z is a real analytic function of Re $\hat{z}$, Im $\hat{z}$, Re $\hat{w}$, Im $\hat{w}$ and we write $x = x(\hat{z}, \hat{w})$. Substituting $x = x(\hat{z}, \hat{w})$ and $w = \hat{w}$ into the right side of (3) would result in a local graphing function for $\widehat{M}$, except that this would still involve the variable $\hat{y} =$ Im $\hat{z}$. A graphing function should only involve $\hat{x}$ and $\hat{w}$. To remedy this, we use the fact that $D\tilde{h}(0, 0, 0) = 0$ and the implicit function theorem to solve for the variable $\hat{y} = \hat{y}(\hat{x}, \hat{w}, \bar{\hat{w}})$ in (3) (with $x = x(\hat{z}, \hat{w})$ and $w = \hat{w}$) as a real analytic function of $\hat{x}$, Re $\hat{w}$, Im $\hat{w}$ near the origin. Substituting $\hat{y}(\hat{x}, \hat{w}, \bar{\hat{w}})$ for $\hat{y} =$ Im $\hat{z}$ in $x(\hat{z}, \hat{w})$ yields a new function which we denote by $x(\hat{x}, \hat{w}, \bar{\hat{w}})$. This function is real analytic in a neighborhood of the origin.

Now substituting $x(\hat{x}, \hat{w}, \bar{\hat{w}})$ for x and $\hat{w}$ for w, (3) becomes

$$\hat{y} = \text{Re}\{\hat{h}(\hat{x}, \hat{w}, \bar{\hat{w}})\} \quad \text{for} \quad (\hat{z}, \hat{w}) \in \widehat{M}$$

where

$$\begin{aligned}\hat{h}(\hat{x}, \hat{w}, \bar{\hat{w}}) = {} & \tilde{h}(x(\hat{x}, \hat{w}, \bar{\hat{w}}), \hat{w}, \bar{\hat{w}}) \\ & - \tilde{h}(\phi(x(\hat{x}, \hat{w}, \bar{\hat{w}}) + i\tilde{h}(x(\hat{x}, \hat{w}, \bar{\hat{w}}), \hat{w}, \bar{\hat{w}}), \hat{w}), \hat{w}, 0).\end{aligned}$$

Therefore, Re $\hat{h}$ is the graphing function for $\widehat{M}$.

It remains to show

$$\frac{\partial^{|\alpha|+|\beta|} \operatorname{Re} \hat{h}}{\partial \hat{x}^{\alpha} \partial \hat{w}^{\beta}}(0) = 0 \qquad \text{and} \qquad \frac{\partial^{|\alpha|+|\beta|} \operatorname{Re} \hat{h}}{\partial \hat{x}^{\alpha} \partial \bar{\hat{w}}^{\beta}}(0) = 0. \tag{4}$$

The second equation follows by conjugating the first, so it suffices to show the first.

The function

$$(\hat{x}, \hat{w}) \longmapsto x(\hat{x}, \hat{w}, \bar{\hat{w}})$$

is real analytic near the origin and therefore can be expressed as a power series in $\hat{x}$, $\hat{w}$, and $\bar{\hat{w}}$. Replacing $\hat{x}$ by $\hat{z} \in \mathbb{C}^d$ and $\bar{\hat{w}}$ by $\hat{\eta} \in \mathbb{C}^{n-d}$, we obtain a map

$$x\colon \mathbb{C}^d \times \mathbb{C}^{n-d} \times \mathbb{C}^{n-d} \to \mathbb{C}^d$$

which is holomorphic in a neighborhood of the origin. By substituting $x(\hat{z}, \hat{w}, \hat{\eta})$ for $x(\hat{x}, \hat{w}, \bar{\hat{w}})$ and $\hat{\eta}$ for $\bar{\hat{w}}$ in the definition of $\hat{h}$, we obtain a function $\hat{h}\colon \mathbb{C}^d \times \mathbb{C}^{n-d} \times \mathbb{C}^{n-d} \to \mathbb{C}^d$ given by

$$\begin{aligned}\hat{h}(\hat{z}, \hat{w}, \hat{\eta}) &= \tilde{h}(x(\hat{z}, \hat{w}, \hat{\eta}), \hat{w}, \hat{\eta}) \\ &\quad -\tilde{h}(\phi(x(\hat{z}, \hat{w}, \hat{\eta}) + i\tilde{h}(x(\hat{z}, \hat{w}, \hat{\eta}), \hat{w}, \hat{\eta}), \hat{w}), \hat{w}, 0)\end{aligned}$$

which is holomorphic for $(\hat{z}, \hat{w}, \hat{\eta})$ in a neighborhood of the origin. To establish the first equation in (4), we must show

$$\hat{h}(\hat{z}, \hat{w}, 0) = 0.$$

From (1) with z replaced by $x(\hat{z}, \hat{w}, 0)$ and w replaced by $\hat{w}$, we obtain

$$\phi(x(\hat{z}, \hat{w}, 0) + i\tilde{h}(x(\hat{z}, \hat{w}, 0), \hat{w}, 0), \hat{w}) = x(\hat{z}, \hat{w}, 0).$$

Substituting this into $\hat{h}$ gives

$$\begin{aligned}\hat{h}(\hat{z}, \hat{w}, 0) &= \tilde{h}(x(\hat{z}, \hat{w}, 0), \hat{w}, 0) - \tilde{h}(x(\hat{z}, \hat{w}, 0), \hat{w}, 0) \\ &= 0\end{aligned}$$

as desired. Therefore (4) holds and the proof of the theorem is complete. ∎

A C^k version of Theorem 1 follows easily from the real analytic version. Suppose $M = \{y = h(x, w)\}$ where h is of class C^k for $k \geq 2$. From a kth order Taylor expansion of h, we have

$$h(x, w) = p(x, w, \bar{w}) + e(x, w)$$

where p is a polynomial of degree k in the variables $x, w, \bar{w}$ and where the Taylor remainder e satisfies

$$\frac{\partial^{|\alpha|+|\beta|+|\gamma|} e(0)}{\partial x^{\alpha} \partial w^{\beta} \partial \bar{w}^{\gamma}} = 0 \qquad \text{for} \qquad |\alpha| + |\beta| + |\gamma| \leq k.$$

Therefore, the manifolds M and $\widehat{M} = \{y = p(x, w, \bar{w})\}$ agree to order k at the origin. Since p is real analytic, Theorem 1 applies to $\widehat{M}$. The images of M and $\widehat{M}$ in the new coordinates still agree to order k at the origin. We have established the following.

THEOREM 2
Suppose M is a generic, CR submanifold of $\mathbb{C}^n$ of class C^k $(k \geq 2)$ with $\dim_{\mathbb{R}} M = 2n - d$ $(1 \leq d \leq n)$. Suppose that p_0 is a point in M. There is a neighborhood U of p_0 in $\mathbb{C}^n$; a biholomorphism $\Phi\colon U \to \Phi\{U\} \subset \mathbb{C}^n$; and a function $h\colon \mathbb{R}^d \times \mathbb{C}^{n-d} \to \mathbb{R}^d$ of class C^k with $h(0) = 0$ and $Dh(0) = 0$ such that

$$\Phi\{M \cap U\} = \{(x + iy, w) \in \Phi\{U\} \subset \mathbb{C}^d \times \mathbb{C}^{n-d}; y = h(x, w)\}$$

Furthermore

$$\frac{\partial^{|\alpha|+|\beta|} h(0)}{\partial x^\alpha \partial w^\beta} = \frac{\partial^{|\alpha|+|\beta|} h(0)}{\partial x^\alpha \partial \bar{w}^\beta} = 0 \qquad \textit{for} \qquad |\alpha| + |\beta| \leq k.$$

Now we turn to the question of finding a canonical local basis for $H^{1,0}(M)$ and $H^{0,1}(M)$. We assume that M is graphed over the origin as in the conclusion of Lemma 1, i.e.,

$$M = \{y = h(x, w)\}$$

where $h\colon \mathbb{R}^d \times \mathbb{C}^{n-d} \to \mathbb{R}^d$ is of class C^k, for $k \geq 2$, and where $h(0) = 0$ and $Dh(0) = 0$. The space $H_0^{0,1}(M)$ is spanned (over $\mathbb{C}$) by $\partial/\partial\bar{w}_1, \ldots, \partial/\partial\bar{w}_{n-d}$ and the space $H_0^{1,0}(M)$ is spanned by $\partial/\partial w_1, \ldots, \partial/\partial w_{n-d}$. It is our desire to extend these vectors to vector fields that locally generate $H^{0,1}(M)$ and $H^{1,0}(M)$, respectively.

THEOREM 3
Suppose $M = \{(x+iy, w) \in \mathbb{C}^d \times \mathbb{C}^{n-d}; y = h(x, w)\}$ where $h\colon \mathbb{R}^d \times \mathbb{C}^{n-d} \to \mathbb{R}^d$ is of class C^k $(k \geq 2)$ with $h(0) = 0$ and $Dh(0) = 0$. A basis for $H^{1,0}(M)$ near the origin is given by $L_1, \ldots, L_{n-d}$ with

$$L_j = \frac{\partial}{\partial w_j} + 2i \sum_{\ell=1}^{d} \left(\sum_{k=1}^{d} \mu_{\ell k} \frac{\partial h_k}{\partial w_j} \frac{\partial}{\partial z_\ell} \right), \qquad 1 \leq j \leq n - d$$

where $\mu_{\ell k}$ is the (ℓ, k)th element of the $d \times d$ matrix

$$\left(I - i \frac{\partial h}{\partial x} \right)^{-1}.$$

A basis for $H^{0,1}(M)$ near the origin is given by $\overline{L}_1, \ldots, \overline{L}_{n-d}$.

Since $Dh(0)=0$, note that $L_j|_0=\partial/\partial w_j$ and $\overline{L}_j|_0 = \partial/\partial \overline{w}_j$, for $1\leq j\leq n-d$.

PROOF Let

$$L_j = \frac{\partial}{\partial w_j} + \sum_{k=1}^{d} A_{kj}\frac{\partial}{\partial z_k} \in T^{1,0}(\mathbb{C}^n), \qquad 1 \leq j \leq n-d$$

where the A_{kj} are smooth functions to be chosen so that $L_j|_M$ belongs to $H^{1,0}(M)$.

Let $\rho_j(z,w) = \text{Im } z_j - h_j(x,w)$, $1 \leq j \leq d$. The manifold M is the common zero set of $\rho_1, \ldots, \rho_d$. By Lemma 2 in Section 7.1, a vector field L in $T^{1,0}(\mathbb{C}^n)|_M$ belongs to $H^{1,0}(M)$, provided

$$\langle \partial\rho_\ell, L\rangle = 0 \quad \text{on} \quad M \quad \text{for} \quad 1 \leq \ell \leq d,$$

where $\langle\ ,\ \rangle$ denotes the pairing between 1-forms and vectors. Inserting $L = L_j$ into this equation, we obtain

$$\frac{1}{2i}A_{\ell j} - \frac{1}{2}\sum_{k=1}^{d}\frac{\partial h_\ell}{\partial x_k}A_{kj} - \frac{\partial h_\ell}{\partial w_j} = 0,$$

for $1 \leq \ell \leq d$ and $1 \leq j \leq n-d$. This can be rewritten in matrix form as

$$\frac{1}{2i}\left[I - i\frac{\partial h}{\partial x}\right]\cdot(A) = \frac{\partial h}{\partial w}.$$

The formula for the L_j given in the theorem now follows. Since $H^{0,1}(M) = \overline{H^{1,0}(M)}$ the set

$$\{\overline{L}_1, \ldots, \overline{L}_{n-d}\}$$

forms a local basis for $H^{0,1}(M)$. ∎

If the graphing function h for M is independent of the variable x, then the local basis in Theorem 3 for $H^{1,0}(M)$ has the following simpler form:

$$L_j = \frac{\partial}{\partial w_j} + 2i\sum_{\ell=1}^{d}\frac{\partial h_\ell}{\partial w_j}\frac{\partial}{\partial z_\ell} \qquad 1 \leq j \leq n-d.$$

We give such a manifold a special name.

DEFINITION 1 *A CR submanifold of the form*

$$M = \{(x+iy, w) \in \mathbb{C}^d \times \mathbb{C}^{n-d}; y = h(w)\}$$

where h: $\mathbb{C}^{n-d} \to \mathbb{R}^d$ is smooth with $h(0) = 0$ and $Dh(0) = 0$ is called rigid.

In Chapter 12, we will give an intrinsic, coordinate-free description of rigidity due to Baouendi, Rothschild, and Treves.

7.3 Quadric submanifolds

By Theorem 2 in the previous section, a smooth, generic, CR submanifold of $\mathbb{C}^n$ has a locally defined graphing function h with no pure terms in its Taylor expansion up to a given order. In particular, the quadratic terms in the Taylor expansion of h contain no terms involving the x-coordinates. Therefore all the second-order information is contained in the term

$$q(w, \bar{w}) = \sum_{j,k=1}^{n-d} \frac{\partial h^2(0)}{\partial w_j \partial \bar{w}_k} w_j \bar{w}_k.$$

By replacing $\bar{w}$ by the independent variable $\eta \in \mathbb{C}^{n-d}$, we obtain a (vector valued) quadratic form.

DEFINITION 1 *A map q: $\mathbb{C}^m \times \mathbb{C}^m \to \mathbb{C}^d$ is a quadratic form if*

(i) *q is bilinear over $\mathbb{C}$*

(ii) *q is symmetric, i.e., $q(w, \eta) = q(\eta, w)$ for $w, \eta \in \mathbb{C}^m$*

(iii) *$\overline{q(w, \eta)} = q(\bar{w}, \bar{\eta})$ for $w, \eta \in \mathbb{C}^m$.*

DEFINITION 2 *A submanifold $M \subset \mathbb{C}^n$ defined by*

$$M = \{(x + iy, w) \in \mathbb{C}^d \times \mathbb{C}^{n-d}; y = q(w, \bar{w})\}$$

where q: $\mathbb{C}^{n-d} \times \mathbb{C}^{n-d} \to \mathbb{C}^d$ is a quadratic form is called a quadric submanifold of $\mathbb{C}^n$.

Requirements (ii) and (iii) of Definition 1 imply that $q(w, \bar{w})$ is a vector in $\mathbb{R}^d$. Hence, the quadric submanifold in Definition 2 is a well-defined submanifold of real dimension $2n - d$.

Quadric submanifolds are rigid since their graphing functions are independent of x. As we will see, the class of quadric submanifolds provides a class of easily studied examples. From the discussion at the beginning of this section, any generic CR submanifold can be approximated to third order at the origin by a quadric submanifold. Therefore, a quadric submanifold often serves as the model for a more general CR submanifold.

Another reason why quadric submanifolds are interesting is that each quadric submanifold has a group structure, which we now describe.

DEFINITION 3 *Let q: $\mathbb{C}^{n-d} \times \mathbb{C}^{n-d} \to \mathbb{C}^d$ be a quadratic form. For (z_1, w_1), $(z_2, w_2) \in \mathbb{C}^d \times \mathbb{C}^{n-d}$, define*

$$(z_1, w_1) \circ (z_2, w_2) = (z_1 + z_2 + 2iq(w_1, \bar{w}_2), w_1 + w_2).$$

LEMMA 1
The operation $\circ$ defines a group structure on $\mathbb{C}^n \times \mathbb{C}^n$ that restricts to a group structure on $M \times M$, where $M = \{y = q(w, \bar{w})\}$.

PROOF The operation $\circ$ is easily shown to be associative. The identity element is the origin. The inverse of the point (z, w) is the point

$$(z, w)^{-1} = (-z + 2iq(w, \bar{w}), -w).$$

It remains to show that $\circ$ restricts to a group structure on $M \times M$. If (z_1, w_1) and (z_2, w_2) belong to M, then Im $z_1 = q(w_1, \bar{w}_1)$ and Im $z_2 = q(w_2, \bar{w}_2)$. Therefore

$$\text{Im}\{z_1 + z_2 + 2iq(w_1, \bar{w}_2)\} = q(w_1, \bar{w}_1) + q(w_2, \bar{w}_2) + 2 \text{ Re } q(w_1, \bar{w}_2).$$

Using the properties of q from Definition 1, this can be rewritten as

$$\text{Im}\{z_1 + z_2 + 2iq(w_1, \bar{w}_2)\} = q(w_1 + w_2, \bar{w}_1 + \bar{w}_2).$$

This shows that $M \times M$ is closed under $\circ$. We leave it to the reader to show that if (z, w) belongs to M then $(z, w)^{-1}$ also belongs to M. The proof of the lemma is now complete. ▮

The lemma implies that a quadric submanifold is a *Lie group*, which means that the group operation $(\circ)$: $M \times M \to M$ is a smooth function. For $p_0 = (z_0, w_0) \in M$, define the map G_{p_0}: $M \to M$

$$\begin{aligned} G_{p_0}(z, w) &= (z, w) \circ (z_0, w_0) \\ &= (z + z_0 + 2iq(w, \bar{w}_0), w + w_0). \end{aligned}$$

$G_{p_0}(z, w)$ is a smooth function in both (z, w) and p_0. For fixed p_0, G_{p_0} is the restriction of a holomorphic map since q is complex linear as a function of the first factor.

From Theorem 3 in Section 7.2, the generators for $H^{1,0}(M)$ are

$$L_j = \frac{\partial}{\partial w_j} + 2i \sum_{\ell=1}^{d} \frac{\partial q_\ell}{\partial w_j} \frac{\partial}{\partial z_\ell} \qquad 1 \leq j \leq n - d$$

where we have written $q = (q_1, \ldots, q_d)$. Likewise, the generators for $H^{0,1}(M)$ are $\overline{L}_1, \ldots, \overline{L}_{n-d}$. These vector fields are globally defined since M is globally presented as the graph of q. Also note that $\partial/\partial x_1, \ldots, \partial/\partial x_d$ are the global generators for the totally real tangent bundle $X(M)$.

The vector fields $L_1, \ldots, L_{n-d}, \overline{L}_1, \ldots, \overline{L}_{n-d}$ and $\partial/\partial x_1, \ldots, \partial/\partial x_d$ have another important property. They are *invariant* under the group action $(\circ)$ for M. Let us explain what this means. The map $(z, w) \mapsto G_{p_0}(z, w)$ sends the origin, 0, to the point $p_0 = (z_0, w_0) \in M$. Therefore, $(G_{p_0})_*(0)$ is a complex linear map from $T_0(M)\otimes\mathbb{C}$ to $T_{p_0}(M)\otimes\mathbb{C}$ as explained in Part I (see Sections 2.3 and 3.1).

DEFINITION 4 *A vector field $L \in T^{\mathbb{C}}(M)$ is said to be (left) invariant for the group action $(\circ)$ for M if*

$$(G_p)_*(0)\{L_0\} = L_p \quad \textit{for each} \quad p \in M$$

where $G_p(z, w) = (z, w) \circ p$.

Since $G_{p_1} \circ G_{p_2} = G_{p_2 \circ p_1}$ and $(G_{p_1} \circ G_{p_2})_* = (G_{p_1})_* \circ (G_{p_2})_*$, an invariant vector field L also satisfies

$$(G_{p_1})_*(p_2)\{L_{p_2}\} = L_{p_2 \circ p_1} \quad \text{for} \quad p_1, p_2 \in M.$$

THEOREM 1
The vector fields

$$\begin{aligned} & L_j = \tfrac{\partial}{\partial w_j} + 2i \textstyle\sum_{\ell=1}^{d} \tfrac{\partial q_\ell}{\partial w_j} \tfrac{\partial}{\partial z_\ell} && 1 \le j \le n-d \\ & \overline{L}_j && 1 \le j \le n-d \\ \textit{and} \quad & \frac{\partial}{\partial x_j} && 1 \le j \le d \end{aligned}$$

are invariant for the group action $(\circ)$ defined on $M = \{y = q(w, \bar{w})\}$.

PROOF If G: $\mathbb{C}^n \to \mathbb{C}^n$ is a smooth map, then

$$G_*(0)\left\{\frac{\partial}{\partial \zeta_j}\right\} = \sum_{k=1}^{n} \frac{\partial G_k}{\partial \zeta_j}(0)\frac{\partial}{\partial \zeta_k} + \frac{\partial \overline{G}_k}{\partial \zeta_j}(0)\frac{\partial}{\partial \bar{\zeta}_k},$$

$$G_*(0)\left\{\frac{\partial}{\partial \bar{\zeta}_j}\right\} = \sum_{k=1}^{n} \frac{\partial G_k}{\partial \bar{\zeta}_j}(0)\frac{\partial}{\partial \zeta_k} + \overline{\left(\frac{\partial G_k}{\partial \zeta_j}(0)\right)}\frac{\partial}{\partial \bar{\zeta}_k}$$

where we have written $\zeta = (\zeta_1, \ldots, \zeta_n)$ as the coordinates for $\mathbb{C}^n$ and we have written the component functions of G as $G = (G_1, \ldots, G_n)$.

In our case, the map $G_p(\zeta) = G_p(z, w)$ is holomorphic in $\zeta = (z, w) \in \mathbb{C}^n$, and so $\partial G_p/\partial \bar{\zeta}_j = 0 = \partial(\overline{G}_p)/\partial \zeta_j$. Using the formula for the L_j and writing $\partial/\partial x_j = \partial/\partial z_j + \partial/\partial \bar{z}_j$, the proof of the theorem reduces to an easy calculation. ∎

Example 1

One of the most important examples of a quadric is the Heisenberg group. This is the real hypersurface in $\mathbb{C}^n$ defined by

$$M = \{(z, w) \in \mathbb{C} \times \mathbb{C}^{n-1};\ \text{Im } z = |w|^2\}.$$

Here, $q(w,\eta) = w \cdot \eta = \sum_{j=1}^{n-1} w_j \eta_j$ for $w = (w_1, \ldots, w_{n-1}) \in \mathbb{C}^{n-1}$ and $\eta = (\eta_1, \ldots, \eta_{n-1}) \in \mathbb{C}^{n-1}$. The group action is given by

$$(z^1, w^1) \circ (z^2, w^2) = (z^1 + z^2 + 2iw^1 \cdot \bar{w}^2, w^1 + w^2)$$

for $(z^1, w^1), (z^2, w^2) \in \mathbb{C} \times \mathbb{C}^{n-1}$.

The invariant generators for $H^{1,0}(M)$ are given by

$$L_j = 2i\overline{w}_j \frac{\partial}{\partial z} + \frac{\partial}{\partial w_j} \qquad 1 \leq j \leq n-1.$$

The invariant vector fields $\overline{L}_1, \ldots, \overline{L}_{n-1}$ generate $H^{0,1}(M)$, and the invariant vector field $\partial/\partial x$ (where $x = \text{Re } z$) generates $X(M)$. The vector fields $L_1, \ldots, L_{n-1}$ have interesting nonsolvability properties which will be discussed at the end of Part IV. □

We conclude this section on quadric surfaces by deriving a normal form for the quadric surfaces of real codimension two in $\mathbb{C}^4$. This normal form will be useful for exhibiting various types of CR extension phenomena in Part III.

Let (z_1, z_2, w_1, w_2) be the coordinates for $\mathbb{C}^4$ and write $z_1 = x_1 + iy_1$ and $z_2 = x_2 + iy_2$. A codimension two quadric in $\mathbb{C}^4$ has the form

$$M = \left\{ \begin{array}{l} y_1 = q_1(w, \bar{w}) \\ y_2 = q_2(w, \bar{w}) \end{array} \right\}$$

where q_1 and q_2 are scalar-valued quadratic forms on $\mathbb{C}^2$. We say that q_1 and q_2 are *linearly independent* over $\mathbb{R}$ if the restrictions of q_1 and q_2 to the set $\{(w, \bar{w}); w \in \mathbb{C}^2\}$ are linearly independent over $\mathbb{R}$ in the usual sense. From the properties of a quadratic form given in Definition 1, it is easy to see that q_1 and q_2 are linearly independent over $\mathbb{R}$ if and only if q_1 and q_2 as functions from $\mathbb{C}^2 \times \mathbb{C}^2 \to \mathbb{C}$ are linearly independent over $\mathbb{C}$.

THEOREM 2

Suppose M is a quadric codimension two submanifold of $\mathbb{C}^4$ defined by

$$M = \left\{ \begin{array}{l} y_1 = q_1(w, \bar{w}) \\ y_2 = q_2(w, \bar{w}) \end{array} \right\}.$$

(a) *If q_1 and q_2 are linearly dependent over $\mathbb{R}$, then there is a nonsingular complex linear change of coordinates in $\mathbb{C}^4$ so that in the new coordinates*

$$M = \left\{ \begin{array}{l} y_1 = q_1(w, \bar{w}) \\ y_2 = 0 \end{array} \right\}$$

where q_1 is a scalar-valued quadratic form.

(b) *If q_1 and q_2 are linearly independent over $\mathbb{R}$ then there is a nonsingular complex linear change of coordinates in $\mathbb{C}^4$ so that in the new coordinates, M has one of the following forms:*

$$\textit{(i)}\quad M=\left\{\begin{array}{l} y_1=|w_1|^2 \\ y_2=|w_2|^2 \end{array}\right\}$$

$$\textit{(ii)}\quad M=\left\{\begin{array}{l} y_1=|w_1|^2 \\ y_2=\ Re(w_1\bar{w}_2) \end{array}\right\}$$

$$\textit{(iii)}\quad M=\left\{\begin{array}{l} y_1=\ Re(w_1\bar{w}_2) \\ y_2=\ Im(w_1\bar{w}_2) \end{array}\right\}.$$

In Chapter 10, we will relate the quadratic forms q_1 and q_2 to the Levi form. Then we will have a more intrinsic way of deciding which of the above normal forms applies to a given quadric submanifold in $\mathbb{C}^4$.

Given a scalar-valued quadratic form q: $\mathbb{C}^2\times\mathbb{C}^2\to\mathbb{C}$, there is a 2×2 matrix with complex entries that represents q. There are complex numbers A, B, C, D with

$$q(w,\eta)=(w_1 \quad w_2)\begin{pmatrix} A & D \\ B & C \end{pmatrix}\begin{pmatrix}\eta_1 \\ \eta_2\end{pmatrix}.$$

Since $q(w,\bar{w})$ is real valued and since q is symmetric, A and C are both real and $B=\overline{D}$. That is, the matrix that represents q is Hermitian symmetric. Throughout the proof of this theorem, we identify q with its matrix.

The proof of part (b) will show that if q_1 or q_2 is positive or negative definite, then the normal form given in (i) can be arranged. The same is true if both q_1 and q_2 have a vanishing eigenvalue. If one of q_1 and q_2 has a vanishing eigenvalue and the other has eigenvalues of opposite sign, then the normal form given in (ii) can be arranged. If both q_1 and q_2 have eigenvalues of opposite sign, then the normal form given in (iii) can be arranged.

PROOF OF THEOREM 2 For part (a), if $q_2=\lambda q_1$ for some $\lambda\in\mathbb{R}$, then we make the following nonsingular complex linear change of variables

$$\begin{aligned}\hat{z}_1&=z_1\\ \hat{z}_2&=z_2-\lambda z_1\\ \hat{w}_1&=w_1\\ \hat{w}_2&=w_2.\end{aligned}$$

In the new coordinates

$$\begin{aligned}\hat{y}_1=\ \text{Im }\hat{z}_1&=\ \text{Im }z_1\\ &=q_1(w,\bar{w}) \quad \text{for} \quad (z,w)\in M\\ &=q_1(\hat{w},\hat{w})\end{aligned}$$

and

$$\begin{aligned}\hat{y}_2 = \text{ Im } \hat{z}_2 &= \text{ Im } z_2 - \lambda \text{ Im } z_1 \\ &= q_2(w, \bar{w}) - \lambda q_1(w, \bar{w}) \quad \text{for} \quad (z, w) \in M \\ &= 0.\end{aligned}$$

Therefore, if $\widehat{M}$ is the image of M under the map $(z, w) \mapsto (\hat{z}, \hat{w})$, then $\widehat{M}$ is defined by

$$\begin{aligned}\hat{y}_1 &= q_1(\hat{w}, \bar{\hat{w}}) \\ \hat{y}_2 &= 0.\end{aligned}$$

This completes the proof of part (a).

For part (b), we start by diagonalizing the matrix of q_1. This can be done by a unitary change of coordinates which affects only w_1 and w_2. After rescaling, there are three cases to consider:

1. $q_1(w, \bar{w}) = |w_1|^2 + |w_2|^2$ (q_1 is positive definite)
2. $q_1(w, \bar{w}) = |w_1|^2 - |w_2|^2$ (q_1 has eigenvalues of opposite sign)
3. $q_1(w, \bar{w}) = |w_1|^2$ (q_1 has one vanishing eigenvalue).

Note that case 1 also includes the case where q_1 is negative definite, for in this case, the change of variables $\hat{z} = -z$ and $\hat{w} = w$ will make the resulting q_1 positive definite.

We let

$$q_2(w, \bar{w}) = A|w_1|^2 + 2 \text{ Re}(\lambda w_1 \bar{w}_2) + C|w_2|^2$$

where A and C are real and λ is complex.

Case 1. (q_1 is positive definite). We first look for complex numbers a and b (with $b \neq 0$) and a real number t so that

$$q_2(w, \bar{w}) + t q_1(w, \bar{w}) = |a w_1 + b w_2|^2.$$

Expanding this equation, we see that a, b, and t must satisfy

$$\begin{aligned}a\bar{b} &= \lambda \\ |a|^2 &= A + t \\ |b|^2 &= C + t.\end{aligned}$$

If $\lambda = 0$, then we proceed as follows. We have $A > C$ or $C > A$; for if $A = C$ and $\lambda = 0$, then q_1 and q_2 are linearly dependent. By switching the roles of w_1 and w_2 if necessary, we assume $C > A$. Set $a = 0$. This forces $t = -A$. This in turn forces $|b|^2 = C - A > 0$. So we may let $b = \sqrt{C - A}$. With these choices, the above equations are satisfied with a nonzero b.

If $\lambda \neq 0$, then we must choose nonzero a and b. From the above equations, we have

$$A + t = |a|^2 = \frac{|\lambda|^2}{|b|^2} = \frac{|\lambda|^2}{C + t}.$$

This can be rearranged into the following quadratic equation in t

$$t^2 + (A + C)t + (AC - |\lambda|^2) = 0.$$

Its discriminant, $(A + C)^2 - 4(AC - |\lambda|^2) = (A - C)^2 + 4|\lambda|^2$, is positive because $\lambda \neq 0$. Therefore, the above quadratic equation has two distinct real roots. Let t be the larger root. Clearly $t + A > 0$ and $t + C > 0$ because $(t + A)(t + C) = |\lambda|^2 > 0$. Let θ be an argument of λ. The above three equations are satisfied by this choice of t and

$$\begin{aligned} a &= (t + A)^{1/2} e^{i\theta} \\ b &= (t + C)^{1/2} > 0. \end{aligned}$$

With this choice of a, b, t, we have

$$\begin{aligned} q_1(w, \bar{w}) &= |w_1|^2 + |w_2|^2 \\ q_2(w, \bar{w}) + tq_1(w, \bar{w}) &= |aw_1 + bw_2|^2. \end{aligned} \tag{1}$$

Define the following linear change of variables:

$$\begin{aligned} \hat{z}_1 &= z_1 & \hat{z}_2 &= z_2 + tz_1 \\ \hat{w}_1 &= w_1 & \hat{w}_2 &= aw_1 + bw_2. \end{aligned}$$

This is a nonsingular linear map since $b \neq 0$. We have

$$\begin{aligned} \text{Im } \hat{z}_1 = {} & \text{Im } z_1 \\ &= q_1(w, \bar{w}) \quad (\text{if } (z, w) \in M) \\ &= |w_1|^2 + |w_2|^2 \\ &= |\hat{w}_1|^2 + \left| \frac{1}{b}(\hat{w}_2 - a\hat{w}_1) \right|^2 \\ &= \alpha|\hat{w}_1|^2 + 2 \text{ Re}(\gamma \hat{w}_1 \bar{\hat{w}}_2) + \beta|\hat{w}_2|^2 \end{aligned}$$

where α and β are positive real numbers and γ is a complex number. Similarly

$$\begin{aligned} \text{Im } \hat{z}_2 = {} & \text{Im } z_2 + t \text{ Im } z_1 \\ &= q_2(w, \bar{w}) + tq_1(w, \bar{w}) \qquad (\text{if } (z, w) \in M) \\ &= |aw_1 + bw_2|^2 \qquad \text{by (1)} \\ &= |\hat{w}_2|^2. \end{aligned}$$

Therefore, if $\widehat{M}$ is the image of M under the linear map $(z, w) \longmapsto (\hat{z}, \hat{w})$, then in the new coordinates, $\widehat{M} = \{\text{Im } \hat{z} = \hat{q}(\hat{w}, \bar{\hat{w}})\}$ with $\hat{q} = (\hat{q}_1, \hat{q}_2)$ where

$$\begin{aligned}\hat{q}_1(\hat{w}, \bar{\hat{w}}) &= \alpha|\hat{w}_1|^2 + 2 \text{ Re}(\gamma \hat{w}_1 \bar{\hat{w}}_2) + \beta |\hat{w}_2|^2 \\ \hat{q}_2(\bar{w}, \bar{\hat{w}}) &= |\hat{w}_2|^2.\end{aligned}$$

Now we complete the square in $\hat{q}_1$. After dropping the $\wedge$, we obtain

$$q_1(w, \bar{w}) = |\alpha^{1/2} w_1 + \bar{\gamma}\alpha^{-1/2} w_2|^2 + (\beta - |\gamma|^2 \alpha^{-1})|w_2|^2$$

(recall that $\alpha > 0$). We make one further linear change of coordinates

$$\begin{aligned}\hat{z}_1 &= z_1 - (\beta - |\gamma|^2\alpha^{-1}) z_2, \quad \hat{z}_2 = z_2 \\ \hat{w}_1 &= \alpha^{1/2} w_1 + \bar{\gamma}\alpha^{-1/2} w_2, \quad \hat{w}_2 = w_2.\end{aligned}$$

This change of coordinates is nonsingular since $\alpha > 0$. Again, let $\widehat{M}$ be the image of M under the map $(z, w) \mapsto (\hat{z}, \hat{w})$. The defining equation for $\widehat{M}$ in the new coordinates is given by

$$\widehat{M} = \left\{ \begin{array}{l} \text{Im } \hat{z}_1 = |\hat{w}_1|^2 \\ \text{Im } \hat{z}_2 = |\hat{w}_2|^2 \end{array} \right\}$$

which is the normal form given in (i) of part (b) in the theorem.

Case 2. $q_1(w, \bar{w}) = |w_1|^2 - |w_2|^2$ (q_1 has eigenvalues of opposite sign). In this case, we make the following nonsingular linear change of coordinates:

$$\begin{aligned} z_1 &= \hat{z}_1 & z_2 &= \hat{z}_2 \\ w_1 &= \tfrac{1}{2}(\hat{w}_1 + \hat{w}_2) & w_2 &= \tfrac{1}{2}(\hat{w}_1 - \hat{w}_2). \end{aligned}$$

In the new coordinates, (the image of) M is given by (after dropping the $\wedge$)

$$M = \left\{ \begin{array}{ll} \text{Im } z_1 = & \text{Re}(w_1 \bar{w}_2) \\ \text{Im } z_2 = & A|w_1|^2 + 2 \text{ Re}(\lambda w_1 \bar{w}_2) + B|w_2|^2 \end{array} \right\}$$

for some choice of $A, B \in \mathbb{R}$, and $\lambda \in \mathbb{C}$. Let

$$\lambda = r + si \quad \text{and} \quad r, s \in \mathbb{R}.$$

After the change of variables $\hat{z}_1 = z_1$, $\hat{z}_2 = z_2 - 2rz_1$, $\hat{w}_1 = w_1$, $\hat{w}_2 = w_2$, M is given by (drop the $\wedge$) $\{\text{Im } z_1 = q_1(w, \bar{w}),\ \text{Im } z_2 = q_2(w, \bar{w})\}$ where

$$\begin{aligned} q_1(w, \bar{w}) &= \text{Re}(w_1 \bar{w}_2) \\ q_2(w, \bar{w}) &= A|w_1|^2 - 2s \text{ Im}(w_1 \bar{w}_2) + B|w_2|^2. \end{aligned}$$

In matrix form, we have

$$q_2(w, \bar{w}) = (\bar{w}_1, \bar{w}_2) \begin{pmatrix} A & -is \\ is & B \end{pmatrix} \begin{pmatrix} w_1 \\ w_2 \end{pmatrix}.$$

If the determinant $AB - s^2$ is positive, then the matrix of q_2 is positive or negative definite and so this falls under Case 1 above with the roles of q_1 and q_2 reversed. This leads to the normal form given in (i) of part (b). So we assume

$$AB - s^2 \leq 0.$$

We first show that we can force the coefficient of $|w_1|^2$ in q_2 to vanish by a change of variables of the form

$$\begin{aligned} z_1 &= \hat{z}_1 \qquad & z_2 &= \hat{z}_2 \\ w_1 &= \hat{w}_1 \qquad & w_2 &= \hat{w}_2 + it\hat{w}_1 \end{aligned}$$

for an appropriate $t \in \mathbb{R}$ to be chosen later. Such a change of variables preserves q_1. We obtain

$$\begin{aligned} q_2(w, \bar{w}) = (A + 2st + Bt^2)|\hat{w}_1|^2 - 2(s + Bt)\ \mathrm{Im}(\hat{w}_1\bar{\hat{w}}_2) \\ + B|\hat{w}_2|^2. \end{aligned}$$

There is a real root t to the quadratic equation $A + 2st + Bt^2 = 0$ because its discriminant $4(s^2 - AB)$ is nonnegative. With this choice of t, the coefficient of $|\hat{w}_1|^2$ vanishes and so we may assume (after dropping the $\wedge$)

$$\begin{aligned} q_1(w, \bar{w}) &= \ \mathrm{Re}(w_1\bar{w}_2) \\ q_2(w, \bar{w}) &= \alpha\ \mathrm{Im}(w_1\bar{w}_2) + \beta|w_2|^2 \end{aligned}$$

where α and β are real numbers.

If $\alpha = 0$ then β must be nonzero, for otherwise $q_2 \equiv 0$. After a rescale in the z_2-variable, M is in the normal form given in (ii) of part (b) with the roles of w_1 and w_2 and the roles of z_1 and z_2 reversed.

If $\alpha \neq 0$, then we can force the coefficient of $|w_2|^2$ to vanish by a change of variables of the form

$$\begin{aligned} z_1 &= \hat{z}_2 \qquad & z_2 &= \hat{z}_2 \\ w_1 &= \hat{w}_1 + it\hat{w}_2 \qquad & w_2 &= \hat{w}_2 \end{aligned}$$

where t is a real number. Again, any change of variables of this form preserves q_1. We have

$$q_2(w, \bar{w}) = \alpha\ \mathrm{Im}(\hat{w}_1\bar{\hat{w}}_2) + (\beta + \alpha t)|\hat{w}_2|^2.$$

Since $\alpha \neq 0$, we may let $t = -\beta\alpha^{-1}$, which forces the coefficient of $|\hat{w}_2|^2$ to vanish. After a rescale in the z_2-variable, M is now in the normal form given in (iii) of part (b).

Case 3. $q_1(w, \overline{w}) = |w_1|^2$ (q_1 has a vanishing eigenvalue). We let

$$q_2(w, \bar{w}) = A|w_1|^2 + 2\,\mathrm{Re}(\lambda w_1 \bar{w}_2) + B|w_2|^2$$

where A and B are real and λ is complex. We make the change of variables

$$\hat{z}_1 = z_1 \qquad \hat{z}_2 = z_2 - Az_1$$

$$\hat{w}_1 = w_1 \qquad \hat{w}_2 = w_2.$$

In the new variables, M is defined by

$$M = \left\{ \begin{array}{l} \mathrm{Im}\ \hat{z}_1 = |\hat{w}_1|^2 \\ \mathrm{Im}\ \hat{z}_2 = 2\,\mathrm{Re}(\lambda \hat{w}_1 \bar{\hat{w}}_2) + B|\hat{w}_2|^2 \end{array} \right\}.$$

Let $q_2(w, \bar{w}) = 2\,\mathrm{Re}(\lambda w_1 \bar{w}_2) + B|w_2|^2$. The matrix that represents q_2 is

$$\begin{pmatrix} 0 & \bar{\lambda} \\ \lambda & B \end{pmatrix}.$$

If $\lambda \neq 0$, then the determinant of this matrix is $-|\lambda|^2$ which is negative. Hence, the eigenvalues of the matrix of q_2 have opposite sign and this falls under Case 2 above with the roles of q_1 and q_2 reversed. If $\lambda = 0$ then, after a rescale, M has the normal form given in (i) of part (b). The proof of the theorem is now complete. ∎

7.4 Abstract CR manifolds

So far, we have been dealing with CR submanifolds of $\mathbb{C}^n$. In this section, we define the concept of an abstract CR manifold which requires no mention of an ambient $\mathbb{C}^n$ or complex manifold.

Let M be an abstract $\mathbb{C}^\infty$ manifold. As defined in Part I, $T^{\mathbb{C}}(M)$ denotes the complexified tangent bundle whose fiber at each point $p \in M$ is $T_p(M) \otimes \mathbb{C}$. If M is a CR submanifold of $\mathbb{C}^n$, then from Lemma 3 in Section 7.1,

(i) $H^{1,0}(M) \cap \overline{H^{1,0}(M)} = \{0\}$

(ii) $H^{1,0}(M)$ and $H^{0,1}(M)$ are involutive.

These two properties make no mention of a complex structure on $\mathbb{C}^n$ other than to define the space $H^{1,0}(M)$. Therefore, we define an abstract CR manifold to be a manifold together with a subbundle of $T^{\mathbb{C}}(M)$ which satisfies the above two properties.

DEFINITION 1 *Let M be a C^∞ manifold and suppose $\mathbb{L}$ is a subbundle of $T^{\mathbb{C}}(M)$. The pair $(M, \mathbb{L})$ is called (an abstract) CR manifold or CR structure if*

(a) $\mathbb{L}_p \cap \overline{\mathbb{L}}_p = \{0\}$ *for each* $p \in M$.

(b) $\mathbb{L}$ *is involutive, that is,* $[L_1, L_2]$ *belongs to* $\mathbb{L}$ *whenever* $L_1, L_2 \in \mathbb{L}$.

It is clear from the above discussion that if M is a CR submanifold of $\mathbb{C}^n$, then the pair $(M, \mathbb{L})$ with $\mathbb{L} = H^{1,0}(M)$ is a CR structure.

By analogy with the imbedded case, we call $\dim_{\mathbb{C}}\{T^{\mathbb{C}}(M)/\mathbb{L} \oplus \overline{\mathbb{L}}\}$ the *CR codimension* of $(M, \mathbb{L})$.

There is a complex structure map J defined on the real subbundle which generates $\mathbb{L}$ so that the eigenspaces of the extension of J to $\mathbb{L} \oplus \overline{\mathbb{L}}$ are $\mathbb{L}$ (for the eigenvalue $+i$) and $\overline{\mathbb{L}}$ (for the eigenvalue $-i$). This follows from Lemma 3 in Section 3.2.

In Section 4.3, we said that a pair $(M, \mathbb{L})$ is an almost complex structure if $\mathbb{L}$ is a subbundle of $T^{\mathbb{C}}(M)$ with $\mathbb{L} \oplus \overline{\mathbb{L}} = T^{\mathbb{C}}(M)$. In particular, $\mathbb{L} \cap \overline{\mathbb{L}} = \{0\}$. Thus, an involutive almost complex structure is an example of a CR structure.

As mentioned in Part I, the Newlander–Nirenberg theorem [NN] states that a manifold with an involutive almost complex structure is a complex manifold. Now since a complex manifold can be locally imbedded into $\mathbb{C}^n$ (by definition), this prompts the analogous question for CR manifolds: if $(M, \mathbb{L})$ is an abstract CR structure, then does there exist a locally defined diffeomorphism Φ: $M \to \mathbb{C}^n$ so that $\Phi(M)$ is a CR submanifold of $\mathbb{C}^n$ with $\Phi_*\{\mathbb{L}\} = H^{1,0}(\Phi\{M\})$? This last requirement for Φ implies that the CR structure for M (namely $\mathbb{L}$) is pushed forward to the CR structure for $\Phi\{M\}$ (namely $H^{1,0}(\Phi\{M\})$. The answer to this question is a qualified yes. If M is real analytic, then there is a real analytic imbedding, as we will show in Section 11.1. If M is only smooth, then the answer, in general, is no, as we will show in Section 11.2, where we present Nirenberg's counterexample. There are further conditions on a smooth CR structure that will guarantee a local imbedding and we will briefly discuss these in Chapter 12.

8

The Tangential Cauchy–Riemann Complex

For a CR submanifold of $\mathbb{C}^n$, there are two ways to define the tangential Cauchy–Riemann complex and both approaches appear in the literature. The first way is an extrinsic approach that uses the $\bar{\partial}$-complex of the ambient $\mathbb{C}^n$. The second way is an intrinsic approach that makes no use of the ambient $\mathbb{C}^n$ and therefore generalizes to abstract CR manifolds. In this chapter, we present both approaches. In the imbedded case, these approaches lead to different tangential Cauchy–Riemann complexes but in Section 8.3, we show they are isomorphic.

8.1 Extrinsic approach

Here, we assume the reader is familiar with the bundle of (p,q)-forms on $\mathbb{C}^n$, denoted $\Lambda^{p,q}T^*(\mathbb{C}^n)$. The space of smooth sections of $\Lambda^{p,q}T^*(\mathbb{C}^n)$ over an open set U in $\mathbb{C}^n$ is denoted $\mathcal{E}^{p,q}(U)$. Basic facts about the bundle of (p,q)-forms and the associated Cauchy–Riemann complex $\bar{\partial}$: $\mathcal{E}^{p,q} \to \mathcal{E}^{p,q+1}$ are given in Section 3.3. As also mentioned in Chapter 3, for each point $p_0 \in \mathbb{C}^n$, the Hermitian inner product on $\Lambda^{p,q}_{p_0}T^*(\mathbb{C}^n)$ is defined by declaring that the set $\{dz^I \wedge d\bar{z}^J;\ |I| = p,\ |J| = q,\ I, J \text{ increasing}\}$ is an orthonormal basis.

Let M be a smooth, generic, CR submanifold of $\mathbb{C}^n$ with real dimension $2n-d$. We define $\Lambda^{p,q}T^*(\mathbb{C}^n)|_M$ to be the restriction of the bundle $\Lambda^{p,q}T^*(\mathbb{C}^n)$ to M, that is, $\Lambda^{p,q}T^*(\mathbb{C}^n)|_M$ is the union of $\Lambda^{p,q}_{p_0}T^*(\mathbb{C}^n)$ where p_0 ranges over M. This space is different from the space $\{j^*\omega;\ \omega \in \Lambda^{p,q}T^*(\mathbb{C}^n)\}$, where j: $M \to \mathbb{C}^n$ is the inclusion map. A smooth section of $\Lambda^{p,q}T^*(\mathbb{C}^n)|_M$ is an element of the form

$$f = \sum_{\substack{|I|=p \\ |J|=q}} f_{IJ}\, dz^I \wedge d\bar{z}^J \in \mathcal{E}^{p,q}(\mathbb{C}^n)$$

whose coefficient functions, f_{IJ}, have been restricted to M.

For $0 \le p, q \le n$, define

$$I^{p,q} = \left\{ \begin{array}{l} \text{the ideal in } \Lambda^{p,q}T^*(\mathbb{C}^n) \text{ which is generated} \\ \text{by } \rho \text{ and } \bar{\partial}\rho \text{ where } \rho\colon \mathbb{C}^n \to \mathbb{R} \text{ is} \\ \text{any smooth function that vanishes on } M \end{array} \right\}.$$

Elements of $I^{p,q}$ are sums of forms of the type

$$\Phi_1\rho + \Phi_2 \wedge \bar{\partial}\rho, \quad \Phi_1 \in \Lambda^{p,q}T^*(\mathbb{C}^n), \quad \Phi_2 \in \Lambda^{p,q-1}T^*(\mathbb{C}^n).$$

If $\{\rho_1, \ldots, \rho_d\}$ is a local defining system for M, then $\{\rho_1, \ldots, \rho_d\}$ locally generates the ideal of all real-valued functions that vanish on M as shown in Lemma 3 of Section 2.2. Therefore, $I^{p,q}$ is the ideal in $\Lambda^{p,q}T^*(\mathbb{C}^n)$ that is locally generated by

$$\rho_1, \ldots, \rho_d, \bar{\partial}\rho_1, \ldots, \bar{\partial}\rho_d.$$

The restriction of $I^{p,q}$ to M, denoted $I^{p,q}|_M$, is the ideal in $\Lambda^{p,q}T^*(\mathbb{C}^n)|_M$ locally generated by $\bar{\partial}\rho_1, \ldots, \bar{\partial}\rho_d$. Since M is CR, the dimension of the fiber $I^{p,q}_{p_0}|_M$ is independent of the point $p_0 \in M$. Thus, $I^{p,q}|_M$ is a subbundle of $\Lambda^{p,q}T^*(\mathbb{C}^n)|_M$.

Let

$$\Lambda^{p,q}T^*(M) = \left\{ \begin{array}{l} \text{the orthogonal complement of } I^{p,q}|_M \text{ in} \\ \Lambda^{p,q}T^*(\mathbb{C}^n)|_M \end{array} \right\}.$$

Elements in $\Lambda^{p,q}_{p_0}T^*(M)$ for $p_0 \in M$ are orthogonal to the ideal in $\Lambda^{p,q}_{p_0}T^*(\mathbb{C}^n)$ generated by $\bar{\partial}\rho_1(p_0), \ldots, \bar{\partial}\rho_d(p_0)$. Let k be the number of linearly independent elements from $\{\bar{\partial}\rho_1(p_0), \ldots, \bar{\partial}\rho_d(p_0)\}$ (i.e., k is the CR codimension of M). Since M is CR, k is independent of the point $p_0 \in M$. Therefore, the dimension of $\Lambda^{p,q}_{p_0}T^*(M)$ is independent of the point $p_0 \in M$. Hence, the space $\Lambda^{p,q}T^*(M)$ is a subbundle of $\Lambda^{p,q}T^*(\mathbb{C}^n)|_M$. Note that $\Lambda^{p,q}T^*(M) = 0$ if either $p > n$ or $q > n - k$. If M is generic, then $k = d$ (the real codimension of M) by Lemma 4 in Section 7.1.

The space $\Lambda^{p,q}T^*(M)$ is *not* intrinsic to M, i.e., it is not a subspace of the exterior algebra generated by the complexified cotangent bundle of M. This is due to the fact that if $\rho : \mathbb{C}^n \mapsto \mathbb{R}$ vanishes on M, then $\bar{\partial}\rho = (1/2)(d\rho + iJ^*d\rho)$ is not orthogonal to the cotangent bundle of M due to the presence of $J^*d\rho$.

For $s \ge 0$, let

$$\Lambda^s_M = \Lambda^{s,0}T^*(M) \oplus \cdots \oplus \Lambda^{0,s}T^*(M)$$

where some of the summands on the right may vanish. The space Λ^s_M is *not* the same as the space $\Lambda^s T^*(M)$. The latter space is intrinsic to M whereas the former is not. As an example, let $M = \{(z, w) \in \mathbb{C}^2;\ \text{Im } z = 0\}$; then $\rho(z, w) = (2i)^{-1}(z - \bar{z})$. We have $\bar{\partial}\rho = -(i/2)d\bar{z}$ and so $\Lambda^{p,q}T^*(M)$ is the space of (p, q)-forms on M that are orthogonal to the ideal generated by

$d\bar{z}$. In particular, $\Lambda^{2,1}T^*(M)$ is generated by the form $dz \wedge dw \wedge d\bar{w}$ and $\Lambda^{1,2}T^*(M) = 0$. Therefore, $\Lambda^3_M = \Lambda^{2,1}T^*(M)$, whereas $\Lambda^3T^*(M)$ is the space generated by $dx \wedge dw \wedge d\bar{w}$ where $x =$ Re z. Note that $\{j^*\omega; \omega \in \Lambda^3_M\} = \Lambda^3T^*(M)$. More will be said about $j^*\{\Lambda^{p,q}T^*(M)\}$ in Section 8.3 where we discuss the relationship between the extrinsic and intrinsic tangential Cauchy–Riemann complexes.

For an open set $U \subset M$, the space of smooth sections of $\Lambda^{p,q}T^*(M)$ over U will be denoted $\mathcal{E}^{p,q}_M(U)$, and $\mathcal{D}^{p,q}_M(U)$ will denote the space of compactly supported elements in $\mathcal{E}^{p,q}_M(U)$. If the open set U is not essential for the discussion, then it will be omitted from the notation.

For $s \geq 0$, we let

$$\mathcal{E}^s_M(U) = \mathcal{E}^{s,0}_M(U) \oplus \cdots \oplus \mathcal{E}^{0,s}_M(U)$$

where some of the summands on the right may vanish. Again, note that $\mathcal{E}^s_M(U) \neq \mathcal{E}^s(U)$.

Let t_M: $\Lambda^{p,q}T^*(\mathbb{C}^n)|_M \to \Lambda^{p,q}T^*(M)$ be the orthogonal projection map. For a form $f \in \Lambda^{p,q}T^*(\mathbb{C}^n)|_M$, we often write f_{t_M} for $t_M(f)$ and call this the *tangential part* of f. If f is a smooth (p,q)-form on $\mathbb{C}^n$, then f_{t_M} is an element of $\mathcal{E}^{p,q}_M$. Conversely, any form $f \in \mathcal{E}^{p,q}_M(U)$ can be extended to an element $\tilde{f} \in \mathcal{E}^{p,q}(\widetilde{U})$, where $\widetilde{U}$ is an open set in $\mathbb{C}^n$ with $\widetilde{U} \cap M = U$. This is accomplished by writing

$$f = \sum_{\substack{|I|=p \\ |J|=q}} f_{IJ} dz^I \wedge d\bar{z}^J \quad \text{with} \quad f_{IJ} \in \mathcal{E}(U)$$

and extending each coefficient function f_{IJ} to an open subset $\widetilde{U}$ of $\mathbb{C}^n$.

We now define the (extrinsic) tangential Cauchy–Riemann complex.

DEFINITION 1 *For an open set $U \subset M$, the tangential Cauchy–Riemann complex $\bar{\partial}_M: \mathcal{E}^{p,q}_M(U) \to \mathcal{E}^{p,q+1}_M(U)$ is defined as follows. For $f \in \mathcal{E}^{p,q}_M(U)$, let $\widetilde{U}$ be an open set in $\mathbb{C}^n$ with $\widetilde{U} \cap M = U$ and let $\tilde{f} \in \mathcal{E}^{p,q}(\widetilde{U})$ with $\tilde{f}_{t_M} = f$ on $\widetilde{U} \cap M = U$. Then*

$$\bar{\partial}_M f = (\bar{\partial}\tilde{f})_{t_M}.$$

The form $\bar{\partial}_M f$ is calculated by extending f ambiently to an open set in $\mathbb{C}^n$, then applying $\bar{\partial}$ and taking the tangential part of the result. Since there are many possible ambient extensions of a given element of $\mathcal{E}^{p,q}_M$, we must show that the definition of $\bar{\partial}_M$ is independent of the ambient extension.

LEMMA 1

$\bar{\partial}_M$ is well defined, that is, if $\tilde{f}_1$ and $\tilde{f}_2$ are elements in $\mathcal{E}^{p,q}(\widetilde{U})$ with $(\tilde{f}_1)_{t_M} = (\tilde{f}_2)_{t_M}$ on $M \cap \widetilde{U}$, then $(\bar{\partial}\tilde{f}_1)_{t_M} = (\bar{\partial}\tilde{f}_2)_{t_M}$ on $M \cap \widetilde{U}$.

PROOF Note that $(\tilde{f}_1 - \tilde{f}_2)$ is an element of $I^{p,q}$. Therefore, it suffices to show that $\bar{\partial}$ maps smooth sections of $I^{p,q}$ to $I^{p,q+1}$. If $\alpha \in \mathcal{E}^{p,q}(\widetilde{U})$ and $\beta \in \mathcal{E}^{p,q-1}(\widetilde{U})$ and if $\rho\colon \widetilde{U} \to \mathbb{R}$ vanishes on $M \cap \widetilde{U}$, then

$$\bar{\partial}(\alpha\rho + \beta \wedge \bar{\partial}\rho) = (\bar{\partial}\alpha)\rho + (\alpha + \bar{\partial}\beta) \wedge \bar{\partial}\rho.$$

The right side is clearly an element of $I^{p,q+1}$. ∎

As already mentioned, the spaces $\Lambda^{p,q}T^*(M)$ are not intrinsic to M. Therefore, the spaces $\mathcal{E}_M^{p,q}$ and the resulting tangential Cauchy–Riemann operator are not intrinsic to M. For this reason, we refer to the above-defined tangential Cauchy–Riemann complex as being *extrinsically defined.*

It is useful to have a procedure for computing $\bar{\partial}_M$. We shall specialize to the case of a real hypersurface.

LEMMA 2
Suppose $M = \{z \in \mathbb{C}^n; \rho(z) = 0\}$ is a real hypersurface in $\mathbb{C}^n$ where $\rho\colon \mathbb{C}^n \to \mathbb{R}$ is smooth with $|d\rho| = 1$ on M. Let $\mathbb{N} = 4(\partial\rho/\partial z) \cdot \partial/\partial\bar{z} = 4\sum_{j=1}^n (\partial\rho/\partial z_j)(\partial/\partial\bar{z}_j)$. Then

$$\phi_{t_M} = \mathbb{N} \lrcorner (\bar{\partial}\rho \wedge \phi) \quad \textit{for} \quad \phi \in \Lambda^{p,q}T^*(\mathbb{C}^n)|_M$$

and

$$\bar{\partial}_M f = \mathbb{N} \lrcorner (\bar{\partial}\rho \wedge \bar{\partial}\tilde{f}) \quad \textit{for} \quad f \in \mathcal{E}_M^{p,q}$$

where $\tilde{f}$ is any ambiently defined smooth (p,q)-form with $\tilde{f}_{t_M} = f$ on M.

Recall that $\lrcorner$ denotes the contraction operator of a vector with a form (see Section 1.5). The hypothesis that $|d\rho| = 1$ on M can easily be arranged by replacing ρ by $\rho/|d\rho|$.

PROOF Since $|d\rho| = 1$, the vector field $\mathbb{N}$ is dual to the form $\bar{\partial}\rho$. From Section 1.5, if $\phi \in \Lambda^{p,q}T^*(\mathbb{C}^n)$ and $\psi \in \Lambda^{p,q-1}T^*(\mathbb{C}^n)$, we have

$$(\mathbb{N} \lrcorner \phi) \cdot \psi = \phi \cdot (\bar{\partial}\rho \wedge \psi)$$

where $(\cdot)$ is the Hermitian inner product on $\Lambda^*T^*(\mathbb{C}^n)$. Therefore

$$\mathbb{N} \lrcorner \phi = 0 \quad \text{on} \quad M \quad \text{if and only if} \quad \phi \in \Lambda^{p,q}T^*(M).$$

From the product rule for $\lrcorner$ (see Lemma 1 in Section 1.5), we have

$$\mathbb{N} \lrcorner (\bar{\partial}\rho \wedge \phi) = (\mathbb{N} \lrcorner \bar{\partial}\rho)\phi - \bar{\partial}\rho \wedge (\mathbb{N} \lrcorner \phi).$$

Since $\mathbb{N} \lrcorner \bar{\partial}\rho = |d\rho|^2 = 1$ on M, this becomes

$$\phi = \mathbb{N} \lrcorner (\bar{\partial}\rho \wedge \phi) + \bar{\partial}\rho \wedge (\mathbb{N} \lrcorner \phi). \tag{1}$$

Now, $\mathbb{N}\lrcorner(\mathbb{N}\lrcorner) \equiv 0$ and so $\mathbb{N}\lrcorner(\bar{\partial}\rho \wedge \phi)|_M$ is an element of $\Lambda^{p,q}T^*(M)$. The form $\bar{\partial}\rho \wedge (\mathbb{N}\lrcorner\phi)$ is an element of $I^{p,q}$ (the ideal generated by ρ and $\bar{\partial}\rho$). Therefore, equation (1) provides an orthogonal decomposition of an element ϕ of $\Lambda^{p,q}T^*(\mathbb{C}^n)|_M$ into its tangential part $\phi_{t_M} \in \Lambda^{p,q}T^*(M)$ and the component of ϕ in $I^{p,q}|_M$. In particular, $\phi_{t_M} = \mathbb{N}\lrcorner(\bar{\partial}\rho \wedge \phi)$, as claimed. The formula for $\bar{\partial}_M f$ follows from the expression for ϕ_{t_M} and the definition of $\bar{\partial}_M$. ∎

The term $\mathbb{N}\lrcorner\phi$ is called the normal component of ϕ and it is denoted by ϕ_{n_M}. Equation (1) then reads

$$\phi = \phi_{t_M} + \bar{\partial}\rho \wedge \phi_{n_M} \quad \text{on} \quad M.$$

This equation provides an orthogonal decomposition of ϕ into its tangential and normal components. If ϕ is an ambiently defined (p, q)-form on $\mathbb{C}^n$, then both ϕ_{t_M} and ϕ_{n_M} are ambiently defined because ρ and hence $\mathbb{N}$ are ambiently defined. The equation $\phi = \phi_{t_M} + \bar{\partial}\rho \wedge \phi_{n_M}$ also holds ambiently provided $|d\rho| = 1$ ambiently. This can be arranged, for example, if ρ is the signed distance function to M.

At this point, the reader may wonder whether or not $j^* \circ \bar{\partial} = \bar{\partial}_M \circ j^*$ where $j: M \to \mathbb{C}^n$ is the inclusion map. However, the right side of this equation does not make sense because the domain of the tangential Cauchy–Riemann operator is $\mathcal{E}_M^{p,q}$ which is not contained in $\Lambda^*T^*(M)$, which is the range of j^*. The right side of this equation does make sense if the tangential Cauchy–Riemann operator is defined intrinsically (see the next section). In Section 9.2, we shall discuss CR maps (such as j) and the validity of commuting their pull backs with the tangential Cauchy–Riemann operator.

The following lemma follows easily from the analogous properties of $\bar{\partial}$ (see Section 3.3).

LEMMA 3

Suppose M is a smooth CR submanifold of $\mathbb{C}^n$.

(a) $\bar{\partial}_M(f \wedge g) = \bar{\partial}_M f \wedge g + (-1)^{p+q} f \wedge \bar{\partial}_M g$ *for* $f \in \mathcal{E}_M^{p,q}$ *and* $g \in \mathcal{E}_M^{r,s}$.

(b) $\bar{\partial}_M \circ \bar{\partial}_M = 0$.

From part (b), if $\bar{\partial}_M f = g$, then $\bar{\partial}_M g = 0$. An important question is to ask whether or not the converse holds: if $\bar{\partial}_M g = 0$, then does there exist a form f with $\bar{\partial}_M f = g$? This solvability question for the tangential Cauchy–Riemann operator will be discussed in Part IV.

It is useful to interpret the equation $\bar{\partial}_M f = g$ in terms of currents (see Chapter 6 for basic facts about currents). If M is a smooth, oriented submanifold of $\mathbb{C}^n$ of real dimension $2n - d$ $(1 \le d \le n)$, then the current "integration over M" is defined by

$$\langle [M], \phi \rangle_{\mathbb{C}^n} = \int_M \phi \quad \text{for} \quad \phi \in \mathcal{D}^{2n-d}(\mathbb{C}^n).$$

$[M]$ is a current of degree d and therefore it splits up into its various bidegrees

$$[M] = [M]^{d,0} + \cdots + [M]^{0,d}.$$

In particular

$$\langle [M]^{0,d}, \phi \rangle_{\mathbb{C}^n} = \int_M \phi^{n,n-d} \quad \text{for} \quad \phi \in \mathcal{D}^{2n-d}(\mathbb{C}^n)$$

where $\phi^{n,n-d}$ indicates the piece of ϕ of bidegree $(n, n-d)$. By Stokes' theorem, we have $d[M] = 0$. Since $\bar{\partial}[M]^{0,d}$ is the piece of $d[M]$ of bidegree $(0, d+1)$, we have

$$\bar{\partial}[M]^{0,d} = 0.$$

LEMMA 4
Let M be an oriented, generic, CR submanifold of $\mathbb{C}^n$ of real dimension $2n - d$, $1 \leq d \leq n$. Suppose $f \in \mathcal{E}^{p,q}(\mathbb{C}^n)$. Then $f_{t_M} = 0$ on M if and only if $[M]^{0,d} \wedge f = 0$ as a current on $\mathbb{C}^n$.

PROOF By a partition of unity argument, it suffices to prove the lemma for forms f with support in a $\mathbb{C}^n$-neighborhood U of a given point $p_0 \in M$. From Section 6.1, the current $[M]$ is given by $\mu_M \alpha d\rho_1 \wedge \ldots \wedge d\rho_d$ where $\{\rho_1, \ldots, \rho_d\}$ is a local defining system for M near p_0 and $\alpha = |d\rho_1 \wedge \ldots \wedge d\rho_d|^{-1}$. Here, μ_M is the distribution given by Hausdorff measure on M (see Section 5.1). The piece of bidegree $(0, d)$ of this current is

$$[M]^{0,d} = \mu_M \alpha \bar{\partial}\rho_1 \wedge \ldots \wedge \bar{\partial}\rho_d.$$

Thus for $f \in \mathcal{D}^{p,q}(U)$, $[M]^{0,d} \wedge f = 0$ if and only if

$$\bar{\partial}\rho_1 \wedge \ldots \wedge \bar{\partial}\rho_d \wedge f = 0 \quad \text{on } M \cap U.$$

Since M is generic, we have $\bar{\partial}\rho_1 \wedge \ldots \wedge \bar{\partial}\rho_d \neq 0$ and

$$\bar{\partial}\rho_1 \wedge \ldots \wedge \bar{\partial}\rho_d \wedge f = \bar{\partial}\rho_1 \wedge \ldots \wedge \bar{\partial}\rho_d \wedge f_{t_M}.$$

It follows that the equation $[M]^{0,d} \wedge f = 0$ is equivalent to $f_{t_M} = 0$ on M. ∎

Suppose f is an element of $\mathcal{E}_M^{p,q}$. Technically, the current $[M]^{0,d} \wedge f$ is not well defined because $[M]^{0,d}$ is a current on $\mathbb{C}^n$ and f is not ambiently defined on $\mathbb{C}^n$. However, we can define $[M]^{0,d} \wedge f$ as $[M]^{0,d} \wedge \tilde{f}$ where $\tilde{f}$ is any smooth, ambiently defined (p, q)-form with $\tilde{f}_{t_M} = f$. Lemma 4 implies that this definition is independent of the extension $\tilde{f}$. With this in mind, we state and prove the following.

LEMMA 5

Suppose M is an oriented, CR, generic submanifold of $\mathbb{C}^n$ of real dimension $2n-d$. Let $f \in \mathcal{E}_M^{p,q}$ and $g \in \mathcal{E}_M^{p,q+1}$. The equation $\bar\partial_M f = g$ on M is equivalent to the current equation

$$\bar\partial(f \wedge [M]^{0,d}) = g \wedge [M]^{0,d} \quad \text{on} \quad \mathbb{C}^n.$$

PROOF From Lemma 4 and the fact that $\bar\partial[M]^{0,d} = 0$, we have

$$\begin{aligned}\bar\partial(f \wedge [M]^{0,d}) &= \bar\partial f \wedge [M]^{0,d} \\ &= (\bar\partial f)_{t_M} \wedge [M]^{0,d}.\end{aligned}$$

Therefore, $\bar\partial_M f = g$ if and only if $\bar\partial(f \wedge [M]^{0,d}) = g \wedge [M]^{0,d}$, as desired. ∎

Both the $\bar\partial$ and d operators satisfy an integration by parts formula. For $f \in \mathcal{E}^{p,q}(\mathbb{C}^n)$ and $g \in \mathcal{D}^{n-p,n-q-1}(\mathbb{C}^n)$, we have

$$\langle \bar\partial f, g\rangle_{\mathbb{C}^n} = (-1)^{p+q+1}\langle f, \bar\partial g\rangle_{\mathbb{C}^n}.$$

The same formula holds (by definition) if f is a current of bidegree (p,q). From this equation, we will derive an integration by parts formula for $\bar\partial_M$. Recall that the current pairing on M is denoted by $\langle\ ,\ \rangle_M$. We extend this current pairing to the spaces $\mathcal{D}_M^r$ (which are not intrinsic to M) by setting

$$\begin{aligned}\langle f, g\rangle_M &= \int_M f \wedge g \\ &= \langle [M] \wedge f, g\rangle_{\mathbb{C}^n}.\end{aligned}$$

for $f \in \mathcal{E}_M^r(M)$ and $g \in \mathcal{D}_M^{2n-d-r}(M)$. For a generic manifold M, $\Lambda^{p,q}T^*(M) = 0$ provided either $p > n$ or $q > n-d$. Therefore, if $f \in \mathcal{E}_M^{p,q}$ and $g \in \mathcal{D}_M^{2n-d-(p+q)}$, then

$$\langle f, g\rangle_M = \langle f, g^{n-p,n-q-d}\rangle_M$$

where $g^{n-p,n-q-d}$ is the piece of g of bidegree $(n-p, n-q-d)$. If $f \in \mathcal{E}_M^{p,q}$ and $g \in \mathcal{D}_M^{n-p,n-q-d}$, then $f \wedge g$ has bidegree $(n, n-d)$ and so

$$\begin{aligned}\langle f, g\rangle_M &= \int_M f \wedge g \\ &= \langle [M]^{0,d} \wedge f, g\rangle_{\mathbb{C}^n}.\end{aligned}$$

LEMMA 6

Suppose M is an oriented, CR, generic submanifold of $\mathbb{C}^n$ with real dimension $2n-d$. Let $f \in \mathcal{E}_M^{p,q}(M)$ and $g \in \mathcal{D}_M^{n-p,n-q-d-1}(M)$. Then

$$\langle \bar\partial_M f, g\rangle_M = (-1)^{p+q+1}\langle f, \bar\partial_M g\rangle_M.$$

PROOF From the discussion preceding the statement of Lemma 6, we have

$$\langle \bar{\partial}_M f, g \rangle_M = \langle [M]^{0,d}, (\bar{\partial} \tilde{f})_{t_M} \wedge \tilde{g} \rangle_{\mathbb{C}^n}$$

where $\tilde{f} \in \mathcal{E}^{p,q}(\mathbb{C}^n)$ and $\tilde{g} \in \mathcal{D}^{n-p,n-q-d-1}(\mathbb{C}^n)$ are any ambient extensions of f and g. We have

$$\begin{aligned}
\langle \bar{\partial}_M f, g \rangle_M &= \langle [M]^{0,d} \wedge (\bar{\partial} \tilde{f})_{t_M}, \tilde{g} \rangle_{\mathbb{C}^n} \\
&= \langle [M]^{0,d} \wedge \bar{\partial} \tilde{f}, \tilde{g} \rangle_{\mathbb{C}^n} \quad \text{(by Lemma 4)} \\
&= (-1)^d \langle \bar{\partial} \{ [M]^{0,d} \wedge \tilde{f} \}, \tilde{g} \rangle_{\mathbb{C}^n} \quad (\text{since } \bar{\partial}[M]^{0,d} \equiv 0) \\
&= (-1)^{p+q+1} \langle [M]^{0,d} \wedge \tilde{f}, \bar{\partial} \tilde{g} \rangle_{\mathbb{C}^n}
\end{aligned}$$

where the last equation follows from the definition of $\bar{\partial}$ applied to a current. By Lemma 4, we have $[M]^{0,d} \wedge \bar{\partial} \tilde{g} = [M]^{0,d} \wedge (\bar{\partial} \tilde{g})_{t_M}$. Therefore

$$\begin{aligned}
\langle \bar{\partial}_M f, g \rangle_M &= (-1)^{p+q+1} \langle [M]^{0,d}, \tilde{f} \wedge (\bar{\partial} \tilde{g})_{t_M} \rangle_{\mathbb{C}^n} \\
&= (-1)^{p+q+1} \langle f, \bar{\partial}_M g \rangle_M
\end{aligned}$$

as desired. ∎

This integration by parts formula allows us to extend the definition of $\bar{\partial}_M$ to currents on M. The space of currents on M of bidimension (p, q) is the dual of the space $\mathcal{D}_M^{p,q}$ and it is denoted by $\{\mathcal{D}_M^{p,q}\}'$. By adapting the proof of Lemma 1 in Section 6.1 to this context, the reader can easily show that $\{\mathcal{D}_M^{p,q}\}'$ is isomorphic to $\mathcal{D}_M'^{n-p,n-q-d}$, which is the space of currents of bidegree $(n-p, n-q-d)$ on M. An element of $\mathcal{D}_M'^{n-p,n-q-d}$ is a form of bidegree $(n-p, n-q-d)$ on M with distribution coefficients.

DEFINITION 2 *Suppose M is an oriented, CR, generic submanifold of $\mathbb{C}^n$ with real dimension $2n - d$. Let $T \in \mathcal{D}_M'^{p,q}$, then $\bar{\partial}_M T \in \mathcal{D}_M'^{p,q+1}$ is the current defined by*

$$\langle \bar{\partial}_M T, g \rangle_M = (-1)^{p+q+1} \langle T, \bar{\partial}_M g \rangle_M, \quad \textit{for} \quad g \in \mathcal{D}_M^{n-p,n-q-d-1}.$$

From Lemma 3, we have $\bar{\partial}_M^2(T) = 0$ for $T \in \mathcal{D}_M'^{p,q}$. In addition, Lemma 5 holds for currents, although we should say a word about the definition of $[M]^{0,d} \wedge T$ when T is an element of $\mathcal{D}_M'^{p,q}$. It suffices by a partition of unity argument to define $[M]^{0,d} \wedge T$ locally. Let $\rho_1, \ldots, \rho_d$ be a local defining system for M. Then, $[M]^{0,d} = \mu_M \overline{\partial} \rho_1 \wedge \ldots \wedge \overline{\partial} \rho_d$ where μ_M is Hausdorff measure on M. A typical element of $\mathcal{D}_M'^{p,q}$ is $T = T_1 \phi$ where T_1 is a distribution on M and where $\phi \in \mathcal{E}_M^{p,q}$. There is an element $\Phi \in \mathcal{E}^{p,q}(\mathbb{C}^n)$ with $\Phi_{t_M} = \phi$ on M. To define $[M]^{0,d} \wedge T$, it suffices to define the distribution $\mu_M \cdot T_1$. By using a (smooth) local coordinate system, we may assume $M = \{(z, w) \in \mathbb{C}^d \times \mathbb{C}^{n-d}; \text{Im} z = 0\}$. In these coordinates, we have $\mu_M = \delta_0(y)$, where $y = \text{Im} z$. A distribution T_1

on M acts in the variables $x = \text{Re}z$ and w. Therefore, we define the distribution $(\mu_M \cdot T_1)(x, y, w) = \delta_0(y) \otimes T_1(x, w)$ where $\otimes$ denotes the tensor product of distributions (see Section 5.2).

The proof of Lemma 5 for currents now proceeds by approximating a current by a sequence of smooth forms and then using Lemma 5 for smooth forms.

If M is not generic, then the CR codimension of M is less than d. In this case, the reader can show

$$[M]^{0,d} = [M]^{1,d-1} = \cdots = [M]^{d-k-1,k+1} = 0$$

where k is the CR codimension of M. Lemmas 4 through 6 hold with $[M]^{0,d}$ replaced by $[M]^{d-k,k}$.

8.2 Intrinsic approach to $\bar{\partial}_M$

Our treatment of the intrinsically defined tangential Cauchy–Riemann complex is similar to that in [PW]. We assume that $(M, \mathbb{L})$ is an abstract CR structure (see Definition 1 in Section 7.4). Therefore, $\mathbb{L}$ is an involutive subbundle of $T^{\mathbb{C}}(M)$ and $\mathbb{L} \cap \overline{\mathbb{L}} = \{0\}$. It will be necessary to choose a complementary subbundle to $\mathbb{L} \oplus \overline{\mathbb{L}}$. In order to do this, we assume that M comes equipped with a Hermitian metric for $T^{\mathbb{C}}(M)$ so that $\mathbb{L}$ is orthogonal to $\overline{\mathbb{L}}$. If M is a submanifold of $\mathbb{C}^n$, then the natural metric to use is the restriction to $T^{\mathbb{C}}(M)$ of the usual Hermitian metric on $T^{\mathbb{C}}(\mathbb{C}^n)$. If M is an abstract manifold, then a metric can be constructed locally by declaring a local basis of vector fields to be orthonormal. This metric can be extended globally by a partition of unity.

For each point $p_0 \in M$, we let X_{p_0} be the orthogonal complement of $\mathbb{L}_{p_0} \oplus \overline{\mathbb{L}}_{p_0}$ in $T_{p_0}(M) \otimes \mathbb{C}$. Clearly, the spaces $\{X_{p_0}; p_0 \in M\}$ fit together smoothly (since the $\mathbb{L}_{p_0} \oplus \overline{\mathbb{L}}_{p_0}$ do) and so the space

$$X(M) = \bigcup_{p_0 \in M} X_{p_0}$$

forms a subbundle of $T^{\mathbb{C}}(M)$.

If M is a CR submanifold of $\mathbb{C}^n$, then $\mathbb{L} = H^{1,0}(M)$ and $\overline{\mathbb{L}} = H^{0,1}(M)$. In this case, $X(M)$ is the totally real part of the tangent bundle.

Define the subbundles

$$T^{0,1}(M) = \overline{\mathbb{L}}$$
$$T^{1,0}(M) = \mathbb{L} \oplus X(M).$$

We emphasize that $T^{1,0}(M)$ is *not* analogous to $H^{1,0}(M)$ for an imbedded CR manifold (unless $X(M) = \{0\}$).

The dual of each of these spaces is denoted $T^{*^{0,1}}(M)$ and $T^{*^{1,0}}(M)$, respectively. Forms in $T^{*^{0,1}}(M)$ annihilate vectors in $T^{1,0}(M)$ and forms in $T^{*^{1,0}}(M)$ annihilate vectors in $T^{0,1}(M)$ (by the definition of dual).

Define the bundle

$$\Lambda^{p,q}T^*(M) = \Lambda^p(T^{*^{1,0}}(M))\widehat{\otimes}\Lambda^q(T^{*^{0,1}}(M)).$$

This is called the space of (p,q)-forms on M. Unlike the extrinsic approach, these spaces are intrinsic to M (in the abstract case, they cannot be anything but intrinsic since there is no ambient space). Let $m = \dim_{\mathbb{C}} \mathbb{L}$ and $d = \dim_{\mathbb{C}} X(M)$. Recall that d is the CR codimension of M. If $p > m+d$ or $q > m$, then $\Lambda^{p,q}T^*(M) = 0$.

The pointwise metric on $T^{\mathbb{C}}(M)$ induces a pointwise dual metric on $T^{*^{\mathbb{C}}}(M)$ in the usual way. Let $\phi_1, \dots, \phi_{m+d}$ be an orthonormal basis for $T^{*^{1,0}}(M)$ and let $\psi_1, \dots, \psi_m$ be an orthonormal basis for $T^{*^{0,1}}(M)$. The metric for $T^{*^{\mathbb{C}}}(M)$ extends to a metric on $\Lambda^{p,q}T^*(M)$ by declaring that the set

$$\{\phi^I \wedge \psi^J;\ |I| = p,\ |J| = q,\ I,\ J \text{ are increasing multiindices}\}$$

is an orthonormal basis. We also declare that $\Lambda^{p,q}T^*(M)$ is orthogonal to $\Lambda^{r,s}T^*(M)$ if either $p \neq r$ or $q \neq s$. We have the following orthogonal decomposition

$$\Lambda^r T^{*^{\mathbb{C}}}(M) = \Lambda^{r,0}T^*(M) \oplus \cdots \oplus \Lambda^{0,r}T^*(M),$$

with the understanding that some of these summands vanish if $r > m$. Let

$$\pi_M^{p,q}\colon \Lambda^r T^{*^{\mathbb{C}}}(M) \longrightarrow \Lambda^{p,q}T^*(M) \quad \text{for} \quad p+q = r$$

be the natural projection map.

The space of smooth r-forms on an open set $U \subset M$ is denoted by $\mathcal{E}_M^r(U)$. The space of smooth sections of $\Lambda^{p,q}T^*(M)$ over U is denoted by $\mathcal{E}_M^{p,q}(U)$, and $\mathcal{D}_M^{p,q}(U)$ is the space of compactly supported elements of $\mathcal{E}_M^{p,q}(U)$. The "U" may be omitted from the notation if it is unimportant for the discussion at hand.

The intrinsic definition of the tangential Cauchy–Riemann operator can now be given in terms of the exterior derivative $d_M\colon \mathcal{E}_M^r \to \mathcal{E}_M^{r+1}$.

DEFINITION 1 *The tangential Cauchy–Riemann operator* $\bar{\partial}_M\colon \mathcal{E}_M^{p,q} \to \mathcal{E}_M^{p,q+1}$ *is defined by* $\bar{\partial}_M = \pi_M^{p,q+1} \circ d_M$.

This definition of $\bar{\partial}_M$ is analogous to the definition of $\bar{\partial}$ on $\mathbb{C}^n$ (or any other complex manifold).

We will show that $\bar{\partial}_M\colon \mathcal{E}_M^{p,q} \to \mathcal{E}_M^{p,q+1}$ is a complex, i.e., $\bar{\partial}_M \circ \bar{\partial}_M = 0$. This will follow from the equation $d_M \circ d_M = 0$ and type considerations. First, we need a preliminary result.

LEMMA 1
If M is a CR manifold, then

$$d_M\{\mathcal{E}_M^{p,q}\} \subset \mathcal{E}_M^{p+2,q-1} \oplus \mathcal{E}_M^{p+1,q} \oplus \mathcal{E}_M^{p,q+1}.$$

Conceivably, the exterior derivative of a (p,q)-form, ϕ, might be a sum of forms of various bidegrees. The point is that the only possible nonvanishing components of $d_M\phi$ have bidegrees $(p+2,q-1)$, $(p+1,q)$ and $(p,q+1)$.

PROOF The case $p = 1$, $q = 0$ will be handled first. We must show $\pi^{0,2}(d_M\phi) = 0$ if $\phi \in \mathcal{E}_M^{1,0}$. This is equivalent to showing

$$\langle d_M\phi, \overline{L}_1 \wedge \overline{L}_2\rangle = 0 \quad \text{for all} \quad \overline{L}_1, \overline{L}_2 \in \overline{\mathbb{L}} = T^{0,1}(M).$$

where $\langle\ ,\ \rangle$ denotes the pairing between forms and vectors. From Lemma 3 in Section 1.4

$$\begin{aligned}\langle d_M\phi, \overline{L}_1 \wedge \overline{L}_2\rangle = \overline{L}_1\{\langle\phi, \overline{L}_2\rangle\} - \overline{L}_2\{\langle\phi, \overline{L}_1\rangle\} \\ - \langle\phi, [\overline{L}_1, \overline{L}_2]\rangle.\end{aligned}$$

Since $\phi \in \mathcal{E}_M^{1,0}$ and $\overline{L}_1, \overline{L}_2 \in T^{0,1}(M)$, we have $\langle\phi, \overline{L}_1\rangle = \langle\phi, \overline{L}_2\rangle = 0$. By the definition of a CR structure, $\overline{\mathbb{L}}$ is involutive, and so $[\overline{L}_1, \overline{L}_2] \in \overline{\mathbb{L}}$. Therefore $\langle\phi, [\overline{L}_1, \overline{L}_2]\rangle = 0$, from which $\langle d_M\phi, \overline{L}_1 \wedge \overline{L}_2\rangle = 0$ follows, as desired. This proves the lemma for the case $p = 1, q = 0$.

Note that the lemma automatically holds for $p = 0$ and $q = 1$. For $p, q \geq 1$, $\mathcal{E}_M^{p,q}$ is generated by the following terms:

$$\phi_1 \wedge \ldots \wedge \phi_p \wedge \psi_1 \wedge \ldots \wedge \psi_q$$

where each ϕ_j is a smooth (1,0)-form and each ψ_j is a smooth $(0,1)$-form. The general case now follows by using the product rule for d_M and the lemma for the case of a (1,0)-form or (0,1)-form. ∎

The key ingredient of the proof is that $T^{0,1}(M) = \overline{\mathbb{L}}$ is involutive. Since $T^{1,0}(M)$ is not necessarily involutive, we cannot conclude that $\pi^{p+2,q-1}(d_M\phi) = 0$ for $\phi \in \mathcal{E}_M^{p,q}$. If M is a complex manifold, then $T^{1,0}(M)$ is involutive and so $\pi^{p+2,q-1}(d_M\phi) = 0$ for $\phi \in \mathcal{E}_M^{p,q}$. This is a key difference between the class of complex manifolds and the class of CR manifolds that are not complex manifolds.

LEMMA 2
If M is a CR manifold, then $\bar{\partial}_M \circ \bar{\partial}_M = 0$.

PROOF Suppose ϕ is a smooth (p,q)-form; then $\bar{\partial}_M \phi = \pi^{p,q+1}(d_M \phi)$. Lemma 1 implies

$$\bar{\partial}_M \phi = d_M \phi - [\pi^{p+2,q-1}(d_M \phi) + \pi^{p+1,q}(d_M \phi)].$$

Therefore

$$\begin{aligned}\bar{\partial}_M \bar{\partial}_M \phi &= \pi^{p,q+2}[d_M(\bar{\partial}_M \phi)] \\ &= -\pi^{p,q+2}[d_M \pi^{p+2,q-1}(d_M \phi) + d_M \pi^{p+1,q}(d_M \phi)]\end{aligned}$$

where the last equality uses $d_M \circ d_M = 0$. From Lemma 1, the term on the right vanishes. Therefore, $\bar{\partial}_M \bar{\partial}_M \phi = 0$, as desired. ∎

The following product rule for $\bar{\partial}_M$ follows from the product rule for the exterior derivative.

LEMMA 3
If $f \in \mathcal{E}_M^{p,q}$ *and* $g \in \mathcal{E}_M^{r,s}$, *then*

$$\bar{\partial}_M (f \wedge g) = (\bar{\partial}_M f) \wedge g + (-1)^{p+q} f \wedge \bar{\partial}_M g.$$

From Stokes' theorem and the product rule for $\bar{\partial}_M$ we obtain an integration by parts formula for $\bar{\partial}_M$.

LEMMA 4
If $f \in \mathcal{E}_M^{p,q}$ *and* g *is any smooth form on* M *with compact support, then*

$$\int_M (\bar{\partial}_M f) \wedge g = (-1)^{p+q+1} \int_M f \wedge \bar{\partial}_M g.$$

PROOF Let m be the dimension of $\mathbb{L}$ (and $\overline{\mathbb{L}}$); let d be the dimension of $X(M)$. Therefore, $\dim_{\mathbb{C}} (\mathbb{L} \oplus X) = \dim_{\mathbb{C}} T^{1,0}(M) = m + d$ and $\dim_{\mathbb{C}} T^{0,1}(M) = m$. $T^{\mathbb{C}}(M)$ has complex dimension $2m+d$ which is the same as the real dimension of M. So a form of top degree on M has bidegree $(m+d, m)$. If $f \in \mathcal{E}_M^{p,q}$, then $\bar{\partial}_M f \in \mathcal{E}_M^{p,q+1}$ and so $\bar{\partial}_M f$ pairs with forms of bidegree $(m+d-p, m-q-1)$. Let $g \in \mathcal{D}_M^{m+d-p,m-q-1}$. From the product rule, we obtain

$$\int_M (\bar{\partial}_M f) \wedge g = \int_M \bar{\partial}_M (f \wedge g) + (-1)^{p+q+1} \int_M f \wedge \bar{\partial}_M g. \tag{1}$$

The bidegree of $f \wedge g$ is $(m+d, m-1)$. Since top degree on M is $(m+d, m)$, we have

$$\bar{\partial}_M (f \wedge g) = d_M (f \wedge g).$$

Since g has compact support on M, we have

$$\int_M d_M(f \wedge g) = 0$$

by Stokes' theorem. This together with (1), establishes the lemma. ∎

Just as with the extrinsic case, the integration by parts formula allows us to extend the definition of $\bar{\partial}_M$ to currents. By definition, the space of currents of bidimension (p,q) on an open set $U \subset M$ is the dual of the space $\mathcal{D}_M^{p,q}(U)$. By adapting the proof of Lemma 1 in Section 6.1, the reader can easily show that this is isomorphic to the space of currents of bidegree $(m+d-p, m-q)$ which is denoted by $\mathcal{D}_M'^{m+d-p,m-q}(U)$. Elements in this space are $(m+d-p,m-q)$-forms on U with coefficients in $\mathcal{D}'(U)$. If the open set U is not essential to the discussion, then it will be omitted from the notation.

DEFINITION 2 *If $T \in \mathcal{D}_M'^{p,q}$, then the current $\bar{\partial}_M T \in \mathcal{D}_M'^{p,q+1}$ is defined by*

$$\langle \bar{\partial}_M T, g \rangle_M = (-1)^{p+q+1} \langle T, \bar{\partial}_M g \rangle_M.$$

An easy argument using this definition together with Lemma 2 shows that $\bar{\partial}_M(\bar{\partial}_M T) = 0$ for a current $T \in \mathcal{D}_M'^{p,q}$.

8.3 The equivalence of the extrinsic and intrinsic tangential Cauchy–Riemann complexes

For a CR submanifold M of $\mathbb{C}^n$, there is a choice of viewpoints for the tangential Cauchy–Riemann complex — the extrinsic and the intrinsic. These two complexes are different, but in this section, we show they are isomorphic. Before we establish this isomorphism, let us precisely define an isomorphism between two complexes.

DEFINITION 1 *(a) A complex is a collection of vector spaces $\mathcal{A} = \{\mathcal{A}_q; q \in \mathbb{Z}, q \geq 0\}$ with maps d_q: $\mathcal{A}^q \to \mathcal{A}^{q+1}$ such that $d_{q+1} \circ d_q = 0$ for $q \geq 0$.*

(b) Suppose $\mathcal{A} = \{\mathcal{A}_q, d_q; q \geq 0\}$ and $A = \{A_q, D_q; q \geq 0\}$ are two complexes. These complexes are isomorphic if there exists a collection of isomorphisms of vector spaces P_q: $A_q \to \mathcal{A}_q$ that intertwine d_q and D_q, i.e.,

$$P_{q+1} \circ D_q = d_q \circ P_q.$$

The following commutative diagram describes part (b) of the definition.

$$\begin{array}{ccccccccc}
\cdots & \to & A_q & \xrightarrow{D_q} & A_{q+1} & \xrightarrow{D_{q+1}} & A_{q+2} & \to & \cdots \\
 & & \downarrow P_q & & \downarrow P_{q+1} & & \downarrow P_{q+2} & & \\
 & \to & \mathcal{A}_q & \xrightarrow{d_q} & \mathcal{A}_{q+1} & \xrightarrow{d_{q+1}} & \mathcal{A}_{q+2} & \to & \cdots
\end{array}$$

As an example, let M and N be smooth manifolds and suppose F: $M \to N$ is a diffeomorphism. The complexes $\{d_M\colon\ \mathcal{E}^r(M) \to \mathcal{E}^{r+1}(M)\}$ and $\{d_N\colon\ \mathcal{E}^r(N) \to \mathcal{E}^{r+1}(N)\}$ are isomorphic and the isomorphism is given by $F^*\colon\ \mathcal{E}^r(N) \to \mathcal{E}^r(M)$.

THEOREM 1

Suppose M is a CR submanifold of $\mathbb{C}^n$. The extrinsic and intrinsic tangential Cauchy–Riemann complexes are isomorphic.

PROOF Fix p with $0 \le p \le n$. For $q \ge 0$, let

$$A_q = \mathcal{E}_M^{p,q} \ -\ \text{via the extrinsic definition}$$

$$\mathcal{A}_q = \mathcal{E}_M^{p,q} \ -\ \text{via the intrinsic definition.}$$

A_q is the space of smooth sections of the extrinsically defined bundle $\Lambda^{p,q}T^*(M)$ which by definition is the orthogonal complement of $I^{p,q}$ in $\Lambda^{p,q}T^*(\mathbb{C}^n)|_M$. On the other hand, $\mathcal{A}_q$ is the space of smooth sections of the intrinsically defined bundle $\Lambda^{p,q}T^*(M)$ which by definition is

$$\Lambda^p\{H^{*1,0}(M) \oplus X^*(M)\}\widehat{\otimes}\Lambda^q\{H^{*0,1}(M)\}.$$

We let

$$D_q\colon\ A_q \to A_{q+1} \text{ be the extrinsically defined } \bar{\partial}_M$$

$$d_q\colon\ \mathcal{A}_q \to \mathcal{A}_{q+1} \text{ be the intrinsically defined } \bar{\partial}_M.$$

The operator D_q is the tangential part of $\bar{\partial}$ (i.e., $t_M \circ \bar{\partial}$), and $d_q = \pi_M^{p,q+1} \circ d_M$ where d_M is the exterior derivative on M and $\pi_M^{p,q+1}$ is the projection of $\Lambda^{p+q+1}T^*(M)$ onto $\Lambda^p\{H^{*^{1,0}}(M) \oplus X^*(M)\}\widehat{\otimes}\Lambda^q\{H^{*^{0,1}}(M)\}$.

Let $j : M \mapsto \mathbb{C}^n$ be the inclusion map. We will show that j^* is the desired isomorphism between the complexes $\{D_q\colon\ A_q \to A_{q+1}\}$ and $\{d_q\colon\ \mathcal{A}_q \to \mathcal{A}_{q+1}\}$. The following two statements must be shown:

(i) The map j^* takes A_q onto $\mathcal{A}_q$ isomorphically.

(ii) $j^* \circ D_q = d_q \circ j^*$.

To prove (i) it suffices to show the following.

LEMMA 1

For each point $p_0 \in M$, j^* *maps the extrinsic* $\Lambda^{p,q}T^*_{p_0}(M)$ *isomorphically onto the intrinsic* $\Lambda^{p,q}T^*_{p_0}(M)$.

PROOF For notational simplicity, we will prove this lemma for the case where M is generic. From Lemma 1 in Section 7.2, we can find an affine complex linear change of coordinates $A : \mathbb{C}^n \mapsto \mathbb{C}^n$ so that the given point $p_0 \in M$ is the origin and

$$M = \{(x + iy, w) \in \mathbb{C}^d \times \mathbb{C}^{n-d}; y = h(x, w)\}$$

where $h\colon \mathbb{R}^d \times \mathbb{C}^{n-d} \to \mathbb{R}^d$ is smooth with $h(0) = 0$ and $Dh(0) = 0$. As mentioned in the remark after the proof of this lemma, A preserves the holomorphic tangent space of M, the totally real tangent space of M and the metric for the real tangent space of M. Therefore, the definition of the intrinsic $\Lambda^{p,q}T^*(M)$ is invariant under this change of coordinates. In addition, the pull back of A commutes with $\bar\partial$. Therefore, the definition of the extrinsic $\Lambda^{p,q}T^*(M)$ is also invariant under this change of coordinates.

The following arguments can be easily modified for the nongeneric case by using the remarks that follow the proof of Lemma 1 in Section 7.2.

A local defining system for M is given by $\{\rho_1, \ldots, \rho_d\}$ where $\rho_j(z, w) = \operatorname{Im} z_j - h_j(\operatorname{Re} z, w)$. Since $Dh(0) = 0$, we have $\bar\partial\rho_j(0) = -(2i)^{-1}d\bar z_j$. By definition, the extrinsic $\Lambda^{p,q}T^*(M)$ is the orthogonal complement in $\Lambda^{p,q}T^*(\mathbb{C}^n)|_M$ of the ideal generated by $\bar\partial\rho_1, \ldots, \bar\partial\rho_d$. Therefore, a basis for the extrinsic $\Lambda^{p,q}T^*(M)$ at the origin is given by

$$\{dz^I \wedge dw^J \wedge d\bar w^K;\quad |I| + |J| = p, |K| = q,\ I,\ J,\ K \text{ increasing}\}.$$

A basis for $\mathbb{L}_0^* = H_0^{*^{1,0}}(M)$ is given by $\{dw_1, \ldots, dw_{n-d}\}$ and a basis for $\overline{\mathbb{L}}_0^* = H_0^{*^{0,1}}(M)$ is given by $\{d\bar w_1, \ldots, d\bar w_{n-d}\}$. Since $X_0^*(M)$ is the orthogonal complement of $\mathbb{L}^* \oplus \overline{\mathbb{L}}^*$ in $T_0^{*\mathbb{C}}(M)$, a basis for $X_0^*(M)$ is given by $\{dx_1, \ldots, dx_d\}$. Therefore, a basis for the intrinsic $\Lambda^{p,q}T_0^*(M)$ is given by

$$\{dx^I \wedge dw^J \wedge d\bar w^K;\quad |I| + |J| = p, |K| = q,\ I,\ J,\ K \text{ increasing}\}.$$

Since $Dh(0) = 0$, the following relations hold at the origin:

$$\begin{array}{ll} j^*(dw_k) = dw_k & 1 \le k \le n-d \\ j^*(d\bar w_k) = d\bar w_k & 1 \le k \le n-d \\ j^*(dy_k) = 0 & 1 \le k \le d \\ j^*(dx_k) = dx_k & 1 \le k \le d. \end{array}$$

In particular, $j^*(dz_k) = dx_k$ for $1 \le k \le d$ and so

$$j^*(dz^I \wedge dw^J \wedge d\bar w^K) = dx^I \wedge dw^J \wedge d\bar w^K.$$

Hence, j^* takes a basis for the extrinsic $\Lambda^{p,q}T^*(M)$ (at 0) to a basis for the intrinsic $\Lambda^{p,q}T^*(M)$ (at 0). This completes the proof of the lemma and therefore statement (i) follows. ∎

To prove statement (ii), first note from the definitions

$$D_q = t_M \circ \bar{\partial} = t_M \circ \pi_{\mathbb{C}^n}^{p,q+1} \circ d_{\mathbb{C}^n}$$

$$d_q = \pi_M^{p,q+1} \circ d_M.$$

Since $j^* \circ d_{\mathbb{C}^n} = d_M \circ j^*$, statement (ii) will follow from the following lemma.

LEMMA 2

For any integers p, q

$$j^* \circ t_M \circ \pi_{\mathbb{C}^n}^{p,q+1} = \pi_M^{p,q+1} \circ j^*$$

as maps from $\{\Lambda^{p,q+1}T^*(\mathbb{C}^n) \oplus \Lambda^{p+1,q}T^*(\mathbb{C}^n)\}|_M$ to the intrinsic $\Lambda^{p,q+1}T^*(M)$.

PROOF For each point $p_0 \in M$, the maps in this lemma take the space

$$\{\Lambda^{p,q+1}T^*_{p_0}(\mathbb{C}^n) \oplus \Lambda^{p+1,q}T^*(\mathbb{C}^n)\}|_M$$

to the intrinsically defined $\Lambda^{p,q+1}T^*_{p_0}(M)$. As with the proof of Lemma 1, we assume M is generic, the point p_0 is the origin, and

$$M = \{(x+iy, w) \in \mathbb{C}^d \times \mathbb{C}^{n-d}; y = h(x,w)\}$$

where h: $\mathbb{R}^d \times \mathbb{C}^{n-d} \to \mathbb{R}^d$ is smooth with $h(0) = 0, Dh(0) = 0$. At the origin, $j^*(dw_k) = dw_k$, $j^* d\bar{w}_k = d\bar{w}_k$ for $1 \le k \le n-d$ and $j^*(dz_k) = j^*(d\bar{z}_k) = dx_k$ for $1 \le k \le d$.

First, suppose ϕ is an element of $\Lambda^{p,q+1}T_0^*(\mathbb{C}^n)$. We have

$$\phi = \sum_{r=0}^{\min(d,p)} \sum_{s=0}^{\min(d,q+1)} \sum_{\substack{|I|=r \\ |J|=s}} \phi_{IJ}^{r,s} \wedge dz^I \wedge d\bar{z}^J$$

where each $\phi_{IJ}^{r,s}$ is a form (at 0) of bidegree $(p-r, q+1-s)$ that *only* involves $dw_1, \ldots, dw_{n-d}, d\bar{w}_1, \ldots, d\bar{w}_{n-d}$. At the origin, t_M annihilates $d\bar{z}_1, \ldots, d\bar{z}_d$. Therefore

$$t_M \circ \pi_{\mathbb{C}^n}^{p,q+1}(\phi) = t_M(\phi) = \sum_{r=0}^{\min(d,p)} \sum_{|I|=r} \phi_I^{r,0} \wedge dz^I.$$

Since $j^* dz_k = dx_k$, we have

$$j^* \circ t_M \circ \pi_{\mathbb{C}^n}^{p,q+1}(\phi) = \sum_{r=0}^{\min(d,p)} \sum_{|I|=r} \phi_I^{r,0} \wedge dx^I. \tag{1}$$

On the other hand

$$\pi_M^{p,q+1} \circ j^*(\phi) = \pi_M^{p,q+1} \left\{ \sum_{r=0}^{\min(d,p)} \sum_{s=0}^{\min(d,q+1)} \sum_{\substack{|I|=r \\ |J|=s}} \phi_{IJ}^{r,s} \wedge dx^I \wedge dx^J \right\}.$$

Each dx_k belongs to the intrinsic $\Lambda^{1,0}T_0^*(M)$ and $\phi_{IJ}^{r,s}$ belongs to the intrinsic $\Lambda^{p-r,q+1-s}\ T_0^*(M)$. So $\phi_{IJ}^{r,s} \wedge dx^I \wedge dx^J$ belongs to the intrinsic $\Lambda^{p+s,q+1-s}T_0^*(M)$. Due to the presence of $\pi_M^{p,q+1}$, the only contributing term to the sum on the right occurs when $s = 0$. Therefore

$$\pi_M^{p,q+1} \circ j^*(\phi) = \sum_{r=0}^{\min(d,p)} \sum_{|I|=r} \phi_I^{r,0} \wedge dx^I.$$

By comparing this with (1), we see that

$$j^* \circ t_M \circ \pi_{\mathbb{C}^n}^{p,q+1} = \pi_M^{p,q+1} \circ j^* \tag{2}$$

on the space $\Lambda^{p,q+1}T^*(\mathbb{C}^n)|_M$. By examining the above argument, we see that both of these maps vanish on $\Lambda^{p+1,q}T^*(\mathbb{C}^n)|_M$ (this is trivial in the case of $j^* \circ t_M \circ \pi_{\mathbb{C}^n}^{p,q+1}$ due to the presence of $\pi_{\mathbb{C}^n}^{p,q+1}$). We conclude that (2) also holds on the space $\Lambda^{p+1,q}T^*(\mathbb{C}^n)|_M$ and so the proof of the lemma is complete.

This completes the proof of statement (ii) and hence the proof of Theorem 1 is also complete. ▮

In view of this theorem and in order to keep the notation to a minimum, we shall not distinguish between the extrinsic and intrinsic $\bar{\partial}_M$-complexes. It will be clear from the context which point of view will be used.

As the final item in this chapter, we make a remark about metrics. Here, a metric is used to choose a complement, $X(M)$, to $\mathbb{L} \oplus \overline{\mathbb{L}}$. A different choice of metric leads to a different $X(M)$ and hence a different $\bar{\partial}_M$-complex. However, given any two metrics, the associated complementary bundles are isomorphic and therefore the two resulting $\bar{\partial}_M$ complexes are isomorphic.

A metric can be avoided by using quotient spaces. For example in the imbedded case, we may let

$$\Lambda^{p,q}T^*(M) = \Lambda^{p,q}T^*(\mathbb{C}^n)/I^{p,q}.$$

In the abstract setting, we let $A^{p,q}(M)$ be the space of forms on M of degree $p+q$ that annihilate any $(p+q)$-vector on M that has more than q-factors contained in $\overline{\mathbb{L}}$. Then we define

$$\Lambda^{p,q}T^*(M) = \frac{A^{p,q}(M)}{A^{p+1,q-1}(M)}.$$

Both of these definitions of $\Lambda^{p,q}T^*(M)$ avoid the use of a metric.

In the imbedded case, the Cauchy–Riemann operator maps the space of smooth sections of $I^{p,q}$ to $I^{p,q+1}$. The tangential Cauchy–Riemann operator can then be defined as the induced map of the Cauchy–Riemann operator on the quotient spaces. Similarly, in the abstract case, the exterior derivative maps the space of smooth sections of $A^{p,q}(M)$ to $A^{p,q+1}(M)$. We leave the verification of this to the reader. In this case, the tangential Cauchy–Riemann operator can be defined as the induced map of the exterior derivative on quotient spaces. Both of these complexes are isomorphic to the tangential Cauchy–Riemann complexes defined earlier in this chapter once a metric has been chosen.

We have chosen not to emphasize this point of view because computations usually require a choice of a metric. The metric point of view will be especially useful in Part IV of this book.

9

CR Functions and Maps

In this chapter, we present the definitions and basic properties of CR functions and CR maps. CR functions are analogous to holomorphic functions on a complex manifold. However, there are important differences. For example, CR functions are not always smooth. There are relationships between CR functions on an imbedded CR manifold and holomorphic functions on the ambient $\mathbb{C}^n$. For example, the restriction of a holomorphic function to a CR submanifold is a CR function. However, CR functions do not always extend to holomorphic functions. In this chapter, we show that real analytic CR functions on a real analytic CR submanifold locally extend to holomorphic functions. A C^∞ version of this is also given. The chapter ends with a discussion of CR maps between CR manifolds.

9.1 CR functions

DEFINITION 1 *Suppose $(M, \mathbb{L})$ is a CR structure. A function f: $M \to \mathbb{C}$ (or distribution) is called a CR function if $\bar{\partial}_M f = 0$ on M.*

For most of this chapter, we shall be dealing with CR functions that are of class C^1.

The above definition applies to any CR manifold — either abstract or imbedded in $\mathbb{C}^n$. We now present other characterizations of a CR function.

LEMMA 1
(a) Suppose $(M, \mathbb{L})$ is a CR structure. A C^1 function f: $M \to \mathbb{C}$ is CR if and only if $\overline{L} f = 0$ on M for all $\overline{L} \in \overline{\mathbb{L}}$. (b) Suppose $M = \{z \in \mathbb{C}^n; \rho_1(z) = \cdots = \rho_d(z) = 0\}$ is a generic, CR submanifold of $\mathbb{C}^n$. A C^1 function f: $M \to \mathbb{C}$ is CR if and only if $\bar{\partial} \tilde{f} \wedge \bar{\partial}\rho_1 \wedge \ldots \wedge \bar{\partial}\rho_d = 0$ on M where $\tilde{f}$: $\mathbb{C}^n \to \mathbb{C}$ is any C^1 extension of f.

PROOF For the proof of (a), recall that $\bar{\partial}_M f = \pi^{0,1} d_M f$. Since $\pi^{0,1}$ is the projection of $T^{*^{\mathbb{C}}}(M)$ onto $T^{*^{0,1}}(M) = \overline{\mathbb{L}}^*$, we have $\pi^{0,1}(d_M f) = 0$ if and only if $\langle d_M f, \overline{L} \rangle = 0$ for all $\overline{L} \in \overline{\mathbb{L}}$. Part (a) now follows from the equation

$$\langle d_M f, \overline{L} \rangle = \overline{L}\{f\}$$

which is the definition of the exterior derivative of a function.

Part (b) follows from the extrinsic definition of $\bar{\partial}_M f$ as the piece of $\bar{\partial}\tilde{f}|_M$ which is orthogonal to the ideal generated by $\{\bar{\partial}\rho;\ \rho\colon\ \mathbb{C}^N \to \mathbb{R}$ is smooth with $\rho = 0$ on $M\}$. ∎

If M is a CR submanifold of $\mathbb{C}^n$ then any holomorphic function on a neighborhood of M in $\mathbb{C}^n$ restricts to a CR function on M by part (b) of the lemma. However, the converse is not true, that is, CR functions do not always extend as holomorphic functions. This is fortunate, for otherwise the study of CR functions would be much less interesting. The following example illustrates this behavior.

Example 1

Let $M = \{(z, w) \in \mathbb{C}^2;\ \text{Im } z = 0\}$. Here, $\overline{\mathbb{L}} = H^{0,1}(M)$ is spanned (over $\mathcal{E}(M)$) by the vector field $\partial/\partial\bar{w}$. A function $f\colon\ M \to \mathbb{C}$ is CR if

$$\frac{\partial f}{\partial \bar{w}}(x, w) = 0 \qquad (x = \text{ Re } z).$$

A CR function on M is a function that is holomorphic in w with $x \in \mathbb{R}$ held fixed. Since there is no condition on the behavior of a CR function in the x-variable, an arbitrary function of x is automatically CR. Therefore any nonanalytic function of x is an example of a CR function that does not extend to a holomorphic function on a neighborhood of M in $\mathbb{C}^2$. □

In this example, a real analytic CR function on $M = \{y = 0\}$ is always the restriction of a holomorphic function defined near M. A real analytic CR function on M can be represented (near the origin) by a power series in x and w (no $\bar{w}$). The holomorphic extension is obtained by replacing x by z in this power series. This idea will be exploited in the existence part of the proof of the next theorem, which is due to Tomassini [Tom].

THEOREM 1

Suppose M is a real analytic, generic CR submanifold of $\mathbb{C}^n$ with real dimension at least n. Suppose $f\colon\ M \to \mathbb{C}$ is a real analytic CR function on M. Then there is a neighborhood U of M in $\mathbb{C}^n$ and a unique holomorphic function $F\colon\ U \to \mathbb{C}$ with $F|_M = f$.

The neighborhood U in this theorem depends on the CR function f. Additional geometric conditions on M can be added to ensure that the neighborhood U is independent of f. With these added conditions, the real analyticity of f is unnecessary. This and other CR extension topics are discussed in Part III.

The uniqueness part of the theorem requires that M be generic but it does not require the real analyticity of M. We present it as a lemma.

LEMMA 2

Suppose M is a smooth, generic CR submanifold of $\mathbb{C}^n$ of real dimension $2n-d$, $0 \le d \le n$. If f is holomorphic in a connected neighborhood of M in $\mathbb{C}^n$ and if f vanishes on M, then f vanishes identically.

PROOF By the identity theorem for holomorphic functions, it suffices to show all the derivatives of f vanish at a fixed point $p_0 \in M$. Near $p_0 \in M$, there is a local basis for $H^{1,0}(M)$ consisting of smooth vector fields $L_1, \dots, L_{n-d}$ (for example, use Theorem 3 in Section 7.2). The collection of vector fields $\{\overline{L}_1, \dots, \overline{L}_{n-d}\}$ forms a local basis for $H^{0,1}(M)$. Let $X_1, \dots, X_d$ be a local basis for the totally real tangent bundle, $X(M)$. By ambiently extending the coefficients, we may assume these vector fields are defined in a neighborhood of p_0 in $\mathbb{C}^n$. The vector fields $N_1 = JX_1, \dots, N_d = JX_d$ (restricted to M) are transverse to M. Since M is generic, a basis for $T^{\mathbb{C}}(\mathbb{C}^n)$ near p_0 is given by

$$\{L_1, \dots, L_{n-d}, \overline{L}_1, \dots, \overline{L}_{n-d}, X_1, \dots, X_d, N_1, \dots, N_d\}.$$

The vector fields $L_1, \dots, L_{n-d}, \overline{L}_1, \dots, \overline{L}_{n-d}, X_1, \dots, X_d$ will be called tangential (since their restrictions to M belong to $T^{\mathbb{C}}(M)$). The vector fields $N_1, \dots, N_d$ will be called transverse.

To prove $D^\alpha f = 0$ near p_0 on M for all differential operators D^α, we use a double induction argument on both the order of the differential operator D^α and the number, m, of transverse vector fields in D^α.

If $m = 0$, then D^α involves only tangential vector fields and so $D^\alpha f = 0$ on M because $f = 0$ on M.

Now we assume by induction that for $m \ge 0$, $D^\alpha f = 0$ on M for all differential operators that involve only m-transverse vector fields. We will show

$$N_{j_1} \dots N_{j_{m+1}}\{f\} = 0 \quad \text{on} \quad M$$

where $j_1, \dots, j_{m+1}$ are any indices from the set $\{1, \dots, d\}$. From the Cauchy–Riemann equations on $\mathbb{C}^n$, we have $(X + iJX)(f) = 0$ for $X \in T(\mathbb{C}^n)$. Therefore, $N_{j_{m+1}}\{f\} = JX_{j_{m+1}}\{f\} = iX_{j_{m+1}} f$ near p_0 in $\mathbb{C}^n$. Thus

$$N_{j_1} \dots N_{j_{m+1}}\{f\} = iN_{j_1} \dots N_{j_m} X_{j_{m+1}}\{f\}.$$

The right side involves a differential operator with only m-transverse vector fields and therefore it vanishes on M as desired.

To complete the double induction, we assume the following: for integers, $N \geq 0$, $m \geq 1$

$$D^\alpha f = 0 \quad \text{on} \quad M$$

for $|\alpha| \leq N$ and where D^α involves only m-transverse vector fields. We also assume $D^\alpha f = 0$ on M for operators of any order that involve at most $(m-1)$-transverse vector fields. We must show that $D^\alpha f = 0$ on M for $|\alpha| = N+1$ and where D^α involves m-transverse vector fields.

We have two cases to consider.

Case (i). $D^\alpha = T \circ D^{\alpha'}$.

Here, T is a tangential vector field and $D^{\alpha'}$ is a differential operator of order $|\alpha'| \leq N$ that involves only m-transverse vector fields. In this case, $D^{\alpha'} f = 0$ on M by the induction hypothesis and so $T\{D^{\alpha'} f\} = 0$ on M, as desired.

Case (ii) $D^\alpha = N_j \circ D^{\alpha'}$.

Here, $N_j = JX_j$ is transverse and $D^{\alpha'}$ is a differential operator of order $|\alpha'| \leq N$ that involves only $(m-1)$-transverse vector fields. In this case

$$\begin{aligned} D^\alpha f &= N_j\{D^{\alpha'} f\} \\ &= D^{\alpha'}\{N_j f\} + [N_j, D^{\alpha'}]\{f\} \end{aligned}$$

where [,] denotes the commutator. The first term equals $iD^{\alpha'}\{X_j f\}$ by the Cauchy–Riemann equations. This term vanishes on M by the induction hypothesis because $D^{\alpha'} X_j$ involves only $(m-1)$-transverse vector fields. The second term is a differential operator of order N and so it vanishes on M, again by the induction hypothesis.

From double induction, it follows that $D^\alpha f = 0$ near p_0 on M for all differential operators and so f vanishes identically. The proof of the lemma is now complete. ∎

For the existence part of Theorem 1, we give two proofs. The first is perhaps simpler but the second can be easily modified to handle a C^∞ version of Theorem 1. Both proofs illustrate important ideas.

FIRST PROOF OF EXISTENCE The first proof treats both ζ and $\bar{\zeta} \in \mathbb{C}^n$ as independent coordinates. We will show that the given real analytic CR function f extends to a holomorphic function on $\mathbb{C}^{2n}$ (of ζ and $\bar{\zeta}$). By using the tangential Cauchy–Riemann equations, we will show that the holomorphic extension of f is independent of the coordinate $\bar{\zeta}$ and so it restricts to a holomorphic function on $\mathbb{C}^n$ which is the desired extension of f.

Now we present the details. It suffices to holomorphically extend the given CR function to a neighborhood of a fixed point $p_0 \in M$. The global extension can then be obtained by piecing together the local extensions. The uniqueness part of Theorem 1 ensures that the local extensions agree on overlaps.

From Lemma 1 in Section 7.2, we may assume the point p_0 is the origin and

$$M = \{(z = x + iy, w) \in \mathbb{C}^d \times \mathbb{C}^{n-d}; y = h(x, w)\}$$

where $h\colon \mathbb{R}^d \times \mathbb{C}^{n-d} \to \mathbb{R}^d$ is real analytic in a neighborhood of the origin and $Dh(0) = 0$. From Theorem 3 in Section 7.2, a local basis for $H^{0,1}(M)$ is given by

$$\overline{L}_j = -2i \sum_{\ell=1}^{d} \sum_{k=1}^{d} \mu_{\ell k} \frac{\partial h_k}{\partial \bar{w}_j} \frac{\partial}{\partial \bar{z}_\ell} + \frac{\partial}{\partial \bar{w}_j}, \qquad 1 \le j \le n - d$$

where $\mu_{\ell k}$ is the (ℓ, k)th entry of the matrix $(I + i(\partial h/\partial x))^{-1}$.

Since h is a real analytic function near the origin, h can be expressed in a power series in the variables $z, \bar{z} \in \mathbb{C}^d$ and $w, \bar{w} \in \mathbb{C}^{n-d}$. By replacing $\bar{z}$ by the independent variable $\zeta \in \mathbb{C}^d$ and $\bar{w}$ by the independent variable $\eta \in \mathbb{C}^{n-d}$ in the power series for h, we obtain a holomorphic function $\tilde{h}\colon \mathbb{C}^d \times \mathbb{C}^d \times \mathbb{C}^{n-d} \times \mathbb{C}^{n-d} \to \mathbb{C}^d$, with $\tilde{h}(z, \bar{z}, w, \bar{w}) = h(x, w)$. Define

$$M_{\mathbb{C}} = \{(z, \zeta, w, \eta) \in \mathbb{C}^d \times \mathbb{C}^d \times \mathbb{C}^{n-d} \times \mathbb{C}^{n-d};\ \frac{1}{2i}(z - \zeta) = \tilde{h}(z, \zeta, w, \eta)\}.$$

Also define $\Delta\colon \mathbb{C}^d \times \mathbb{C}^{n-d} \to \mathbb{C}^d \times \mathbb{C}^d \times \mathbb{C}^{n-d} \times \mathbb{C}^{n-d}$ by

$$\Delta(z, w) = (z, \bar{z}, w, \bar{w}).$$

$M_{\mathbb{C}}$ is a complex submanifold of $\mathbb{C}^{2n}$ with complex dimension $2n - d$. Moreover, $\Delta\{M\}$ is imbedded as a totally real submanifold of $M_{\mathbb{C}}$.

If $f\colon M \to \mathbb{C}$ is a real analytic function on M, then the above procedure of replacing $\bar{z}$ by ζ and $\bar{w}$ by η in the power series expansion of f produces a holomorphic function $\tilde{f}\colon M_{\mathbb{C}} \to \mathbb{C}$ with $\tilde{f} \circ \Delta = f$. Similarly, the real analytic coefficients of $\overline{L}_j$ can be holomorphically extended to vector fields $\tilde{L}_1, \ldots, \tilde{L}_{n-d} \in T^{1,0}(M_{\mathbb{C}})$ with

$$\tilde{L}_j = -2i \sum_{\ell=1}^{d} \sum_{k=1}^{d} \mu_{\ell k}(z, \zeta, w, \eta) \frac{\partial \tilde{h}_k}{\partial \eta_j} \frac{\partial}{\partial \zeta_\ell} + \frac{\partial}{\partial \eta_j}, \qquad 1 \le j \le n - d. \qquad (1)$$

Since $\tilde{f}$ is holomorphic, we have $(\tilde{L}_j \tilde{f}) = \Delta_* \overline{L}_j\{\tilde{f}\}$ on $\Delta\{M\}$. If f is CR, then

$$\begin{aligned}(\tilde{L}_j \tilde{f}) \circ \Delta &= \overline{L}_j\{\tilde{f} \circ \Delta\} \\ &= \overline{L}_j f = 0 \quad \text{on} \quad M.\end{aligned}$$

Since $\Delta\{M\}$ is a $2n-d$ real dimensional generic (totally real) submanifold of $M_{\mathbb{C}}$, we have

$$\tilde{L}_j \tilde{f} \equiv 0 \tag{2}$$

on $M_{\mathbb{C}}$ by Lemma 2.

Each vector $\partial/\partial\zeta_j$ is transverse to $M_{\mathbb{C}}$ because $D\tilde{h}(0)=0$. Define $\widetilde{F}\colon \mathbb{C}^d \times \mathbb{C}^d \times \mathbb{C}^{n-d} \times \mathbb{C}^{n-d} \to \mathbb{C}$ to be the extension of $\tilde{f}\colon M_{\mathbb{C}} \to \mathbb{C}$ that is independent of ζ. So

$$\frac{\partial \widetilde{F}}{\partial \zeta_j} = 0 \qquad \text{near the origin,} \qquad 1 \le j \le d. \tag{3}$$

We also claim

$$\frac{\partial \widetilde{F}}{\partial \eta_j} = 0 \quad \text{near the origin,} \qquad 1 \le j \le n-d. \tag{4}$$

To see this, note

$$\frac{\partial}{\partial \zeta_j}\left(\frac{\partial \widetilde{F}}{\partial \eta_j}\right) = \frac{\partial}{\partial \eta_j}\left(\frac{\partial \widetilde{F}}{\partial \zeta_j}\right) = 0.$$

In view of (1) and using $\partial\widetilde{F}/\partial\zeta_j = 0$, we obtain

$$\begin{aligned} \frac{\partial \widetilde{F}}{\partial \eta_j} &= \tilde{L}_j \widetilde{F} \qquad \text{on} \qquad M_{\mathbb{C}}, \quad 1 \le j \le n-d \\ &= 0 \qquad \text{on} \qquad M_{\mathbb{C}} \quad \text{(by (2))}. \end{aligned}$$

Since $\partial/\partial\zeta_j$ is transverse to $M_{\mathbb{C}}$, the previous two sets of equations imply that $\partial\widetilde{F}/\partial\eta_j = 0$, as claimed.

Finally, the holomorphic extension of f on $\mathbb{C}^n = \mathbb{C}^d \times \mathbb{C}^{n-d}$ is obtained by setting $F = \widetilde{F} \circ \Delta$. From (3) and (4), $\widetilde{F}$ is independent of ζ and η and so the power series of $F = \widetilde{F} \circ \Delta$ is independent of $\bar{z}$ and $\bar{w}$. Therefore, F is holomorphic on a neighborhood of the origin in $\mathbb{C}^n$. Moreover $F|_M = f$ because

$$\begin{aligned} F|_M &= \widetilde{F} \circ \Delta|_M \\ &= \tilde{f} \circ \Delta|_M \\ &= f. \end{aligned}$$

∎

SECOND PROOF OF EXISTENCE This second proof is based on ideas in a paper by Baouendi, Jacobowitz, and Treves [BJT]. We again start with M presented near the origin as

$$M = \{(x+iy, w) \in \mathbb{C}^d \times \mathbb{C}^{n-d}; y = h(x,w)\},$$

where h is real analytic and $h(0) = 0$, $Dh(0) = 0$. If M is flat (i.e., $h \equiv 0$) then a real analytic CR function near the origin is a convergent power series in x and w (no $\bar{w}$). The desired holomorphic extension is obtained by replacing x with $z = x + iy$ in its power series. We want to mimic this procedure as much as possible for the general case. The problem is that in general, a real analytic CR function will depend on $\bar{w}$. However, its dependence on $\bar{w}$ is closely linked with the dependence of the power series of h on $\bar{w}$. Instead of letting $\bar{w}$ be an independent complex coordinate as in the first proof, we shall change the complex structure for $\mathbb{C}^n$ so that h becomes holomorphic.

Now we present the details. In the power series of h, we replace x by z. So h is defined on $\mathbb{C}^d \times \mathbb{C}^{n-d}$ and $h(z, w)$ is holomorphic in z near the origin. Define $H: \mathbb{C}^d \times \mathbb{C}^{n-d} \to \mathbb{C}^d \times \mathbb{C}^{n-d}$ by

$$H(z, w) = (z + ih(z, w), w).$$

Let $H_1, \dots, H_n$ be the component functions for H. We use $H = (H_1, \dots, H_n)$ as a coordinate chart to define a new complex structure for $\mathbb{C}^n = \mathbb{C}^d \times \mathbb{C}^n$. A function g is holomorphic with respect to the new complex structure if there exists a holomorphic function G in the usual sense with $g = G \circ H$. This complex structure agrees with the usual complex structure in the z-variables since H is holomorphic in $z \in \mathbb{C}^d$ in the usual sense.

The $T^{0,1}$-vector fields for this new complex structure are those vector fields that annihilate the coordinate functions $H_1, \dots, H_n$.

LEMMA 3

A local basis for the bundle $T^{0,1}(\mathbb{C}^n)$ for the new complex structure is given by

$$\Lambda_j = -i \sum_{k,\ell=1}^{d} \mu_{\ell,k}(z, w) \frac{\partial h_k}{\partial \bar{w}_j}(z, w) \frac{\partial}{\partial z_\ell} + \frac{\partial}{\partial \bar{w}_j} \quad 1 \le j \le n-d, \quad \frac{\partial}{\partial \bar{z}_j}, \quad 1 \le j \le d$$

where $\mu_{\ell k}$ is the (ℓ, k)th entry in the $d \times d$ matrix $[I + i(\partial h / \partial z)]^{-1}$.

PROOF This lemma follows by showing that for $1 \le \ell \le n$, $\Lambda_j\{H_\ell\} = 0$ for $1 \le j \le n - d$, and $(\partial / \partial \bar{z}_j)\{H_\ell\} = 0$ for $1 \le j \le d$. Also note that these vector fields are linearly independent near the origin because $Dh(0) = 0$. ∎

Let $H_0 = H|_{\{\mathrm{Im}\, z=0\}}$. The map H_0: $\mathbb{R}^d \times \mathbb{C}^{n-d} \to M$ is a parameterization for M. Let $\pi: \mathbb{C}^d \times \mathbb{C}^{n-d} \to \mathbb{R}^d \times \mathbb{C}^{n-d}$ be the projection map given by $\pi(x + iy, w) = (x, w)$. Clearly, $\pi|_M$ is the inverse of H_0.

From Theorem 3 in Section 7.2, a local basis for $H^{0,1}(M)$ is given by

$$\overline{L}_j = -2i \sum_{k,\ell=1}^{d} \mu_{\ell k} \frac{\partial h_k}{\partial \bar{w}_j} \frac{\partial}{\partial \bar{z}_\ell} + \frac{\partial}{\partial \bar{w}_j} \qquad 1 \le j \le n - d$$

where $\mu_{\ell k}$ is the (ℓ, k)th entry of the matrix $[I + i(\partial h/\partial x)]^{-1}$. A C^1 function $f\colon M \to \mathbb{C}$ is CR on M if and only if $\overline{L}_j f = 0$, $1 \leq j \leq n-d$. This is equivalent to $\pi_* \overline{L}_j \{f \circ H_0\} = 0$ on $\mathbb{R}^d \times \mathbb{C}^{n-d}$ because $H_0 \circ \pi$ is the identity map on M. The vector field $\pi_* \overline{L}_j$ can be computed by using $\pi_*(\partial/\partial y_j) = 0, \pi_*(\partial/\partial x_j) = \partial/\partial x_j$ and $\pi_*(\partial/\partial \bar{w}_j) = \partial/\partial \bar{w}_j$. We have

$$\pi_* \overline{L}_j = -i \sum_{k,\ell=1}^{d} \mu_{\ell k} \frac{\partial h_k}{\partial \bar{w}_j} \frac{\partial}{\partial x_\ell} + \frac{\partial}{\partial \bar{w}_j}, \qquad 1 \leq j \leq n-d.$$

Suppose $f\colon M \to \mathbb{C}$ is a real analytic CR function. Let $f_0 = f \circ H_0\colon \mathbb{R}^d \times \mathbb{C}^{n-d} \to \mathbb{C}$. The function f_0 is a real analytic function of $x \in \mathbb{R}^d$ and $w \in \mathbb{C}^{n-d}$. Let $F_0\colon \mathbb{C}^d \times \mathbb{C}^{n-d} \to \mathbb{C}$ be the extension of f_0 obtained by replacing x by $z = x + iy$ in the power series expansion of f_0 about the origin. Since F_0 and h are holomorphic in $z \in \mathbb{C}^d$, we have $\partial h/\partial z_k = \partial h/\partial x_k, \partial F_0/\partial z_k = \partial F_0/\partial x_k$. Comparing the expressions for Λ_j and $\pi_* \overline{L}_j$, we obtain

$$\begin{aligned} \Lambda_j F_0|_{\{\mathrm{Im}\ z=0\}} &= \pi_* \overline{L}_j \{f_0\} \\ &= 0 \quad \text{(since } f \text{ is CR).} \end{aligned}$$

Since $\Lambda_j F_0$ is holomorphic in $z \in \mathbb{C}^d$ and vanishes on $\{\mathrm{Im}\ z = 0\}$, $\Lambda_j F_0$ must vanish identically.

Since both $\Lambda_j F_0 = 0$ and $\partial F_0/\partial \bar{z}_k = 0$, the function F_0 is holomorphic with respect to the new complex structure. There is a function $F\colon \mathbb{C}^n \to \mathbb{C}$ that is holomorphic in the usual sense defined in a neighborhood of the origin with $F_0 = F \circ H$. The restriction of F to M is f because $f \circ H_0 = F \circ H_0$ on $\{\mathrm{Im} z = 0\}$. This completes the second proof of the existence part of Theorem 1. ∎

We have shown by example that CR functions of class C^∞ are not necessarily the restrictions of holomorphic functions. However, smooth CR functions are the restrictions of functions that satisfy the ambient Cauchy–Riemann equations on M. This is presented in the next theorem.

THEOREM 2

Suppose M is a C^∞, generic CR submanifold of $\mathbb{C}^n$ with real dimension $2n-d$, $1 \leq d \leq n$. If f is a C^∞, CR function on M, then there is a C^∞ function F defined on $\mathbb{C}^n$ such that $\bar{\partial} F$ vanishes on M to infinite order and $F|_M = f$. F is unique modulo the space of functions that vanish to infinite order on M.

If M or f is only of class C^k, $k \geq 2$, then an easy modification of the proof produces a C^k extension, F, such that $\bar{\partial} F$ vanishes on M to order $k-1$.

The key ingredient in the proof of Theorem 1 is the fact that a real analytic function $\phi\colon \mathbb{R}^d \to \mathbb{C}$ is locally the restriction of a holomorphic function $\Phi\colon \mathbb{C}^d \to \mathbb{C}$. In the proof of Theorem 2, we must replace this idea with the following.

LEMMA 4

(a) Suppose ϕ: $\mathbb{R}^d \to \mathbb{C}$, $d \geq 1$, *is a* C^∞ *function. Then there exists a* C^∞ *function* Φ: $\mathbb{C}^d \to \mathbb{C}$ *such that* $\bar{\partial}\Phi$ *vanishes to infinite order on* $\{\text{Im } z = 0\}$ *and* $\Phi = \phi$ *on* $\{\text{Im } z = 0\}$.

(b) Φ *in part (a) is unique modulo the space of smooth functions that vanish to infinite order on* $\{\text{Im } z = 0\}$.

PROOF Let $z = x + iy$. The requirement that $\Phi = \phi$ on $\{y = 0\}$ determines all the x-derivatives of Φ on $\{y = 0\}$. The requirement that $\bar{\partial}\Phi$ vanish to infinite order on $\{y = 0\}$ is equivalent to the requirement

$$\frac{\partial}{\partial y^\alpha}\left\{\frac{\partial \Phi}{\partial y_j} - i\frac{\partial \Phi}{\partial x_j}\right\} = 0 \quad \text{on} \quad \{y = 0\}, \quad 1 \leq j \leq d \tag{5}$$

for all indices α. This equation inductively determines all the y-derivatives of Φ on $\{y = 0\}$. Part (a) now follows from the Whitney extension theorem (see Theorem 2 in Section 5.3).

For part (b), if Φ vanishes on $\{y = 0\}$, then all x-derivatives of Φ also vanish on $\{y = 0\}$. If $\bar{\partial}\Phi$ vanishes to infinite order on $\{y = 0\}$ then from (5) and induction, it follows that all derivatives (in x and y) of Φ vanish on $\{y = 0\}$. ∎

PROOF OF THEOREM 2 The proof of the uniqueness part of Theorem 2 is similar to the proof of Lemma 2. The arguments there show that if $F = 0$ on M and if $\bar{\partial}F$ vanishes to infinite order on M, then all derivatives of F must also vanish on M.

The proof of the existence part of Theorem 2 is analogous to the second proof of the existence part of Theorem 1. We sketch the ideas. The graphing function h for M (and hence H) can be extended to $\mathbb{C}^d \times \mathbb{C}^{n-d}$ as smooth functions so that $\bar{\partial}_z h$ and $\bar{\partial}_z H$ vanish to infinite order on $\{\text{Im } z = 0\}$ by Lemma 4. As before, we use $H = (H_1, \ldots, H_n)$ to define a new complex structure for $\mathbb{C}^n$. Lemma 3, which exhibits a local basis $\{\Lambda_j, \partial/\partial\bar{z}_j\}$ for the $T^{0,1}$ bundle of this new complex structure, is still valid. For a smooth CR function f on M, let $f_0 = f \circ H|_{\{\text{Im } z=0\}}$. Extend f_0 to F_0: $\mathbb{C}^d \times \mathbb{C}^{n-d} \to \mathbb{C}$ so that $\bar{\partial}_z F_0$ vanishes to infinite order on $\{\text{Im } z = 0\}$. The computation that $\Lambda_j F_0$ vanishes on $\{\text{Im } z = 0\}$ is still valid. Since $\bar{\partial}_z h$ and $\bar{\partial}_z F_0$ vanish to infinite order on $\{\text{Im } z = 0\}$, $\bar{\partial}_z\{\Lambda_j F_0\}$ also vanishes to infinite order on $\{\text{Im } z = 0\}$. By part (b) of Lemma 4, $\Lambda_j F_0$ vanishes to infinite order on $\{\text{Im } z = 0\}$. Hence, both $\Lambda_j F_0$ and $\partial F_0/\partial\bar{z}_j$ vanish to infinite order on $\{\text{Im } z = 0\}$.

Since H: $\mathbb{C}^d \times \mathbb{C}^{n-d} \to \mathbb{C}^d \times \mathbb{C}^{n-d}$ is a diffeomorphism near the origin, we can define

$$F = F_0 \circ H^{-1}.$$

As before, $F|_M = f$ because $F \circ H|_{\{\text{Im } z=0\}} = f_0 = f \circ H|_{\{\text{Im } z=0\}}$. It remains to show that $\bar{\partial}F$ vanishes to infinite order on M. This is equivalent to showing $\overline{L}F$ vanishes to infinite order on M for all $\overline{L}$ belonging to $T^{0,1}(\mathbb{C}^n)$ for the usual complex structure for $\mathbb{C}^n$. We already know that $\overline{L}F_0$ vanishes to infinite order on $\{\text{Im}z = 0\}$ for all $\overline{L}$ in the $T^{0,1}$ bundle for the new complex structure for $\mathbb{C}^n$. Since $H : \mathbb{C}^n \mapsto \mathbb{C}^n$ is the coordinate chart for the new complex structure for $\mathbb{C}^n$, the push forward map under H^{-1} sends the $T^{0,1}$ bundle for the usual complex structure for $\mathbb{C}^n$ to the $T^{0,1}$ bundle for the new complex structure for $\mathbb{C}^n$. Thus, $(H_*^{-1}\overline{L})F_0$ vanishes to infinite order on $\{\text{Im}z = 0\}$ for all $\overline{L}$ in the $T^{0,1}$ bundle for the usual complex structure for $\mathbb{C}^n$. Since $F = F_0 \circ H^{-1}$, we conclude that $\overline{L}F$ vanishes to infinite order on M for all $\overline{L}$ in the $T^{0,1}$ bundle for the usual complex structure for $\mathbb{C}^n$, as desired. The proof of Theorem 2 is now complete. ∎

9.2 CR maps

Suppose M and N are CR manifolds and f: $M \to N$ is a C^1 map. If the target space N is a CR submanifold of $\mathbb{C}^m$, then f has component functions $(f_1, \dots, f_m)$. In this case, it is reasonable to call f a *CR map* if each of the component functions is a CR function. This definition needs to be modified in the case where N is not imbedded in $\mathbb{C}^m$. To motivate the abstract definition, let us more closely examine the case where both M and N are CR submanifolds of $\mathbb{C}^n$ and $\mathbb{C}^m$. Using Theorem 2 in the previous section, we can extend $f = (f_1, \dots, f_m)$: $M \to N$ to a function $F = (F_1, \dots, F_m)$ defined on $\mathbb{C}^n$ so that $\bar{\partial}F_j = 0$ on M, $1 \le j \le m$ (here, all we need is that $\bar{\partial}F_j$ vanish to first order on M). So for $p \in M$, $F_*(p)$, maps $T_p^{1,0}(\mathbb{C}^n)$ into $T_{F(p)}^{1,0}(\mathbb{C}^n)$ and $T_p^{0,1}(\mathbb{C}^n)$ into $T_{F(p)}^{0,1}(\mathbb{C}^n)$. In addition, $F_*(p)$ maps $T_p^{\mathbb{C}}(M)$ to $T_{F(p)}^{\mathbb{C}}(N)$ because F maps M to N. Therefore, for $p \in M, F_*(p)$ maps $H_p^{1,0}(M)$ into $H_{F(p)}^{1,0}(N)$ and $H_p^{0,1}(M)$ into $H_{F(p)}^{0,1}(N)$. For an abstract CR structure $(M, \mathbb{L})$, the subbundle $\mathbb{L}$ takes the place of $H^{1,0}(M)$. This motivates the following definition.

DEFINITION 1 *Suppose $(M, \mathbb{L}_M)$ and $(N, \mathbb{L}_N)$ are CR structures. A C^1 map f: $M \to N$ is called a CR map if $f_*\{\mathbb{L}_M\} \subset \mathbb{L}_N$.*

The extension of f_* to $T^{\mathbb{C}}(M)$ satisfies $f_*(\overline{L}) = \overline{f_*(L)}$ for any $L \in T^{\mathbb{C}}(M)$ (this is true for any C^1 map). Therefore, Definition 1 is equivalent to the requirement that $f_*\{\overline{\mathbb{L}}_M\} \subset \overline{\mathbb{L}}_N$. In particular, $f_*\{\mathbb{L}_M \oplus \overline{\mathbb{L}}_M\} \subset \mathbb{L}_N \oplus \overline{\mathbb{L}}_N$.

LEMMA 1
Suppose $(M, \mathbb{L})$ is a CR structure. A C^1 map $f = (f_1, \dots, f_m) : M \mapsto \mathbb{C}^m$ is a CR map if and only if each f_j is a CR function.

PROOF Here, the target space is the CR structure $(\mathbb{C}^m, T^{1,0}(\mathbb{C}^m))$. The map f is CR if and only if $f_*(\overline{L})$ belongs to $T^{0,1}(\mathbb{C}^m)$ for each $\overline{L} \in \overline{\mathbb{L}}$. For $1 \leq j \leq m$, let z_j be the jth coordinate function for $\mathbb{C}^m$. We have

$$\begin{aligned}(f_*(\overline{L})\{z_j\}) \circ f &= \overline{L}\{z_j \circ f\} \\ &= \overline{L}\{f_j\}.\end{aligned}$$

It follows that $f_*(\overline{L})$ belongs to $T^{0,1}(\mathbb{C}^m)$ if and only if $\overline{L}\{f_j\} = 0$ for $1 \leq j \leq m$. The proof of the lemma now follows from part (a) of Lemma 1 in Section 9.1. ∎

Define the subbundle

$$H(M) = \{L + \overline{L};\ L \in \mathbb{L}_M\}.$$

This is a subbundle of the real tangent bundle to M. From Lemma 3 in Section 3.2, there is a complex structure map J: $H(M) \to H(M)$ so that the extension of J to $H^{\mathbb{C}}(M) = \mathbb{L} \oplus \overline{\mathbb{L}}$ has eigenspaces $\mathbb{L}$ and $\overline{\mathbb{L}}$ corresponding to the eigenvalues $+i$ and $-i$. The following theorem gives an alternative characterization of a CR map in terms of the action of f_* on $H(M)$ and the J map.

THEOREM 1
Suppose $(M, \mathbb{L}_M)$ and $(N, \mathbb{L}_N)$ are CR structures. Let J_M: $H(M) \to H(M)$ and J_N: $H(N) \to H(N)$ be the associated complex structure maps. A C^1 map f: $M \to N$ is a CR map if and only if for each $p \in M$, $f_*(p)\{H_p(M)\} \subset H_{f(p)}(N)$ and

$$J_N \circ f_*(p) = f_*(p) \circ J_M \quad \text{on} \quad H_p(M).$$

One of the characterizations of a holomorphic mapping between two complex manifolds is that the derivative commutes with the complex structures. The point of Theorem 1 is that the analogous characterization holds for CR maps between CR manifolds. Theorem 1 is sometimes summarized by saying that $f_*(p)$ is a complex linear map from $H_p(M)$ to $H_{f(p)}(N)$. The reader should not confuse this meaning of complex linear with the concept of a complex linear map between two complex vector spaces such as $T_p^{\mathbb{C}}(M)$ and $T_{f(p)}^{\mathbb{C}}(N)$. The push forward of *any* C^1 map $f : M \mapsto N$ is complex linear as a map from $T_p^{\mathbb{C}}(M)$ to $T_{f(p)}^{\mathbb{C}}(N)$ (see Section 3.1).

PROOF OF THEOREM 1 Let us first assume that f is a CR map as in Definition 1. For $L \in \mathbb{L}_M$

$$f_*(L + \overline{L}) = f_*(L) + \overline{f_*(L)}.$$

Since $f_*(L) \in \mathbb{L}_N$ (by Definition 1), clearly $f_*(L+\overline{L})$ is an element of $H(N)$. In addition, $\mathbb{L}_M$ and $\overline{\mathbb{L}}_M$ are the $+i$ and $-i$ eigenspaces for J_M. Therefore

$$\begin{aligned} f_*(J_M(L+\overline{L})) &= f_*(iL - i\overline{L}) \\ &= i(f_*(L) - \overline{f_*(L)}). \end{aligned}$$

Since $f_*(L)$ is an element of $\mathbb{L}_N$, which is the $+i$ eigenspace of J_N, the above equation becomes

$$f_*(J_M(L+\overline{L})) = J_N(f_*(L+\overline{L})).$$

Thus, $f_* \circ J_M = J_N \circ f_*$ on $H(M)$, as desired.

For the converse, note that each element L in $\mathbb{L}_M$ can be written as

$$L = X - iJ_M X$$

where $X = 1/2(L+\overline{L}) \in H(M)$. We have

$$\begin{aligned} f_*(L) &= f_*(X) - if_*(J_M X) \\ &= f_*(X) - iJ_N f_*(X). \end{aligned}$$

$\mathbb{L}_N$ is generated by vectors of the form $Y - iJ_N Y$ for $Y \in H(N)$. Since $f_*(X)$ belongs to $H(N)$, the above equation shows that $f_*(L)$ belongs to $\mathbb{L}_N$, as desired. The proof of the theorem is complete. ∎

Now we turn our attention from the push forward of vectors to the pull back of forms via a CR map. If f is a holomorphic map between two complex manifolds then f^* commutes with $\bar{\partial}$. This must be modified for CR maps with the tangential Cauchy–Riemann operator. If $(M, \mathbb{L})$ is a CR structure, then the tangential Cauchy–Riemann complex involves the totally real part of the cotangent bundle, $X^*(M)$ as well as $\mathbb{L}^*$ and $\overline{\mathbb{L}}^*$. However, the definition of a CR map makes no requirement about the behavior of f_* on $X(M)$. So unlike the case for holomorphic maps, the pull back operator of a CR map, f^*, does *not* preserve bidegree, i.e., $f^*\{\mathcal{E}_N^{p,q}\}$ is not necessarily contained in $\mathcal{E}_M^{p,q}$ (see Lemma 2 below). Therefore, f^* does not quite commute with the tangential Cauchy–Riemann operator. For example, let $M = \{(z = x+iy, w) \in \mathbb{C}^2;\ y = 0\}$. Suppose $f : M \mapsto M$ is given by $f(x,w) = (x, w+x)$. Clearly, f is a CR map because the component functions of f are holomorphic in w. Note that $f^*\overline{\partial}_M(\overline{w}) = f^* d\overline{w} = d\overline{w} + dx$. We also have $\overline{\partial}_M f^*(\overline{w}) = \overline{\partial}_M(\overline{w}+x) = d\overline{w}$, and so $f^*\overline{\partial}_M \overline{w} \neq \overline{\partial}_M f^*(\overline{w})$. Instead, the following equation holds: $\overline{\partial}_M f^*(\overline{w}) = \pi_M^{0,1} f^* \overline{\partial}_M(\overline{w})$. More generally, we have the following theorem.

THEOREM 2
Suppose M and N are CR manifolds and f: $M \to N$ is a CR map. Then

$$\bar{\partial}_M \circ \pi_M^{p,q} \circ f^* = \pi_M^{p,q+1} \circ f^* \circ \bar{\partial}_N$$

as maps from $\mathcal{E}_N^{p,q}$ to $\mathcal{E}_M^{p,q+1}$.

We remark that if the metric-free version of the tangential Cauchy–Riemann complex (defined at the end of Chapter 8) is used, then the pull back operator via a CR map preserves the bidegree. In this case, the tangential Cauchy–Riemann operator commutes with the pull back operator via a CR map. We leave the verification of this to the reader. As mentioned in Chapter 8, computations usually require the choice of a metric. For this reason, we have chosen to show how the tangential Cauchy–Riemann complex defined via a metric behaves with respect to pull backs of CR maps.

As an example, let M be a CR submanifold of $\mathbb{C}^n$ and suppose $j : M \mapsto \mathbb{C}^n$ is the inclusion map. Since $H^{1,0}(M) \subset T^{1,0}(\mathbb{C}^n)$, j is a CR map. From Theorem 2, we have

$$\overline{\partial}_M \circ \pi_M^{p,q} \circ j^* = \pi_M^{p,q+1} \circ j^* \circ \overline{\partial}.$$

This equation relates the intrinsically defined tangential Cauchy–Riemann operator to the ambient $\overline{\partial}$-operator.

The proof of Theorem 1 requires a preliminary result.

LEMMA 2
Suppose F: $M \to N$ is a CR map between CR structures $(M, \mathbb{L}_M)$ and $(N, \mathbb{L}_N)$. Then

(a) $F^*\{T^{*^{1,0}}(N)\} \subset T^{*^{1,0}}(M)$

(b) *For $p, q \geq 0$*

$$F^*\{\Lambda^{p,q}T^*(N)\} \subset \Lambda^{p,q}T^*(N) \oplus \cdots \oplus \Lambda^{p+r,q-r}T^*(N)$$

where $r = \min\{q, n-p\}$ with $n = \dim_{\mathbb{C}} T^{*^{1,0}}(M)$.

PROOF For the proof of part (a), let $\phi \in T^{*^{1,0}}(N)$. We must show

$$\langle F^*\phi, \overline{L} \rangle = 0 \quad \text{for all} \quad \overline{L} \in T^{0,1}(M) = \overline{\mathbb{L}}_M$$

where $\langle\ ,\ \rangle$ denotes the pairing between forms and vectors. Since F is CR, $F_*\overline{L}$ is an element of $\overline{\mathbb{L}}_N = T^{0,1}(N)$ and so $0 = \langle \phi, F_*\overline{L} \rangle = \langle F^*\phi, \overline{L} \rangle$, as desired.

Part (b) follows by writing a typical term in $\Lambda^{p,q}T^*(N)$ as

$$\phi = \phi_1 \wedge \ldots \wedge \phi_p \wedge \psi_1 \wedge \ldots \wedge \psi_q$$

where $\phi_i \in T^{*^{1,0}}(N)$ and $\psi_j \in T^{*^{0,1}}(N)$ and then by using part (a) for $F^*\phi_i$. Note that $F^*\psi_j$ generally has nontrivial components of type (1,0) and (0,1). ∎

PROOF OF THEOREM 2 Let ϕ be an element of $\mathcal{E}_N^{p,q}$. From Lemma 2, we have

$$\pi^{p,q}F^*\phi = F^*\phi - [\pi_M^{p+1,q-1}(F^*\phi) + \cdots + \pi_M^{p+r,q-r}(F^*\phi)].$$

From the definition of $\bar{\partial}_M$, we have

$$\begin{aligned}\bar{\partial}_M(\pi^{p,q}F^*\phi) &= \pi_M^{p,q+1}d_M F^*\phi \\ &\quad -\pi_M^{p,q+1}\left\{\sum_{j=1}^{r} d_M(\pi_M^{p+j,q-j}F^*\phi)\right\}.\end{aligned}$$

In view of Lemma 1 in Section 8.2, $d_M\{\pi_M^{p+j,q-j}F^*\phi\}$ belongs to $\mathcal{E}_M^{p+j+2,q-j-1} \oplus \mathcal{E}_M^{p+j+1,q-j} \oplus \mathcal{E}_M^{p+j,q-j+1}$. Since $j \geq 1$, the sum on the right must vanish. Therefore

$$\bar{\partial}_M(\pi_M^{p,q}F^*\phi) = \pi_M^{p,q+1}(d_M F^*\phi).$$

Using the fact that d_M commutes with F^*, we obtain

$$\bar{\partial}_M(\pi_M^{p,q}F^*\phi) = \pi^{p,q+1}(F^*d_N\phi). \tag{1}$$

From the definition of $\bar{\partial}_N$ and Lemma 1 in Section 8.2, we have

$$d_N\phi = \bar{\partial}_N\phi + [\pi_N^{p+1,q}(d_N\phi) + \pi_N^{p+2,q-1}(d_N\phi)]. \tag{2}$$

By Lemma 2, $F^*(\pi_N^{p+1,q}(d_N\phi))$ is a sum of terms of type $(p+1+j, q-j)$ for $0 \leq j \leq \min(q, n-p-1)$. In particular

$$\pi_M^{p,q+1}(F^*\pi_N^{p+1,q}(d_N\phi)) = 0.$$

Similarly

$$\pi_M^{p,q+1}(F^*\pi_N^{p+2,q-1}(d_N\phi)) = 0.$$

These two equations together with (1) and (2) yield

$$\begin{aligned}\bar{\partial}_M\pi^{p,q}F^*\phi &= \pi^{p,q+1}F^*d_N\phi \\ &= \pi^{p,q+1}F^*\bar{\partial}_N\phi.\end{aligned}$$

The proof of Theorem 2 is complete. ∎

We discuss two corollaries. First, we give the extrinsic version of Theorem 2 for imbedded submanifolds.

COROLLARY 1
Suppose M and N are CR submanifolds of $\mathbb{C}^n$ and $\mathbb{C}^m$, respectively. Suppose $f\colon M \to N$ is a CR map. Let $F\colon \mathbb{C}^n \to \mathbb{C}^m$ be an extension of f with $\bar{\partial}F = 0$ on M. Then

$$\bar{\partial}_M \circ t_M \circ F^* = t_M \circ F^* \circ \bar{\partial}_N$$

as maps from the extrinsically defined $\mathcal{E}_N^{p,q}$ to the extrinsically defined $\mathcal{E}_M^{p,q+1}$.

Here, $\overline{\partial}_M$ and $\overline{\partial}_N$ refer to the extrinsically defined tangential Cauchy–Riemann operators. Elements of $\mathcal{E}_N^{p,q}$ are not intrinsic to N. Rather, they are smooth sections of $\Lambda^{p,q}T^*(N) \subset \Lambda^{p,q}T^*(\mathbb{C}^n)|_N$. For this reason, it is necessary to have an ambient extension (F) of the CR map (f) for the statement of the corollary.

PROOF One approach to the proof is to show that F^* maps $I^{p,q}|_N$ to $I^{p,q}|_M$ (see the definitions of these spaces in Section 8.2). Then, the corollary will follow from the definition of the extrinsic version of the tangential Cauchy–Riemann complex. The other approach to the proof involves reducing the statement given in the corollary to Theorem 2. We will give the details of the latter approach and leave the details of the former approach as an exercise.

Let $j_M : M \mapsto \mathbb{C}^n$ and j_N: $N \to \mathbb{C}^n$ be the inclusion maps. By Section 8.3, j_M^* and j_N^* are isomorphisms between the extrinsic and intrinsic tangential Cauchy–Riemann complexes of M and N. Therefore, the statement of the corollary is equivalent to

$$j_M^* \circ \bar{\partial}_M \circ t_M \circ F^* = j_M^* \circ t_M \circ F^* \circ \bar{\partial}_N$$

as operators on the extrinsically defined $\mathcal{E}_N^{p,q}(U)$. Since $\bar{\partial} F = 0$ on M, F^* preserves bidegree for elements of $\Lambda^{p,q}T^*(\mathbb{C}^n)|_M$ and so this equation is equivalent to

$$j_M^* \circ \bar{\partial}_M \circ t_M \circ \pi_{\mathbb{C}^n}^{p,q} \circ F^* = j_M^* \circ t_M \circ \pi_{\mathbb{C}^n}^{p,q+1} \circ F^* \circ \bar{\partial}_N.$$

From Theorem 1 in Section 8.3, $j_M^* \circ \bar{\partial}_M = \bar{\partial}_M \circ j_M^*$ and $\overline{\partial}_N = (j_N^*)^{-1} \circ \overline{\partial}_N \circ j_N^*$ where the $\bar{\partial}_M$ and $\bar{\partial}_N$ on the left are extrinsic and the $\bar{\partial}_M$ and $\bar{\partial}_N$ on the right are intrinsic. In addition, $j_M^* \circ t_M \circ \pi_{\mathbb{C}^n}^{p,q} = \pi_M^{p,q} \circ j_M^*$ from Lemma 2 in Section 8.3. Therefore, the above equation is equivalent to

$$\bar{\partial}_M \circ \pi_M^{p,q} \circ (F \circ j_M)^* = \pi_M^{p,q+1} \circ (F \circ j_M)^* \circ (j_N^*)^{-1} \circ \bar{\partial}_N \circ j_N^*.$$

Here, $\bar{\partial}_M$ and $\bar{\partial}_N$ are now the intrinsic tangential Cauchy–Riemann operators. Finally, we use the fact that $F \circ j_M = j_N \circ f$ (since $F = f$ on M) to see that this equation is equivalent to

$$(\overline{\partial}_M \circ \pi_M^{p,q} \circ f^*) \circ j_N^* = (\pi_M^{p,q+1} \circ f^* \circ \overline{\partial}_N) \circ j_N^*.$$

Since j_N^* is an isomorphism between the extrinsic and intrinsic $\mathcal{E}_M^{p,q}$, the proof of the corollary now follows from Theorem 2. ∎

We say that the CR structures $(M, \mathbb{L})$ and $(N, \mathbb{L}_N)$ are *CR equivalent* if there is a CR diffeomorphism between M and N. Using Theorem 2, we will show (in Corollary 2) that if M and N are CR equivalent, then the tangential Cauchy–Riemann complex on M is solvable if and only if the same is true for N. To be precise, we say that the tangential Cauchy–Riemann complex is *solvable at bidegree* (p, q) if for any form $f \in \mathcal{E}_M^{p,q}(M)$ with $\bar{\partial}_M f = 0$, there is a form $u \in \mathcal{E}_M^{p,q-1}(M)$ with $\bar{\partial}_M u = f$.

COROLLARY 2

Suppose $(M, \mathbb{L}_M)$ and $(N, \mathbb{L}_N)$ are equivalent CR structures. The tangential Cauchy–Riemann complex is solvable at bidegree (p, q) on M if and only if the same is true for N.

Since an open subset of a CR manifold is also a CR manifold, the above corollary applies to CR equivalent open subsets of CR manifolds.

PROOF Suppose F: $M \to N$ is a CR diffeomorphism and suppose f is an element of $\mathcal{E}_N^{p,q}(N)$ with $\bar{\partial}_N f = 0$. By Theorem 2, we have

$$\begin{aligned}\bar{\partial}_M(\pi_M^{p,q} F^* f) &= \pi_M^{p,q+1} F^* \bar{\partial}_N f \\ &= 0 \quad \text{on} \quad M.\end{aligned}$$

If the $\bar{\partial}_M$-complex is solvable at bidegree (p, q) on M, then there is a form $u \in \mathcal{E}_M^{p,q-1}(M)$ with

$$\bar{\partial}_M u = \pi_M^{p,q} F^* f.$$

Applying $\pi_N^{p,q} \circ F^{-1^*}$ to this equation and using Theorem 2 with M replaced by N and F replaced by F^{-1}, we obtain

$$\bar{\partial}_N\{\pi_N^{p,q-1}(F^{-1^*} u)\} = \pi_N^{p,q}(F^{-1^*} \pi_M^{p,q} F^* f). \tag{3}$$

From Lemma 2, we have

$$(\pi_M^{p,q} F^* f) = F^* f - \left(\sum_{j=1}^{r} \pi_M^{p+j,q-j}(F^* f) \right)$$

where $r = \min(q, n-p)$. Applying $\pi_N^{p,q} \circ F^{-1^*}$ to this equation and using Lemma 2 with F^{-1} instead of F, we obtain

$$\begin{aligned}\pi_N^{p,q}(F^{-1^*}(\pi_M^{p,q} F^* f)) &= \pi_N^{p,q} F^{-1^*} F^* f \\ &= \pi_N^{p,q}(f) \\ &= f \quad (\text{since } f \in \mathcal{E}_N^{p,q}).\end{aligned}$$

Substituting this equation into the right side of (3) yields

$$\bar{\partial}_N\{\pi_N^{p,q-1}(F^{-1^*} u)\} = f \quad \text{on} \quad N.$$

Hence, the solvability of $\bar{\partial}_M$ implies the solvability of $\bar{\partial}_N$. The converse is established the same way. ∎

10

The Levi Form

In previous chapters, concepts such as the tangential Cauchy–Riemann complex are introduced first for imbedded CR manifolds and then later for abstract CR manifolds. In this chapter, we take the opposite approach. First, we give the definition of the Levi form for the case of an abstract CR structure and then proceed to give more concrete representations of the Levi form in the case of an imbedded CR manifold. The Levi form for the case of a real hypersurface in $\mathbb{C}^n$ is discussed in some detail. In particular, the relationship between the Levi form and the first fundamental form of a hypersurface is presented.

10.1 Definitions

One of the defining properties of an abstract CR structure $(M, \mathbb{L})$ is that $\mathbb{L}$ is involutive (i.e., $[L_1, L_2] \in \mathbb{L}$ whenever $L_1, L_2 \in \mathbb{L}$). The subbundle $\mathbb{L} \oplus \overline{\mathbb{L}} \subset T^{\mathbb{C}}(M)$ is not necessarily involutive. In fact, the Levi form for M is defined so that it measures the degree to which $\mathbb{L} \oplus \overline{\mathbb{L}}$ fails to be involutive.

For $p \in M$, let

$$\pi_p \colon T_p(M) \otimes \mathbb{C} \longrightarrow \{T_p(M) \otimes \mathbb{C}\}/(\mathbb{L}_p \oplus \overline{\mathbb{L}}_p)$$

be the natural projection map.

DEFINITION 1 *The Levi form at a point $p \in M$ is the map $\mathcal{L}_p \colon \mathbb{L}_p \longmapsto \{T_p(M) \otimes \mathbb{C}\}/(\mathbb{L}_p \oplus \overline{\mathbb{L}}_p)$ defined by*

$$\mathcal{L}_p(L_p) = \frac{1}{2i}\pi_p\{[\overline{L}, L]_p\} \quad \textit{for} \quad L_p \in \mathbb{L}_p$$

where L is any vector field in $\mathbb{L}$ that equals L_p at p.

The vector field $[\overline{L}, L]$ lies in $T^{\mathbb{C}}(M)$ since $T^{\mathbb{C}}(M)$ is involutive. So the Levi form measures the piece of $(1/2i)[\overline{L}, L]_p$ that lies "outside" of $\mathbb{L}_p \oplus \overline{\mathbb{L}}_p$. The factor $1/2i$ is introduced to make the Levi form real valued, i.e., $\overline{\mathcal{L}}_p = \mathcal{L}_p$.

In order to show that the Levi form is well defined, we must show that its definition is independent of the $\mathbb{L}$-vector field extension of the vector $L_p \in \mathbb{L}_p$.

LEMMA 1

Suppose L and Z are two vector fields in $\mathbb{L}$ with $L_p = Z_p$, then

$$\pi_p[\overline{L}, L]_p = \pi_p[\overline{Z}, Z]_p.$$

PROOF Fix $p \in M$ and let $\{L_1, \dots, L_m\}$ be a basis for $\mathbb{L}$ that is defined near p. For some unique collection of smooth functions $a_1, \dots, a_m$ and $b_1, \dots, b_m$, we have

$$L = \sum_{j=1}^{m} a_j L_j$$

$$Z = \sum_{j=1}^{m} b_j L_j \qquad \text{near} \qquad p.$$

The assumption that $L_p = Z_p$ means that $a_j(p) = b_j(p)$ for $1 \le j \le m$. Expanding the Lie bracket, we obtain

$$\begin{aligned}[\overline{L}, L] &= \left[\sum_{j=1}^{m} \bar{a}_j \overline{L}_j, \sum_{k=1}^{m} a_k L_k\right] \\ &= \sum_{j,k=1}^{m} \bar{a}_j a_k [\overline{L}_j, L_k] \bmod (\mathbb{L} \oplus \overline{\mathbb{L}}).\end{aligned}$$

Therefore

$$\pi_p[\overline{L}, L]_p = \sum_{j,k=1}^{m} \bar{a}_j(p) a_k(p) \pi_p[\overline{L}_j, L_k]_p.$$

In a similar manner, we have

$$\pi_p[\overline{Z}, Z]_p = \sum_{j,k=1}^{m} \bar{b}_j(p) b_k(p) \pi_p[\overline{L}_j, L_k]_p.$$

Since $a_j(p) = b_j(p)$, the proof of the lemma is complete. ∎

If F: $M \to N$ is a CR map between the CR structures $(M, \mathbb{L}_M)$ and $(N, \mathbb{L}_N)$, then $F_*(p)$ maps $(\mathbb{L}_M \oplus \overline{\mathbb{L}}_M)_p$ to $(\mathbb{L}_N \oplus \overline{\mathbb{L}}_N)_{F(p)}$. Therefore, $F_*(p)$ induces a map on the quotient spaces

$$[F_*(p)]\colon \{T_p(M) \otimes \mathbb{C}\}/(\mathbb{L}_M \oplus \overline{\mathbb{L}}_M)_p \longrightarrow \{T_{F(p)}(N) \otimes \mathbb{C}\}/(\mathbb{L}_N \oplus \overline{\mathbb{L}}_N)_{F(p)}.$$

LEMMA 2
Suppose $(M, \mathbb{L}_M)$ *and* $(N, \mathbb{L}_N)$ *are CR structures and let* $\mathcal{L}^M$ *and* $\mathcal{L}^N$ *be their respective Levi forms. If* $F\colon M \to N$ *is a CR diffeomorphism, then for* $p \in M$

$$[F_*(p)] \circ \mathcal{L}^M_p = \mathcal{L}^N_{F(p)} \circ F_*(p)$$

as maps from $\{\mathbb{L}_M\}_p$ *to* $\{T_{F(p)}(N) \otimes \mathbb{C}\}/(\mathbb{L}_N \oplus \overline{\mathbb{L}}_N)_{F(p)}$.

PROOF The proof follows from the definitions and the observation that $F_*[\overline{L}, L] = [\overline{F_*(L)}, F_*(L)]$. ∎

We say a CR structure $(M, \mathbb{L})$ is *Levi flat* if the Levi form of M vanishes at each point in M. For example, let $M = \{(z, w) \in \mathbb{C} \times \mathbb{C}^{n-1};\ \text{Im } z = 0\}$. A (global) basis for $\mathbb{L} = H^{1,0}(M)$ is given by $\partial/\partial w_1, \ldots, \partial/\partial w_{n-1}$. Since $[\partial/\partial w_j, \partial/\partial \bar{w}_k] = 0$, M is Levi flat. Also note that M is foliated by the complex manifolds

$$M_x = \{(x, w); w \in \mathbb{C}^{n-1}\} \qquad for \quad x \in \mathbb{R}.$$

The complexified tangent bundle of each M_x is given by $\mathbb{L} \oplus \overline{\mathbb{L}}$. We have the following more general result.

THEOREM 1
Suppose $(M, \mathbb{L})$ *is a Levi flat CR structure. Then* M *is locally foliated by complex manifolds whose complexified tangent bundle is given by* $\mathbb{L} \oplus \overline{\mathbb{L}}$.

PROOF The idea of the proof is to show that $\mathbb{L} \oplus \overline{\mathbb{L}}$ and its underlying real bundle are involutive. Then the foliation is obtained by the real Frobenius theorem. The Newlander–Nirenberg theorem will then be used to show that the submanifolds in the foliation are complex manifolds.

Since $\mathbb{L}$ and $\overline{\mathbb{L}}$ are involutive by the definition of a CR manifold, $\mathbb{L} \oplus \overline{\mathbb{L}}$ is involutive if and only if $[\overline{L}_1, L_2]$ is an element of $\mathbb{L} \oplus \overline{\mathbb{L}}$ whenever L_1 and L_2 belong to $\mathbb{L}$. Since M is Levi flat, $[\overline{L}_1, L_1]$, $[\overline{L}_2, L_2]$, and $[\overline{L}_1 + \overline{L}_2, L_1 + L_2]$ belong to $\mathbb{L} \oplus \overline{\mathbb{L}}$. After expanding $[\overline{L}_1 + \overline{L}_2, L_1 + L_2]$, we see that

$$A = [\overline{L}_1, L_2] - \overline{[\overline{L}_1, L_2]}$$

belongs to $\mathbb{L} \oplus \overline{\mathbb{L}}$. A similar computation involving $[\overline{L}_1 + \overline{iL_2}, L_1 + iL_2]$ implies that the vector field

$$B = [\overline{L}_1, L_2] + \overline{[\overline{L}_1, L_2]}$$

also belongs to $\mathbb{L} \oplus \overline{\mathbb{L}}$. Adding A and B, we see that $[\overline{L}_1, L_2]$ belongs to $\mathbb{L} \oplus \overline{\mathbb{L}}$ and so $\mathbb{L} \oplus \overline{\mathbb{L}}$ is involutive.

The underlying real bundle for $\mathbb{L} \oplus \overline{\mathbb{L}}$ is the space

$$H(M) = \{L + \overline{L};\ L \in \mathbb{L}\}.$$

Since $\mathbb{L}\oplus\overline{\mathbb{L}}$ is involutive, $H(M)$ is an involutive real subbundle of $T(M)$. The real Frobenius theorem (Section 4.1) implies that M is foliated by submanifolds, $\{M'\}$, such that $T_p(M') = H_p(M)$ for each $p \in M'$. The complexified tangent space for M' at p is given by $H_p(M) \otimes \mathbb{C} = \mathbb{L}_p \oplus \overline{\mathbb{L}}_p$. Since $\mathbb{L}$ is involutive, $(M', \mathbb{L})$ forms an involutive almost complex structure. By the Newlander–Nirenberg theorem, there is a complex manifold structure for M' so that the resulting $T^{1,0}(M')$-bundle is $\mathbb{L}$. This completes the proof of the theorem.

The reader should note that if M is a Levi flat CR submanifold of $\mathbb{C}^n$, then the easier imbedded version of the Newlander–Nirenberg theorem (Theorem 2 in Section 4.3) can be used to show that each leaf of the foliation, M', is a complex submanifold of $\mathbb{C}^n$. ∎

10.2 The Levi form for an imbedded CR manifold

Computations with the Levi form are facilitated by identifying the quotient space $\{T_p(M)\otimes\mathbb{C}\}/(\mathbb{L}_p\oplus\overline{\mathbb{L}}_p)$ with a subspace of $T_p(M)\otimes\mathbb{C}$. This is accomplished by choosing a metric for $T^{\mathbb{C}}(M)$ and then by identifying the quotient space $\{T_p(M) \otimes \mathbb{C}\}/\mathbb{L}_p \oplus \overline{\mathbb{L}}_p$ with the orthogonal complement of $\mathbb{L}_p \oplus \overline{\mathbb{L}}_p$ (denoted $X_p(M)$ in Section 8.2).

For an imbedded CR manifold M, a natural metric exists — namely the restriction of the Euclidean metric on $T^{\mathbb{C}}(\mathbb{C}^n)$ to $T^{\mathbb{C}}(M)$. In this case, $\mathbb{L} = H^{1,0}(M)$, $\overline{\mathbb{L}} = H^{0,1}(M)$ and the quotient space

$$T^{\mathbb{C}}(M)/\mathbb{L}\oplus\overline{\mathbb{L}}$$

is identified with the complexified totally real part of the tangent bundle. As mentioned earlier, the Levi form is real valued, i.e., $\overline{\mathcal{L}_p(L_p)} = \mathcal{L}_p(L_p)$ and so the image of the Levi form is contained in $X_p(M)$, which is the totally real part of the real tangent space of M at p.

With this identification, the Levi form of a CR submanifold, M, at a point $p \in M$ is the map

$$\mathcal{L}_p\colon\ H^{1,0}_p(M) \longrightarrow X_p(M)$$

given by

$$\mathcal{L}_p(L_p) = \frac{1}{2i}\pi_p[\overline{L}, L]_p, \quad L_p \in H^{1,0}_p(M)$$

where $\pi_p\colon T_p(M) \to X_p(M)$ is the orthogonal projection and where L is any $H^{1,0}(M)$-vector field extension of the vector L_p.

Sometimes, it is convenient to think of the Levi form of an imbedded CR manifold as a map into the normal space of M at p, denoted $N_p(M)$, which is the orthogonal complement of $T_p(M)$ in $T_p(\mathbb{R}^{2n})$. This is accomplished by composing $\mathcal{L}_p$ with J and then projecting onto $N_p(M)$. Let

$$\tilde{\pi}_p : T_p(\mathbb{C}^n) \mapsto N_p(M)$$

be the orthogonal projection map.

DEFINITION 1 *The extrinsic Levi form of M at p is the map $\widetilde{\mathcal{L}}_p$: $H_p^{1,0}(M) \to N_p(M)$ given by $\tilde{\pi}_p \circ J \circ \mathcal{L}_p$.*

Since $H_p^{1,0}(M)$ and $H_p^{0,1}(M)$ are J-invariant, we have

$$\widetilde{\mathcal{L}}_p(L_p) = \frac{1}{2i}\tilde{\pi}_p(J[L, \overline{L}]_p)$$

where L is any $H^{1,0}(M)$-vector field extension of L_p.

Now, we develop a formula for the Levi form in terms of the complex hessian of a set of defining functions for M.

THEOREM 1

Suppose $M = \{\zeta \in \mathbb{C}^n; \rho_1(\zeta) = \cdots = \rho_d(\zeta) = 0\}$ is a smooth CR submanifold of $\mathbb{C}^n$, with $1 \leq d \leq n$. Let p be a point in M and suppose $\{\nabla\rho_1(p), \ldots, \nabla\rho_d(p)\}$ is an orthonormal basis for $N_p(M)$. Then the extrinsic Levi form is given by

$$\widetilde{\mathcal{L}}_p(W) = -\sum_{\ell=1}^{d}\left(\sum_{j,k=1}^{n} \frac{\partial^2\rho_\ell(p)}{\partial\zeta_j\partial\bar{\zeta}_k} w_j\overline{w}_k\right)\nabla\rho_\ell(p)$$

for $W = \sum_{k=1}^n w_k(\partial/\partial\zeta_k) \in H_p^{1,0}(M)$.

PROOF We start with the definition of $\widetilde{\mathcal{L}}_p$

$$\widetilde{\mathcal{L}}_p(W_p) = \frac{1}{2i}\tilde{\pi}_p\{J[\overline{W}, W]_p\}$$

for $W \in H^{1,0}(M)$. Since $\{\nabla\rho_1(p), \ldots, \nabla\rho_d(p)\}$ is an orthonormal basis for $N_p(M)$, the projection $\tilde{\pi}_p$: $T_p(\mathbb{C}^n) \to N_p(M)$ is given by

$$\tilde{\pi}_p(v) = \sum_{\ell=1}^{d}\langle d\rho_\ell(p), v\rangle\nabla\rho_\ell(p)$$

where $\langle\ ,\ \rangle$ denotes the pairing between one-forms and vectors. So the ℓth component of $\widetilde{\mathcal{L}}_p(W_p)$ is given by

$$\frac{1}{2i}\langle d\rho_\ell(p), J[\overline{W}, W]_p\rangle.$$

Using the dual of J, denoted J^*: $T_p^*(\mathbb{C}^n) \to T_p^*(\mathbb{C}^n)$, we obtain

$$\ell\text{th component of } \widetilde{\mathcal{L}}_p(W_p) = \frac{1}{2i}\langle J^* d\rho_\ell(p), [\overline{W}, W]_p\rangle.$$

Recall that $d = \partial + \bar{\partial}$; $J^* \circ \partial = i\partial$ and $J^* \circ \bar{\partial} = -i\bar{\partial}$. Therefore

$$\ell\text{th component of } \widetilde{\mathcal{L}}_p(W_p) = \frac{1}{2}\langle(\partial - \bar{\partial})\rho_\ell(p), [\overline{W}, W]_p\rangle. \tag{1}$$

Now we use the formula for the exterior derivative of a one-form in terms of its action on vectors (see Lemma 3 in Section 1.4). For a one-form ϕ and vector fields L_1, L_2, we have

$$\langle d\phi, L_1 \wedge L_2\rangle = L_1\{\langle\phi, L_2\rangle\} - L_2\{\langle\phi, L_1\rangle\} - \langle\phi, [L_1, L_2]\rangle. \tag{2}$$

We apply this formula with $\phi = (1/2)(\partial - \bar{\partial})\rho_\ell$ and $L_1 = \overline{W} \in H^{0,1}(M)$, $L_2 = W \in H^{1,0}(M)$. We have

$$\langle\phi, L_2\rangle = \frac{1}{2}\langle\partial\rho_\ell, W\rangle - \frac{1}{2}\langle\bar{\partial}\rho_\ell, W\rangle.$$

Since W is of type (1,0) and $\bar{\partial}\rho_\ell$ is a form of the bidegree (0,1), the second term on the right vanishes. The first term on the right also vanishes in view of Lemma 2 in Section 7.1. Therefore, we have $\langle\phi, L_2\rangle = 0$. Similarly, we have $\langle\phi, L_1\rangle = 0$. Equation (2) becomes

$$\langle d(\partial - \bar{\partial})\rho_\ell, \overline{W} \wedge W\rangle = -\langle(\partial - \bar{\partial})\rho_\ell, [\overline{W}, W]\rangle.$$

Comparing this with (1) yields

$$\begin{aligned}\ell\text{th component of } \widetilde{\mathcal{L}}_p(W_p) &= -\frac{1}{2}\langle d(\partial - \bar{\partial})\rho_\ell, \overline{W} \wedge W\rangle_p \\ &= -\langle\bar{\partial}\partial\rho_\ell, \overline{W} \wedge W\rangle_p \quad (\text{since } d = \partial + \bar{\partial}) \\ &= -\sum_{j,k=1}^{n} \frac{\partial^2 \rho_\ell(p)}{\partial\zeta_j \partial\bar{\zeta}_k} w_j \bar{w}_k,\end{aligned}$$

for $W_p = \sum_{k=1}^{n} w_k(\partial/\partial\zeta_k)$. The proof of the theorem is now complete. ∎

Theorem 1 is often used in conjunction with Lemma 1 in Section 7.2. In that lemma, the point $p \in M$ is the origin and coordinates $\zeta = (z = x + iy, w) \in \mathbb{C}^d \times \mathbb{C}^{n-d}$ are chosen so that the defining functions for M are $\rho_\ell(z, w) = y_\ell - h_\ell(x, w)$, $1 \leq \ell \leq d$, with $h_\ell(0) = 0$ and $Dh_\ell(0) = 0$. Note that $\nabla\rho_\ell(0) = \partial/\partial y_\ell$ and so $\{\nabla\rho_1(0), \ldots, \nabla\rho_d(0)\}$ is an orthonormal basis for $N_0(M)$. We identify $N_0(M)$ with $\mathbb{R}^d$ via the map $y = (y_1, \ldots, y_d) \longmapsto \sum_{\ell=1}^{d} y_\ell(\partial/\partial y_\ell) \in$

$N_0(M)$. We also identify $H_0^{1,0}(M)$ with $\mathbb{C}^{n-d}$ by the map

$$w = (w_1, \ldots, w_{n-d}) \longmapsto \sum_{k=1}^{n-d} w_k \frac{\partial}{\partial w_k}.$$

With these identifications, the restriction of the action of the complex hessian of ρ_ℓ to the directions in $H_0^{1,0}(M)$ is the same as the action of the $(w, \bar{w})$-hessian of ρ_ℓ on $\mathbb{C}^{n-d}$. We obtain the following corollary.

COROLLARY 1

Suppose $M = \{(x+iy, w) \in \mathbb{C}^d \times \mathbb{C}^{n-d}; y = h(x, w)\}$ where h: $\mathbb{R}^d \times \mathbb{C}^{n-d} \to \mathbb{R}^d$ is smooth and $h(0) = 0, Dh(0) = 0$. Then the extrinsic Levi form at 0 is given by

$$\widetilde{\mathcal{L}}_0(W) = (y_1, \ldots, y_d) \in \mathbb{R}^d$$

$$y_\ell = \sum_{j,k=1}^{n-d} \frac{\partial^2 h_\ell(0)}{\partial w_j \partial \bar{w}_k} w_j \bar{w}_k$$

for $W = (w_1, \ldots, w_{n-d}) \in \mathbb{C}^{n-d}$.

This corollary can also be established by expanding $[\overline{L}_k, L_j]$ where $L_1, \ldots, L_{n-d}$ is the local basis for $H^{1,0}(M)$ given in Theorem 3 in Section 7.2.

Let us specialize to the case of a quadric submanifold $M = \{y = q(w, \bar{w})\}$, where q: $\mathbb{C}^{n-d} \times \mathbb{C}^{n-d} \to \mathbb{C}^d$ is a quadratic form (Definition 1 in Section 7.3). From Corollary 1, the Levi form of M at the origin is given by

$$w = (w_1, \ldots, w_{n-d}) \longmapsto \sum_{j,k=1}^{n-d} \frac{\partial^2 q(0)}{\partial w_j \partial \bar{w}_k} w_j \bar{w}_k = q(w, \bar{w}) \in \mathbb{R}^d.$$

In other words, the Levi form at the origin of a quadric submanifold is the associated quadratic form q (restricted to $\{(w, \bar{w}); w \in \mathbb{C}^{n-d}\}$).

Theorem 2 in Section 7.3, which describes a normal form for codimension two quadrics in $\mathbb{C}^4$, can now be interpreted in terms of the Levi form. In this case, $N_0(M)$ is a copy of $\mathbb{R}^2$ and $H_0^{1,0}(M)$ is a copy of $\mathbb{C}^2$. In part (a) of that theorem, $M = \{y_1 = q_1(w, \bar{w}), y_2 = 0\}$ and the image of the Levi form of M at 0 is contained in a one-dimensional line (the y_1 axis) in $\mathbb{R}^2$. In all three cases in part (b), the image of the Levi form is a two-dimensional cone in $\mathbb{R}^2$. In case (i), $M = \{y_1 = |w_1|^2, y_2 = |w_2|^2\}$ and the image of the Levi form is the closed quadrant $\{y_1 \geq 0, y_2 \geq 0\}$. In case (ii), $M = \{y_1 = |w_1|^2, y_2 = \operatorname{Re}(w_1\bar{w}_2)\}$ and the image of the Levi form is the open half space $\{y_1 > 0\}$ together with the origin. In case (iii), $M = \{y_1 = \operatorname{Re}(w_1\bar{w}_2), y_2 = \operatorname{Im}(w_1\bar{w}_2)\}$ and the image of the Levi form is all of $\mathbb{R}^2 \simeq N_0(M)$. In all of these cases, the image of the Levi form is a convex cone in $N_0(M)$. In general, the image of the Levi form is a cone in $N_p(M)$. However, the image of the Levi form is not always convex, as we shall see in example (v) in Section 14.3.

10.3 The Levi form of a real hypersurface

One of the best studied classes of CR manifolds is the class of real hypersurfaces in $\mathbb{C}^n$. So we shall devote a section to the study of the Levi form for a real hypersurface in $\mathbb{C}^n$. Let $M = \{z \in \mathbb{C}^n; \rho(z) = 0\}$, where $\rho\colon \mathbb{C}^n \to \mathbb{R}$ is smooth. If p is a point on M with $|\nabla\rho(p)| = 1$, then from Theorem 1 in Section 10.2, the extrinsic Levi form is given by

$$\widetilde{\mathcal{L}}_p(W) = -\left(\sum_{j,k=1}^{n} \frac{\partial^2 \rho}{\partial\zeta_j \partial\bar{\zeta}_k}(p) w_j \bar{w}_k\right) \nabla\rho(p)$$

for $W = \sum_{k=1}^n w_k(\partial/\partial\zeta_k) \in H_p^{1,0}(M)$. In this case, $N_p(M)$ is isomorphic to a real line via the map $t \mapsto t\nabla\rho(p), t \in \mathbb{R}$. For this reason, $\nabla\rho(p)$ is often dropped and the Levi form is then identified with the restriction of the complex hessian of ρ to $H_p^{1,0}(M)$.

The above formula requires $|\nabla\rho(p)| = 1$ which can always be arranged by multiplying ρ by a suitable scalar. However, it is important to note that if $\tilde{\rho}$ is another defining equation for M with $d\tilde{\rho} \neq 0$ on M, then the map

$$W = (w_1, \ldots, w_n) \longmapsto \sum_{j,k=1}^{n} \frac{\partial^2 \tilde{\rho}(p)}{\partial\zeta_j \partial\bar{\zeta}_k} w_j \bar{w}_k \quad \text{for} \quad W \in H_p^{1,0}(M)$$

is a nonzero multiple of the Levi form at p. To see this, first note that $\tilde{\rho} = \alpha\rho$ for some smooth function $\alpha\colon \mathbb{C}^n \to \mathbb{R}$, which is nonzero near M (see Lemma 3 in Section 2.2). Therefore

$$\begin{aligned}\sum_{j,k=1}^{n} \frac{\partial^2 \tilde{\rho}}{\partial\zeta_j \partial\bar{\zeta}_k}(p) w_j \bar{w}_k = \alpha(p) \sum_{j,k=1}^{n} \frac{\partial^2 \rho(p)}{\partial\zeta_j \partial\bar{\zeta}_k} w_j \bar{w}_k & \\ +2\,\mathrm{Re}\left\{\left(\sum_{j=1}^{n} \frac{\partial\rho(p)}{\partial\zeta_j} w_j\right)\left(\sum_{k=1}^{n} \frac{\partial\alpha(p)}{\partial\bar{\zeta}_k} \bar{w}_k\right)\right\} & \\ +\rho(p) \sum_{j,k=1}^{n} \frac{\partial^2 \alpha(p)}{\partial\zeta_j \partial\bar{\zeta}_k} w_j \bar{w}_k. &\end{aligned}$$

The third term on the right vanishes because $\rho(p) = 0$ for $p \in M$. The second term also vanishes because $\sum_{j=1}^n (\partial\rho/\partial\zeta_j) w_j = 0$ for $W = \sum_{j=1}^n w_j(\partial/\partial\zeta_j) \in H^{1,0}(M)$ (by Lemma 2 in Section 7.1). Therefore, the complex hessian of $\tilde{\rho}$

applied to the vector $W = \sum_{j=1}^{n} w_j(\partial/\partial\zeta_j) \in H_p^{1,0}(M)$ differs from the Levi form of M at p by the factor $\alpha(p)$. In particular, information about the Levi form such as the number of nonzero eigenvalues can be determined by examining the complex hessian of any defining function for M.

A special case worth examining occurs when the Levi form is definite.

DEFINITION 1 *A real hypersurface M is called strictly pseudoconvex at a point $p \in M$ if the Levi form at p is either positive or negative definite, i.e., if there exists a defining function ρ for M so that*

$$\sum_{j,k=1}^{n} \frac{\partial^2 \rho}{\partial\zeta_j \partial\bar{\zeta}_k}(p) w_j \bar{w}_k > 0$$

for all $W = \sum_{j=1}^{n} w_j(\partial/\partial\zeta_j) \in H_p^{1,0}(M)$.

The above inequality is an open condition. Furthermore, it is invariant under a local biholomorphic change of coordinates. This follows by explicitly computing the complex hessian of $\rho \circ F$ where F is a biholomorphism or by using Lemma 2 in Section 10.1.

A real hypersurface M is called *strictly pseudoconvex* if M is strictly pseudoconvex at each point $p \in M$.

THEOREM 1

Suppose $M \subset \mathbb{C}^n$ is a smooth real hypersurface that is strictly pseudoconvex at a point $p \in M$. Then there is a biholomorphic map F defined on a neighborhood U of p in $\mathbb{C}^n$ so that $F\{M \cap U\}$ is a strictly convex hypersurface in $F\{U\} \subset \mathbb{C}^n$.

PROOF The idea of the proof is to holomorphically change variables so that the *real* hessian of the defining function in the new variables is positive definite.

First, we choose coordinates $(z, w) \in \mathbb{C} \times \mathbb{C}^{n-1}$ as in Lemma 1 in Section 7.2 so that p is the origin and

$$M = \{(z = x + iy, w) \in \mathbb{C} \times \mathbb{C}^{n-1}; y = h(x, w)\}$$

where $h\colon \mathbb{R} \times \mathbb{C}^{n-1} \to \mathbb{R}$ is smooth and $h(0) = 0, Dh(0) = 0$. By Theorem 2 in Section 7.2 (with $k = 2$), we may assume there are no second-order pure terms in the expansion of h about the origin, i.e.,

$$\frac{\partial^2 h(0)}{\partial w_j \partial w_k} = \frac{\partial^2 h(0)}{\partial x_j \partial w_k} = 0.$$

Let $\rho(z, w) = y - h(x, w)$. We have

$$\rho(z, w) = y + \sum_{j,k=1}^{n-1} \frac{\partial^2 \rho(0)}{\partial w_j \partial \bar{w}_k} w_j \bar{w}_k + \mathcal{O}(3)$$

where $\mathcal{O}(3)$ denotes terms that vanish to third order in x and w (i.e., $|\mathcal{O}(3)| \le C(|x|^3 + |w|^3)$. We may assume the quadratic expression in w and $\bar{w}$ is positive definite (the negative definite case is similar).

Now, we modify ρ and make a holomorphic change of coordinates so that the quadratic piece of the defining function in the new coordinates is positive definite in $z, \bar{z}$ as well as $w, \bar{w}$. Let

$$\tilde{\rho} = \rho + 2\rho^2.$$

Note that $\tilde{\rho}$ is also a defining function for M. We have

$$\begin{aligned}\tilde{\rho}(z,w) &= y + 2y^2 + \sum_{j,k=1}^{n-1} \frac{\partial^2 \rho(0)}{\partial w_j \partial \bar{w}_k} w_j \bar{w}_k + \mathcal{O}(3) \\ &= y - \operatorname{Re}(z^2) + |z|^2 + \sum_{j,k=1}^{n-1} \frac{\partial^2 \rho(0)}{\partial w_j \partial \bar{w}_k} w_j \bar{w}_k + \mathcal{O}(3). \qquad (1)\end{aligned}$$

Define the following change of variables, $(\hat{z}, \hat{w}) = F(z,w)$

$$\hat{z} = z - iz^2, \qquad z \in \mathbb{C}$$

$$\hat{w} = w, \qquad w \in \mathbb{C}^{n-1}$$

F is a local biholomorphism which preserves the set $\{(0,w); w \in \mathbb{C}^{n-1}\}$. If $\hat{z} = \hat{x} + i\hat{y}$, then

$$\begin{aligned}\hat{y} &= y - \operatorname{Re}(z^2) \\ |\hat{z}|^2 &= |z|^2 + \mathcal{O}(3). \qquad (2)\end{aligned}$$

Let $\widehat{M} = F\{M\}$ and let $\hat{\rho}(\hat{z}, \hat{w}) = \tilde{\rho}(F^{-1}(\hat{z},\hat{w})) = \tilde{\rho}(z,w)$. A defining equation for $\widehat{M}$ is given by $\hat{\rho}(\hat{z},\hat{w}) = 0$. Using (1) and (2), we obtain

$$\hat{\rho}(\hat{z},\hat{w}) = \hat{y} + |\hat{z}|^2 + \sum_{j,k=1}^{n-1} \frac{\partial^2 \rho(0)}{\partial w_j \partial \bar{w}_k} \hat{w}_j \bar{\hat{w}}_k + \mathcal{O}(3).$$

Clearly, the quadratic piece of $\hat{\rho}$ is positive definite in $\hat{z}, \hat{w}$. Therefore, $\widehat{M} = \{(\hat{z},\hat{w}); \hat{\rho}(\hat{z},\hat{w}) = 0\}$ is strictly convex in a neighborhood of the origin. ∎

If the complex hessian of the defining equation of M is only positive semidefinite on $H_p^{1,0}(M)$ for each $p \in$M, then the hypersurface is called *pseudoconvex*. The analogue of Theorem 1 does not hold for pseudoconvex hypersurfaces. There is an example (see [KN]) of a real hypersurface that is strictly pseudoconvex everywhere except at one point p_0 which is not biholomorphic near p_0 to any (weakly) convex hypersurface in $\mathbb{C}^n$.

We conclude this chapter with the comparison of the Levi form of a real hypersurface in $\mathbb{C}^n$ with its second fundamental form. Our presentation is similar to that in [Tai]. First, we review the definition of the second fundamental form. Suppose M is a real hypersurface in $\mathbb{R}^N$ that locally separates $\mathbb{R}^N$ in two open sets D and $\mathbb{R}^N - \overline{D}$. Let $\mathbb{N}$ be the outward pointing unit normal vector field to D on M. We assume M is locally oriented according to $\mathbb{N}$, which means that a collection of vectors $X_1, \ldots, X_{N-1}$ in $T_p(M)$ is considered positively oriented if $\{\mathbb{N}_p, X_1, \ldots, X_{N-1}\}$ has the same orientation as $\mathbb{R}^N$.

Suppose $W = \sum w_j(\partial/\partial x_j)$ is a vector field on $\mathbb{R}^N$ and let V_p be an element of $T_p(\mathbb{R}^N)$. Define the vector $\nabla_{V_p} W \in T_p(\mathbb{R}^N)$ by

$$\nabla_{V_p} W = \sum_{j=1}^{N} V_p\{w_j\} \frac{\partial}{\partial x_j}.$$

In other words, $\nabla_{V_p} W$ is the derivative of W in the direction of V_p. If both V and W are vector fields, then $[V, W]_p = \nabla_{V_p} W - \nabla_{W_p} V$.

DEFINITION 2 *The second fundamental form is the map $\mathbb{I}_p$: $T_p(M) \times T_p(M) \to \mathbb{R}$ defined by*

$$\mathbb{I}_p(V_p, W_p) = -(\nabla_{V_p}\mathbb{N}) \cdot W_p \quad \textit{for} \quad V_p, W_p \in T_p(M)$$

where $(\cdot)$ is the Euclidean inner product on $\mathbb{R}^N$.

In the next lemma, we derive a formula for $\mathbb{I}_p$ in terms of the real hessian of the following defining function for M

$$\rho(x) = \begin{cases} -\text{dist}(x, M) & \text{if } x \in D \\ \text{dist}(x, M) & \text{if } x \in \mathbb{R}^N - D. \end{cases}$$

If M is C^∞ then ρ is C^∞ near M in $\mathbb{R}^N$ and $\nabla\rho = \mathbb{N}$ on M.

LEMMA 1

Let $p \in M$, and suppose V_p and W_p are elements in $T_p(M)$. Suppose V and W are any $T(M)$-vector field extensions of V_p and W_p respectively. Then

(a) $\mathbb{I}_p(V_p, W_p) = (\nabla_{V_p} W) \cdot N_p$

(b) *If $V_p = \sum_{j=1}^N v_j(\partial/\partial x_j)$ and $W_p = \sum_{k=1}^N w_k(\partial/\partial x_k)$, then*

$$\mathbb{I}_p(V_p, W_p) = -\sum_{j,k=1}^{N} \frac{\partial^2 \rho(p)}{\partial x_j \partial x_k} v_j w_k.$$

PROOF For part a), note that $\mathbb{N} \cdot W = 0$, since $W \in T(M)$ and $\mathbb{N}$ is the unit normal. From the product rule, we have

$$0 = V\{\mathbb{N} \cdot W\} = (\nabla_V \mathbb{N}) \cdot W + \mathbb{N} \cdot (\nabla_V W)$$

and part (a) follows. For (b), write $\mathbb{N} = \nabla \rho = \sum (\partial \rho / \partial x_k)(\partial / \partial x_k)$. Then

$$\nabla_{V_p} \mathbb{N} = \sum_{j,k=1}^{N} \frac{\partial \rho^2(p)}{\partial x_j \partial x_k} v_j \frac{\partial}{\partial x_k}.$$

Taking the inner product of this vector with W_p yields the formula in part (b). ∎

Note that part (b) shows that $\mathbb{I}_p(\cdot, \cdot)$ is a symmetric bilinear form.

To compare the Levi form with the second fundamental form, the first problem to overcome is that $\mathbb{I}_p$ is defined on the real tangent space whereas $\widetilde{\mathcal{L}}_p$ is defined on $H_p^{1,0}(M)$ which is a subspace of the complexified tangent space of M. If W_p is an element of $H_p^{1,0}(M)$ then $W_p = X_p - iJX_p$ where $X_p = 1/2(W_p + \overline{W}_p) \in H_p(M) \subset T_p(M)$. If X is a $H(M)$-vector field extension of X_p, then $W = X - iJX$ is a $H^{1,0}(M)$-vector field extension of W_p. We have

$$\begin{aligned}
\widetilde{\mathcal{L}}_p(W_p) &= \tilde{\pi}_p \left\{ \frac{1}{2i} J[\overline{W}, W]_p \right\} \\
&= \tilde{\pi}_p \left\{ \frac{1}{2i} J[X + iJX, X - iJX]_p \right\} \\
&= -\tilde{\pi}_p \{ J[X, JX]_p \}.
\end{aligned}$$

So for the purposes of comparing $\widetilde{\mathcal{L}}_p$ and $\mathbb{I}_p$, let us identify X_p with $W_p = X_p - iJX_p$. Then $\widetilde{\mathcal{L}}_p$ is identified with the map $\widetilde{\mathcal{L}}_p(X_p) = -\tilde{\pi}_p\{J[X, JX]_p\}$, for $X_p \in H_p(M)$.

The projection $\tilde{\pi}_p \colon T_p(\mathbb{C}^n) \to N_p(M)$ is given by $\tilde{\pi}_p(V) = (V \cdot \mathbb{N}_p)\mathbb{N}_p$. Therefore

$$\widetilde{\mathcal{L}}_p(X_p) = -J[X, JX]_p \cdot \mathbb{N}_p.$$

The Lie bracket $[X, JX]$ can be expressed as $\nabla_X(JX) - \nabla_{JX}(X)$. We also have $J(\nabla_X Y) = \nabla_X JY$ by explicit computation. Therefore

$$\begin{aligned}
\widetilde{\mathcal{L}}_p(X_p) &= -J\{\nabla_{X_p}(JX) - \nabla_{JX_p}(X)\} \cdot \mathbb{N}_p \\
&= (\nabla_{X_p}(X)) \cdot \mathbb{N}_p + (\nabla_{JX_p}(JX)) \cdot \mathbb{N}_p \\
&= \mathbb{I}_p(X_p, X_p) + \mathbb{I}_p(JX_p, JX_p) \quad \text{(from Lemma 1)}.
\end{aligned}$$

Note that if X_p is a vector in the J-invariant subspace $H_p(M)$, then JX_p also belongs to $H_p(M) \subset T_p(M)$ and therefore $\mathbb{I}_p(JX_p, JX_p)$ is well defined. We have established the following theorem.

THEOREM 2

$\widetilde{\mathcal{L}}_p(X_p) = \mathbb{I}_p(X_p, X_p) + \mathbb{I}_p(JX_p, JX_p)$, for $X_p \in H_p(M)$, where $X_p \in H_p(M)$ is identified with $W_p = X_p - iJX_p \in H_p^{1,0}(M)$ in the definition of the Levi form.

If M is convex, then the second fundamental form is positive (or negative) semidefinite. If $\widetilde{\mathcal{L}}_p(X_p) = 0$ then by Theorem 2, both $\mathbb{I}_p(X_p, X_p)$ and $\mathbb{I}_p(JX_p, JX_p)$ must vanish. In this case, both X_p and JX_p are null vectors (i.e., in the 0-eigenspace) for the second fundamental form. Therefore

$$\mathbb{I}_p(X_p, Y_p) = \mathbb{I}_p(JX_p, Y_p) = 0 \quad \text{for all} \quad Y_p \in T_p(M).$$

We have

$$\begin{aligned} 0 &= -\mathbb{I}_p(X_p, Y_p) \\ &= (\nabla_{X_p}\mathbb{N}) \cdot Y_p \end{aligned} \tag{3}$$

for all $Y_p \in T_p(M)$. Now for any $X_p \in T_p(M)$, $\nabla_{X_p}\mathbb{N}$ is an element of $T_p(M)$ because

$$0 = X_p\{1\} = X_p\{\mathbb{N} \cdot \mathbb{N}\} = 2(\nabla_{X_p}\mathbb{N}) \cdot \mathbb{N}.$$

Therefore (3) implies $\nabla_{X_p}\mathbb{N} = 0$. Likewise, we have $\nabla_{JX_p}\mathbb{N} = 0$. This fact has the following geometric interpretation for a convex hypersurface: if X_p is a null vector for the Levi form then the derivatives of the unit normal vector field in the directions of X_p and JX_p both vanish.

11

The Imbeddability of CR Manifolds

In Section 11.1, we show that any abstract real analytic CR structure is locally CR equivalent — via a real analytic CR diffeomorphism — to a generic, real analytic CR submanifold of $\mathbb{C}^n$. Nirenberg's C^∞ counterexample presented in Section 11.2 shows that without additional hypothesis, the corresponding theorem for C^∞ CR structures is false. Additional imbedding results will be discussed in Chapter 12.

11.1 The real analytic imbedding theorem

Recall that the CR codimension of M is the number $d = \dim_{\mathbb{C}}\{T^{\mathbb{C}}(M)/\mathbb{L}\oplus\overline{\mathbb{L}}\}$. If M is a generic CR submanifold of $\mathbb{C}^n$, then the CR codimension of M is the same as the real codimension of M. To say that a CR structure $(M, \mathbb{L})$ is *real analytic* means that M is a real analytic manifold and that $\mathbb{L}$ is a real analytic subbundle of $T^{\mathbb{C}}(M)$, i.e., $\mathbb{L}$ is locally generated by real analytic vector fields. Now we state the imbedding theorem, which first appeared in [AnHi1].

THEOREM 1
Suppose $(M, \mathbb{L})$ is an abstract real analytic CR structure with CR codimension $= d \geq 1$. Given any point $p_0 \in M$, there is a neighborhood U of p_0 in M so that $(M \cap U, \mathbb{L})$ is CR equivalent via a real analytic CR map to a generic real analytic CR submanifold of complex Euclidean space with codimension d.

One of the defining properties of a CR structure $(M, \mathbb{L})$ is that $\mathbb{L}$ is an involutive subbundle of $T^{\mathbb{C}}(M)$. One might be tempted to think that this theorem should follow from the real Frobenius theorem (which does not require the real analyticity of $\mathbb{L}$). However, the real Frobenius theorem requires the underlying real subbundle of $\mathbb{L}$ (i.e., $H(M) = \{L + \overline{L}; L \in \mathbb{L}\}$) to be involutive. If $H(M)$ is involutive, then $\mathbb{L} \oplus \overline{\mathbb{L}}$ is involutive. This is equivalent to saying that M is

Levi flat which is not assumed here. Instead, we shall use the real analyticity of M and $\mathbb{L}$ to complexify $\mathbb{L}$ and then use the complex analytic version of the Frobenius theorem.

PROOF Suppose that $m = \dim_{\mathbb{C}} \mathbb{L} = \dim_{\mathbb{C}} \overline{\mathbb{L}}$. Near p_0 in M, $\overline{\mathbb{L}}$ is generated by m-real analytic vector fields $\{\overline{L}_1, \ldots, \overline{L}_m\}$. Since d is the CR codimension of $(M, \mathbb{L})$, we have $2m + d = \dim_{\mathbb{C}} T^{\mathbb{C}}(M) = \dim_{\mathbb{R}}(M)$. Using a local real analytic coordinate system for M, we may assume that M is an open subset of $\mathbb{R}^{2m+d}$ containing the origin and that each $\overline{L}_j$ is a real analytic vector field in $T^{\mathbb{C}}(\mathbb{R}^{2m+d})$. Denote the coordinates of $\mathbb{R}^{2m+d}$ by $(u_1, \ldots, u_{2m+d})$. We write

$$(\overline{L}_j)_u = \sum_{k=1}^{2m+d} a_{jk}(u) \frac{\partial}{\partial u_k} \qquad 1 \le j \le m$$

where each a_{jk} is a real analytic, complex-valued function of $u \in \mathbb{R}^{2m+d}$. Since $\{\overline{L}_1, \ldots, \overline{L}_m\}$ is linearly independent, the matrix $(a_{jk}(0))$ $1 \le j \le m$, $1 \le k \le 2m + d$, has complex rank m. By reordering the coordinates if necessary, we may assume the $m \times m$ block $A = (a_{jk})_{j,k=1}^{m}$ is nonsingular in a neighborhood U of the origin in $\mathbb{R}^{2m+d}$. Let $u = (t, x), t \in \mathbb{R}^m, x \in \mathbb{R}^{m+d}$ be the coordinates for $\mathbb{R}^{2m+d}$. By multiplying the coefficients of $\{\overline{L}_1, \ldots, \overline{L}_m\}$ by A^{-1}, we obtain another basis for $\overline{\mathbb{L}}$ (over U) of the form $\{\overline{L}_1, \ldots, \overline{L}_m\}$ where

$$\overline{L}_j = \frac{\partial}{\partial t_j} + \sum_{k=1}^{m+d} \lambda_{jk}(t, x) \frac{\partial}{\partial x_k} \qquad 1 \le j \le m.$$

Here, each λ_{jk} is a real analytic, complex-valued function of $u = (t, x) \in U \subset \mathbb{R}^{2m+d}$.

The Lie bracket $[\overline{L}_j, \overline{L}_k]$ has no $(\partial/\partial t_\ell)$-component. On the other hand, $[\overline{L}_j, \overline{L}_k]$ is a linear combination of $\{\overline{L}_1, \ldots, \overline{L}_m\}$ because $\overline{\mathbb{L}}$ is involutive. Any nontrivial linear combination of $\{\overline{L}_1, \ldots, \overline{L}_m\}$ contains a nontrivial linear combination of $\partial/\partial t_1, \ldots, \partial/\partial t_m$. Therefore, we conclude

$$[\overline{L}_j, \overline{L}_k] = 0 \qquad 1 \le j, k \le m.$$

Now we complexify each $\overline{L}_j$. Let $\zeta \in \mathbb{C}^m$ and $z \in \mathbb{C}^{m+d}$ be the complexifications of $t \in \mathbb{R}^m$ and $x \in \mathbb{R}^{m+d}$, respectively (so Re $\zeta = t$ and Re $z = x$). By replacing t and x by ζ and z in the power series expansion of λ_{jk} (about 0), we obtain functions $\tilde{\lambda}_{jk}$: $\mathbb{C}^{2m+d} \to \mathbb{C}$ which are holomorphic in a neighborhood $\widetilde{U}$ of the origin in $\mathbb{C}^{2m+d}$ with $\tilde{\lambda}_{jk}(t, x) = \lambda_{jk}(t, x)$. Define

$$\tilde{L}_j = \frac{\partial}{\partial \zeta_j} + \sum_{k=1}^{m+d} \tilde{\lambda}_{jk}(\zeta, z) \frac{\partial}{\partial z_k} \qquad 1 \le j \le m.$$

Since $\tilde{\lambda}_{jk}(\zeta, z)$ is holomorphic in ζ and z, we have

$$\frac{\partial \tilde{\lambda}_{jk}(t,x)}{\partial \zeta_\ell} = \frac{\partial \lambda_{jk}(t,x)}{\partial t_\ell} \quad \text{and} \quad \frac{\partial \tilde{\lambda}_{jk}(t,x)}{\partial z_\ell} = \frac{\partial \lambda_{jk}(t,x)}{\partial x_\ell}.$$

Together with $[\overline{L}_j, \overline{L}_k] = 0$, we obtain

$$[\tilde{L}_j, \tilde{L}_k]_{(t,x)} = 0 \quad \text{for} \quad (t,x) \in U \subset \mathbb{R}^{2m+d}.$$

From the identity theorem for holomorphic functions, we obtain

$$[\tilde{L}_j, \tilde{L}_k] \equiv 0 \quad \text{on} \quad \widetilde{U} \subset \mathbb{C}^{2m+d}.$$

From Lemma 1 in Section 4.2 (with $n = 2m+d$), there is a holomorphic map $\widetilde{Z} = (\widetilde{Z}_1, \dots, \widetilde{Z}_{m+d})\colon \mathbb{C}^m \times \mathbb{C}^{m+d} \to \mathbb{C}^{m+d}$ which is defined on a possibly smaller neighborhood $\widetilde{U}$ of the origin in $\mathbb{C}^{2m+d}$, so that

$$\begin{aligned} &\tilde{L}_j \widetilde{Z}_k \equiv 0 \quad \text{on} \quad \widetilde{U} \quad 1 \le j \le m, \quad 1 \le k \le m+d \\ &\widetilde{Z}(0,z) = z \quad \text{for} \quad (0,z) \in \widetilde{U}. \end{aligned}$$

Here, we are only using the last $(m+d)$-components of the map $Z\colon \mathbb{C}^m \times \mathbb{C}^{m+d} \to \mathbb{C}^{2m+d}$ given in this lemma. Define the real analytic map $Z\colon \widetilde{U} \cap \{\mathbb{R}^m \times \mathbb{R}^{m+d}\} \to \mathbb{C}^{m+d}$ by

$$Z(t,x) = \widetilde{Z}(t,x) \quad \text{for} \quad (t,x) \in \mathbb{R}^m \times \mathbb{R}^{m+d}.$$

We have

$$\overline{L}_j Z_k = \tilde{L}_j \widetilde{Z}_k = 0 \quad \text{on} \quad \widetilde{U} \cap \{\mathbb{R}^m \times \mathbb{R}^{m+d}\}.$$

In addition

$$Z(0,x) = \widetilde{Z}(0,x) = x \quad \text{for} \quad (0,x) \in \widetilde{U}.$$

We claim that Z is our desired imbedding.

Since $\overline{L}_j Z_k = 0$ for $1 \le j \le m$, $1 \le k \le m+d$, Z is a CR map by Lemma 1 in Section 9.2 and Lemma 1 in Section 9.1. So, it suffices to show that the real derivative at the origin of Z as a map from $\mathbb{R}^{2m+d}$ to $\mathbb{C}^{m+d} \simeq \mathbb{R}^{2m+2d}$ has maximal rank. Write $Z(t,x) = U(t,x) + iV(t,x)$. We must show that the matrix

$$\mathcal{M} = \begin{pmatrix} \frac{\partial U}{\partial t}(0) & \frac{\partial U}{\partial x}(0) \\ \frac{\partial V}{\partial t}(0) & \frac{\partial V}{\partial x}(0) \end{pmatrix}$$

has real rank $2m + d$. Since $x = Z(0,x) = U(0,x) + iV(0,x)$, we have

$$\frac{\partial U}{\partial x}(0) = I \quad \frac{\partial V}{\partial x}(0) = 0.$$

The imaginary part of the equation $\overline{L}_j\{Z_k\} = 0$ can be described in the following matrix form:

$$\frac{\partial V}{\partial t} + \left(\frac{\partial U}{\partial x}\right)(\text{Im }\lambda)^t + \left(\frac{\partial V}{\partial x}\right)(\text{Re }\lambda)^t = 0.$$

Here, Re λ and Im λ are the $m \times (m + d)$ matrices whose (j, k)th entry is Re $\lambda_{jk}(t, x)$ and Im $\lambda_{jk}(t, x)$, respectively. By evaluating this equation at the origin and using $(\partial U/\partial x)(0) = I$ and $(\partial V/\partial x)(0) = 0$, we obtain

$$\frac{\partial V}{\partial t}(0) = -(\text{Im }\lambda)^t(0).$$

Therefore, the matrix $\mathcal{M}$ has maximal rank if and only if Im $\lambda(0)$ has rank m. It is here that we use the assumption that $\mathbb{L} \cap \overline{\mathbb{L}} = \{0\}$ (from the definition of a CR structure). This implies that the set $\{L_1 - \overline{L}_1, \ldots, L_m - \overline{L}_m\}$ is linearly independent over $\mathbb{C}$. For $1 \leq j \leq m$, the term

$$\frac{1}{2i}(\overline{L}_j - L_j) = \sum_{k=1}^{m+d} \text{Im }\lambda_{jk}\frac{\partial}{\partial x_k}$$

is essentially the jth row of the matrix Im λ. It follows that Im $\lambda(0)$ has rank m and so $\mathcal{M}$ has maximal rank, as desired. Therefore, there is a neighborhood U of the origin in $\mathbb{R}^{2m+d}$ such that

$$M' = Z\{U\}$$

is a real analytic submanifold of $\mathbb{C}^{m+d}$. The real codimension of M' in $\mathbb{C}^{m+d}$ is d because $\dim_{\mathbb{R}} M' = \dim_{\mathbb{R}} U = 2m + d$. The proof of Theorem 1 is now complete. ∎

11.2 Nirenberg's nonimbeddable example

In this section, we present Nirenberg's example [Nir] which shows that the imbedding theorem fails for C^∞ CR structures. Our treatment of this example is taken from [JT]. Nirenberg's example is a three-dimensional C^∞ CR structure that cannot be imbedded into $\mathbb{C}^2$. Defining a three-dimensional CR structure is equivalent to specifying a vector field $L \in T^{\mathbb{C}}(\mathbb{R}^3)$ so that L and $\overline{L}$ are linearly independent. A CR map $Z = (Z_1, Z_2)$: $\mathbb{R}^3 \to \mathbb{C}^2$ must satisfy $\overline{L}Z_1 = 0, \overline{L}Z_2 = 0$. We will construct $\overline{L}$ so that any two functions Z_1, Z_2: $\mathbb{R}^3 \to \mathbb{C}$ with $\overline{L}Z_1 = \overline{L}Z_2 = 0$ must also satisfy $dZ_1 \wedge dZ_2 = 0$ at the origin and therefore the map $Z = (Z_1, Z_2)$: $\mathbb{R}^3 \to \mathbb{C}^2$ cannot be an imbedding of any neighborhood of the origin of $\mathbb{R}^3$.

The desired vector field $\overline{L}$ will be a perturbation of the generator of the CR structure for the Heisenberg group. Recall the Heisenberg group in $\mathbb{C}^2$ is given

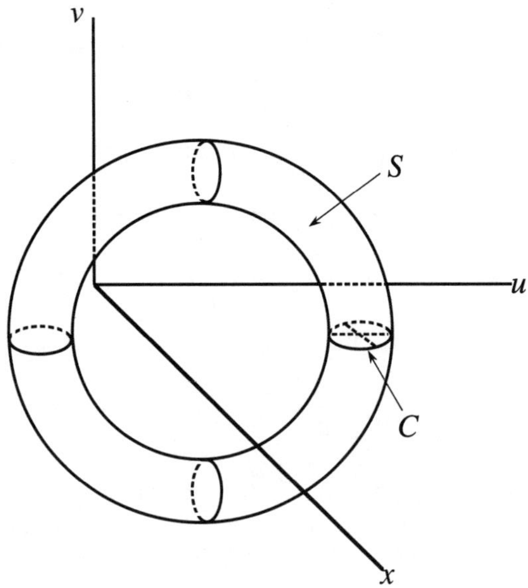

FIGURE 11.1

by

$$M = \{(z = x + iy, w) \in \mathbb{C}^2; y = |w|^2\}.$$

The graphing function for M is given by H: $\mathbb{R}^3 \to \mathbb{C}^2$

$$H(x, w) = (x + i|w|^2, w).$$

The real tangent space of M at 0 is the copy of $\mathbb{R}^3$ with the coordinates (x, u, v) where $w = u + iv$. Let C be a circle in the copy of $\mathbb{R}^2$ given by $\{v = 0\}$. The circle C is constructed so that the u-coordinate of each point in C is positive. Each point $p_0 = (x_0, u_0, 0) \in C$ is contained in the unique circle

$$C_{p_0} = \{(x_0, u, v); u^2 + v^2 = u_0^2\}.$$

Let

$$S = \bigcup_{p \in C} C_p.$$

S is a two-dimensional torus in $\mathbb{R}^3$. We define T to be the open solid torus in $\mathbb{R}^3$ whose boundary is S. T is obtained by filling in the circle C with a disc, D, and sweeping it around the torus.

With $w = u + iv$, let

$$\widetilde{C} = H\{C\} = \{(x + i|w|^2, w);\ (x, w) \in C\} \subset M$$

$$\widetilde{S} = H\{S\} = \{(x + i|w|^2, w);\ (x, w) \in S\} \subset M$$

$$\widetilde{T} = H\{T\} = \{(x + i|w|^2, w);\ (x, w) \in T\} \subset M.$$

Also let π_1: $\mathbb{C}^2 \to \mathbb{C}$ be the projection $\pi_1(z, w) = z$. Define

$$C_1 = \pi_1\{\widetilde{C}\} \subset \mathbb{C}$$

$$S_1 = \pi_1\{\widetilde{S}\} \subset \mathbb{C}$$

$$T_1 = \pi_1\{\widetilde{T}\} \subset \mathbb{C}.$$

Note that $\widetilde{C}$, $\widetilde{S}$, and $\widetilde{T}$ are circled in the w-variable, which means that if the point (z_0, w_0) belongs to $\widetilde{S}$ (or $\widetilde{C}$ or $\widetilde{T}$), then the circle $\{(z_0, w); |w| = |w_0|\}$ is also contained in $\widetilde{S}$ (or $\widetilde{C}$ or $\widetilde{T}$). Therefore, C_1 is a simple closed curve in $\mathbb{C}$, $S_1 = C_1$, and T_1 is the open set in $\mathbb{C}$ whose boundary is C_1.

LEMMA 1
$\pi_1\{M - \widetilde{T}\}$ is a connected subset of $\{z \in \mathbb{C};\ Im\ z \geq 0\}$ which contains $\{Im\ z = 0\}$ (the x-axis).

PROOF Note that $M - \widetilde{T}$ is circled in the w-variable and so $\pi_1\{M - \widetilde{T}\} = \{z \in \mathbb{C};\ \text{Im } z \geq 0\} - T_1$, which is connected. C does not intersect the x-axis and so neither does C_1. Thus, the x-axis is contained in $\pi_1\{M - \widetilde{T}\}$, as desired. ∎

We remark that if $\widetilde{T}_i$ is a countable sequence of nonoverlapping tori constructed as above, then the lemma still holds with $\widetilde{T}$ replaced by $\cup_i \widetilde{T}_i$.

From Theorem 3 in Section 7.2, the generator for $H^{0,1}(M)$ is given by

$$\overline{\widetilde{L}}_1 = -2iw\frac{\partial}{\partial \bar{z}} + \frac{\partial}{\partial \bar{w}}.$$

Let π: $\mathbb{C}^2 \to \mathbb{R}^3$ be the projection given by $\pi(x + iy, w) = (x, w)$. Define

$$\overline{L}_1 = \pi_*\{\overline{\widetilde{L}}_1\} = -iw\frac{\partial}{\partial x} + \frac{\partial}{\partial \bar{w}}.$$

A CR function on M can be viewed either as a function $\tilde{f}$: $M \to \mathbb{C}$ with $\overline{\widetilde{L}}_1 \tilde{f} = 0$ on M or as a function $f = \tilde{f} \circ H$: $\mathbb{R}^3 \to \mathbb{C}$ with $\overline{L}_1 f = 0$ on $\mathbb{R}^3$.

For $z \in \mathbb{C}$ with Im $z \geq 0$, let $\widetilde{\Gamma}(z)$ be the circle $\{(z, w); |w|^2 = \text{ Im } z\}$. So $\widetilde{\Gamma}(z) = \pi_1^{-1}\{z\} \cap M$.

LEMMA 2
Suppose $\tilde{f}$: $M \to \mathbb{C}$ is a C^1 function.

(a) *If $\overline{\tilde{L}}_1 \tilde{f} = 0$ on an open set $D \subset M$, then the function*

$$z \mapsto F(z) = \int_{\widetilde{\Gamma}(z)} \tilde{f} dw$$

is holomorphic on the set $\{z \in \mathbb{C}; \widetilde{\Gamma}(z) \subset D\}$.

(b) *If $\overline{\tilde{L}}_1 \tilde{f} = 0$ on $M - \widetilde{T}$ and $\widetilde{\Gamma}(z) \subset M - \widetilde{T}$, then*

$$\int_{\widetilde{\Gamma}(z)} \tilde{f} dw = 0.$$

(c) *If $\overline{\tilde{L}}_1 \tilde{f} = 0$ on $M - \widetilde{T}$, then*

$$\iint_{\widetilde{S}} \tilde{f} dw \wedge dz = 0.$$

PROOF For the proof of part (a), we parameterize the circle $\widetilde{\Gamma}(z) = \{(z, w); |w| = \sqrt{\text{Im } z}\}$ by $w = w(z, \phi) = \sqrt{\text{Im } z} e^{i\phi}$ for $0 \le \phi \le 2\pi$. Using Theorem 2 in Section 9.1, we can extend $\tilde{f}$ ambiently so that $\bar{\partial} \tilde{f} = 0$ on $D \subset M$. Here, all we need is that $\bar{\partial} \tilde{f}$ vanish on $D \subset M$ to first order. We have

$$F(z) = \int_{\widetilde{\Gamma}(z)} \tilde{f} dw = \int_0^{2\pi} \tilde{f}(z, w(z, \phi)) w_\phi(z, \phi) d\phi.$$

Since $\partial \tilde{f} / \partial \bar{z} = \partial \tilde{f} / \partial \bar{w} = 0$ on $D \subset M$, we have

$$\begin{aligned} \frac{\partial F}{\partial \bar{z}} &= \int_0^{2\pi} (\tilde{f}_w(z, w) \cdot w_{\bar{z}} w_\phi + \tilde{f}(z, w) w_{\phi \bar{z}}) d\phi \\ &= \int_0^{2\pi} (\tilde{f}(z, w) w_{\bar{z}})_\phi d\phi \\ &= 0 \quad \text{for} \quad z \text{ with } \widetilde{\Gamma}(z) \subset D. \end{aligned}$$

For (b), if $\overline{\tilde{L}}_1 \tilde{f} = 0$ on $M - \widetilde{T}$, then part (a) implies that F is holomorphic on the interior of the set $\{z \in \mathbb{C}; \widetilde{\Gamma}(z) \subset M - \widetilde{T}\}$. F is also continuous on the closure of this set. By Lemma 1, this set is connected and contains $\{\text{Im } z = 0\}$. When $\text{Im } z = 0, \widetilde{\Gamma}(z)$ degenerates to a point and so $F(z) = 0$. Therefore, $F \equiv 0$ on $\{z; \widetilde{\Gamma}(z) \subset M - \widetilde{T}\}$, as desired.

Part (c) will follow from the fact that $\widetilde{S}$ is foliated by curves $\widetilde{\Gamma}(z)$ which satisfy part (b). We parameterize $C_1 = \pi_1\{H(C)\}$ by $\theta \mapsto z(\theta), 0 \le \theta \le 2\pi$. Therefore, $\widetilde{S}$ is parameterized by $(\phi,\theta) \mapsto (z(\theta), w(\phi,\theta))$ with $w(\phi,\theta) = \sqrt{\text{Im } z(\theta)}e^{i\phi}$. We have

$$\iint_{\widetilde{S}} \tilde{f} dw \wedge dz = \int_0^{2\pi}\int_0^{2\pi} \tilde{f}(z(\theta), w(\phi,\theta))z'(\theta)w_\phi(\phi,\theta)d\phi d\theta$$

$$= \int_0^{2\pi}\left(\int_0^{2\pi} \tilde{f}(z(\theta), w(\phi,\theta))w_\phi(\phi,\theta)d\phi\right) z'(\theta)d\theta$$

$$= \int_0^{2\pi}\left(\int_{\widetilde{\Gamma}(z(\theta))} \tilde{f}dw\right) z'(\theta)d\theta.$$

Since $\widetilde{\Gamma}(z(\theta)) \subset \widetilde{S} \subset M - \widetilde{T}$, this last integral vanishes by part (b), as desired. ∎

Lemma 2 is valid if $\widetilde{T}$ and $\widetilde{S}$ are replaced by a nonoverlapping union of such tori. In this case, part (c) is valid for each torus in this union.

LEMMA 3

Suppose D is an open set with smooth boundary in M. Let $\tilde{f}$: $\overline{D} \to \mathbb{C}$ be a C^1 function. With $w = u + iv$

$$\iint_{\partial D} \tilde{f} dw \wedge dz = 2i \iiint_D \overline{L}_1 \tilde{f} dx du dv$$

where $\overline{L}_1 = (\partial/\partial\bar{w}) - 2iw(\partial/\partial\bar{z})$ is the generator for $H^{0,1}(M)$.

PROOF We use Stokes' theorem to obtain

$$\iint_{\partial D} \tilde{f} dw \wedge dz = \iiint_D (d\tilde{f}) \wedge dw \wedge dz.$$

We have

$$d\tilde{f} \wedge dw \wedge dz = \bar{\partial}\tilde{f} \wedge dw \wedge dz$$

$$= \left(\frac{\partial \tilde{f}}{\partial \bar{z}} d\bar{z} + \frac{\partial \tilde{f}}{\partial \bar{w}} d\bar{w}\right) \wedge dw \wedge dz.$$

We also have $z = x + i|w|^2$ on M and so $dz = dx + iwd\bar{w} + i\bar{w}dw$ on M.

Therefore

$$d\tilde{f} \wedge dw \wedge dz = \left(\frac{\partial \tilde{f}}{\partial \bar{w}} - 2iw \frac{\partial \tilde{f}}{\partial \bar{z}} \right) dx \wedge d\bar{w} \wedge dw.$$

The lemma now follows after rewriting $d\bar{w} \wedge dw$ as $2idu \wedge dv$. ∎

Now we construct Nirenberg's nonimbeddable example. Let $\{T_i\}$ be a sequence of nonoverlapping tori in $\mathbb{R} \times \mathbb{C}$ which converge to the origin as $i \to \infty$. Each T_i is constructed like the torus, T, given at the beginning of this section. Let g: $\mathbb{R} \times \mathbb{C} \to \mathbb{R}$ be a nonnegative C^∞ function that is positive on each T_i. The function g must vanish to infinite order at the origin. Let $\overline{L} = \overline{L}_1 + g(\partial/\partial x)$, where $\overline{L}_1 = (\partial/\partial \bar{w}) - iw(\partial/\partial x)$. The vector field $\overline{L}$ agrees with $\overline{L}_1$ to infinite order at the origin. Clearly, L and $\overline{L}$ are linearly independent (over $\mathbb{C}$) at each point in a neighborhood U of the origin in $\mathbb{R} \times \mathbb{C}$. Since $[L, L] \equiv 0$, the subbundle $\mathbb{L}$ generated by L is involutive. Therefore, $(U, \mathbb{L})$ defines a three-dimensional CR structure on $U \subset \mathbb{R} \times \mathbb{C}$.

This CR structure is not CR equivalent to any three-dimensional CR submanifold of $\mathbb{C}^n$, $n \geq 1$, as the next theorem shows.

THEOREM 1
Suppose $\overline{L} = \overline{L}_1 + g(\partial/\partial x)$, where g is constructed as above. Suppose Z_1, Z_2: $\mathbb{R} \times \mathbb{C} \mapsto \mathbb{C}$ are C^1 functions with $\overline{L}Z_1 = \overline{L}Z_2 = 0$ near the origin in $\mathbb{R} \times \mathbb{C}$. Then $dZ_1 \wedge dZ_2 = 0$ at the origin. In particular, the derivative at the origin of any CR map Z: $\mathbb{R} \times \mathbb{C} \to \mathbb{C}^n$ must have real rank at most 2 and therefore Z cannot be an imbedding of any neighborhood of the origin of $\mathbb{R} \times \mathbb{C}$.

PROOF Let H: $\mathbb{R} \times \mathbb{C} \to M$ be the graphing function for the Heisenberg group $(H(x, w) = (x + i|w|^2, w))$. As in the beginning of this section, we let $\tilde{\overline{L}}_1 = H_*(\overline{L}_1)$. Define

$$\tilde{\overline{L}} = H_*(\overline{L}) = \tilde{\overline{L}}_1 + \tilde{g} \frac{\partial}{\partial x}$$

where $\tilde{g}$: $M \to \mathbb{R}$ is defined by the equation $\tilde{g} \circ H = g$. For a function f: $\mathbb{R} \times \mathbb{C} \to \mathbb{C}$, define $\tilde{f}$: $M \to \mathbb{C}$ by $\tilde{f} \circ H = f$. The equation $\overline{L}f = 0$ on a neighborhood U of $\mathbb{R} \times \mathbb{C}$ is equivalent to the equation $\tilde{\overline{L}}\tilde{f} = 0$ on $H\{U\} \subset M$. So if $\overline{L}f = 0$, then

$$\tilde{\overline{L}}_1 \tilde{f} = -\tilde{g} \frac{\partial}{\partial x} \tilde{f}.$$

Since g vanishes on $\mathbb{R} \times \mathbb{C} - \{\cup_i T_i\}$, we have $\tilde{\overline{L}}_1 \tilde{f} = 0$ on $M - \{\cup_i \widetilde{T}_i\}$. Therefore

$$\iint_{\widetilde{S}_i} \tilde{f} dw \wedge dz = 0$$

by part (c) of Lemma 2. Using Lemma 3, we obtain

$$0 = \iint_{\widetilde{S}_i} \tilde{f} dw \wedge dz = 2i \iiint_{\widetilde{T}_i} \tilde{\overline{L}}_1 \tilde{f} dx \wedge du \wedge dv$$

$$= -2i \iiint_{\widetilde{T}_i} \tilde{g} \frac{\partial \tilde{f}}{\partial x} dx \wedge du \wedge dv.$$

Since $\tilde{g} > 0$ on each $\widetilde{T}_i$, each of the functions $\mathrm{Re}(\partial \tilde{f}/\partial x)$, $\mathrm{Im}(\partial \tilde{f}/\partial x)$ must vanish at a point in each $\widetilde{T}_i$. Equivalently, each of the functions $\mathrm{Re}(\partial f/\partial x)$ and $\mathrm{Im}(\partial f/\partial x)$ must vanish at a point in T_i. Since the T_i converge to $\{0\}$ as $i \mapsto \infty$, $\partial f/\partial x$ must vanish at the origin. From the equations $(\overline{L}_1 + g(\partial/\partial x))f = 0$ and $(\overline{L}_1)_0 = \partial/\partial\bar{w}$ we have $\partial f(0)/\partial \bar{w} = 0$. Since both $(\partial f/\partial \bar{w})(0) = 0$ and $(\partial f/\partial x)(0) = 0$, we obtain

$$df(0) = \frac{\partial f}{\partial w}(0) dw.$$

Therefore, if $\overline{L} Z_1 = \overline{L} Z_2 = 0$ near the origin in $\mathbb{R} \times \mathbb{C}$, then

$$dZ_1(0) \wedge dZ_2(0) = \left(\frac{\partial Z_1(0)}{\partial w}\right)\left(\frac{\partial Z_2(0)}{\partial w}\right) dw \wedge dw$$
$$= 0.$$

This completes the proof of Theorem 1. ∎

12

Further Results

In this chapter, we discuss further results, whose proofs are too involved to include in a book of reasonable length. Instead, we refer the reader to the references. We start this chapter with some refinements of the normal form given in section 7.2. Then we discuss the Levi form in more detail. The chapter ends with some facts concerning nongeneric CR submanifolds.

12.1 Bloom–Graham normal form

Suppose that M is a real hypersurface in $\mathbb{C}^n$. If the Levi form of M at a point $p \in M$ is not identically zero, then the totally real direction of the tangent space at p can be expressed as the projection of a Lie bracket of the form $[\overline{L}, L]_p$ for some $L \in H^{1,0}(M)$. If the Levi form of M at p vanishes identically, then it still may be the case that the totally real direction of the tangent space can be expressed as some higher order Lie bracket generated by vector fields in $H^{\mathbb{C}}(M) = H^{1,0}(M) \oplus H^{0,1}(M)$. This leads us to define the concept of a point of higher type.

To make this concept precise, we need some definitions. For $j \geq 2$ and a point $p \in M$, we call an operator of the form

$$\left[L_1, [L_2, \ldots [L_{j-1}, L_j] \ldots]\right]_p, \quad L_k \in H^{\mathbb{C}}(M)$$

a *Lie bracket of length j at p generated by* $H^{\mathbb{C}}(M)$. Let $\mathcal{L}_p^1(M)$ be $H_p^{\mathbb{C}}(M)$. For $j \geq 2$, we define $\mathcal{L}_p^j(M)$ as the vector subspace of $T_p^{\mathbb{C}}(M)$ spanned (over $\mathbb{C}$) by $H_p^{\mathbb{C}}(M)$ and all Lie brackets of length $k \leq j$ at p generated by $H^{\mathbb{C}}(M)$. Note that $\mathcal{L}_p^j(M) \subset \mathcal{L}_p^{j+1}(M)$. The dimension of $\mathcal{L}_p^j(M)$ typically varies with p and so the union (over p) of these spaces typically cannot be considered as a subbundle of $T^{\mathbb{C}}(M)$. If M is a real hypersurface, then we say that $p \in M$ is a point of *type* m if $\mathcal{L}_p^{m-1}(M) = H_p^{\mathbb{C}}(M)$ and $\mathcal{L}_p^m(M) = T_p^{\mathbb{C}}(M)$. In other words, a point $p \in M$ is a point of type m if the totally real direction of the tangent space of M at p can be expressed as the projection of a Lie bracket of length m at p generated by $H^{\mathbb{C}}(M)$ but not by any Lie bracket of length less than m. We should mention that there is a different notion of type defined by D'Angelo [D] which is defined in terms of the order of contact of complex analytic varieties.

Now we express the type of a point in terms of coordinates. Suppose the point p is the origin and suppose M is graphed over its tangent space at the origin in the manner described by Theorem 2 in Section 7.2. Coordinates for $\mathbb{C}^n$ are given by (z, w) where $z = x + iy \in \mathbb{C}$ and $w \in \mathbb{C}^{n-1}$. The defining equation for M is given by $y = h(x, w)$ where $h : \mathbb{R} \times \mathbb{C}^{n-1} \mapsto \mathbb{R}$ is smooth (say C^∞) and where there are no pure terms through order $m + 1$ in the Taylor expansion of h at the origin. Let $L_1, \ldots, L_{n-1}$ be the local basis for $H^{1,0}(M)$ given in Theorem 3 in Section 7.2. If the origin is a point of type m then the $(\partial/\partial x)$-component of any Lie bracket of the L'^s and the $\overline{L}'^s$ of length less than m at 0 vanishes. Together with the fact that there are no pure terms in the expansion of h, we can inductively show that all the w and $\overline{w}$ derivatives of h through order $m - 1$ at the origin vanish. We have the following expansion of h:

$$h(x, w) = p(w, \overline{w}) + e(x, w)$$

where p is a homogeneous polynomial of degree m in w and $\overline{w}$ with no pure terms and where e is a smooth function that satisfies the estimate

$$|e(x, w)| \leq C(|w|^{m+1} + |x||w| + |x|^2)$$

for some uniform constant $C > 0$. The variable w is the coordinate for $H^{1,0}(M)$ and we assign the *weight* 1 to w and $\overline{w}$. The variable x is the coordinate for the totally real tangent space direction of M at the origin. Since this direction can be expressed as the projection of a Lie bracket at the origin of length m generated by $H^{\mathbb{C}}(M)$, we assign the *weight* m to the variable x. Likewise, we also assign the weight m to the normal variable y and to the complex coordinates $z = x + iy$ and $\overline{z} = x - iy$. The weight of a monomial $w^\alpha \overline{w}^\beta z^j \overline{z}^k$ is by definition the number $|\alpha| + |\beta| + m(j + k)$. By definition, the weight of a smooth function is the minimal weight of all of the monomials appearing in its formal Taylor expansion about the origin. So the homogeneous polynomial p has weight m whereas e has weight greater than m. Note that in an unweighted sense, the polynomial p may vanish at the origin to higher order than does e. For example, suppose

$$h(x, w) = |w|^2 \text{Re}\{w^2\} + |w|^2 x_1.$$

Then the origin is a point of type 4 and the term $e(x, w) = |w|^2 x_1$ has weight 6. In an unweighted sense, e vanishes at the origin to third order.

The above discussion for hypersurfaces generalizes to submanifolds of $\mathbb{C}^n$ of codimension $d > 1$. In this case, we follow Bloom and Graham [BG] and say that a point $p \in M$ has type $(m_1, \ldots, m_d)$ (with $m_j \leq m_k$ for $j < k$) if the following conditions hold:

(a) $\mathcal{L}_p^j(M) = H_p^{\mathbb{C}}(M)$ for $j < m_1$ and thus $\dim_{\mathbb{C}} \mathcal{L}_p^j(M) = 2n - 2d$.

(b) $\dim_{\mathbb{C}} \mathcal{L}_p^j(M) = 2n - 2d + i$ for all j such that $m_i \leq j < m_{i+1}$.

(c) $\dim_{\mathbb{C}} \mathcal{L}_p^j(M) = 2n - d = \dim_{\mathbb{C}} T_p^{\mathbb{C}}(M)$ for all j such that $j \geq m_d$.

The numbers $m_1, \ldots, m_d$ are also called the *Hörmander numbers.* Condition (b) is vacuous if $m_i = m_{i+1}$. Note that we allow the case where $m_j = \infty$. We say that the point p has *finite type* if all the m_j are finite.

Suppose the point $p \in M$ is the origin and suppose the defining equation for M is $y = h(x, w)$ where $h : \mathbb{R}^d \times \mathbb{C}^{n-d} \mapsto \mathbb{R}^d$ is smooth. If M is rigid (i.e., h is independent of the variable x) then the type of the origin can be determined by examining the order of vanishing (in w) of the components of h. For example, if $M = \{(z_1, z_2, w) \in \mathbb{C}^3;\ x_1 = |w|^2,\ x_2 = |w|^2 \text{Re} w\}$, then the type of the origin is $(2, 3)$. For another example, suppose

$$M = \left\{(z_1, z_2, z_3, w) \in \mathbb{C}^4;\ x_1 = |w|^4,\ x_2 = |w|^2 \text{Re}\{w^2\},\ x_3 = |w|^6\right\}.$$

In this case, the type of the origin is $(4, 4, 6)$.

The nonrigid case is more complicated. We assign weights to the coordinates as follows. As with the hypersurface case, we assign weight 1 to the variables w and $\overline{w}$. Let $h = (h_1, \ldots, h_d)$. We assume $h_1, \ldots, h_d$ are ordered so that $h_1(0, w)$ has the smallest weight among the functions $h_1(0, w), \ldots, h_d(0, w)$. Let l_1 be the weight of $h_1(0, w)$. We assign weight l_1 to the variable z_1 (and to $\overline{z}_1$ and x_1, y_1). If this collection of functions vanishes to infinite order, then we assign the weight of ∞ to z_1 (and to $\overline{z}_1, x_1, y_1$). Next, we order $h_2, \ldots, h_d$ so that $h_2(x_1, 0, \ldots, w)$ has the smallest weight among the functions $h_2(x_1, 0, 0 \ldots, w), \ldots, h_d(x_1, 0, 0 \ldots, w)$. Let l_2 be the weight of $h_2(x_1, 0, 0 \ldots, w)$. We assign the weight l_2 to the variable z_2 (and to $\overline{z}_2$, x_2, y_2). Continuing in this way, we assign weights $(l_1, \ldots, l_d)$ to all the coordinates $(z_1, \ldots, z_d)$. We have $l_i \leq l_{i+1}$. Unlike the rigid case, the weights of these coordinates, in general, are *not* the same as the numbers $(m_1, \ldots, m_d)$ coming from the type of the origin. In fact, Bloom and Graham show that if the origin is a point of type $(m_1, \ldots, m_d)$ and if the weights $(l_1, \ldots, l_d)$ are assigned as above, then $m_i \geq l_i$ for $1 \leq i \leq d$.

Suppose the weights of the z-coordinates are $(l_1, \ldots, l_d)$. We say that the manifold $M = \{y = h(x, w)\}$ is presented in *Bloom–Graham normal form* if the following three conditions on h are met:

(a) For $1 \leq i \leq d$

$$h_i(x, w) = p_i(x_1, \ldots, x_{i-1}, w, \overline{w}) + e_i(x, w)$$

where p_i is a homogeneous polynomial of weight l_i and the weight of e_i is greater than l_i.

(b) There are no pure terms in the polynomial p_i (i.e., there are no terms of the form $x^\alpha w^\beta$ or $x^\alpha \overline{w}^\beta$).

(c) For $1 \leq j < i$ and nonnegative integers $\alpha_1, \ldots, \alpha_j$

$$\left(\frac{\partial}{\partial x_1}\right)^{\alpha_1} \cdots \left(\frac{\partial}{\partial x_j}\right)^{\alpha_j} \pi_j(p_i)(0) = 0$$

where π_j is the differential operator $p_j(\partial/\partial x, \partial/\partial w, \partial/\partial \overline{w})$ (i.e., just replace x by $\partial/\partial x$ etc. in the expression for p_j).

In requirement (a), p_i is not allowed to depend on the variables $x_i, \ldots, x_d$. Requirement (c) means that there are no terms of the form $x_1^{\alpha_1} \ldots x_j^{\alpha_j} p_j$ contained in p_i. For example, if

$$M = \{(z_1, z_2, w) \in \mathbb{C}^3;\ y_1 = |w|^2,\ y_2 = x_1|w|^2\}$$

then the defining functions for M satisfy requirements (a) and (b) but not (c). (because $p_2 = x_1 p_1$).

The main theorem in Bloom and Graham's paper is that if a point $p \in M$ has type $(m_1, \ldots, m_d)$ with all the m_j finite, then there is a local biholomorphic change of coordinates so that in the new coordinates, the defining equations satisfy Bloom and Graham's normal form with $l_i = m_i$ for $1 \leq i \leq d$. Conversely, they show that if M is already presented in their normal form, then the origin is a point of type $(m_1, \ldots, m_d) = (l_1, \ldots, l_d)$.

Example 1

Suppose $M = \{(z_1, z_2, w) \in \mathbb{C}^3;\ y_1 = |w|^2,\ y_2 = x_1|w|^2 \mathrm{Re} w\}$. Then M is presented in Bloom–Graham normal form. The weight of the (z_1, z_2) coordinates is $(2, 5)$ and the type of the origin is also $(2, 5)$. □

Example 2

Suppose $M = \{(z_1, z_2, w) \in \mathbb{C}^3;\ y_1 = |w|^2,\ y_2 = x_1|w|^2\}$. M is *not* presented in normal form since condition (c) is not satisfied as mentioned above. Since $y_1 = |w|^2$ on M, we can replace p_2 by the expression $x_1 y_1$ which is the imaginary part of $(1/2)z_1^2$. We can make the change of variables $\hat{z}_1 = z_1$ and $\hat{z}_2 = z_2 - (1/2)z_1^2$. In the new variables (drop the hat), the defining equations for the manifold M become $y_1 = |w|^2$ and $y_2 = 0$. Thus, the type of the origin is $(2, \infty)$. This example illustrates the importance of condition (c) in the Bloom–Graham normal form. A naive count of the weights of the original defining functions for M might lead one to erroneously think that the type of the origin is $(2, 4)$. This example also illustrates some of the ideas used to prove Bloom–Graham's theorem mentioned above. Another illustration is provided in the following example. □

Example 3

(See Section 6 in [BG].) Suppose

$$M = \{(z_1, z_2, z_3, w) \in \mathbb{C}^4;\ y_1 = |w|^6,\ y_2 = x_1|w|^2,\ y_3 = x_2^2 x_1 |w|^2\}.$$

This manifold is not presented in Bloom–Graham normal form because $p_3 = x_2^2 p_2$. We note that p_3 can be replaced by $x_2^2 y_2$ and furthermore

$$\frac{1}{3}\mathrm{Im} z_2^3 = x_2^2 y_2 - \frac{y_2^3}{3}.$$

We change variables by setting $\hat{z}_1 = z_1$ and $\hat{z}_2 = z_2$ and $\hat{z}_3 = z_3 - (1/3)z_2^3$. In the new variables (drop the hat), the defining equations for M become

$$y_1 = |w|^6,\ \ y_2 = x_1|w|^2,\ \ y_3 = \frac{y_2^3}{3} = \frac{x_1^3}{3}|w|^6.$$

The manifold is still not presented in Bloom–Graham normal form because $p_3 = (1/3)x_1^3 p_1$. We apply the same procedure again. We have $(1/12)\mathrm{Im} z_1^4 = (1/3)(x_1^3 y_1 - x_1 y_1^3)$. We let $\hat{z}_3 = z_3 - (1/12)z_1^4$. After dropping the hat, we see that in the new coordinates, the defining equations for M become

$$y_1 = |w|^6,\ \ y_2 = x_1|w|^2,\ \ y_3 = \frac{1}{3}x_1|w|^{18}.$$

The manifold M is now presented in Bloom–Graham normal form. The type of the origin is $(6, 8, 24)$. $\square$

12.2 Rigid and semirigid submanifolds

At the end of Section 7.2, we said that a submanifold is rigid if its defining equation has the form $y = h(w)$. As before, the coordinates for $\mathbb{C}^n$ are given by (z, w) where $z = x + iy \in \mathbb{C}^d$ and $w \in \mathbb{C}^{n-d}$ (here, d is the real codimension of M). The point is that for a rigid submanifold, the graphing function h is independent of the variable $x \in \mathbb{R}^d$.

In [BRT], Baouendi, Rothschild, and Treves present a coordinate invariant description of rigidity. To describe this condition, we make a few definitions. First, we say that a vector subspace $\mathcal{G}$ (over $\mathbb{C}$) of the space of smooth sections of the tangent bundle $T^{\mathbb{C}}(M)$ to a manifold M is a *Lie subalgebra* if $\mathcal{G}$ is involutive (i.e., closed under the Lie bracket $[\ ,\ ]$ operation). A subalgebra is called *abelian* if $[L_1, L_2] = 0$ for $L_1, L_2 \in \mathcal{G}$. We say that an abstract CR structure $(M, \mathbb{L})$ is *invariant under a transversal Lie group action* if there is a finite dimensional Lie subalgebra, $\mathcal{G}$, of $T^{\mathbb{C}}(M)$ with the following properties:

$$\mathbb{L} \oplus \overline{\mathbb{L}} \oplus \mathcal{G} = T^{\mathbb{C}}(M)$$

$$[\overline{L}, G] \in \overline{\mathbb{L}} \quad \text{for } \overline{L} \in \overline{\mathbb{L}} \text{ and } G \in \mathcal{G}.$$

If M is a rigid submanifold of $\mathbb{C}^n$, then the vector fields

$$\overline{L}_j = \frac{\partial}{\partial \overline{w}_j} - 2i \sum_{l=1}^{d} \frac{\partial h_l}{\partial \overline{w}_j} \frac{\partial}{\partial \overline{z}_l} \qquad 1 \leq j \leq n-d.$$

generate $H^{0,1}(M)$ by Theorem 3 in Section 7.2. Since the coefficients of each $\overline{L}_j$ are independent of x, the Lie bracket of each $\overline{L}_j$ with the vector field $\partial/\partial x_k$ vanishes. Therefore, if M is a rigid submanifold of $\mathbb{C}^n$, then $(M, H^{1,0}(M))$ is invariant under a transversal Lie group action given by the abelian subalgebra generated by linear combinations (over $\mathbb{C}$) of the vector fields $\partial/\partial x_1, \ldots, \partial/\partial x_d$.

In [BRT], Baouendi, Rothschild, and Treves establish the converse: if an abstract CR structure $(M, \mathbb{L})$ is invariant under a transversal abelian Lie group action, then the manifold can be locally imbedded into $\mathbb{C}^n$ with coordinates (z, w), $z = x + iy \in \mathbb{C}^d$ where d is the CR codimension of M and where the defining equation for M is given by $y = h(w)$.

Another class of CR manifolds of interest is the class of semirigid manifolds. If the CR manifold is imbedded into $\mathbb{C}^n$ and presented in Bloom–Graham normal form, then we say that M is *semirigid* if each p_i is independent of the variable x. The graphing function $h = (h_1, \ldots, h_d)$ for M is allowed to depend on x but the terms of lowest weight in each h_i are required to be independent of x.

In [BR2], Baouendi and Rothschild give a coordinate invariant description of semirigidity. To describe it, we need some more definitions. For an abstract CR structure $(M, \mathbb{L})$, let $X(M)$ be a complement to $\mathbb{L} \oplus \overline{\mathbb{L}}$ as explained in Section 8.2. If M is imbedded into $\mathbb{C}^n$, then $X(M)$ is the totally real tangent bundle to M. Let $X^*(M)$ be the space of one-forms that are dual to $X(M)$. For $p \in M$ and $\phi \in X^*(M)$, we let $m(p, \phi)$ be the smallest integer m for which there exists a Lie bracket of the form

$$L^m = \big[M_1, [M_2, \ldots, [M_{m-1}, M_m] \ldots], \big] \quad M_j \in \mathbb{L} \oplus \overline{\mathbb{L}} \tag{1}$$

with $\langle \phi, L^m \rangle_p \neq 0$. If no such finite integer m exists then we set $m(p, \phi) = \infty$. As usual, the notation $\langle \ , \ \rangle$ denotes the pairing between one-forms and vectors.

The abstract CR structure $(M, \mathbb{L})$ is defined to be *semirigid* at a point $p \in M$ if for each $\phi \in X_p^*(M)$, we have $\langle \phi, [L^k, L^j] \rangle_p = 0$ for all commutators L^j and L^k of the form (1) with $j \geq 2$, $k \geq 2$, and $j + k \leq m(p, \phi)$. Baouendi and Rothschild's result states that if an abstract CR structure is semirigid, then near any point of finite type, it can be locally imbedded into $\mathbb{C}^n$ so that in its Bloom–Graham normal form, the polynomials $p_1, \ldots, p_d$ are independent of the variable x. They also show that the class of semirigid manifolds contain (1) the class of all real hypersurfaces in $\mathbb{C}^n$; (2) manifolds in which the largest Hörmander number is at most 3; (3) the class of CR structures $(M, \mathbb{L})$ where the complex dimension of $\mathbb{L}$ is one and the largest Hörmander number is at most 4; and (4) the class of manifolds where all the Hörmander numbers are the same or, more generally, if the difference between any two Hörmander numbers m_i is at most one.

12.3 More on the Levi form

In Section 10.1, we show that a Levi flat CR manifold $(M, \mathbb{L})$ is locally foliated by complex submanifolds. Furthermore, for each $p \in M$, $\mathbb{L}_p$ is the complex tangent space at p of the leaf of the foliation that passes through p. This result can be strengthened. First, let us extend the Levi form to a bilinear map

$$\mathcal{L}_p : \{\mathbb{L}_p \oplus \overline{\mathbb{L}}_p\} \times \{\mathbb{L}_p \oplus \overline{\mathbb{L}}_p\} \mapsto \frac{\{T_p(M) \otimes \mathbb{C}\}}{\{\mathbb{L}_p \oplus \overline{\mathbb{L}}_p\}}$$

by

$$\mathcal{L}_p(X_p, Y_p) = \frac{1}{2i}\pi_p[X, Y]_p$$

where π_p is the projection

$$\pi_p : T_p(M) \otimes \mathbb{C} \mapsto \frac{\{T_p(M) \otimes \mathbb{C}\}}{\{\mathbb{L}_p \oplus \overline{\mathbb{L}}_p\}}.$$

Here, X and $Y \in \mathbb{L} \oplus \overline{\mathbb{L}}$ are vector field extensions of the vectors X_p and Y_p. The Levi form defined in Chapter 10 is the restriction of the above map to the case where $Y_p \in \mathbb{L}_p$ and $X_p = \overline{Y}_p \in \overline{\mathbb{L}}_p$. As with the usual Levi form, this definition of $\mathcal{L}_p(X_p, Y_p)$ is independent of the vector field extensions $X, Y \in \mathbb{L} \oplus \overline{\mathbb{L}}$.

For $p \in M$, let

$$\mathbb{N}_p = \{X_p \in \mathbb{L}_p \oplus \overline{\mathbb{L}}_p;\ \mathcal{L}_p(X_p, Y_p) = 0 \text{ for every } Y_p \in \mathbb{L}_p \oplus \overline{\mathbb{L}}_p\}.$$

$\mathbb{N}_p$ is called the *Levi null set at p*. As a simple example, let $M = \{(z, w_1, w_2) \in \mathbb{C}^3; \text{Re}\, z = |w_1|^2\}$. In this case, a basis for $\overline{\mathbb{L}}$ is given by $\partial/\partial\overline{w}_2$ and $L_1 = (\partial/\partial\overline{w}_1) - 2iw_1(\partial/\partial\overline{z})$. Since the coefficients of L_1 are independent of w_2, we have $[\partial/\partial w_2, L_1] = [\partial/\partial w_2, \overline{L}_1] = 0$. Therefore, the Levi null set at any point $p \in M$ is spanned (over $\mathbb{C}$) by the vectors $\partial/\partial\overline{w}_2$, $\partial/\partial w_2$. Note that M is foliated by one-dimensional complex manifolds that are translates of the w_2-axis.

We can generalize the above example to more general CR structures $(M, \mathbb{L})$. If the dimension of $\mathbb{N}_p$ is independent of $p \in M$, then $\mathbb{N} = \cup_p \mathbb{N}_p$ forms a subbundle of $\mathbb{L} \oplus \overline{\mathbb{L}}$. We have

$$\mathbb{N} = \{X \in \mathbb{L} \oplus \overline{\mathbb{L}};\ [X,Y] \in \mathbb{L} \oplus \overline{\mathbb{L}} \ \text{ for each } Y \in \mathbb{L} \oplus \overline{\mathbb{L}}\}. \tag{1}$$

Under this constant rank assumption on $\mathbb{N}$, M can be foliated by complex manifolds such that for each $p \in M$, $\mathbb{N}_p$ is the complex tangent space to the leaf passing through p. The proof of this result (due to M. Freeman [Fr]) uses similar arguments to those in the proof of Theorem 1 in Section 10.1. The key idea is to show that the subbundle $\mathbb{N}$ is involutive, i.e., $[\mathbb{N},\mathbb{N}] \subset \mathbb{N}$. To see that $\mathbb{N}$ is involutive, let X and Y belong to $\mathbb{N}$ and let Z be an element of $\mathbb{L} \oplus \overline{\mathbb{L}}$. Then from (1), $[X,Z]$ and $[Y,Z]$ belong to $\mathbb{L} \oplus \overline{\mathbb{L}}$. Hence, $[[X,Z],Y]$ and $[[Y,Z],X]$ also belong to $\mathbb{L} \oplus \overline{\mathbb{L}}$. From the Jacobi identity

$$[[X,Y],Z] + [[Z,X],Y] + [[Y,Z],X] = 0$$

we conclude that $[[X,Y],Z]$ also belongs to $\mathbb{L} \oplus \overline{\mathbb{L}}$. This means $[X,Y]$ belongs to $\mathbb{N}$ and so $\mathbb{N}$ is involutive. It follows that the underlying real subbundle which generates $\mathbb{N}$ is involutive and Freeman's result follows from the real Frobenius theorem and the Newlander–Nirenberg theorem.

We should mention that the Frobenius theorem and the Levi form can also be viewed from the differential form point of view rather than the vector field point of view taken earlier in this book. Suppose M is a manifold and $\mathbb{L}$ is a subbundle of the real tangent bundle, $T(M)$. Let $\mathcal{I} \subset T^*(M)$ be the annihilator of $\mathbb{L}$, i.e.,

$$\mathcal{I} = \{\phi \in T^*(M);\ \langle \phi, L\rangle = 0 \ \text{ for all } L \in \mathbb{L}\}.$$

$\mathcal{I}$ is a subbundle of $T^*(M)$. The condition that $\mathbb{L}$ is involutive is equivalent to the condition that $d\phi \in \mathcal{I} \wedge T^*(M)$ for each $\phi \in \mathcal{I}$. This is easily seen from the equation

$$\langle d\phi, L_1 \wedge L_2\rangle = L_1\{\langle \phi, L_2\rangle\} - L_2\{\langle \phi, L_1\rangle\} - \langle \phi, [L_1, L_2]\rangle$$

(see Lemma 3 in Section 1.4). The differential form version of the Frobenius theorem is the following: if $\mathcal{I}$ is a k-(real) dimensional subbundle of one-forms with $d\{\mathcal{I}\} \subset \mathcal{I} \wedge T^*(M)$, then there exist smooth functions $u_1, \ldots, u_k$ such that $du_1, \ldots du_k$ locally generate $\mathcal{I}$ over the ring of smooth functions on M. This version follows from Theorem 1 in Section 4.1.

For a CR structure $(M, \mathbb{L})$, the annihilator of $\mathbb{L} \oplus \overline{\mathbb{L}}$ is $X^*(M)$. The Levi form is designed to measure the degree to which $\mathbb{L} \oplus \overline{\mathbb{L}}$ fails to be involutive. Therefore from the differential form point of view, the Levi form at p is the map $\phi_p \mapsto \pi_p(d\phi)$ for $\phi_p \in \mathcal{I}_p$, where π_p is the projection

$$\pi_p : \Lambda^2 T_p^{*\mathbb{C}}(M) \mapsto \frac{\{\Lambda^2 T_p^{*\mathbb{C}}(M)\}}{\{\mathcal{I}_p \wedge T_p^{*\mathbb{C}}(M)\}}.$$

where $\phi \in \mathcal{I}$ is any extension of ϕ_p. Note that if this Levi form vanishes for all $p \in M$, then $d\{\mathcal{I}\} \subset \mathcal{I} \wedge T^{*\mathbb{C}}(M)$ and so $\mathbb{L} \oplus \overline{\mathbb{L}}$ is involutive. In this case, M is Levi flat.

12.4 Kuranishi's imbedding theorem

Nirenberg's example of a nonimbeddable abstract CR manifold shows that the real analytic imbedding theorem in Chapter 11 does not hold, in general, for the class of C^∞ CR manifolds. In [Ku], Kuranishi shows that with the additional assumption of strict pseudoconvexity and a restriction on the dimension, then a given C^∞ CR structure can be locally imbedded into $\mathbb{C}^n$. The dimension restriction requires that the underlying manifold M have real dimension $2n-1$ with $n \geq 4$ and that the CR codimension of the CR structure $(M, \mathbb{L})$ is one. In general, the image of the Levi form at a point $p \in M$ (see Definition 1 in Section 10.1) is a real subspace of $(T_p(M) \otimes \mathbb{C}/\{\mathbb{L}_p \oplus \overline{\mathbb{L}}_p\})$. If the CR codimension is one, then the complex dimension of $(T_p(M) \otimes \mathbb{C}/\{\mathbb{L}_p \oplus \overline{\mathbb{L}}_p\})$ is one, and therefore the Levi form can be considered a real valued map. A CR structure $(M, \mathbb{L})$ is then said to be *strictly pseudoconvex at p* if the Levi form is either always positive or always negative. Under these assumptions, Kuranishi's result states that the CR structure is locally CR equivalent to an imbedded hypersurface in $\mathbb{C}^n$. Kuranishi's result only handles CR structures of dimension at least 7. Nirenberg's nonimbeddable example is a strictly pseudoconvex CR structure of real dimension 3. It is still unknown whether or not a strictly pseudoconvex CR structure of real dimension 5 can be locally imbedded as a real hypersurface in $\mathbb{C}^3$.

12.5 Nongeneric and non-CR manifolds

Within the class of imbedded CR manifolds, most of our concern in this book is with the class of generic submanifolds. One reason for this is that "most" submanifolds of $\mathbb{C}^n$ are generic. Another reason is that a nongeneric CR submanifold can always be locally presented as the graph of a CR map over a generic CR submanifold of some lower dimensional complex Euclidean space. This is noted in [HT2].

To see this, suppose M is a nongeneric submanifold of $\mathbb{C}^n$. Suppose d is the real codimension of M and let k be the CR codimension of M. As explained after the proof of Lemma 1 in Section 7.2, $d-k$ must be a positive even integer which we denote by $2j$. Furthermore, there exist coordinates (z, ζ, w) for $\mathbb{C}^n$ where $z \in \mathbb{C}^j$, $\zeta = x + iy \in \mathbb{C}^k$, and $w \in \mathbb{C}^{n-d+j}$ such that

$$M = \{(z, \zeta, w) \in \mathbb{C}^n;\ z = H(x, w),\ y = h(x, w)\}.$$

Here, h and H are smooth functions defined on $\mathbb{R}^k \times \mathbb{C}^{n-d+j}$, where h has values in $\mathbb{R}^k$ and H has values in $\mathbb{C}^j$. Both h and H along their first derivatives vanish at the origin. Let M_0 be the submanifold of $\mathbb{C}^{n-j} = \mathbb{C}^k \times \mathbb{C}^{n-d+j}$ defined by

$$M_0 = \{(\zeta = x + iy, w) \in \mathbb{C}^k \times \mathbb{C}^{n-d+j};\ y = h(x, w)\}.$$

By an easy dimension count, M_0 is a generic submanifold of $\mathbb{C}^{n-j}$. Moreover, M is the graph over M_0 of the map $G : M_0 \mapsto M$ given by

$$p = (x + ih(x, w), w) \mapsto G(p) = (H(x, w), x + ih(x, w), w).$$

The inverse of G is the projection map $\pi|_M : M \mapsto M_0$ given by $\pi(z, \zeta, w) = (\zeta, w)$. Since π is a holomorphic map, the restriction of π to M is a CR map. Therefore, the graphing map G is also a CR map, as desired.

Let us briefly turn our attention to the class of compact CR submanifolds of $\mathbb{C}^n$. We mention a result of Wells [W] that states that if M is a compact, oriented, generic, CR submanifold of $\mathbb{C}^n$, then the Euler characteristic of M must vanish. For example, the equator of the unit sphere in $\mathbb{C}^n$ is *not* a CR submanifold. In fact in Section 7.1, we show that the equator of the unit sphere has two points where the real tangent space becomes complex linear (i.e., invariant under the J map). The above-mentioned result of Wells shows that any submanifold of $\mathbb{C}^n$ that is homeomorphic to the equator of the unit sphere in $\mathbb{C}^n$ cannot be a CR manifold.

Analysis near points where the dimension of the holomorphic tangent space changes is typically messy. For example, the proof of Kuranishi's imbedding theorem involves the solution of the tangential Cauchy–Riemann complex with estimates on a domain whose boundary is homeomorphic to the equator of the unit sphere. The hard work involves obtaining the right estimates near points on the boundary where the dimension of the complex tangent space changes.

To make sense of the notion of CR function and other "CR" objects near points where a submanifold is not CR also requires a lot of work. We refer the reader to the work of Harris [Har] for research along these lines.

Part III

The Holomorphic Extension of CR Functions

As shown in Part II, every holomorphic function on $\mathbb{C}^n$ restricts to a CR function on a CR submanifold M of $\mathbb{C}^n$. However, not all CR functions are the restrictions of holomorphic functions. In Part III, we examine geometric conditions on M that guarantee that CR functions on M extend as holomorphic functions on some open set Ω in $\mathbb{C}^n$. Under certain geometric conditions on M, the open set Ω contains an open subset of M, and under other geometric conditions, Ω lies to one side of M.

We are especially interested in CR extension to an open set that is function independent. That is, we ask: given an open set ω in M, does there exist an open set Ω in $\mathbb{C}^n$ such that each CR function on ω extends to a holomorphic function on Ω? This is a different question than the one answered by Theorem 1 in Section 9.1, which shows that any real analytic CR function on a real analytic CR submanifold of $\mathbb{C}^n$ holomorphically extends to an open set in $\mathbb{C}^n$ that may depend on the CR function. If there are no further geometric conditions on the CR submanifold, then CR extension to an open set that is function independent is impossible — even when the CR submanifold is real analytic. For example, let $M = \{(z, w) \in \mathbb{C}^2;\ \text{Im } z = 0\}$. Each real analytic CR function on an open set $\omega \subset M$ extends to a holomorphic function on an open set $\Omega \subset \mathbb{C}^2$ with $\Omega \cap M = \omega$. On the other hand, we have

$$\omega = \bigcap_{\varepsilon > 0} \Omega_\varepsilon$$

where

$$\Omega_\epsilon = \{(z, w) \in \mathbb{C}^2;\ (\text{Re} z,\ w) \in \omega \text{ and } |\text{Im } z| < \epsilon\}.$$

If ω is convex, then each Ω_ϵ is convex and hence a domain of holomorphy. From the theory of several complex variables, a holomorphic function $f_\epsilon : \Omega_\epsilon \to \mathbb{C}$ exists that cannot be analytically continued past any part of the boundary of

Ω_ϵ. The restriction of f_ϵ to ω is an example of a real analytic CR function on ω that cannot be analytically extended past any part of the boundary of Ω_ϵ. Since $\epsilon > 0$ is arbitrary, we conclude that there does not exist a single open set $\Omega \subset \mathbb{C}^2$ to which all CR functions on ω holomorphically extend.

In the above example, M is Levi flat. A similar construction of the Ω_ϵ can be carried out for any Levi flat submanifold in $\mathbb{C}^n$. This suggests that the Levi form might play a role in CR extension. As we show in the subsequent chapters, the Levi form is a key geometric object that determines whether or not CR extension to a fixed open set is possible. For a real hypersurface M in $\mathbb{C}^n$, we present Hans Lewy's theorem [L1] which states that if the Levi form at a point is not identically zero, then CR functions holomorphically extend to a fixed open set lying to one side of M. If the Levi form of M has eigenvalues of opposite sign, then CR functions holomorphically extend to a fixed open set containing both sides of M. We then generalize this theorem to the case where M has higher codimension.

We present two approaches to the proof of the CR extension theorem since the techniques of the proof are as important as the result. The first approach involves the use of analytic discs, which is an idea pioneered by Lewy and Bishop. The second more recent approach uses the Fourier transform. This technique was first applied to the CR extension problem by Baouendi and Treves. Both techniques are used in today's research problems.

13

An Approximation Theorem

Both the analytic disc and Fourier transform approaches to CR extension currently use an approximation theorem by Baouendi and Treves. This theorem roughly states that CR functions on a submanifold of $\mathbb{C}^n$ can be locally approximated by entire functions on $\mathbb{C}^n$. This approximation theorem has an interesting history. There are earlier versions of this theorem that are weaker (and some with errors in their proofs). All of these earlier versions have some convexity assumptions on the Levi form. Then around 1980, Baouendi and Treves caught the CR extension community by surprise. They showed that no convexity assumptions are needed. Moreover, their proof is a simple but clever adaptation of the proof of the classical Weierstrass theorem on approximating continuous functions on $\mathbb{R}^n$ by polynomials (see Section 1.1).

The statement of the approximation theorem of Baouendi and Treves allows very general systems of partial differential equations. However, we state and prove their theorem for the case of the tangential Cauchy–Riemann equations on a CR submanifold of $\mathbb{C}^n$. This version of their theorem has a simpler proof.

THEOREM 1 [BT1]
Suppose p is a point in a generic CR submanifold M of $\mathbb{C}^n$ of class C^2 with $\dim_{\mathbb{R}} M = 2n - d$, $0 \leq d \leq n$. Given an open neighborhood ω_1 of p in M, there exists an open set ω_2 in M with $p \in \omega_2 \subset \omega_1$ so that each CR function of class C^1 on ω_1 can be uniformly approximated on ω_2 by a sequence of entire functions in $\mathbb{C}^n$.

The proof requires two lemmas which we present after some notation.

It will be convenient to choose holomorphic coordinates for $\mathbb{C}^n$ (as in Lemma 1 of Section 7.2) so that the given point p is the origin and

$$M = \{(z = x + iy, w) \in \mathbb{C}^d \times \mathbb{C}^{n-d};\ y = h(x, w)\}$$

where h: $\mathbb{R}^d \times \mathbb{C}^{n-d} \to \mathbb{R}^d$ is of class C^2 with $h(0) = 0$ and $Dh(0) = 0$. Let $w = u + iv \in \mathbb{C}^{n-d}; t = (x, u) \in \mathbb{R}^d \times \mathbb{R}^{n-d} = \mathbb{R}^n$, and $s = (y, v) \in$

$\mathbb{R}^d \times \mathbb{R}^{n-d} = \mathbb{R}^n$. Note that coordinates for $\mathbb{C}^n$ can be written as $\zeta = t + is$.

Define $H\colon \mathbb{R}^n \times \mathbb{R}^{n-d} \to \mathbb{R}^d \times \mathbb{R}^{n-d}$ by

$$H(t,v) = (h(x, u+iv), v) \quad (\text{with } t = (x,u)).$$

Since $h(0) = 0$ and $Dh(0) = 0$, we have $H(0) = 0$ and $\partial H(0)/\partial t = 0$. We are only concerned with a small neighborhood of the origin and so we multiply h by a suitable cutoff function and assume that h has support in a neighborhood $U_1 \times U_2$ in $\mathbb{R}^n \times \mathbb{R}^{n-d}$ of the origin. By choosing U_1 and U_2 small enough, we can arrange

$$\left|\frac{\partial H}{\partial t}(t,v)\right| \leq \frac{1}{2} \quad \text{for all} \quad (t,v) \in \mathbb{R}^n \times \mathbb{R}^{n-d}.$$

Here, $|A|$ denotes the norm of the matrix A, i.e., $|A| = \sup\limits_{|v|=1} |A(v)|$.

By the mean value theorem, we have

$$|H(t_1,v) - H(t_2,v)| \leq \frac{1}{2}|t_1 - t_2| \tag{1}$$

for all $(t_1,v),(t_2,v) \in \mathbb{R}^n \times \mathbb{R}^{n-d}$.

Near the origin, M is parameterized by H, i.e.,

$$M = \{t + iH(t,v); (t,v) \in U_1 \times U_2 \subset \mathbb{R}^n \times \mathbb{R}^{n-d}\}.$$

M is also foliated by the n-dimensional slices M_v for $v \in U_2$ where

$$M_v = \{(t + iH(t,v);\ t \in U_1 \subset \mathbb{R}^n\}.$$

For $\zeta = (\zeta_1, \ldots, \zeta_n)$, let

$$[\zeta]^2 = \zeta_1^2 + \cdots + \zeta_n^2$$
$$d\zeta = d\zeta_1 \wedge \ldots \wedge d\zeta_n.$$

LEMMA 1

Let $g \in \mathcal{D}(U_1)$ with $g \equiv 1$ on a neighborhood U_1' with $0 \in U_1' \subset\subset U_1 \subset \mathbb{R}^n$. Extend g to $\mathbb{C}^n$ so that $g(\zeta)$ is independent of Im ζ, for $\zeta \in \mathbb{C}^n$. Let $f\colon M \to \mathbb{C}$ be a continuous function. Then for $\zeta \in M_v \cap \{U_1' \times U_2\}$

$$f(\zeta) = \lim_{\epsilon \to 0} \pi^{-n/2}\epsilon^{-n} \int\limits_{\zeta' \in M_v} g(\zeta')f(\zeta')e^{-\epsilon^{-2}[\zeta-\zeta']^2}d\zeta'.$$

Moreover, this limit is uniform for $v \in U_2$ and $\zeta \in M_v \cap \{U_1' \times U_2\}$.

This lemma and its proof should remind the reader of the first part of the proof of the Weierstrass theorem in Section 1.1 where it is shown that $f * e_\epsilon \to f$ with $e_\epsilon(t) = \pi^{-n/2}\epsilon^{-n}e^{-\epsilon^{-2}|t|^2}$.

Even though the integrand in the lemma is an entire function of $\zeta \in \mathbb{C}^n$, this lemma alone is not enough to prove the approximation theorem. This is because the domain of integration, M_v, depends on the variable ζ (since ζ is required to belong to M_v). In the next lemma, we show that if f is CR, then a domain of integration can be found that is independent of ζ and the approximation theorem will follow.

PROOF Define ζ: $\mathbb{R}^n \times \mathbb{R}^{n-d} \to \mathbb{C}^n$ by

$$\zeta(t,v) = t + iH(t,v).$$

For fixed $v \in U_2 \subset \mathbb{R}^{n-d}$, the map $t' \mapsto \zeta(t',v)$ for $t' \in U_1$ parameterizes M_v. Pulling back the integral in the lemma via this map, we see that it suffices to show

$$f(\zeta(t,v)) =$$
$$\lim_{\epsilon\to 0} \pi^{-n/2}\epsilon^{-n} \int_{t'\in\mathbb{R}^n} g(t') f(\zeta(t',v)) e^{-\epsilon^{-2}[\zeta(t,v)-\zeta(t',v)]^2} \det\left(\frac{\partial\zeta}{\partial t'}(t',v)\right) dt'$$

where the limit is uniform in $(t,v) \in U_1' \times U_2$. Here, we write $t' = (t_1', \ldots, t_n') \in \mathbb{R}^n$; $dt' = dt_1' \ldots dt_n'$ and $(\partial\zeta/\partial t')(t',v)$ is the $n \times n$ complex matrix with entries $(\partial\zeta_j/\partial t_k')(t',v)$ for $1 \le j,k \le n$.

We replace t' by $t - \epsilon s$ (for t fixed), and the above integral becomes

$$\pi^{-n/2} \int_{s\in\mathbb{R}^n} g(t-\epsilon s) f(\zeta(t-\epsilon s, v)) e^{-\epsilon^{-2}[\zeta(t,v)-\zeta(t-\epsilon s,v)]^2} \det\left(\frac{\partial\zeta}{\partial t}(t-\epsilon s, v)\right) ds. \tag{2}$$

If $t \in U_1'$, then $g(t-\epsilon s) \to 1$ as $\epsilon \to 0$. Moreover

$$\zeta(t,v) - \zeta(t-\epsilon s, v) = \frac{-\partial\zeta}{\partial t}(t,v)\cdot(\epsilon s) + \mathcal{O}(\epsilon^2).$$

Therefore, we have

$$\epsilon^{-2}[\zeta(t,v) - \zeta(t-\epsilon s,v)]^2 = \left[\frac{\partial\zeta}{\partial t}(t,v)\cdot s\right]^2 + \mathcal{O}(\epsilon).$$

This means that pointwise in s (but uniformly in $t \in U_1'$ and $v \in U_2$), the integrand in (2) converges, as $\epsilon \to 0$, to

$$f(\zeta(t,v)) e^{-[\frac{\partial\zeta}{\partial t}(t,v)\cdot s]^2} \det\left(\frac{\partial\zeta}{\partial t}(t,v)\right).$$

To show that the corresponding integrals converge, we must dominate the integrand in (2) by an integrable function of $s \in \mathbb{R}^n$. Certainly $g \cdot f \cdot \partial\zeta/\partial t$ is globally bounded. So it suffices to dominate the exponential term. We have

$$\left| e^{-\epsilon^{-2}[\zeta(t,v)-\zeta(t-\epsilon s,v)]^2} \right| = e^{-\epsilon^{-2}\,\mathrm{Re}\{[\zeta(t,v)-\zeta(t-\epsilon s,v)]^2\}}.$$

Since $\zeta(t,v) = t + iH(t,v)$, we obtain

$$\mathrm{Re}\{[\zeta(t,v) - \zeta(t-\epsilon s, v)]^2\} = \epsilon^2|s|^2 - [H(t,v) - H(t-\epsilon s, v)]^2.$$

From (1), we have

$$\mathrm{Re}\{[\zeta(t,v) - \zeta(t-\epsilon s, v)]^2\} \geq \frac{3}{4}\epsilon^2|s|^2.$$

So the exponential term is dominated from above by $e^{-3/4|s|^2}$ which is an integrable function of $s \in \mathbb{R}^n$. By the dominated convergence theorem, the integral in (2) converges (as $\epsilon \to 0$) to

$$\pi^{-n/2} f(\zeta(t,v)) \int\limits_{s\in\mathbb{R}^n} e^{-[\frac{\partial\zeta}{\partial t}(t,v)\cdot s]^2} \det\left(\frac{\partial\zeta}{\partial t}(t,v)\right) ds$$

and the limit is uniform in $t \in U_1'$ and $v \in U_2$.

We have $(\partial\zeta/\partial t)(t,v) = (\partial/\partial t)(t + iH(t,v)) = I + i(\partial H/\partial t)(t,v)$. Moreover $|(\partial H/\partial t)(t,v)| \leq 1/2$. Therefore, the proof of the lemma will be complete after we show

$$\pi^{-n/2} \int\limits_{s\in\mathbb{R}^n} e^{-[A\cdot s]^2}(\det A)ds = 1 \tag{3}$$

for all matrices A in the set

$$\mathcal{A} = \left\{ \begin{array}{l} n \times n \text{ complex matrices, } A, \text{ such that } |\mathrm{Im}\ A| \leq \frac{1}{2}|\mathrm{Re}\ A| \\ \text{and } \mathrm{Re}\ A \text{ is nonsingular} \end{array} \right\}.$$

The integral in (3) is a holomorphic function of the entries in $A \in \mathcal{A}$. In addition, if $\mathrm{Im}\ A = 0$ then (3) follows by the change of variables $s' = A \cdot s \in \mathbb{R}^n$ and a standard polar coordinate calculation. Therefore, (3) must hold for all matrices $A \in \mathcal{A}$ by the identity theorem for holomorphic functions. The proof of the lemma now follows by applying (3) to the matrix $A = (\partial\zeta/\partial t)(t,v)$. ∎

LEMMA 2

Suppose f is a CR function of class C^1 on a neighborhood ω_1 of the origin in M. Then, there is a neighborhood ω_2 in M with $0 \in \omega_2 \subset\subset \omega_1 \subset M$ such that for $\zeta \in \omega_2$

$$f(\zeta) = \lim_{\epsilon\to 0} \pi^{-n/2}\epsilon^{-n} \int\limits_{\zeta'\in M_0} g(\zeta')f(\zeta')e^{-\epsilon^{-2}[\zeta-\zeta']^2} d\zeta'.$$

Moreover, this limit is uniform in $\zeta \in \omega_2$.

PROOF OF THEOREM 1 ASSUMING LEMMA 2 The point of Lemma 2 is that if f is CR, then the domain of integration (M_0) is independent of the variable ζ.

Since the integrand is an entire function of $\zeta \in \mathbb{C}^n$, the proof of Theorem 1 is complete. ∎

PROOF OF LEMMA 2 For $\zeta \in \zeta\{U_1 \times U_2\} \subset M$, we have

$$\zeta = t + iH(t, v)$$

for some unique $t \in U_1$ and $v \in U_2$. For fixed $v \in U_2$, let

$$\widetilde{M}_v = \{\zeta(t', \lambda v) = t' + iH(t', \lambda v) \in M; t' \in U_1 \subset \mathbb{R}^n \text{ and } 0 \leq \lambda \leq 1\}.$$

$\widetilde{M}_v$ is an $(n+1)$-dimensional submanifold of M and its boundary is the union of M_0, M_v and the set $\{\zeta(t', \lambda v); t' \in \partial U_1 \text{ and } 0 \leq \lambda \leq 1\}$. We will only be concerned with functions whose t'-support is contained in U_1. So the only contributing components of the boundary of $\widetilde{M}_v$ are M_0 and M_v. We use Stokes' theorem to transfer the integral in Lemma 1 over M_v into a sum of an integral over M_0 and an integral over $\widetilde{M}_v$. For $\zeta \in \zeta\{U_1' \times U_2\}$, we obtain

$$f(\zeta) = \lim_{\epsilon \to 0} \pi^{-n/2} \epsilon^{-n} \int\limits_{\zeta' \in M_0} g(\zeta') f(\zeta') e^{-\epsilon^{-2}[\zeta - \zeta']^2} d\zeta'$$
$$+ \lim_{\epsilon \to 0} \pi^{-n/2} \epsilon^{-n} \int\limits_{\zeta' \in \widetilde{M}_v} d_{\zeta'}\{g(\zeta') f(\zeta') e^{-\epsilon^{-2}[\zeta - \zeta']^2}\} d\zeta'.$$

The integral over $\widetilde{M}_v$ involves the exterior derivative $d_{\zeta'} = \partial_{\zeta'} + \bar{\partial}_{\zeta'}$, but due to the presence of $d\zeta' = d\zeta_1' \wedge \ldots \wedge d\zeta_n'$, the only contributing term comes from $\bar{\partial}_{\zeta'}$. From Theorem 2 in Section 9.1, we may assume that the given CR function f is extended ambiently to $\mathbb{C}^n$ so that $\bar{\partial} f = 0$ on $\omega_1 \subset M$. Assuming that U_1 and U_2 in Lemma 1 are chosen small enough so that $\zeta\{U_1 \times U_2\} \subset \omega_1$, we obtain

$$f(\zeta) = \lim_{\epsilon \to 0} \pi^{-n/2} \epsilon^{-n} \int\limits_{\zeta' \in M_0} g(\zeta') f(\zeta') e^{-\epsilon^{-2}[\zeta - \zeta']^2} d\zeta'$$
$$+ \lim_{\epsilon \to 0} \pi^{-n/2} \epsilon^{-n} \int\limits_{\zeta' \in \widetilde{M}_v} (\bar{\partial}_{\zeta'} g(\zeta')) f(\zeta') e^{-\epsilon^{-2}[\zeta - \zeta']^2} d\zeta'. \quad (4)$$

The proof of the lemma will be complete after we show that the second limit on the right is zero.

Recall that $\zeta = t + iH(t, v)$. For $\zeta' \in \widetilde{M}_v$, there is a unique λ with $0 \leq \lambda \leq 1$ such that

$$\zeta' = \zeta(t', \lambda v) = t' + iH(t', \lambda v).$$

To estimate $|e^{-\epsilon^{-2}[\zeta-\zeta']^2}|$, we need to estimate the real part of the exponent. We have

$$\mathrm{Re}\{[\zeta-\zeta']^2\} = |t-t'|^2 - |H(t,v) - H(t',\lambda v)|^2. \tag{5}$$

From the mean value theorem and the estimate $|\partial H/\partial t| \leq 1/2$, we have

$$\begin{aligned} |H(t,v) - H(t',\lambda v)| &\leq |H(t,v) - H(t',v)| + |H(t',v) - H(t',\lambda v)| \\ &\leq \frac{1}{2}|t-t'| + C|v-\lambda v| \end{aligned}$$

where C is some uniform constant that depends only on the C^1-norm of H. Squaring the above inequality and using the inequality $2ab \leq (a^2+b^2)$ for $a, b \geq 0$, we obtain

$$|H(t,v) - H(t',\lambda v)|^2 \leq \frac{1}{2}|t-t'|^2 + 2C^2|v-\lambda v|^2.$$

After noting that $|v - \lambda v| \leq |v|$ (since $0 \leq \lambda \leq 1$), the above inequality together with (5) yields

$$\mathrm{Re}\{[\zeta-\zeta']^2\} \geq \frac{1}{2}|t-t'|^2 - 2C^2|v|^2.$$

Now $g(\zeta')$ only depends on $t' = \mathrm{Re}\ \zeta'$. Also, $\bar{\partial}_{\zeta'} g$ vanishes on a neighborhood of the origin (since $g \equiv 1$ on U_1'). So if $\zeta' = t' + iH(t', \lambda v)$ belongs to the support of $\bar{\partial} g$, then $|t'|$ must be bounded away from 0. In addition, $\zeta = t + iH(t,v) \to 0$ if and only if both $t \to 0$ and $v \to 0$. Together with the above inequality, we see that there are constants $r_1, r_2 > 0$ so that if $\zeta \in M_v$ with $|\zeta| < r_2$ and if $\zeta' \in$ supp $\bar{\partial} g$, then

$$\mathrm{Re}\{[\zeta-\zeta']^2\} \geq r_1.$$

Therefore

$$\left| \epsilon^{-n} \int_{\zeta' \in \widetilde{M}_v} (\bar{\partial}_{\zeta'} g(\zeta')) f(\zeta') e^{-\epsilon^{-2}[\zeta-\zeta']^2} d\zeta' \right| \leq C\epsilon^{-n} e^{-\epsilon^{-2} r_1}$$

for all $\zeta \in M_v \subset M$ with $|\zeta| < r_2$. Here, C is a uniform constant that is independent of ϵ, v, and ζ. This inequality shows that the second integral on the right side of (4) converges to 0 as $\epsilon \to 0$, provided $\zeta \in M$ with $|\zeta| < r_2$. The proof of the lemma is now complete. ∎

For the uniqueness part of the CR extension theorem stated in the next chapter, we need to know that holomorphic functions on a wedge in $\mathbb{C}^n$ can be approximated by a sequence of entire functions. By definition, a wedge is an open set of the form $\omega + \{\Gamma \cap B_\epsilon\}$ where $\omega \subset M$ is an open neighborhood of a given point $p \in M$, where Γ is a cone in $N_p(M)$ (the space of vectors at p which are orthogonal to $T_p(M)$) and where B_ϵ is the ball in $N_p(M)$ of radius ϵ centered at p. The proofs of Lemmas 1 and 2 can easily be modified to prove the following approximation result.

THEOREM 2

Let ω_1 be a neighborhood of a point $p \in M$ and let $\epsilon_1 > 0$ be given. Let Γ be a cone in $N_p(M)$. There exist $\epsilon_2 > 0$ and a neighborhood ω_2 of p in M with $\omega_2 \subset \omega_1$ such that if F is holomorphic on $\omega_1 + \{\Gamma \cap B_{\epsilon_1}\}$ and continuous up to ω_1, then F is the uniform limit on $\omega_2 + \{\Gamma \cap B_{\epsilon_2}\}$ of a sequence of entire functions.

14

The Statement of the CR Extension Theorem

We start this chapter by stating Hans Lewy's CR extension theorem for hypersurfaces. The generalization of this theorem to CR submanifolds of higher codimension is then given. This is followed by a number of examples. The proofs of these theorems will follow in Chapters 15 and 16.

14.1 Lewy's CR extension theorem for hypersurfaces

Let $M = \{z \in \mathbb{C}^n; \rho(z) = 0\}$ be a hypersurface in $\mathbb{C}^n$ where $\rho\colon \mathbb{C}^n \to \mathbb{R}$ is smooth with $d\rho \neq 0$ on M. If ρ is scaled so that $|\nabla\rho(p)| = 1$, then from Theorem 1 in Section 10.2, the Levi form of M at p is the map

$$W \mapsto \left(- \sum_{j,k=1}^{n} \frac{\partial^2 \rho(p)}{\partial\zeta_j \partial\bar{\zeta}_k} w_j \bar{w}_k \right) \nabla\rho(p) \quad \text{for} \quad W = \sum_{j=1}^{n} w_j \frac{\partial}{\partial\zeta_j} \in H_p^{1,0}(M).$$

When we speak of the eigenvalues of the Levi form of M at p, we are referring to the eigenvalues of the matrix that represents the Levi form (i.e., the restriction of $((\partial^2\rho/\partial\zeta_j\partial\bar{\zeta}_k)(p))$ to $H_p^{1,0}(M)$).

Let $\Omega^+ = \{z \in \mathbb{C}^n; \rho(z) > 0\}$ and $\Omega^- = \{z \in \mathbb{C}^n; \rho(z) < 0\}$. Hans Lewy's theorem can now be stated.

THEOREM 1 HANS LEWY'S CR EXTENSION THEOREM FOR HYPERSURFACES [L1]

Let M be a real hypersurface in $\mathbb{C}^n, n \geq 2$ of class $C^k, 3 \leq k \leq \infty$, and let p be a point in M.

(a) *If the Levi form of M at p has at least one positive eigenvalue then for each open set ω in M with $p \in \omega$, there is an open set U in $\mathbb{C}^n$ with $p \in U$ such that for each CR function f of class C^1 on ω, there is a* *unique function F which is holomorphic on $U \cap \Omega^+$ and continuous on $U \cap \overline{\Omega^+}$ such that $F|_{U \cap M} = f$.*

(b) *If the Levi form of M at p has at least one negative eigenvalue, then the conclusion of part (a) holds with Ω^+ replaced by Ω^-.*

(c) *If the Levi form of M at p has eigenvalues of opposite sign, then for each open set ω in M with $p \in \omega$, there is an open set U in $\mathbb{C}^n$ with $p \in U$ such that each CR function of class C^1 on ω is the restriction (on $U \cap \omega$) of a unique holomorphic function defined on U.*

Parts (a) and (b) describe one-sided CR extension results whereas part (c) describes a two-sided CR extension result. Note that the quantifiers are arranged so that the open set U depends only on ω and not on the CR function defined there.

It is easier to see the geometric meaning of this theorem by choosing coordinates (as in Lemma 1 of Section 7.2) so that the point p is the origin and so that the defining function ρ has the form $\rho(x + iy, w) = y - h(x, w)$ for $z = x + iy \in \mathbb{C}$ and $w \in \mathbb{C}^{n-1}$. From Corollary 1 in Section 10.2, the Levi form can be identified with the map

$$W = (w_1, \ldots, w_{n-1}) \longmapsto \sum_{j,k=1}^{n-1} \frac{\partial^2 h(0)}{\partial w_j \partial \bar{w}_k} w_j \bar{w}_k .$$

Since we can arrange coordinates so that all the second-order pure terms in the expansion of h (about 0) vanish, the Levi form of M describes the second-order concavity of M near the origin. A positive eigenvalue of the Levi form indicates that M is locally concave up (here, up means toward positive y) along a direction in $H_0^{1,0}(M)$ (i.e., in one of the w-directions). In this case, Lewy's theorem states that CR functions holomorphically extend above M (i.e., to $\{y > h(x, \omega)\}$). A negative eigenvalue of the Levi form means that M is locally concave down along one of the w-directions. In this case, CR functions holomorphically extend below M. If the Levi form has eigenvalues of opposite sign, then the origin is a saddle point for M and CR extension to both sides of M is possible.

Since holomorphic functions are real analytic, part (c) of Theorem 1 implies the following regularity result for CR functions.

THEOREM 2

Suppose M is a hypersurface in $\mathbb{C}^n$ of class $C^k (3 \le k \le \infty)$ and suppose p is a point in M where the Levi form has eigenvalues of opposite sign. Then each CR function on M that is a priori C^1 in a neighborhood of p must be of class C^k in a neighborhood of p. If in addition M is real analytic, then an a priori C^1 CR function defined near p must be real analytic near p.

Theorems 1 and 2 are also valid for CR distributions. This will be briefly discussed in Chapter 17.

14.2 The CR extension theorem for higher codimension

The Levi form is also the key geometric object that governs CR extension from a CR submanifold of higher codimension. If M is a generic CR submanifold of real dimension $2n-d$, $1 \leq d \leq n-1$, then the normal space of M at a point $p \in M$ (denoted $N_p(M)$) is isomorphic to $\mathbb{R}^d$ with p as the origin in $N_p(M)$. We consider the extrinsic Levi form at p, $\tilde{\mathcal{L}}_p$: $H_p(M) \to N_p(M)$. The definition of $\tilde{\mathcal{L}}_p$, along with a coordinate description of $\tilde{\mathcal{L}}_p$ in terms of an appropriate system of defining functions, is given in Section 10.2. As with the hypersurface case, the image of the Levi form at p provides information about the second-order concavity of M near p.

For $p \in M$, let

$$\Gamma_p = \{\text{the convex hull of the image of } \tilde{\mathcal{L}}_p\} \subset N_p(M).$$

Γ_p is a cone, i.e., if v is an element Γ_p, then λv also belongs to Γ_p for all $\lambda \geq 0$.

If M is a real hypersurface, then $N_p(M) \simeq \mathbb{R}$ and Γ_p is either $\{0\}$ (if $\tilde{\mathcal{L}}_p \equiv 0$) or a ray (if $\tilde{\mathcal{L}}_p$ is positive or negative semidefinite) or all of $\Gamma_p \simeq \mathbb{R}$ (if $\tilde{\mathcal{L}}_p$ has eigenvalues of opposite sign). The translation of Lewy's hypersurface theorem into these terms is the following: if Γ_p is a ray, then CR extension is possible to one side of M; if Γ_p is all of $N_p(M) \simeq \mathbb{R}$ then CR extension is possible to both sides of M.

If $d =$ $\operatorname{codim}_{\mathbb{R}} M$ is greater than one, then $\tilde{\mathcal{L}}_p$ is vector valued and so Γ_p is more complicated. As we shall see in Theorem 1 below, Γ_p determines the shape and size of the open set to which CR functions holomorphically extend.

To state the theorem, we need some additional notation. For two cones Γ_1 and Γ_2 in $N_p(M)$ we say that Γ_1 is *smaller than* Γ_2 (and write $\Gamma_1 < \Gamma_2$) if $\Gamma_1 \cap S_p$ is a compact subset of the interior of $\{\Gamma_2\} \cap S_p$, where S_p is the unit sphere in $N_p(M)$. For example, if the codimension of M is two, then $N_p(M)$ is a copy of $\mathbb{R}^2$. In this case, if Γ_1 and Γ_2 are convex cones with $\Gamma_1 < \Gamma_2$, then either $\Gamma_1 = \Gamma_2 = N_p(M)$ or else $\Gamma_1 \subset \Gamma_2$ and the angle formed by the boundary rays for Γ_1 is smaller than the corresponding angle for Γ_2. Note that $N_p(M)$ is always a smaller cone than itself.

For $\epsilon > 0$, let B_ϵ denote the open ball in $N_p(M)$ centered at the origin $= p$ of radius ϵ. For two sets A and B in $\mathbb{C}^n$, we let $A+B = \{a+b; a \in A \text{ and } b \in B\}$.

THEOREM 1 CR EXTENSION FOR HIGHER CODIMENSION [BP]
Suppose M is a generic, CR submanifold of $\mathbb{C}^n$ of class $C^k(4 \leq k \leq \infty)$ with $\dim_{\mathbb{R}} M = 2n-d$, $1 \leq d \leq n-1$. *Let p be a point in M so that Γ_p has*

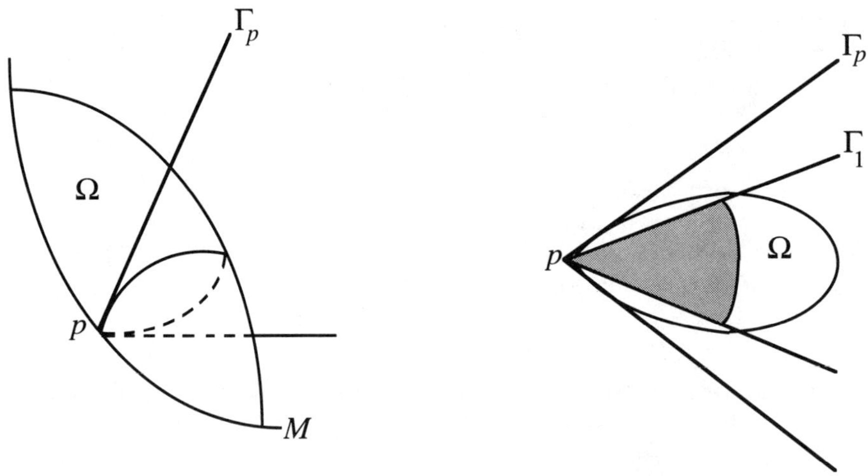

FIGURE 14.1

nonempty interior (with respect to $N_p(M)$). Then for every neighborhood ω of p in M, there is an open set ω' in M and an open set Ω in $\mathbb{C}^n$ such that

(a) $p \in \omega' \subset \overline{\Omega} \cap M \subset \omega$

(b) *for each open cone $\Gamma_1 < \Gamma_p$, there is a connected neighborhood ω_1 of p in M and an $\epsilon > 0$ so that*

$$\omega_1 + \{\Gamma_1 \cap B_\epsilon\} \subset \Omega$$

(c) *for each CR function f of class C^1 on $\omega \subset M$, there is a unique holomorphic function F defined on Ω and continuous on $\Omega \cup \omega'$ with $F = f$ on ω'.*

As with the hypersurface case, the set Ω depends only on ω and Γ_p and not on the CR function defined on ω. Part (b) of the theorem conveys the following picture of Ω which we draw in the case $d = \operatorname{codim}_{\mathbb{R}} M = 2$.

In Figure 14.1, the picture on the right is a side view with M going into the page. $\Gamma_1 \cap B_\epsilon$ is represented by the shaded region. The quantifiers imply that ω_1 and ϵ depend on the cone Γ_1. The closer Γ_1 gets to Γ_p, the smaller ϵ and ω_1 usually get as shown by the examples in the next section. By allowing Γ_1 to approach Γ_p, we see that the tangent cone of Ω at p is spanned by Γ_p and the real tangent space of M at p.

In Theorem 1, if $\Gamma_p = N_p(M)$, then we may let $\Gamma_1 = N_p(M)$ because $N_p(M)$ is smaller than itself. In this case, Ω contains an open set $\{\omega_1 + B_\epsilon\}$

which is an open neighborhood of p in $\mathbb{C}^n$. Therefore, if $\Gamma_p = N_p(M)$, then each CR function near p is locally the restriction of a holomorphic function defined in a neighborhood of p. This is analogous to the two-sided CR extension result (part c) of Lewy's theorem for a hypersurface.

THEOREM 2
Suppose M is a generic CR submanifold of $\mathbb{C}^n$ of class $C^k, 4 \leq k \leq \infty$, and suppose p is a point in M with $\Gamma_p = N_p(M)$. Then for each neighborhood ω of p in M, there is an open set Ω in $\mathbb{C}^n$ with $p \in \Omega$ such that each CR function which is of class C^1 on ω is the restriction of a unique holomorphic function defined on Ω.

Since holomorphic functions are real analytic and hence C^∞, Theorem 2 implies the following regularity result for CR functions.

THEOREM 3
Suppose M is a generic CR submanifold of $\mathbb{C}^n$ of class $C^k(4 \leq k \leq \infty)$ and suppose p is a point in M with $\Gamma_p = N_p(M)$. Then each CR function that is a priori C^1 in a neighborhood of p in M must be of class C^k in a neighborhood of p. If in addition M is real analytic, then each CR function that is a priori C^1 in a neighborhood of p must be real analytic in a neighborhood of p.

Theorems 1, 2, and 3 hold for CR distributions as well as C^1-CR functions. This will be discussed briefly at the end of Part III.

In Theorem 1, the hypothesis that the interior of Γ_p in $N_p(M)$ is nonempty imposes some restrictions on the codimension of M. For example, if $\dim_{\mathbb{C}} H_p^{1,0}(M) = 1$ then the image of the Levi form is contained in a one (real) dimensional subspace of $N_p(M)$. Therefore if $\operatorname{codim}_{\mathbb{R}} M \geq 2$ and if $\dim_{\mathbb{C}} H_p^{1,0}(M) = 1$, then the hypothesis of Theorems 1, 2, or 3 is never satisfied. By using the bilinearity and conjugate symmetry of the Levi form, it is an easy exercise to show that if the interior of Γ_p in $N_p(M)$ is nonempty, then $m(m+1) \geq 2d$ where $m = n - d = \dim_{\mathbb{C}} H^{1,0}(M)$ and $d = \operatorname{codim}_{\mathbb{R}} M$. The reader should not get the impression that CR extension is impossible if $m(m+1) < 2d$. However, CR extension in this case requires an analysis of higher order commutators from $H^{\mathbb{C}}(M)$. Results along these lines will be mentioned at the end of Part III.

14.3 Examples

As already mentioned, the above CR extension theorem for a real hypersurface is simple to interpret. If the theorem applies, then CR functions either locally extend to one or both sides of the hypersurface depending on the sign of the eigenvalues of the Levi form.

The next easiest class of submanifolds to analyze is the class of quadric submanifolds of $\mathbb{C}^4$ with real codimension two. In Section 7.3, we established that any such quadric is biholomorphic to one of the following four normal forms. Here, the coordinates for $\mathbb{C}^4$ are given by (z_1, z_2, w_1, w_2) with $z_j = x_j + iy_j$, $j = 1, 2$.

(i) $M = \{y_1 = q(w, \bar{w}),\ y_2 = 0\}$ where q: $\mathbb{C}^2 \times \mathbb{C}^2 \to \mathbb{C}$ is a scalar-valued quadratic form

(ii) $M = \{y_1 = |w_1|^2,\ y_2 = |w_2|^2\}$

(iii) $M = \{y_1 = |w_1|^2,\ y_2 = \operatorname{Re}(w_1\bar{w}_2)\}$

(iv) $M = \{y_1 = \operatorname{Re}(w_1\bar{w}_2),\ y_2 = \operatorname{Im}(w_1\bar{w}_2)\}$

Analyzing the CR extension phenomenon for each of these four examples will characterize the local CR extension phenomenon on all codimension two quadrics in $\mathbb{C}^4$.

Recall that the Levi form at the origin of the quadric $\{y = q(w, \bar{w})\}$ can be identified with the vector valued quadratic form $w \mapsto q(w, \bar{w})$.

Example 1

Suppose $M = \{y_1 = q(w, \bar{w}),\ y_2 = 0\}$, and $p = 0$ (the origin). The image of the Levi form $(w \mapsto (q(w, \bar{w}), 0))$ is contained in the line $\{(y_1, 0); y_1 \in \mathbb{R}\}$. In this case, the interior of Γ_0 with respect to $N_0(M) \simeq \mathbb{R}^2$ is empty and so the CR extension theorem does not apply. In fact, CR extension to a function independent open set is impossible because M is contained in $\cap_{\epsilon>0}\Omega_\epsilon$ where each $\Omega_\epsilon = \{|y_2| < \epsilon\}$ is a domain of holomorphy. $\square$

Example 2

Suppose $M = \{y_1 = |w_1|^2, y_2 = |w_2|^2\}$. The image of the Levi form $(w_1, w_2) \mapsto \tilde{\mathcal{L}}_0(w_1, w_2) = (|w_1|^2, |w_2|^2)$ is the quadrant $\Gamma_0 = \{y_1 \geq 0, y_2 \geq 0\}$ which is already convex. The CR extension theorem states that CR functions on a neighborhood of the origin in M extend to holomorphic functions on an open set Ω in $\mathbb{C}^4$ whose normal cross section (at the origin) contains sets of the type $B_\epsilon \cap \Gamma_1$ where Γ_1 is any smaller subcone than $\{y_1 \geq 0, y_2 \geq 0\}$ in $N_0(M) \simeq \mathbb{R}^2$ and where $\epsilon > 0$ depends on Γ_1. Note that M is the intersection of the convex boundaries $\{y_1 = |w_1|^2\}$ and $\{y_2 = |w_2|^2\}$. Therefore, CR functions on M cannot holomorphically extend past $\{y_1 \geq |w_1|^2,\ y_2 \geq |w_2|^2\}$. In particular, this shows that in general, we cannot take $\Gamma_1 = \Gamma_0$ in the conclusion of the CR extension theorem (because in this example, $B_\epsilon \cap \Gamma_0$ is not contained in $\{y_1 \geq |w_1|^2, y_2 \geq |w_2|^2\}$). This example also shows that in general, ϵ must depend on the cone Γ_1; for if $\epsilon > 0$ is fixed, then $B_\epsilon \cap \Gamma_1$ is not contained

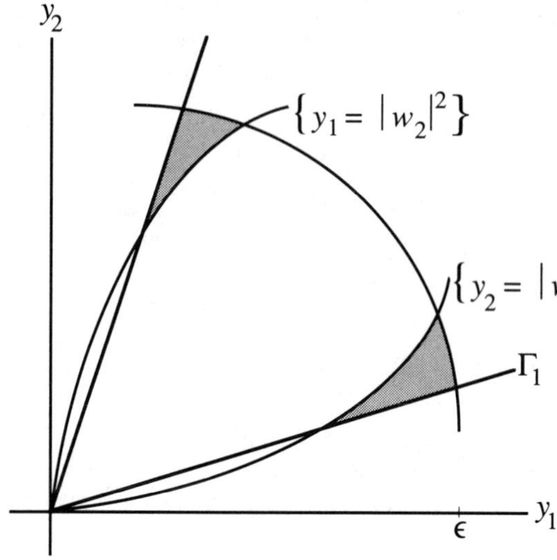

FIGURE 14.2

in $\{y_1 \geq |w_1|^2,\ y_2 \geq |w_2|^2\}$ provided Γ_1 is sufficiently close to $\Gamma_0 = \{y_1 \geq 0,\ y_2 \geq 0\}$. □

Example 3

Suppose $M = \{y_1 = |w_1|^2,\ y_2 = \ \mathrm{Re}(w_1\bar{w}_2)\}$. The image of the Levi form $\tilde{\mathcal{L}}_0(w_1, w_2) = (|w_1|^2,\ \mathrm{Re}(w_1\bar{w}_2))$ is the half space $\{y_1 > 0\}$ together with the origin. The image of $\tilde{\mathcal{L}}_0$ is convex but not closed. The CR extension theorem states that CR functions on an open subset of the origin holomorphically extend to an open set Ω in $\mathbb{C}^4$ whose normal cross section at the origin contains sets of the type $B_\epsilon \cap \Gamma_1$ where Γ_1 is any smaller cone than $\{y_1 > 0\}$. The tangent cone of Ω at the origin contains the half space $\{y_1 > 0\}$. □

Example 4

Suppose $M = \{y_1 = \ \mathrm{Re}(w_1\bar{w}_2),\ y_2 = \ \mathrm{Im}(w_1\bar{w}_2)\}$. The image of the Levi form $\tilde{\mathcal{L}}_0(w_1, w_2) = (\mathrm{Re}(w_1\bar{w}_2),\ \mathrm{Im}(w_1\bar{w}_2))$ is all of $N_0(M) \simeq \mathbb{R}^2$. This can be seen by setting $w_2 = 1$ and letting w_1 range over the complex numbers. Therefore, Theorems 2 and 3 from the previous section apply to this example. A CR function in a neighborhood of the origin in M is the restriction of a holomorphic function defined on a neighborhood of the origin in $\mathbb{C}^4$. In addition, an a priori C^1-CR function near the origin must be real analytic near the origin. □

In the above four examples, the image of the Levi form is convex. This is not always the case for submanifolds of $\mathbb{C}^n$ for $n > 4$ as illustrated by the following example.

Example 5

Let $(z_1, z_2, z_3, z_4, w_1, w_2)$ be the coordinates for $\mathbb{C}^6$. Let

$$M = \{\text{Im } z_1 = |w_1|^2,\ \text{Im } z_2 = |w_2|^2,\ \text{Im } z_3 = \text{Re}(w_1\bar{w}_2),\ \text{Im } z_4 = \text{Im}(w_1\bar{w}_2)\}.$$

Here, M has codimension four in $\mathbb{C}^6$. The image of the Levi form at the origin in $N_0(M) \simeq \mathbb{R}^4$ is the cone

$$\{y \in \mathbb{R}^4; y_1 \geq 0,\ y_2 \geq 0 \quad \text{and} \quad y_3^2 + y_4^2 = y_1 y_2\}.$$

This set is not convex and it has no interior in $\mathbb{R}^4$. However, its convex hull is the set

$$\{y \in \mathbb{R}^4; y_1 \geq 0,\ y_2 \geq 0 \quad \text{and} \quad y_3^2 + y_4^2 \leq y_1 y_2\}$$

which does have nonempty interior in $\mathbb{R}^4$, and so the CR extension theorem applies to this example. □

15

The Analytic Disc Technique

In this chapter, we present the proof of the CR extension theorem using the technique of analytic discs. The rough idea is the following. In Chapter 13, we showed (without any assumption on the Levi form) that a CR function on a CR submanifold M can be uniformly approximated on an open set $\omega \subset M$ by a sequence of entire functions. To extend a given CR function to an open set Ω in $\mathbb{C}^n$, it is natural to try to show that this approximating sequence of entire functions is uniformly convergent on the compact subsets of Ω. This can be accomplished by the use of analytic discs. Let D be the unit disc in $\mathbb{C}$. An *analytic disc* is a continuous map A: $\overline{D} \to \mathbb{C}^n$ which is holomorphic on D. The *boundary* of the analytic disc A is by definition the restriction of A to the unit circle $S^1 = \partial D$. Often in the literature, the analytic disc and its boundary are identified with their images in $\mathbb{C}^n$. Suppose that $\{F_j\}$ is a sequence of entire functions that is uniformly convergent to a given CR function f on the open set $\omega \subset M$. Let us say we wish to show that $\{F_j\}$ also converges on an open set $\Omega \subset \mathbb{C}^n$. The idea behind analytic discs is to show that each point in Ω is contained in (the image of) an analytic disc whose boundary image is contained in ω. From the maximum principle for analytic functions, the sequence of entire functions $\{F_j\}$ must also converge uniformly on Ω. So our CR extension theorem is reduced to a theorem about analytic discs, which we state in Section 15.1. In Section 15.2, this analytic disc theorem is established for hypersurfaces. The proof for hypersurfaces involves an easy slicing argument and thus we obtain an easy proof of Hans Lewy's original CR extension theorem. In Section 15.3, we prove the analytic disc theorem for quadric submanifolds. The proof here is harder than for hypersurfaces but it is still relatively easy since the analytic discs can be explicitly described. The construction of analytic discs for the general case requires the solution of a nonlinear integral equation (Bishop's equation). This is discussed in Section 15.4. In Section 15.5, we complete the proof of the analytic disc theorem for the general case.

15.1 Reduction to analytic discs

The key result concerning analytic discs is the following.

THEOREM 1 ANALYTIC DISCS
Suppose M is a generic CR submanifold of $\mathbb{C}^n$ of class C^k, $4 \leq k \leq \infty$ with $\dim_{\mathbb{R}} M = 2n - d$, $1 \leq d \leq n-1$*. Let p be a point in M such that the interior of Γ_p in $N_p(M)$ is nonempty. Then for each neighborhood ω of p in M and for each cone $\Gamma < \Gamma_p$, there is a neighborhood $\omega_\Gamma \subset \omega$ and a positive number ϵ_Γ such that each point in $\omega_\Gamma + \{\Gamma \cap B_{\epsilon_\Gamma}\}$ is contained in the image of an analytic disc whose boundary image is contained in ω.*

PROOF OF THE CR EXTENSION THEOREM FROM THE ANALYTIC DISC THEOREM
Suppose $p \in \omega \subset M$ is the given point in the CR extension theorem and let f be a CR function on the open set ω. By Theorem 1 in Chapter 13, there is a sequence of entire functions F_j, $j = 1, 2, \ldots$ which converges to f on some open set ω_2 with $p \in \omega_2 \subset \omega \subset M$. Now we apply the analytic disc theorem with $\omega = \omega_2$. Let

$$\Omega = \bigcup_{\Gamma < \Gamma_p} \{\omega_\Gamma + \{\Gamma \cap B_{\epsilon_\Gamma}\}\}$$

$$\omega' = \bigcup_{\Gamma < \Gamma_p} \omega_\Gamma .$$

From the maximum principle, the sequence F_j converges uniformly on each $\omega_\Gamma + \{\Gamma \cap B_{\epsilon_\Gamma}\}$. Therefore, this sequence of entire functions converges uniformly on each compact subset of Ω. The limit of this sequence on Ω is the desired holomorphic extension of f.

For the uniqueness part of the CR extension theorem, we need the following approximation result for wedges stated at the end of Chapter 13: let ω_1 be a neighborhood of p in M and let $\epsilon > 0$ be given; there exist $\epsilon_2 > 0$ and a neighborhood ω_2 of p in M with $\omega_2 \subset \omega_1$ such that if F is holomorphic on $\omega_1 + \{\Gamma_1 \cap B_\epsilon\}$ and continuous up to ω_1, then F is the uniform limit on $\omega_2 + \{\Gamma_1 \cap B_{\epsilon_2}\}$ of a sequence of entire functions $F_n, n = 1, 2, \ldots$.

By the identity theorem for holomorphic functions, it suffices to show the following: suppose F is holomorphic on $\omega_1 + \{\Gamma_1 \cap B_\epsilon\}$ and continuous up to ω_1; if $F = 0$ on ω_1, then $F \equiv 0$ on an open subset of $\omega_1 + \{\Gamma_1 \cap B_\epsilon\}$. So we can assume that the approximating sequence F_n from the previous paragraph converges uniformly to zero on ω_2. From Theorem 1, it follows that there is an open subset U of $\omega_2 + \{\Gamma_1 \cap B_\epsilon\}$ such that each point in U is contained in the image of an analytic disc whose boundary image is contained in ω_2. The maximum principle implies that F_n converges to zero at each point in U. Therefore, $F \equiv 0$ on U, as desired. ▮

The proof of the uniqueness part of Theorem 2 in Section 14.2 is easier. Here, the open set Ω contains an open subset of M. Therefore, uniqueness follows from Lemma 2 in Section 15.1.

15.2 Analytic discs for hypersurfaces

In this section, we prove the analytic disc theorem (and hence Lewy's CR extension theorem) for hypersurfaces. The proof is particularly simple in this case since we can obtain the analytic discs by an elementary slicing argument. Using Theorem 2 in Section 7.2, we can arrange coordinates so that the given point $p \in M$ is the origin and

$$M = \{(z = x + iy, w) \in \mathbb{C} \times \mathbb{C}^{n-1};\ y = h(x, w)\}$$

where $h\colon \mathbb{R} \times \mathbb{C}^{n-1} \to \mathbb{R}$ is of class C^3 with no pure terms in its Taylor expansion through order 2. From a Taylor expansion of h about the origin, we have

$$h(x, w) = \sum_{j,k=1}^{n-1} q_{jk} w_j \bar{w}_k + \mathcal{O}(3)$$

where $q_{jk} = (\partial^2 h(0)/\partial w_j \partial \bar{w}_k)$, $1 \leq j, k \leq n-d$ is the matrix for the Levi form of M at the origin. Here, $\mathcal{O}(3)$ denotes terms depending on both w and x which vanish to third order at the origin. Since $Q = (q_{jk})$ is a Hermitian symmetric matrix, the w coordinates for $\mathbb{C}^{n-1}$ can be chosen so that Q is diagonalized. This is accomplished by finding a unitary matrix U so that ${}^t\overline{U}QU$ is diagonal and then letting $\hat{w} = U \cdot w$.

The hypersurface M divides a neighborhood $\widetilde{\Omega}$ of the origin in $\mathbb{C}^n$ into two sets

$$\widetilde{\Omega}^+ = \{(z = x + iy, w) \in \widetilde{\Omega};\ y > h(x, w)\}$$
$$\widetilde{\Omega}^- = \{(z = x + iy, w) \in \widetilde{\Omega};\ y < h(x, w)\}.$$

Suppose the Levi form (q_{jk}) has at least one positive eigenvalue. This corresponds to the case where $\{y \geq 0\} \subset \Gamma_0$. By reordering the w-coordinates if necessary, we may assume $q_{11} > 0$. From the Taylor expansion of h, we have

$$h(0, w_1, 0) = q_{11}|w_1|^2 + \mathcal{O}(|w_1|^3).$$

Let ω be an open subset of M which contains the origin. Since q_{11} is positive, any small translate of the complex line $\{(0, w_1, 0); w_1 \in \mathbb{C}\}$ in the positive y direction will intersect the open set $\widetilde{\Omega}^+$ in a simply connected open subset of this translated complex line whose boundary is contained in ω. By continuity,

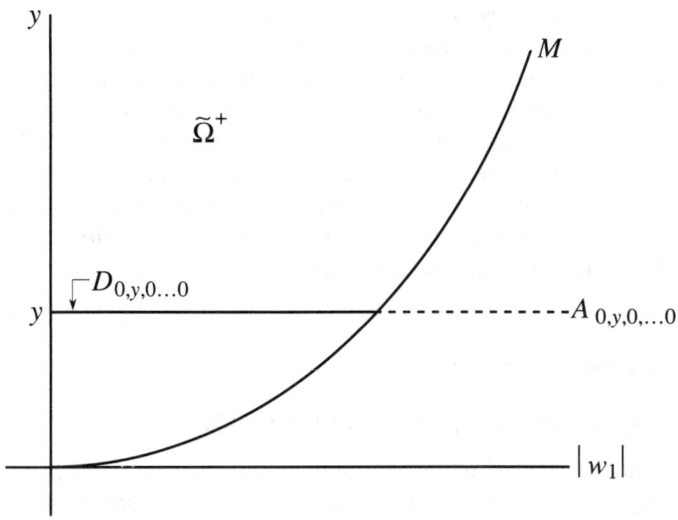

FIGURE 15.1

the same can be said for small translations of this complex line in the x and $w_2, \ldots, w_{n-1}$ directions. More precisely, there are positive numbers $\delta, \epsilon > 0$ such that if $|y| < \epsilon$ and $|x|$, $|w_2|, \ldots, |w_{n-1}| < \delta$, then the complex line

$$A_{x,y,w_2,\ldots,w_{n-1}} = \{(x + iy, \zeta, w_2, \ldots, w_{n-1});\ \zeta \in \mathbb{C}\}$$

intersects $\widetilde{\Omega}^+$ in a simply connected open subset of $A_{x,y,w_2,\ldots,w_{n-1}}$ which we denote by $D_{x,y,w_2,\ldots,w_{n-1}}$ whose boundary lies in ω. The union of the $D_{x,y,w_2,\ldots,w_{n-1}}$ for $|y| < \epsilon$ and $|x|, |w_2|, \ldots, |w_{n-1}| < \delta$ contains an open subset Ω^+ of $\widetilde{\Omega}^+$. From the Riemann mapping theorem, each $D_{x,y,w_2,\ldots,w_{n-1}}$ is biholomorphic to the unit disc in $\mathbb{C}$. This biholomorphism provides the desired analytic disc, and the proof of the analytic disc theorem for the case $\Gamma_0 = \{y \geq 0\}$ is complete.

If instead, $q_{11} < 0$ (i.e., $\Gamma_0 = \{y \leq 0\}$), then the same arguments can be used to construct an open set $\Omega^- \subset \widetilde{\Omega}^-$ which is foliated by the images of analytic discs whose boundaries are contained in ω. If the matrix (q_{jk}) has eigenvalues of opposite sign, then the union of Ω^+, Ω^- together with $\overline{\Omega^+} \cap M$ forms a set Ω in $\mathbb{C}^n$ which contains an open neighborhood of the origin and which is foliated by analytic discs whose boundaries are contained in ω. This completes the proof of the analytic disc theorem for the case of a real hypersurface in $\mathbb{C}^n$.

The proof of Lewy's CR extension theorem for hypersurfaces does not require the Riemann mapping theorem. Instead, one may use the maximum principle directly to each $D_{x,y,w_2,\dots,w_{n-1}}$ to conclude that the approximating sequence of entire functions converges on Ω^+ (or Ω^- or Ω).

There are other proofs of Hans Lewy's theorem which do not use the approximation theorem. Indeed, Lewy did not have the approximation theorem at his disposal. Previous proofs of Lewy's CR extension theorem still arrange coordinates so that geometry of M looks like the picture in Figure 15.1. From there, two approaches can be used; one involves the Cauchy integral formula on each $D_{x,y,w_2,\dots,w_{n-1}}$; the other involves the solution of a $\bar{\partial}$-problem on the complex lines $A_{x,y,w_2,\dots,w_{n-1}}$ (see Hörmander's proof in his book [Ho]). The analytic disc proof presented here more easily generalizes to higher codimension.

15.3 Analytic discs for quadric submanifolds

The slicing argument used in the case of a hypersurface won't work for submanifolds of higher codimension, for if $\text{codim}_{\mathbb{R}} M > 1$, then complex lines usually have empty intersection with M. However, if M is a quadric, then the proof of the analytic disc theorem is still relatively easy because the required analytic discs can be explicitly described.

From Section 7.3, a quadric submanifold M is given by

$$M = \{(z = x + iy, w) \in \mathbb{C}^d \times \mathbb{C}^{n-d};\ y = q(w, \bar{w})\}$$

where q: $\mathbb{C}^{n-d} \times \mathbb{C}^{n-d} \to \mathbb{C}^d$ is a quadratic form. We assume the point $p \in M$ given in the analytic disc theorem is the origin. There is no loss in assuming this because a neighborhood of any point $p \in M$ can be biholomorphically mapped to a neighborhood of the origin by using the group structure on M (see Section 7.3).

We wish to fill out an open set Ω with images of analytic discs whose boundaries are contained in a given open neighborhood of the origin in M. Let us start with a given analytic disc W: $\overline{D} \to \mathbb{C}^{n-d}$. We want to find an analytic disc G: $\overline{D} \to \mathbb{C}^d$ so that the analytic disc given by $A(\zeta) = (G(\zeta), W(\zeta))$ $(\zeta \in \overline{D})$ has boundary image in M. We will then show that by varying W, the images of the corresponding analytic discs, A, will fill out the desired open set Ω.

The analytic disc W: $\overline{D} \to \mathbb{C}^{n-d}$ is given by a convergent power series

$$W(\zeta) = \sum_{j=0}^{\infty} a_j \zeta^j,\ a_j \in \mathbb{C}^{n-d},\ \zeta \in \overline{D}.$$

In our application, all but a finite number of the parameters $\{a_0, a_1, \dots\}$ will vanish. In order for the set $\{A(\zeta) = (G(\zeta), W(\zeta));\ |\zeta| = 1\}$ to be contained in M, the analytic disc G: $\overline{D} \to \mathbb{C}^d$ must satisfy

$$\text{Im}\, G(\zeta) = q(W(\zeta), \overline{W(\zeta)}) \quad \text{for} \quad |\zeta| = 1.$$

After substituting $W(\zeta) = \sum a_j \zeta^j$ and expanding the right side using the bilinearity and symmetry of q, this equation becomes

$$\begin{aligned} \text{Im } G(\zeta) &= q\left(\sum_{j=0}^{\infty} a_j \zeta^j, \sum_{k=0}^{\infty} \bar{a}_k \zeta^{-k}\right) \quad (\text{since } \bar{\zeta} = \zeta^{-1} \text{ on } |\zeta| = 1) \\ &= \sum_{j=0}^{\infty} q(a_j, \bar{a}_j) + 2 \text{ Re}\left\{ \sum_{0 \le k < j} q(a_j, \bar{a}_k)\zeta^{j-k} \right\}. \end{aligned}$$

The term within the brackets on the right is analytic in $\zeta \in D$ and so the analytic disc

$$G(\zeta) = x + i\sum_{j=0}^{\infty} q(a_j, \bar{a}_j) + 2i \sum_{0 \le k < j} q(a_j, \bar{a}_k)\zeta^{j-k}$$

(with $x \in \mathbb{R}^d$) satisfies the equation Im $G(\zeta) = q(W(\zeta), \overline{W(\zeta)})$ for $|\zeta| = 1$. Note that $G(\zeta = 0) = x + i \sum q(a_j, \bar{a}_j)$. We summarize this discussion in the following lemma.

LEMMA 1
Suppose $\{a_0, a_1, \ldots\}$ is a given sequence of vectors in $\mathbb{C}^{n-d}$ with $\sum_{j=0}^{\infty} |a_j|^2 < \infty$. Let $x \in \mathbb{R}^d$ be given. The analytic disc A: $\overline{D} \to \mathbb{C}^n$, $A(\zeta) = (G(\zeta), W(\zeta))$ given by

$$W(\zeta) = \sum_{j=0}^{\infty} a_j \zeta^j$$

$$G(\zeta) = x + i\sum_{j=0}^{\infty} q(a_j, \bar{a}_j) + 2i \sum_{0 \le k < j} q(a_j, \bar{a}_k)\zeta^{j-k}$$

satisfies

$$A(\zeta) \in M \quad \text{for} \quad |\zeta| = 1$$

$$A(\zeta = 0) = \left(x + i\sum_{j=0}^{\infty} q(a_j, \bar{a}_j),\ a_0 \right).$$

The vectors $x \in \mathbb{R}^d$ and $a_0, a_1, \ldots \in \mathbb{C}^{n-d}$ are parameters at our disposal. Each parameter generates an analytic disc A with boundary image in M. Now we will show that by varying these parameters, the images of these discs will sweep out the desired open set Ω for the analytic disc theorem. We consider the center of the disc, $A(\zeta = 0)$. It will be convenient to let $a_0 = w \in \mathbb{C}^{n-d}$.

From Lemma 1, we have

$$A(\zeta = 0) = (x, q(w, \bar{w}), w) + \left(0, \sum_{j=1}^{\infty} q(a_j, \bar{a}_j), 0\right). \tag{1}$$

Here, we have written the coordinates of a point in $\mathbb{C}^n$ as $(x, y, w) \in \mathbb{R}^d \times \mathbb{R}^d \times \mathbb{C}^{n-d}$. The first term on the right describes a point in M. We claim the second term on the right lies in the closure of the convex hull of the image of the Levi form. To see this, recall that the Levi form of the quadric submanifold M at the origin is the map $y = q(w, \bar{w})$ for $w \in \mathbb{C}^{n-d}$. The convex hull of the image of q is the cone

$$\Gamma_0 = \left\{ \sum_{j=1}^{N} t_j q(\alpha_j, \bar{\alpha}_j);\ N \geq 1,\ 0 \leq t_j \leq 1,\ \sum_{j=1}^{N} t_j = 1 \text{ and } \alpha_j \in \mathbb{C}^{n-d} \right\}.$$

Since $tq(\alpha, \bar{\alpha}) = q(\sqrt{t}\alpha, \sqrt{t}\bar{\alpha})$ for $t \geq 0$, we have

$$\Gamma_0 = \left\{ \sum_{j=1}^{\infty} q(a_j, \bar{a}_j);\ a_j \in \mathbb{C}^{n-d} \text{ and } a_j = 0 \text{ for all but a finite number of } j \right\}.$$

Equation (1) now shows that the set of all centers $\{A(\zeta = 0)\}$, where A is given as in Lemma 1, contains the set $M + \Gamma_0$.

The proof of the analytic disc theorem requires a local version of the above analysis. The parameters x, w, and $a_1, a_2, \ldots$ must be restricted so that the boundary of the corresponding analytic disc A is contained in the open subset $\omega \subset M$ given in the analytic disc theorem. In return, we must settle for sweeping out an open set Ω which contains subsets of $M + \Gamma_0$ of the form $\omega_\Gamma + \{B_{\epsilon_\Gamma} \cap \Gamma\}$ for $\Gamma < \Gamma_0$.

For vectors $X_1, \ldots, X_N \in N_0(M) \simeq \mathbb{R}^d$, let

$$\begin{aligned} \Gamma_{X_1 \ldots X_N} &= \text{the convex cone generated by } X_1, \ldots, X_N \\ &= \left\{ \sum_{j=1}^{N} t_j X_j;\ t_j \geq 0 \right\}. \end{aligned}$$

If $\Gamma < \Gamma_0$ is the cone given in the analytic disc theorem, then $X_1, \ldots, X_N$ can be chosen in the image of the Levi form (q) so that

$$\Gamma < \Gamma_{X_1, \ldots, X_N} \subseteq \Gamma_0.$$

For each $j = 1, \ldots, N$, there is a vector $\alpha_j \in \mathbb{C}^{n-d}$ with

$$X_j = q(\alpha_j, \bar{\alpha}_j).$$

Let $a_0 = w \in \mathbb{C}^{n-d}$; let $a_j = t_j \alpha_j$ $(t_j \in \mathbb{R})$ for $1 \leq j \leq N$ and let $a_k = 0$ for $k > N$. From Lemma 1, we obtain a family of analytic discs $A(t, x, w)\colon \overline{D} \to$

$\mathbb{C}^n$ which depend continuously on the parameters $t = (t_1, \ldots, t_N) \in \mathbb{R}^N$ and $x \in \mathbb{R}^d, w \in \mathbb{C}^{n-d}$. From the formula for A, we have

$$|A(t,x,w)(\zeta)| \le C(|t| + |x| + |w|) \quad \text{for} \quad |\zeta| \le 1.$$

Here, C is a uniform constant that is independent of the parameters t, x, w and the variable $\zeta \in \overline{D}$. Suppose ω is the open neighborhood of the origin in M given in the analytic disc theorem. From the above inequality, there is a $\delta = \delta(\Gamma, \omega) > 0$ so that if $|t|, |x|, |w| < \delta$, then the boundary of $A(t,x,w)(\cdot)$ is contained in ω. From (1) and the equation $t_j^2 X_j = q(a_j, \bar{a}_j)$, $1 \le j \le N$, we obtain

$$A(t,x,w)(\zeta = 0) = (x, q(w,\bar{w}), w) + \left(0, \sum_{j=1}^{N} t_j^2 X_j, 0\right). \tag{2}$$

The set $\omega_\Gamma = \{(x, q(w,\bar{w}), w);\ |x|, |w| < \delta\}$ is an open neighborhood of the origin in M. In addition, the set $\{\sum_{j=1}^N t_j^2 X_j;\ |t| = |(t_1, \ldots, t_N)| < \delta\}$ contains the set $\Gamma_{X_1 \ldots X_N} \cap B_\epsilon$ for some suitably small $\epsilon = \varepsilon(\delta) > 0$. Therefore, the set

$$\Omega = \{A(t,x,w)(\zeta = 0);\ |t|, |x|, |w| < \delta\}$$

contains the set $\omega_\Gamma + \{B_\epsilon \cap \Gamma\}$ (since $\Gamma \subset \Gamma_{X_1 \ldots X_N}$), and moreover, each point in Ω is contained in the image of an analytic disc (namely $A(t,x,w)$) whose boundary is contained in ω. This completes the proof of the analytic disc theorem (and hence the CR extension theorem) for the case of a quadric submanifold.

The discussion immediately following (1) suggests that the analytic disc theorem for quadric submanifolds has a global version. That is, we showed that the set $M + \Gamma_0$ can be realized as the union of centers of analytic discs whose boundary images lie in M. So it is not surprising that the CR extension theorem for quadrics has a global version as well.

THEOREM 1
Suppose M is a quadric submanifold of $\mathbb{C}^n$. If the interior of $\{\Gamma_0\}$ is nonempty, then for each CR function f that is of class C^1 on M, there is a function F that is holomorphic on $\Omega = M +$ interior $\{\Gamma_0\}$ and continuous on $\Omega \cup M$ with $F|_M = f$.

PROOF In general, there is no global version of the approximation theorem. Therefore, we use the local version of the CR extension theorem together with the group structure on M to prove Theorem 1. Let Γ be any smaller subcone of Γ_0. From the local CR extension theorem, there is an open set ω in M containing the origin and an $\epsilon_\Gamma > 0$ such that the given CR function f on M extends holomorphically to $\omega_\Gamma + \{\Gamma \cap B_{\epsilon_\Gamma}\}$. The group structure on M is

$$(z_1, w_1) \circ (z_2, w_2) = (z_1 + z_2 + 2iq(w_1, \bar{w}_2), w_1 + w_2)$$

for $(z_1, w_1), (z_2, w_2) \in M$. Note that the group structure preserves $M + \Gamma$ for any convex cone $\Gamma \subset N_0(M)$. Let p be any point on M. The map $g_p(z, w) = (z, w) \circ p$ is a holomorphic map which takes $\omega_\Gamma + \{\Gamma \cap B_{\epsilon_\Gamma}\}$ onto the open set $g_p\{\omega_\Gamma\} + \{\Gamma \cap B_{\epsilon_\Gamma}\}$. By letting p range over M, we see that the CR function f extends to a holomorphic function on $M + \{\Gamma \cap B_{\epsilon_\Gamma}\}$. This argument can be applied to any translate of M lying in $M + \{\Gamma \cap B_{\epsilon_\Gamma}\}$. Continuing in this way, f extends holomorphically to $M + \Gamma$. The proof of this theorem is completed by applying this argument to each smaller subcone $\Gamma < \Gamma_0$. ∎

15.4 Bishop's equation

For a quadric submanifold $M = \{(z = x + iy, w);\ y = q(w, \bar{w})\}$, the construction of the analytic discs is easy. Given an analytic disc $W\colon \overline{D} \to \mathbb{C}^{n-d}$, an explicit formula is given for the analytic disc $G\colon \overline{D} \to \mathbb{C}^d$ (see Lemma 1 in the previous section) so that the boundary of the analytic disc $A = (G, W)\colon \overline{D} \to \mathbb{C}^n$ is contained in M. In this case, the explicit formula for G is possible because the graphing function for M (namely q) is independent of the variable x. However, the graphing function of a more general CR submanifold of $\mathbb{C}^n$ depends on both x and w and the corresponding construction of analytic discs requires the solution of a nonlinear integral equation (Bishop's equation [Bi]) which we now discuss.

As usual, we may assume that coordinates have been chosen so that the given point $p \in M$ is the origin and

$$M = \{(z = x + iy, w) \in \mathbb{C}^d \times \mathbb{C}^{n-d};\ y = h(x, w))\}$$

where $h\colon \mathbb{R}^d \times \mathbb{C}^{n-d} \to \mathbb{R}^d$ is smooth (class C^4) and $h(0) = 0, Dh(0) = 0$. Given an analytic disc $W\colon \overline{D} \to \mathbb{C}^{n-d}$, we wish to find an analytic disc $G\colon \overline{D} \to \mathbb{C}^d$ so that the boundary of the disc $A = (G, W)\colon \overline{D} \to \mathbb{C}^n$ is contained in M. This means that G must satisfy

$$\text{Im} G(\zeta) = h(\text{Re } G(\zeta), W(\zeta)) \quad \text{for} \quad |\zeta| = 1.$$

This equation involves both $u =$ Re G and $v =$ Im G, whereas in the quadric case, the corresponding equation only involves $v =$ Im G. The above equation will be easier to solve by eliminating either u or v. To do this, we use the Hilbert transform which is defined as follows. Let S^1 be the unit circle in $\mathbb{C}$. If $u\colon S^1 \to \mathbb{R}^d$ is a smooth function, then u extends to a unique harmonic function on the unit disc D. This harmonic function has a unique harmonic conjugate in $\overline{D}$ (denoted by v) which vanishes at the origin. The *Hilbert transform* of u (denoted $Tu\colon S^1 \to \mathbb{R}^d$) is defined to be $v|_{S^1}$. If $G = u + iv\colon \overline{D} \to \mathbb{C}^d$ is analytic and continuous up to S^1, then $T(u|_{S^1}) = v|_{S^1} + C$ where $C = -v(\zeta = 0)$. The function $-iG = v - iu$ is also analytic and so $T(v|_{S^1}) = -u + x$ where

$x = u(\zeta = 0)$. Conversely, if $u, v\colon S^1 \to \mathbb{R}^d$ are continuous functions with $u = -Tv + x$, then $u + iv\colon S^1 \to \mathbb{C}^d$ is the boundary values of a unique analytic disc $G\colon \overline{D} \to \mathbb{C}^d$ with Re $G(\zeta = 0) = x$.

Suppose $u + iv = G\colon \overline{D} \to \mathbb{C}^d$ is an analytic disc with $v(e^{i\phi}) = h(u(e^{i\phi}), W(e^{i\phi}))$ for $0 \le \phi \le 2\pi$. We apply $-T$ to both sides of this equation and obtain

$$u(e^{i\phi}) = -T(h(u, W))(e^{i\phi}) + x, \qquad 0 \le \phi \le 2\pi \tag{1}$$

where $x \in \mathbb{R}^d$ is the value of u at $\zeta = 0$. The above equation will be referred to as *Bishop's equation.* Conversely, suppose the analytic disc $W\colon \overline{D} \to \mathbb{C}^{n-d}$ and the vector $x \in \mathbb{R}^d$ are given, and suppose $u\colon S^1 \to \mathbb{R}^d$ is a solution to Bishop's equation. From the above discussion, the function

$$\phi \mapsto u(e^{i\phi}) + ih(u(e^{i\phi}), W(e^{i\phi}))$$

is the boundary values of a unique analytic disc $G\colon \overline{D} \to \mathbb{C}^d$. Since Re $G(e^{i\phi}) = u(e^{i\phi})$, the boundary of the analytic disc $A = (G, W)\colon \overline{D} \to \mathbb{C}^n$ is contained in M. Furthermore, Re $G(\zeta = 0) = x$. We summarize this discussion in the following lemma.

LEMMA 1

Suppose $W\colon \overline{D} \to \mathbb{C}^{n-d}$ is an analytic disc and $x \in \mathbb{R}^d$. If $u\colon S^1 \to \mathbb{R}^d$ is a continuous function that satisfies Bishop's equation

$$u(e^{i\phi}) = -T(h(u, W))(e^{i\phi}) + x \qquad 0 \le \phi \le 2\pi$$

then there is a unique analytic disc $G\colon \overline{D} \to \mathbb{C}^d$ that satisfies

(a) *Re $G(\zeta = 0) = x$*

(b) *Re $G(e^{i\phi}) = u(e^{i\phi})$, $0 \le \phi \le 2\pi$*

(c) *the boundary of the analytic disc $A = (G, W)\colon \overline{D} \to \mathbb{C}^n$ is contained in M.*

Now, we discuss the solution to Bishop's equation. This requires no convexity assumption on the Levi form. In the next section, we shall use the convexity assumption on the Levi form to choose the right family of analytic discs $W\colon \overline{D} \to \mathbb{C}^{n-d}$ so that the associated family of discs $A = (G, W)\colon \overline{D} \to \mathbb{C}^n$ (from Lemma 1) sweeps out the desired open set Ω for the analytic disc theorem.

To solve Bishop's equation, we must set up certain Banach spaces. The Hilbert transform is *not* a continuous linear map on the space of continuous functions on S^1 with the sup-norm. Instead, we consider the space of Hölder continuous functions, which is defined as follows. Suppose L is any normed linear space (usually either $\mathbb{R}^d$ or $\mathbb{C}^d$). A continuous function $f\colon S^1 \to L$ is said to be *Hölder continuous with exponent* α $(0 < \alpha \le 1)$ if there is a finite positive number M so that $|f(e^{i\phi_1}) - f(e^{i\phi_2})| \le M|\phi_1 - \phi_2|^\alpha$ for $0 \le \phi_1, \phi_2 \le 2\pi$. The set of all such functions is denoted by $C^\alpha(S^1, L)$. If the space L is unimportant for the discussion at hand, then it will be omitted from the notation. The space $C^\alpha(S^1)$ is a Banach space under the norm

$$\|f\|_\alpha = |f|_\infty + \sup_{0\leq\phi_1,\phi_2\leq 2\pi} \frac{|f(e^{i\phi_1}) - f(e^{i\phi_2})|}{|\phi_1 - \phi_2|^\alpha}$$

where $|f|_\infty$ is the usual sup-norm. The Hilbert transform is a continuous map from $C^\alpha(S^1, \mathbb{R}^d)$ to itself. To show this, we need an integral formula for Tu. It suffices to consider the case $d = 1$. If u: $S^1 \to \mathbb{R}$ is a continuous function, then its harmonic extension is given by Poisson's integral formula

$$u(re^{i\theta}) = \frac{1}{2\pi}\int_0^{2\pi} \frac{(1-r^2)u(e^{i\phi})d\phi}{1 - 2r\cos(\phi - \theta) + r^2} \qquad 0 \leq r < 1,\ 0 \leq \theta \leq 2\pi.$$

Letting $z = re^{i\theta}$ and $\zeta = e^{i\phi}$, this becomes

$$\begin{aligned} u(z) &= \frac{1}{2\pi i}\int_{|\zeta|=1} \left(\frac{1-|z|^2}{|\zeta - z|^2}\right) u(\zeta)\frac{d\zeta}{\zeta} \\ &= \mathrm{Re}\left\{\frac{1}{2\pi i}\int_{|\zeta|=1}\left(\frac{\zeta+z}{\zeta-z}\right)u(\zeta)\frac{d\zeta}{\zeta}\right\}. \end{aligned}$$

Since the quantity inside the brackets is holomorphic in z for $|z| < 1$, a harmonic conjugate of u is given by

$$v(z) = \mathrm{Im}\left\{\frac{1}{2\pi i}\int_{|\zeta|=1}\left(\frac{\zeta+z}{\zeta-z}\right)u(\zeta)\frac{d\zeta}{\zeta}\right\}.$$

Note that $v(0) = \mathrm{Im}\{(1/2\pi)\int_0^{2\pi} u(e^{i\phi})d\phi\} = 0$ because u is real valued. Therefore, Tu is the boundary values of v on S^1.

The function $(\zeta + z)/\zeta$ is smooth for $|\zeta| = 1$ and $|z| \leq 1$. Therefore, the continuity of T: $C^\alpha(S^1) \to C^\alpha(S^1)$ will follow from the continuity of the Cauchy kernel K: $C^\alpha(S^1) \to C^\alpha(S^1)$, where Ku is the boundary values on S^1 of the function

$$(Ku)(z) = \frac{1}{2\pi i}\int_{|\zeta|=1} \frac{u(\zeta)d\zeta}{\zeta - z}, \qquad |z| < 1.$$

LEMMA 2
Let $0 < \alpha < 1$. *Then*

(a) $K: C^\alpha(S^1) \to C^\alpha(S^1)$ *is a continuous linear map.*
(b) $T: C^\alpha(S^1) \to C^\alpha(S^1)$ *is a continuous linear map.*

This lemma holds for a wide class of singular integral operators which includes both K and T. However, the special properties of the Cauchy kernel can be used to give an easy proof of the lemma for K (and hence for T).

PROOF To prove the continuity of K on $C^\alpha(S^1)$, we first extend u to a function $E(u)$: $\overline{D} \to \mathbb{R}$ so that $\|E(u)\|_{C^\alpha(\overline{D})} \leq C\|u\|_{C^\alpha(S^1)}$ for some uniform constant C that is independent of u. Here, $C^\alpha(\overline{D})$ is defined the same way as $C^\alpha(S^1)$ except that the domain of definition is $\overline{D}$ rather than S^1. To construct $E(u)$, we extend u to be constant on any line that is normal to S^1, and then we multiply this function by a suitable cutoff function. Therefore to prove part (a) it suffices to show K: $C^\alpha(\overline{D}) \to C^\alpha(S^1)$ is continuous.

First, we estimate $\|Ku\|_\infty$. For $z \in D$, Cauchy's integral formula yields

$$Ku(z) = \frac{1}{2\pi i} \int\limits_{|\zeta|=1} \frac{u(\zeta) - u(z)}{\zeta - z} d\zeta + u(z).$$

Therefore

$$|Ku(z)| \leq \frac{1}{2\pi} \left(\int\limits_{|\zeta|=1} |\zeta - z|^{\alpha - 1} |d\zeta| \right) \|u\|_\alpha + \|u\|_\infty.$$

Since $\alpha > 0$, we have $(2\pi)^{-1} \int_{|\zeta|=1} |\zeta - z|^{\alpha-1}|d\zeta| \leq C_\alpha$ for $|z| \leq 1$, where C_α is some finite positive constant that is independent of z. It follows that

$$\|Ku\|_\infty \leq C_\alpha \|u\|_\alpha + \|u\|_\infty.$$

Next, we estimate $Ku(z_1) - Ku(z_2)$ for $z_1, z_2 \in D$. By Cauchy's integral formula, we have

$$K(u)(z_1) - K(u)(z_2) =$$
$$\frac{1}{2\pi i} \int\limits_{|\zeta|=1} \left(\frac{u(\zeta) - u(z_1)}{\zeta - z_1} \right) - \left(\frac{u(\zeta) - u(z_2)}{\zeta - z_2} \right) d\zeta + u(z_1) - u(z_2).$$

Let $\epsilon = |z_1 - z_2|$ and $B_{2\epsilon} = \{\zeta \in \mathbb{C};\ |\zeta - z_1| < 2\epsilon\}$. We rewrite this equation as

$$K(u)(z_1) - K(u)(z_2) = \frac{1}{2\pi i} \int_{\zeta \in S^1 \cap B_{2\epsilon}} \left(\frac{u(\zeta) - u(z_1)}{\zeta - z_1}\right) - \left(\frac{u(\zeta) - u(z_2)}{\zeta - z_2}\right) d\zeta$$

$$+ \frac{1}{2\pi i} \int_{\zeta \in S^1 - B_{2\epsilon}} \frac{u(z_2) - u(z_1)}{\zeta - z_2} d\zeta$$

$$+ \frac{1}{2\pi i} \int_{\zeta \in S^1 - B_{2\epsilon}} \left[\frac{1}{\zeta - z_1} - \frac{1}{\zeta - z_2}\right] (u(\zeta) - u(z_1)) d\zeta + u(z_1) - u(z_2).$$

By Cauchy's integral formula, the second integral on the right is

$$u(z_2) - u(z_1) + \frac{1}{2\pi i} \int_{\zeta \in S^1 \cap B_{2\epsilon}} \frac{u(z_1) - u(z_2)}{\zeta - z_2} d\zeta.$$

Substituting this for the second integral and simplifying, we have

$$K(u)(z_1) - K(u)(z_2) = \frac{1}{2\pi i} \int_{\zeta \in S^1 \cap B_{2\epsilon}} \left(\frac{u(\zeta) - u(z_1)}{\zeta - z_1}\right) - \left(\frac{u(\zeta) - u(z_2)}{\zeta - z_2}\right) d\zeta$$

$$+ \frac{1}{2\pi i} \int_{\zeta \in S^1 \cap B_{2\epsilon}} \frac{u(z_1) - u(z_2)}{\zeta - z_2} d\zeta$$

$$+ \frac{1}{2\pi i} \int_{\zeta \in S^1 - B_{2\epsilon}} \frac{(z_1 - z_2)(u(\zeta) - u(z_1))}{(\zeta - z_1)(\zeta - z_2)} d\zeta.$$

Let A_1, A_2, A_3 be the first, second, and third integrals on the right, respectively. For A_1, we have

$$|A_1| \le \frac{1}{2\pi} \int_{\zeta \in S^1 \cap B_{2\epsilon}} (|\zeta - z_1|^{\alpha - 1} + |\zeta - z_2|^{\alpha - 1}) |d\zeta| \cdot \|u\|_\alpha.$$

Parameterizing the unit circle by $\zeta = (z_1/|z_1|)e^{it}$, this estimate becomes

$$|A_1| \le C\|u\|_\alpha \int_0^{3\epsilon} t^{\alpha - 1} dt$$

where C is a uniform constant. Recalling that $\epsilon = |z_1 - z_2|$ and $\alpha > 0$, we obtain

$$|A_1| \le C_\alpha \|u\|_\alpha |z_1 - z_2|^\alpha$$

where $C_\alpha = C3^\alpha/\alpha$.

In a similar manner, we have

$$
\begin{aligned}
|A_3| &\leq \frac{1}{2\pi}\left(\int\limits_{\zeta\in S^1-B_{2\epsilon}} |\zeta-z_1|^{\alpha-1}|\zeta-z_2|^{-1}|d\zeta|\right)\|u\|_\alpha|z_1-z_2| \\
&\leq C\left(\int\limits_{\epsilon}^{2\pi} t^{\alpha-2}dt\right)\|u\|_\alpha|z_1-z_2| \\
&\leq C_\alpha\|u\|_\alpha|z_1-z_2|^\alpha
\end{aligned}
$$

where $C_\alpha = (1+(2\pi)^{\alpha-1})(1-\alpha)^{-1}C$.

For A_2, we use Cauchy's theorem to deform the contour of integration to obtain

$$
A_2 = \frac{1}{2\pi i}\int\limits_{\zeta\in\partial B_{2\epsilon}-D} \frac{u(z_1)-u(z_2)}{\zeta-z_2}d\zeta \qquad (\text{for } z_1, z_2 \in D).
$$

Since $|\zeta - z_2| > \epsilon$ for $\zeta \in \partial B_{2\epsilon}$, we have

$$
\begin{aligned}
|A_2| &\leq \frac{1}{2\pi}\frac{\|u\|_\alpha|z_1-z_2|^\alpha\cdot 2\pi\epsilon}{\epsilon} \\
&= \|u\|_\alpha|z_1-z_2|^\alpha.
\end{aligned}
$$

By summing the three estimates for A_1, A_2, A_3, we obtain

$$
\frac{|K(u)(z_1)-K(u)(z_2)|}{|z_1-z_2|^\alpha} \leq C_\alpha\|u\|_\alpha \quad \text{for} \quad z_1, z_2 \in D
$$

where C_α is a constant depending only on α. This estimate is uniform for $z_1, z_2 \in D$. Therefore, this estimate holds for $z_1, z_2 \in \overline{D}$. Together with the estimate given above for $\|Ku\|_\infty$, the proof of the lemma is complete. ∎

For functions u: $S^1 \to \mathbb{R}^d$ and W: $S^1 \to \mathbb{C}^{n-d}$, we define $H(u,W)$: $S^1 \to \mathbb{R}^d$ by

$$
H(u,W)(e^{i\phi}) = h(u(e^{i\phi}), W(e^{i\phi})) \quad \text{for} \quad 0 \leq \phi \leq 2\pi.
$$

Suppose h is of class C^1 and let $0 \leq \alpha \leq 1$. It is an easy exercise to show that

$$
H\colon\ C^\alpha(S^1,\mathbb{R}^d)\times C^\alpha(S^1,\mathbb{C}^{n-d}) \to C^\alpha(S^1,\mathbb{R}^d)
$$

is a continuous (nonlinear) map. If h is of class C^2, then H is a C^1 map in the sense of Banach spaces. Furthermore, we have

$$
(D_uH)(u,W)(v) = \frac{\partial h}{\partial x}(u,W)\cdot v.
$$

Here, $D_u H$ is the Banach space derivative with respect to u which is defined for $u, v \in C^\alpha(S^1, \mathbb{R}^d)$ by

$$(D_u H)(u, W)(v) = \lim_{t \to 0} \frac{H(u + tv, W) - H(u, W)}{t}$$

where the limit on the right is taken with respect to the topology on the space $C^\alpha(S^1, \mathbb{R}^d)$. Note there is a slight loss of differentiability (i.e., if h is C^2 then H is only C^1). This is because the norm on $C^\alpha(S^1, \mathbb{R}^d)$ involves the estimate of the fractional difference quotient.

In view of Lemma 2, T: $C^\alpha(S^1, \mathbb{R}^d) \to C^\alpha(S^1, \mathbb{R}^d)$ is a continuous *linear* map. Therefore, T is also differentiable (in fact C^∞) and T is its own derivative, i.e.,

$$(DT)(u)(v) = T(v) \quad \text{for} \quad u, v \in C^\alpha(S^1, \mathbb{R}^d).$$

Bishop's equation can be rewritten

$$u + T(H(u, W)) - x = 0.$$

Given the analytic disc W: $\overline{D} \to \mathbb{C}^{n-d}$ and the vector $x \in \mathbb{R}^d$, we wish to find the solution u: $S^1 \to \mathbb{R}^d$ to Bishop's equation.

THEOREM 1 SOLUTION TO BISHOP'S EQUATION

Fix $0 < \alpha < 1$. Suppose h: $\mathbb{R}^d \times \mathbb{C}^{n-d} \to \mathbb{R}^d$ is of class C^2 with $h(0) = 0, Dh(0) = 0$. There is a $\delta > 0$ such that if $|x| < \delta$ and $W \in C^\alpha(S^1, \mathbb{C}^{n-d})$ with $\|W\|_\alpha < \delta$, then there is a unique element $u = u(W, x) \in C^\alpha(S^1, \mathbb{R}^d)$ that solves Bishop's equation with $u(W = 0, x = 0) = 0$. In addition, if h is of class C^k $(k \geq 2)$, then there is a $\delta > 0$ such that u: $C^\alpha(S^1, \mathbb{C}^{n-d}) \times \mathbb{R}^d \to C^\alpha(S^1, \mathbb{R}^d)$ depends in a C^{k-1} fashion on $x \in \mathbb{R}^d$ and $W \in C^\alpha(S^1, \mathbb{C}^{n-d})$ with $|x| < \delta$ and $\|W\|_\alpha < \delta$.

PROOF The most efficient proof of this theorem involves the Banach space version of the implicit function theorem. This is just like the usual implicit function theorem except that the domain and range are Banach spaces and norms are used instead of absolute values. We observe that the map

$$\mathcal{F}: C^\alpha(S^1, \mathbb{R}^d) \times C^\alpha(S^1, \mathbb{C}^{n-d}) \times \mathbb{R}^d \to C^\alpha(S^1, \mathbb{R}^d)$$

$$\mathcal{F}(u, W, x) = u + T(H(u, W)) - x$$

is of class C^1 in a neighborhood of the origin in $C^\alpha(S^1, \mathbb{R}^d) \times C^\alpha(S^1, \mathbb{C}^{n-d}) \times \mathbb{R}^d$. Since $Dh(0, 0) = 0$, the Banach space derivative of $\mathcal{F}$ with respect to u evaluated at $u = 0$, $W = 0$, $x = 0$ is the identity map from $C^\alpha(S^1, \mathbb{R}^d)$ to itself. Since $\mathcal{F}(0, 0, 0) = 0$, the existence of $u(W, x)$ follows from the implicit function theorem.

Since $\mathcal{F}$ is of class C^1, the solution u depends in a C^1 fashion on W and x in a neighborhood of the origin in $C^\alpha(S^1, \mathbb{R}^d) \times \mathbb{R}^d$. If h is of class C^k, $k \geq 2$, then

$\mathcal{F}$ is of class C^{k-1} and an induction argument shows that u depends in a C^{k-1} fashion on x and W in a neighborhood of the origin (since $(\partial \mathcal{F}/\partial u)(0,0,0) = I$, differentiating the equation $\mathcal{F}(u,W,x) = 0$ allows us to solve for a jth derivative of u in terms of lower order derivatives of u).

If the reader is queasy with the Banach space implicit function theorem, then the reader can use the following outline to fashion his or her own proof using the contraction mapping principle (which is the key tool in the proof of the implicit function theorem). Fix $x \in \mathbb{R}^d$ and $W \in C^\alpha(S^1, \mathbb{C}^{n-d})$ near the origin. Since Dh is small near the origin, the map

$$\widetilde{\mathcal{F}}(x,W)\colon\ C^\alpha(S^1,\mathbb{R}^d) \to C^\alpha(S^1,\mathbb{R}^d)$$
$$\widetilde{\mathcal{F}}(x,W)(u) = x - T(H(u,W))$$

is a contraction on a neighborhood of the origin in $C^\alpha(S^1,\mathbb{R}^d)$ (i.e., $\|\widetilde{\mathcal{F}}(x,W)(u_1) - \widetilde{\mathcal{F}}(x,W)(u_2)\|_{C^\alpha} \le \lambda\|u_1 - u_2\|_{C^\alpha}$ for some fixed λ with $0 \le \lambda < 1$). The contraction mapping principle implies that $\widetilde{\mathcal{F}}(x,W)$ has a fixed point $u = u(x,W)$, which in turn is the solution of Bishop's equation. Now, $\widetilde{\mathcal{F}}$ is uniformly continuous in its dependence on the parameters $x \in \mathbb{R}^d$ and $W \in C^\alpha(S^1,\mathbb{C}^{n-d})$. Since $\widetilde{\mathcal{F}}(x,W)$ is a contraction, u also depends continuously on the parameters $x \in \mathbb{R}^d$ and $W \in C^\alpha(S^1,\mathbb{C}^{n-d})$ (near the origin).

In the next section, the analytic disc W depends smoothly on various real parameters. Since u depends continuously on x and W, u also depends continuously on these real parameters. We need to know that if h is of class C^k, then u depends on these real parameters in a C^{k-1} fashion. This can be established (without the use of Banach space derivatives) by differentiating the equation $u = \widetilde{\mathcal{F}}(x,W)(u)$ with respect to these real parameters together with an induction argument. Details of this approach are left to the reader. ∎

15.5 The proof of the analytic disc theorem for the general case

We first arrange coordinates as in Theorem 2 in Section 7.2 (with $k = 2$) so that the given point p is the origin and

$$M = \{(z = x + iy, w) \in \mathbb{C}^d \times \mathbb{C}^{n-d};\ y = h(x,w)\}$$

where $h\colon \mathbb{R}^d \times \mathbb{C}^{n-d} \to \mathbb{R}^d$ is smooth (class C^4) with $h(0) = 0$, $Dh(0) = 0$ and

$$\frac{\partial^2 h(0)}{\partial x^\alpha \partial w^\beta} = 0 \quad \text{for} \quad |\alpha| + |\beta| = 2.$$

As before, $H_0^{1,0}(M)$ is identified with $\{(0,0,w); w \in \mathbb{C}^{n-d}\}$ and $N_0(M)$ is identified with $\{(0, y, 0); y \in \mathbb{R}^d)\}$. Define the bilinear form $\tilde{\mathcal{L}}$: $\mathbb{C}^{n-d} \times \mathbb{C}^{n-d} \to \mathbb{C}^d$ by

$$\tilde{\mathcal{L}}(w, \hat{w}) = \sum_{j,k=1}^{n-d} \frac{\partial^2 h(0)}{\partial w_j \partial \bar{w}_k} w_j \hat{w}_k, \quad w, \hat{w} \in \mathbb{C}^{n-d}.$$

From Corollary 1 in Section 10.2, the extrinsic Levi form $w \mapsto \tilde{\mathcal{L}}_0(w)$ can be identified with the map $w \mapsto \tilde{\mathcal{L}}(w, \bar{w}) \in \mathbb{R}^d$, for $w \in \mathbb{C}^{n-d}$. By definition, Γ_0 is the convex hull of the image of this map. Suppose $\Gamma < \Gamma_0$ is the cone given in the statement of the analytic disc theorem. As in Section 15.3, we can find vectors $X_1, \ldots, X_N$ which lie in the image $\tilde{\mathcal{L}}_0$ with

$$\Gamma < \Gamma_{X_1 \ldots X_N} \subseteq \Gamma_0$$

where $\Gamma_{X_1 \ldots X_N}$ is the convex cone generated by $X_1, \ldots, X_N$.

Since each X_j lies in the image of $\tilde{\mathcal{L}}_0$, there are vectors $\alpha_j \in \mathbb{C}^{n-d}$ with

$$X_j = \tilde{\mathcal{L}}(\alpha_j, \bar{\alpha}_j) \qquad 1 \le j \le N.$$

For $t \in \mathbb{R}^N$ and $w \in \mathbb{C}^{n-d}$, define the analytic disc

$$W(t, w)(\zeta) = w + \sum_{j=1}^{N} t_j \alpha_j \zeta^j.$$

W depends smoothly on the parameters $t \in \mathbb{R}^N$ and $w \in \mathbb{C}^{n-d}$ and we have $W(0, 0)(\zeta) = 0$. From Theorem 1 in the previous section, there is a solution $u(W(t, w), x)$ to Bishop's equation. For simplicity, we write $u(t, x, w)$ for $u(W(t, w), x)$. We let $A(t, x, w)$: $\overline{D} \to \mathbb{C}^n$ be the resulting analytic disc (see Lemma 1 in the previous section). Now h is assumed to be of class C^4. From the regularity part of the solution to Bishop's equation, we see that u (and hence A) is a map of class C^3 from a neighborhood of the origin in $\mathbb{R}^N \times \mathbb{R}^d \times \mathbb{C}^{n-d}$ to $C^\alpha(S^1)$ (here, α is any fixed number with $0 < \alpha < 1$). The uniqueness part of the solution to Bishop's equation implies that $u(t = 0, x = 0, w = 0)(\cdot) = 0$ and hence $A(t = 0, x = 0, w = 0)(\cdot) = 0$. By suitably restricting the parameters t, x, w, we can ensure that the boundary of the analytic disc $A(t, x, w)(\cdot)$ is contained in the given open set ω of the analytic disc theorem. From Lemma 1 in the previous section, we have Re $G(t, x, w)(\zeta = 0) = x$ where G is the analytic disc with $u =$ Re G. We summarize this discussion in the following lemma. Let $\mathcal{O}^\alpha(D, \mathbb{C}^n)$ be the set of all analytic discs with values in $\mathbb{C}^n$ whose boundaries are elements of $C^\alpha(S^1, \mathbb{C}^n)$.

LEMMA 1

Given ω an open neighborhood of the origin in M, there is a $\delta > 0$ and a map

$$A\colon \{(t, x, w) \in \mathbb{R}^N \times \mathbb{R}^d \times \mathbb{C}^{n-d};\ |t|, |x|, |w| < \delta\} \to \mathcal{O}^\alpha(D, \mathbb{C}^n)$$

of class C^3 such that the boundary of each $A(t,x,w)$ is contained in $\omega \subset M$ for $|t|$, $|x|$, $|w| < \delta$. Furthermore, $A(t,x,w)(\zeta) = (G(t,x,w)(\zeta), W(t,w)(\zeta))$ and Re $G(t,x,w)(\zeta = 0) = x$.

As in the proof for quadric submanifolds, we wish to show that the set of centers

$$\Omega = \{A(t,x,w)(\zeta = 0);\ |t|, |x|, |w| < \delta\}$$

contains a set of the form $\omega_\Gamma + \{\Gamma \cap B_\epsilon\}$ for some $\epsilon = \epsilon_\Gamma > 0$. To do this, we examine a Taylor expansion of $A(t,x,w)(\zeta = 0)$ in t for x and w fixed with $|x|, |w| < \delta$. In the following lemma, we identify a point $(z,w) \in \mathbb{C}^d \times \mathbb{C}^{n-d}$ with $(x,y,w) \in \mathbb{R}^d \times \mathbb{R}^d \times \mathbb{C}^{n-d}$, where $z = x + iy$.

LEMMA 2
Given $\eta > 0$ there are constants $\delta > 0, \epsilon' > 0$ such that for $|x|, |w| < \delta$ and $|t| < \epsilon'$

$$A(t,x,w)(\zeta = 0) = (x, h(x,w), w) + \left(0, \sum_{j=1}^{N} t_j^2 X_j, 0\right) + (0, \xi(t,x,w), 0)$$

where $\xi\colon \mathbb{R}^N \times \mathbb{R}^d \times \mathbb{C}^{n-d} \to \mathbb{R}^d$ is of class C^3 and

$$|\xi(t,x,w)| \leq \eta |t|^2.$$

The sum of the first two terms on the right side of the expansion of $A(t,x,w)(\zeta = 0)$ is exactly the expression for the center of the analytic disc in the quadric case (compare with (2) in Section 15.3). This is not surprising since M can be approximated to third order at the origin by a quadric submanifold. However, the error term in the above lemma is not just any third-order error term. It is crucial for what follows that ξ vanish to second order in t with coefficients that are small in (x,w,t) (for example, we cannot allow a term such as $x_j^2 t_k$ to be part of ξ).

PROOF Write $G(t,x,w)(\zeta) = u(t,x,w)(\zeta) + iv(t,x,w)(\zeta)$. From Lemma 1

$$W(t,w)(\zeta = 0) = w, \qquad u(t,x,w)(\zeta = 0) = x.$$

Therefore

$$A(t,x,w)(\zeta = 0) = (x, v(t,x,w)(\zeta = 0), w).$$

It suffices to examine a Taylor expansion of $v(t,x,w)(\zeta = 0)$ in t. Since $v(t,x,w)(\zeta)$ is harmonic in ζ, the mean value theorem yields

$$v(t,x,w)(\zeta = 0) = \frac{1}{2\pi}\int_0^{2\pi} v(t,x,w)(e^{i\phi})d\phi.$$

Since the boundary of A is contained in M, we have $v(t,x,w)(e^{i\phi}) = h(u(t,x,w)(e^{i\phi}), W(t,w)(e^{i\phi}))$. Therefore

$$v(t,x,w)(\zeta = 0) = \frac{1}{2\pi}\int_0^{2\pi} h(u(t,x,w)(e^{i\phi}), W(t,w)(e^{i\phi}))d\phi. \tag{1}$$

Note that $W(t = 0, w)(\zeta) = w$ (a constant) and so $u(t = 0, x, w)(e^{i\phi}) = x$ is the unique (constant) solution to Bishop's equation in this case. Therefore, we have

$$v(t = 0, x, w)(\zeta = 0) = \frac{1}{2\pi}\int_0^{2\pi} h(x,w)d\phi = h(x,w). \tag{2}$$

This is the constant term (in t) in the expansion of $v(t,x,w)(\zeta = 0)$.

For the linear term, we differentiate (1) with respect to t; evaluate this at $t = 0$ and use the fact that $u = x$ and $W = w$ at $t = 0$. We obtain

$$\begin{aligned}\frac{\partial v}{\partial t_j}(t = 0, x, w)(\zeta = 0) &= \frac{1}{2\pi}\int_0^{2\pi} \frac{\partial h}{\partial x}(x,w)\cdot\frac{\partial u}{\partial t_j}(e^{i\phi})d\phi\bigg|_{t=0} \\ &+\frac{1}{2\pi}\int_0^{2\pi} 2\,\mathrm{Re}\left\{\frac{\partial h}{\partial w}(x,w)\cdot\frac{\partial W}{\partial t_j}(e^{i\phi})\right\}d\phi.\end{aligned}$$

To save space, we have written $(\partial u/\partial t_j)(e^{i\phi})$ for $(\partial u/\partial t_j)(t,x,w)(e^{i\phi})$ and $(\partial W/\partial t_j)(e^{i\phi})$ for $(\partial W/\partial t_j)(t,w)(e^{i\phi})$. Now $(\partial u/\partial t_j)(t = 0, x, w)(\zeta)$ is harmonic in ζ for $|\zeta| \le 1$ (since $u =$ Re G). Furthermore, $u(t,x,w)(\zeta = 0) = x$ and so $(\partial u/\partial t_j)(t,x,w)(\zeta = 0) = 0$. Therefore, the first integral on the right vanishes by the mean value theorem for harmonic functions. The same argument shows that the second integral on the right vanishes since $(\partial W/\partial t_j)(t,w)(\zeta = 0) = 0$. Therefore, we have

$$\frac{\partial v}{\partial t_j}(t = 0, x, w)(\zeta = 0) = 0 \qquad 1 \le j \le N. \tag{3}$$

The second-order part of the Taylor expansion of $v(t,x,w)(\zeta=0)$ in t is obtained by differentiating the right side of (1). We have

$$\begin{aligned}\frac{t_jt_k}{2}\frac{\partial^2 v}{\partial t_j\partial t_k}&(t=0,x,w)(\zeta=0)\\&=\frac{t_jt_k}{2\pi}\int_0^{2\pi}\sum_{\alpha,\beta=1}^{n-d}\frac{\partial^2 h(x,w)}{\partial w_\alpha\partial\bar{w}_\beta}\frac{\partial W_\alpha}{\partial t_j}(e^{i\phi})\frac{\partial\overline{W}_\beta}{\partial t_k}(e^{i\phi})d\phi+\xi_{jk}(t,x,w)\end{aligned}$$

where $\xi_{jk}(t,x,w)$ is a quadratic term in t whose coefficients involve the second-order pure terms from the expansion of h, i.e., $(\partial h/\partial x_\alpha)(x,w)$, $(\partial h/\partial w_\alpha)(x,w)$ $(\partial^2 h/\partial x_\alpha\partial x_\beta)(x,w)$, $\mathrm{Re}\{(\partial^2 h/\partial x_\alpha\partial w_\beta)(x,w)\}$, and $\mathrm{Re}\{(\partial^2 h/\partial w_\alpha\partial w_\beta)(x,w)\}$. These pure terms vanish at $x=0, w=0$ by our choice of local coordinates (from Theorem 2 in Section 7.2). Therefore, $\xi_{jk}(t,x,w)$ can be absorbed into the error term, $\xi(t,x,w)$, with the estimate stated in the conclusion of the lemma.

The term $(\partial^2 h/\partial w_\alpha\partial\bar{w}_\beta)(x,w)$ can be written

$$\frac{\partial^2 h(0,0)}{\partial w_\alpha\partial\bar{w}_\beta}+\left(\frac{\partial^2 h(x,w)}{\partial w_\alpha\partial\bar{w}_\beta}-\frac{\partial^2 h(0,0)}{\partial w_\alpha\partial\bar{w}_\beta}\right).$$

The term in parentheses can be made as small as desired by suitably restricting x and w. Therefore

$$\begin{aligned}\frac{t_jt_k}{2}\frac{\partial^2 v}{\partial t_j\partial t_k}(t=0,x,w)(\zeta=0)&=\frac{t_jt_k}{2\pi}\int_0^{2\pi}\tilde{\mathcal{L}}\left(\frac{\partial W}{\partial t_j}(e^{i\phi}),\overline{\frac{\partial W}{\partial t_k}(e^{i\phi})}\right)d\phi\\&\quad+\tilde{\xi}_{jk}(t,x,w)\end{aligned}\tag{4}$$

where $\tilde{\mathcal{L}}$ is the quadratic form that generates the Levi form at the origin and where

$$|\tilde{\xi}_{jk}(t,x,w)|\le\eta|t|^2$$

provided (t,x,w) belongs to a suitably small neighborhood of the origin.

We have

$$\frac{\partial W}{\partial t_j}(\zeta)=\alpha_j\zeta^j.$$

Substituting this into $\tilde{\mathcal{L}}$ and integrating $\zeta=e^{i\phi}$ over $0\le\phi\le 2\pi$, we obtain

$$\int_0^{2\pi}\tilde{\mathcal{L}}\left(\frac{\partial W}{\partial t_j}(e^{i\phi}),\overline{\frac{\partial W}{\partial t_k}(e^{i\phi})}\right)d\phi=0\quad\text{for}\quad j\neq k.$$

When $j = k$, we have

$$\begin{aligned}\tilde{\mathcal{L}}(\alpha_j\zeta^j, \overline{\alpha_j\zeta^j}) &= \tilde{\mathcal{L}}(\alpha_j, \bar{\alpha}_j) \quad (\text{for } |\zeta| = 1)\\ &= X_j.\end{aligned}$$

Therefore, from (4), we have

$$\frac{t_j t_k}{2}\frac{\partial^2 v}{\partial t_j \partial t_k}(t=0,x,w)(\zeta=0) = \begin{cases} \tilde{\xi}_{jk}(t,x,w) & \text{if } j \neq k \\ t_j^2 X_j + \tilde{\xi}_{jj}(t,x,w) & \text{if } j = k. \end{cases} \tag{5}$$

Since the third-order Taylor remainder in t can be absorbed into the error term, the proof of the lemma follows from (2), (3), and (5). ∎

Let us summarize where we stand. We have shown (Lemma 1) that each point in the set

$$\Omega = \{A(x,w,t)(\zeta = 0);\ |t|, |x|, |w| < \delta\}$$

belongs to the image of an analytic disc whose boundary is contained in the given open set ω for the analytic disc theorem. Furthermore, the Taylor expansion of $A(x,w,t)(\zeta = 0)$ in t is given in Lemma 2. The constant term in this expansion is $(x, h(x,w), w)$ and the set $\omega_\Gamma = \{(x,h(x,w),w); |x|, |w| < \delta\}$ is an open subset of M that contains the origin. Therefore, the proof of the analytic disc theorem will be complete once we show that for fixed x, w with $|x|, |w| < \delta$, the map

$$t \mapsto f(x,w)(t) = \sum_{j=1}^{N} t_j^2 X_j + \xi(t,x,w) \qquad t \in \mathbb{R}^N, |t| < \epsilon'$$

parameterizes a set that contains an open neighborhood of the origin in the cone Γ (here, ϵ' and δ are as in Lemma 2). Now the map $t \mapsto \sum_{j=1}^N t_j^2 X_j$, $|t| < \epsilon'$, parameterizes an open neighborhood of the origin of the cone $\Gamma_{X_1 \dots X_N}$ that contains the cone Γ by the choice of $X_1, \dots, X_N$. The hope is that since $|\xi(t,x,w)|$ is small relative to $|t|^2$ (Lemma 2), the image of $t \mapsto f(x,w)(t)$ will also contain the desired neighborhood of Γ.

To carry out the details, we replace t_j by $\sqrt{t_j}$, for $t_j \geq 0$. We define two maps

$$\hat{F}(t_1, \dots, t_N) = \sum_{j=1}^{N} t_j X_j$$

and

$$\begin{aligned}F(x,w)(t_1,\dots,t_N) &= f(x,w)(\sqrt{t_1},\dots,\sqrt{t_N})\\ &= \hat{F}(t_1,\dots,t_N) + e(t,x,w)\end{aligned}$$

where the error term $e(t,x,w) = \xi(\sqrt{t_1},\dots,\sqrt{t_N},x,w)$ is continuous (but not differentiable at $t = 0$) and satisfies the estimate

$$|e(t, x, w)| \leq \eta |t|$$

provided $|x|, |w| < \delta$ and $|t| < ((\epsilon')^2/N)$. Here, $\eta > 0$ can be chosen as small as desired and δ depends only on η. We wish to show that δ can be chosen so that if $|x|, |w| < \delta$, then the image of the set $\{t \in \mathbb{R}^N;\ t_j \geq 0\ |t| < ((\epsilon')^2/N)\}$ contains an ϵ-neighborhood of the origin of Γ. Recall that $\Gamma < \Gamma_{X_1 \ldots X_N}$. Therefore, for each $v \in \overline{\Gamma} \cap S$ (S is the unit sphere in $\mathbb{R}^d$), there is a conical neighborhood of v, denoted Γ_v, and a collection of d-linearly independent vectors $X_{i_1}, \ldots, X_{i_d}$ with

$$v \in \Gamma_v < \Gamma_{X_{i_1} \ldots X_{i_d}}.$$

Since $\overline{\Gamma} \cap S$ is a compact subset of $\Gamma_{X_1 \ldots X_N}$, we can cover Γ by a finite number of such Γ_v. Therefore, it suffices to show the following lemma.

LEMMA 3
Suppose $X_1, \ldots, X_d$ are linearly independent vectors in $\mathbb{R}^d$. Suppose $\hat{F}(t) = \sum_{j=1}^{d} t_j X_j$. Suppose $\Gamma < \Gamma_{X_1 \ldots X_d}$ and $\epsilon' > 0$ are given. Then there exist η, $\epsilon > 0$ such that if $F : \mathbb{R}^d \mapsto \mathbb{R}^d$ is a continuous map with $|F(t) - \hat{F}(t)| \leq \eta |t|$ for $t = (t_1, \ldots, t_d)$ with $t_j \geq 0$, $1 \leq j \leq d$, and $|t| \leq \epsilon'$, then the image of $\{t = (t_1, \ldots, t_d);\ t_j \geq 0\ |t| < \epsilon'\}$ under F contains $B_\epsilon \cap \Gamma$.

PROOF In our context, F is differentiable except at $t = 0$. We could use the inverse function theorem to examine the images of little balls which are contained in the set $\Gamma_{X_1, \ldots, X_d}$. However, the proof of the above lemma does not require any differentiability assumptions on F. So we offer this purely topological proof.

We may assume that the vectors $X_1, \ldots, X_d$ are the standard basis vectors in $\mathbb{R}^d$. Therefore, we have $\Gamma_{X_1 \ldots X_d} = \{(x_1, \ldots, x_d);\ x_j \geq 0\}$. Given a cone $\Gamma < \Gamma_{X_1 \ldots X_d}$ there exists $\eta' > 0$ and there exists $0 < \epsilon < \epsilon'$ such that if $x \in \Gamma \cap B_\epsilon$ then the Euclidean distance from x to $\partial\{\Gamma_{X_1 \ldots X_d} \cap B_{\epsilon'}\}$ is greater than $\eta'|x|$.

Since $X_1, \ldots, X_d$ are the standard basis vectors, $\hat{F}$ is just the identity map. Suppose $|t - F(t)| \leq \eta |t|$. If η, $\epsilon > 0$ are chosen small relative to η' and ϵ', then the line segment between $t \in \partial\{\Gamma_{X_1 \ldots X_d} \cap B_{\epsilon'}\}$ and $F(t)$ does not intersect $\Gamma \cap B_\epsilon$ (see Figure 15.2). Now suppose the point $x \in \Gamma \cap B_\epsilon$ is not in the image of $\Gamma_{X_1 \ldots X_d} \cap B_{\epsilon'}$ under F. By the above discussion, the restrictions of F and the identity map to $\partial\{\Gamma_{X_1 \ldots X_d} \cap B_{\epsilon'}\}$ are homotopic in $\mathbb{R}^d - \{x\}$. Since $\partial\{\Gamma_{X_1 \ldots X_d} \cap B_{\epsilon'}\}$ is homeomorphic to a $(d-1)$-dimensional sphere in $\mathbb{R}^d$ which encloses x, the homology class of the image of $\partial\{\Gamma_{X_1 \ldots X_d} \cap B_{\epsilon'}\}$ under F is nontrivial in the $(d-1)$st dimensional homology group of $\mathbb{R}^d - \{x\}$. On the other hand, $\Gamma_{X_1 \ldots X_d} \cap B_{\epsilon'}$ is contractible to the origin in $\mathbb{R}^d$. Since F

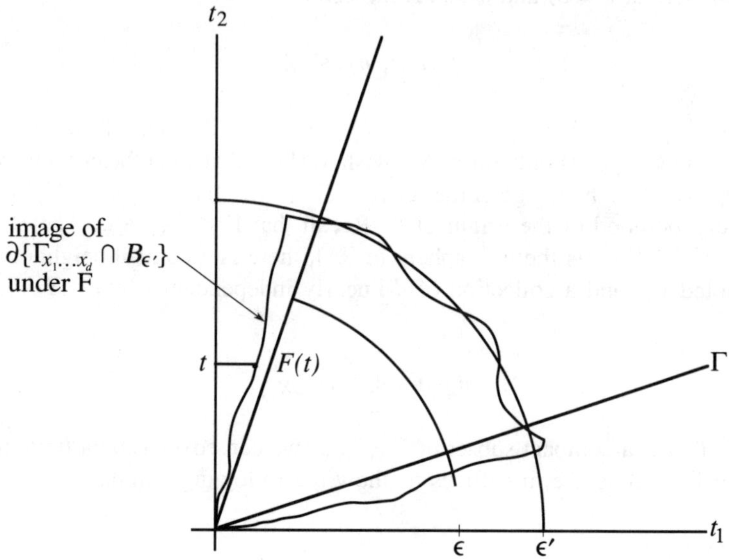

FIGURE 15.2

is continuous and x is not in the image of F, the image of $\partial\{\Gamma_{X_1\ldots X_d} \cap B_{\epsilon'}\}$ under F is also contractible in $\mathbb{R}^d - \{x\}$. This means that the homology class of $F\{\partial\{\Gamma_{X_1\ldots X_d} \cap B_{\epsilon'}\}\}$ is trivial in the $(d-1)$st dimensional homology group of $\mathbb{R}^d - \{x\}$. This contradiction proves the lemma. ∎

As stated earlier, Lemmas 2 and 3 complete the proof of the analytic disc theorem, which in turn completes the proof of the CR extension theorem.

16

The Fourier Transform Technique

In this chapter we present a Fourier transform approach to the proof of the CR extension theorem. This technique has the advantage in that it can be more easily adapted to the holomorphic extension of CR distributions, which will be discussed at the end of Part III. However, the goal of this chapter is to introduce the technique rather than to prove the most general theorem. Therefore, to avoid some of the cumbersome technicalities of more general results, we shall assume the given CR function is sufficiently smooth (class C^{d+2} where $d =$ $\text{codim}_{\mathbb{R}} M$ will suffice). In addition, we shall assume that the submanifold M is *rigid*, which means that near a given point $p \in M$ there is a local biholomorphic change of coordinates so that p is the origin and

$$M = \{(z = x + iy, w) \in \mathbb{C}^d \times \mathbb{C}^{n-d};\ y = h(w)\}$$

where h: $\mathbb{C}^{n-d} \to \mathbb{R}^d$ is smooth (say class C^{d+2}) with $h(0) = 0$ and $Dh(0) = 0$. The point is that the graphing function, h, for a rigid submanifold is independent of the variable $x \in \mathbb{R}^d$. The modifications required to handle the more general case where h depends on both x and w will be mentioned at the end of Part III.

The basic idea of this technique is to use the approximation theorem given in Chapter 13 to derive a modified Fourier inversion formula for smooth functions on M. We will then show that this modified inverse Fourier transform of a given CR function is the restriction of a holomorphic function provided the modified Fourier transform of the CR function is exponentially decreasing. Finally, we show that the set of directions in which the modified Fourier transform of a CR function is exponentially decreasing is related to the convex hull of the image of the Levi form and the CR extension theorem will follow.

The ideas presented in this chapter are due to Baouendi, Treves, Rothschild, Sjöstrand, et al. (see [BCT], [BRT], [BR2], [Sj]). Our presentation is closest to that in [BRT].

16.1 A Fourier inversion formula

Our desired Fourier inversion formula will be derived after we present three lemmas. The first of these lemmas is analogous to Lemma 1 in Chapter 13 for the approximation theorem. Instead of an integral over totally real n-dimensional slices of M as in Chapter 13, we integrate over the following d-dimensional slices. For $p = (z, w) \in M$ let

$$M_p = M_w = \{(z', w) \in M; z' \in \mathbb{C}^d\}.$$

Our analysis will be a local one about the origin. So as in Chapter 13, we assume the graphing function h: $\mathbb{C}^{n-d} \to \mathbb{R}^d$ is suitably cutoff so that $|\nabla h| \leq 1/2$ on $\mathbb{C}^{n-d}$.

LEMMA 1
Suppose U_1 and U_1' are neighborhoods of the origin in $\mathbb{C}^d$ with $U_1' \subset\subset U_1$ and suppose U_2 is a neighborhood of the origin in $\mathbb{C}^{n-d}$. Let g be an element of $\mathcal{D}(U_1)$ with $g \equiv 1$ on U_1'. Suppose f: $\{U_1 \times U_2\} \cap M \to \mathbb{C}$ is a continuous function. Then for $(z, w) \in \{U_1' \times U_2\} \cap M$,

$$f(z, w) = \lim_{\epsilon \to 0} (\pi\epsilon)^{-d/2} \int\limits_{z' \in M_w} g(z') f(z', w) e^{-\epsilon^{-1}[z-z']^2} dz'$$

where $dz' = dz_1' \wedge \ldots \wedge dz_d'$ and $[z - z']^2 = (z_1 - z_1')^2 + \cdots + (z_d - z_d')^2$. Moreover, this limit is uniform in $(z, w) \in \{U_1' \times U_2\} \cap M$.

Note the function f is *not* assumed to be a CR function. We will not assume f is CR until the next section. The purpose of the cutoff function g is to ensure that the integrand has compact support. The reason we do not assume f has compact support is that we will apply this result in the next section to CR functions, which typically do not have compact support.

PROOF The proof is the same as the proof of Lemma 1 in Chapter 13. The only difference is the dimension of the slice of M over which we are integrating. This changes the constants involved but otherwise has no effect on the proof. Also note that ϵ here plays the role of ϵ^2 in Chapter 13. ∎

LEMMA 2
Suppose U_1, U_1', U_2 and g, f are as in Lemma 1. Then for $(z, w) \in \{U_1' \times U_2\} \cap M$

$$f(z, w) = \lim_{\epsilon \to 0} (2\pi)^{-d} \int\limits_{\xi \in \mathbb{R}^d} \int\limits_{z' \in M_w} g(z') f(z', w) e^{i\xi \cdot [z-z'] - \epsilon|\xi|^2} dz' d\xi$$

where the limit is uniform for $(z, w) \in \{U_1' \times U_2\} \cap M$.

Here, $w \cdot u = \sum_{j=1}^{d} w_j u_j$ for $w = (w_1, \ldots, w_d), u = (u_1, \ldots, u_d) \in \mathbb{C}^d$.

PROOF Let us replace ϵ by 4ϵ in the statement of Lemma 1. We obtain

$$f(z,w) = \lim_{\epsilon \to 0} (4\epsilon\pi)^{-d/2} \int_{z' \in M_w} g(z') f(z',w) e^{-\frac{1}{4\epsilon}[z-z']^2} dz'.$$

Lemma 2 will follow once we have established the following identity:

$$e^{-\frac{1}{4\epsilon}[z-z']^2} = \left(\frac{\epsilon}{\pi}\right)^{d/2} \int_{\xi \in \mathbb{R}^d} e^{i\xi \cdot [z-z'] - \epsilon|\xi|^2} d\xi. \tag{1}$$

To show (1), first note that if w is a vector in $\mathbb{R}^d$ (rather than $\mathbb{C}^d$), then by a translation, we have

$$\int_{\xi \in \mathbb{R}^d} e^{-[w+\sqrt{\epsilon}\xi]^2} d\xi = \int_{\xi \in \mathbb{R}^d} e^{-\epsilon|\xi|^2} d\xi. \tag{2}$$

The left side of this equation is an entire function of $w \in \mathbb{C}^d$ which agrees with the right side when $w \in \mathbb{R}^d$. Therefore, equation (2) holds for all $w \in \mathbb{C}^d$ by the identity theorem for holomorphic functions.

Now we complete the square in the exponent to obtain

$$\begin{aligned}
&\int_{\xi \in \mathbb{R}^d} e^{\frac{1}{4\epsilon}[z-z']^2 + i\xi \cdot [z-z'] - \epsilon|\xi|^2} d\xi \\
&\qquad = \int_{\xi \in \mathbb{R}^d} e^{-[\frac{-i}{2\sqrt{\epsilon}}(z-z') + \sqrt{\epsilon}\xi]^2} d\xi \\
&\qquad = \int_{\mathbb{R}^d} e^{-\epsilon|\xi|^2} d\xi \qquad \text{(by letting } w = \tfrac{-i}{2\sqrt{\epsilon}}(z-z') \text{ in (2))} \\
&\qquad = (\pi/\epsilon)^{d/2}.
\end{aligned}$$

The last equality follows from a standard polar coordinate calculation. Multiplying this equation by $(\epsilon/\pi)^{d/2} e^{-(1/4\epsilon)[z-z']^2}$ yields (1) and so the proof of the lemma is complete. ▮

For a continuous function f: $M \to \mathbb{C}$ and for $z \in \mathbb{C}^d$, $w \in \mathbb{C}^{n-d}$, and $\xi \in \mathbb{R}^d$, define

$$\widehat{I}f(z,w,\xi) = (2\pi)^{-d} \int_{z' \in M_w} g(z') f(z',w) e^{i\xi \cdot [z-z']} \, dz'.$$

Lemma 2 can now be rewritten

$$f(z,w)=\lim_{\epsilon\to 0}\int\limits_{\xi\in\mathbb{R}^d}\widehat{I}f(z,w,\xi)e^{-\epsilon|\xi|^2}d\xi.$$

We will show in Lemma 3 below that if f is of class C^{d+1} on M, then $|\widehat{I}f(z,w,\xi)|\leq C(|\xi|+1)^{-(d+1)}$ for some uniform constant C. Since $(|\xi|+1)^{-(d+1)}$ is integrable in $\xi\in\mathbb{R}^d$, the dominated convergence theorem implies

$$f(z,w)=\int\limits_{\xi\in\mathbb{R}^d}\widehat{I}f(z,w,\xi)d\xi \quad \text{for } (z,w)\in\{U_1'\times U_2\}\cap M.$$

This is analogous to the standard Fourier inversion formula for Euclidean space. The only difference is that $\widehat{I}f$ involves an integral over a slice of M (rather than Euclidean space). Also, it is customary in Euclidean space not to include the factor $e^{i\xi\cdot z}$ in the integrand of the Fourier transform as we have done in $\widehat{I}f$. The factor $e^{i\xi\cdot z}$ would then reappear in the Fourier inversion formula.

For technical reasons which will become clear, we wish to modify the above Fourier transform to one with a term of the form $|\xi|[z-z']^2$ in the exponent. For $z\in\mathbb{C}^d$, $w\in\mathbb{C}^{n-d}$, and $\xi\in\mathbb{R}^d$ define

$$If(z,w,\xi)=(2\pi)^{-d}\int\limits_{z'\in M_w} g(z')f(z',w)e^{i\xi\cdot[z-z']-|\xi|[z-z']^2}\Delta(z-z',\xi)dz'$$

where $\Delta(u,\xi)=(1+i(u\cdot\xi/|\xi|))$ for $u,\xi\in\mathbb{C}^d$.

LEMMA 3

There exist neighborhoods U_1 and U_1' of the origin in $\mathbb{C}^d$ with $U_1'\subset\subset U_1$ and a neighborhood U_2 of the origin in $\mathbb{C}^{n-d}$ such that if $g\in\mathcal{D}(U_1)$ with $g=1$ on U_1', then the following holds.

(a) *If $f\colon M\cap\{U_1\times U_2\}\to\mathbb{C}$ is a function of class C^{d+1}, then there is a constant $C>0$ such that*

$$|(\widehat{I}f)(z,w,\xi)|\leq C(|\xi|+1)^{-(d+1)} \quad \textit{for all} \quad \xi\in\mathbb{C}^d$$

$$|(If)(z,w,\xi)|\leq C(|\xi|+1)^{-(d+1)} \quad \textit{for all} \quad \xi\in\mathbb{R}^d$$

and for $(z,w)=(x+ih(w),w)\in M\cap\{U_1'\times U_2\}$.

(b) *More generally, if $f\colon M\cap\{U_1\times U_2\}\to\mathbb{C}$ is a function of class C^N, $N>0$, and if D^j is a jth order derivative $(0\leq j\leq N)$ in $z,\bar z,w,\bar w$ on $\mathbb{C}^n$, then there is a constant $C>0$ such that*

$$|D^j(If)(z,w,\xi)|\leq C(|\xi|+1)^{j-N}e^{-v\cdot\xi+|v|^2|\xi|}$$

for all $\xi\in\mathbb{R}^d$ and $w\in U_2$, $z=x+i(h(w)+v)\in U_1'$ (here $v\in\mathbb{R}^d$).

We will need the above estimate in part (a) on $\widehat{If}(z,w,\xi)$ for $\xi \in \mathbb{C}^d$ (rather than just $\xi \in \mathbb{R}^d$) because in the proof of the next theorem, we will transform the ξ-integral over $\mathbb{R}^d$ in the Fourier inversion formula for $\widehat{If}$ into an integral over a contour in $\mathbb{C}^d$ which will yield a Fourier inversion formula for If. In the next section, we will need the more general result in (b), which estimates the derivatives of If for (z,w) in a $\mathbb{C}^n$-neighborhood of the origin.

PROOF The proof of the estimate for If in part (a) is a special case of part (b) (with $j=0$, $N=d+1$, and $v=0$). Therefore, we first prove the estimate in (b). This estimate clearly holds on the compact set $\{\xi \in \mathbb{R}^d; |\xi| \leq 1\}$. So it suffices to show

$$|D^j\{(If)(z,w,\xi)\}| \leq C|\xi|^{j-N} \quad \text{for} \quad |\xi| \geq 1.$$

For fixed $w \in \mathbb{C}^{n-d}$, the set M_w is parameterized by the map $z' = x' + ih(w)$ for $x' \in \mathbb{R}^d$. For $z = x + i(h(w)+v)$ with $v \in \mathbb{R}^d$, we have

$$\begin{aligned}&(If)(z,w,\xi) = \\ &(2\pi)^{-d} \int\limits_{x'\in\mathbb{R}^d} \tilde{g}(x',w)\tilde{f}(x',w)e^{i\xi\cdot(x-x'+iv)-|\xi|[x-x'+iv]^2}\Delta(x-x'+iv,\xi)dx'\end{aligned}$$

where $\tilde{g}(x',w) = g(x'+ih(w))$ and $\tilde{f}(x',w) = f(x'+ih(w),w)$. If D^j is a jth order derivative in z, $\bar{z}$, w, $\bar{w}$, then

$$D^j\{(If)(z,w,\xi)\} = \sum_{k=0}^{j} \int\limits_{x'\in\mathbb{R}^d} G_k(f)(x',z,w)(\xi)e^{i\xi\cdot(x-x'+iv)-|\xi|[x-x'+iv]^2}dx' \tag{3}$$

where $G_k(f)$ is an expression involving kth order derivatives of $gf\Delta$. Moreover, $G_k(f)(x',z,w)(\xi)$ has compact x'-support and is a polynomial in ξ and $|\xi|$ of degree $j-k$ (as a result of the derivatives of order $j-k$ of the exponential term).

Now the idea is to integrate by parts with the vector field

$$L = \frac{i\xi \cdot D_{x'}}{|\xi|} = \frac{i}{|\xi|}\sum_{j=1}^{d} \xi_j \frac{\partial}{\partial x'_j}.$$

As we will see, each integration by parts will yield a factor of $|\xi|^{-1}$. Iterating this procedure $N-k$ times will yield the lemma.

To carry out the details, note

$$L\{i\xi\cdot(x-x'+iv) - |\xi|[x-x'+iv]^2\} = |\xi| + \eta(x,x',v)(\xi)$$

where

$$\eta(x,x',v)(\xi) = 2i\xi\cdot(x-x'+iv).$$

The term η is homogeneous of degree one in ξ and satisfies the estimate

$$|\eta(x, x', v)(\xi)| \le C(|x| + |x'| + |v|)|\xi|$$

for some uniform constant $C > 0$. Choose neighborhoods $U_1 \subset \mathbb{C}^d$ and $U_2 \subset \mathbb{C}^{n-d}$ of the origin so that if $w \in U_2$ and if $z = x + i(h(w) + v)$ and $z' = x' + ih(w)$ belong to U_1, then

$$|\eta(x, x', v)(\xi)| \le \frac{|\xi|}{2}.$$

We have

$$||\xi| + \eta(x, x', v)(\xi)| \ge \frac{1}{2}|\xi|. \tag{4}$$

Let $U_1' \subset\subset U_1$ be a neighborhood of the origin in $\mathbb{C}^d$. Let $g \in \mathcal{D}(U_1)$ with $g = 1$ on U_1'. We have

$$\frac{L\{e^{i\xi\cdot(x-x'+iv)-|\xi|[x-x'+iv]^2}\}}{(|\xi| + \eta(x, x', v)(\xi))} = e^{i\xi\cdot(x-x'+iv)-|\xi|[x-x'+iv]^2}. \tag{5}$$

Substituting the left side for the right and integrating by parts, we obtain

$$\int_{x'\in\mathbb{R}^d} G_k(f)(x', z, w)(\xi) e^{i\xi\cdot(x-x'+iv)-|\xi|[x-x'+iv]^2} dx'$$

$$= - \int_{x'\in\mathbb{R}^d} L\left\{\frac{G_k(f)(x', z, w)(\xi)}{|\xi| + \eta(x, x', v)(\xi)}\right\} e^{i\xi\cdot(x-x'+iv)-|\xi|[x-x'+iv]^2} dx'.$$

Using (4) together with the fact that $G_k(f)(x', z, w)(\xi)$ is a polynomial in ξ and $|\xi|$ of degree $j - k$, we have

$$\left|\frac{G_k(f)(x', z, w)(\xi)}{|\xi| + \eta(x, x', v)(\xi)}\right| \le C|\xi|^{j-k-1} \quad \text{for} \quad |\xi| \ge 1$$

where C is some uniform constant. Since L is a differential operator in x' whose coefficients are homogeneous of degree zero in ξ, we have

$$\left|L\left\{\frac{G_k(f)(x', z, w)(\xi)}{|\xi| + \eta(x, x', v)(\xi)}\right\}\right| \le C|\xi|^{j-k-1} \quad \text{for} \quad |\xi| \ge 1.$$

Recall that $|G_k(f)(x', z, w)(\xi)|$ is homogeneous of degree $j - k$ in $|\xi|$, whereas the above term is homogeneous of degree $j - k - 1$ in $|\xi|$. If f is of class C^N, then $G_k(f)$ is of class C^{N-k} and we may iterate the above procedure starting with (5) $N-k$ times to obtain

$$\int\limits_{x'\in\mathbb{R}^d} G_k(f)(x',z,w)(\xi)e^{i\xi\cdot(x-x'+iv)-|\xi|[x-x'+iv]^2}dx'$$
$$= \int\limits_{x'\in\mathbb{R}^d} \widetilde{G}_k(f)(x',z,w)(\xi)e^{i\xi\cdot(x-x'+iv)-|\xi|[x-x'+iv]^2}dx' \tag{6}$$

where $|\widetilde{G}_k(f)(x',z,w)(\xi)|$ is homogeneous of degree $j-N$ in $|\xi|$. Therefore

$$|\widetilde{G}_k(f)(x',z,w)(\xi)| \leq C|\xi|^{j-N} \quad \text{for} \quad |\xi| \geq 1 \tag{7}$$

where C is a constant independent of $w \in U_2$ and $z = x + i(h(w)+v) \in U_1'$.

For the exponential term, we have

$$\begin{aligned}|e^{i\xi\cdot(x-x'+iv)-|\xi|[x-x'+iv]^2}| &= e^{\mathrm{Re}\{i\xi\cdot(x-x'+iv)-|\xi|[x-x'+iv]^2\}}\\ &= e^{-\xi\cdot v-|\xi||x-x'|^2+|\xi||v|^2}\\ &\leq e^{-\xi\cdot v+|\xi||v|^2}.\end{aligned}$$

This estimate together with (6), (7), and (3) yield part (b) of the lemma.

As already mentioned, the estimate on If given in part (a) is a special case of part (b). The estimate on $\widehat{I}f$ in part (a) is proved in a similar manner. In fact, this estimate is easier since the exponent occurring in $\widehat{I}f$ is simpler. Therefore, it will be left to the reader. ∎

We remark that the estimate in part (b) also holds for ξ in a $\mathbb{C}^d$-conical neighborhood of $\mathbb{R}^d$.

We now state and prove the Fourier inversion formula for the transform If.

THEOREM 1

There exist neighborhoods $U_1' \subset\subset U_1$ of the origin in $\mathbb{C}^d$ and a neighborhood U_2 of the origin in $\mathbb{C}^{n-d}$ such that if $g \in \mathcal{D}(U_1)$ with $g \equiv 1$ on U_1' and if $f\colon M \to \mathbb{C}$ is of class C^{d+1}, then

$$f(z,w) = \int\limits_{\xi\in\mathbb{R}^d} If(z,w,\xi)d\xi$$

for $(z,w) \in \{U_1' \times U_2\} \cap M$.

PROOF From Lemma 2 and Lemma 3 part (a) and the dominated convergence theorem, we have

$$f(z,w) = \int\limits_{\xi \in \mathbb{R}^d} (\widehat{I}f)(z,w,\xi)d\xi$$

$$= \lim_{R\to\infty} \int\limits_{-R}^{R} \cdots \int\limits_{-R}^{R} (\widehat{I}f)(z,w,\xi)d\xi_1 \ldots d\xi_d. \tag{8}$$

For $(z,w) \in M$ near the origin, we write $z = x + ih(w)$ for some $x \in \mathbb{R}^d$. The slice M_w is parameterized by $z' = x' + ih(w)$, for $x' \in \mathbb{R}^d$. Therefore

$$(\widehat{I}f)(z,w,\xi) = (2\pi)^{-d} \int\limits_{x' \in \mathbb{R}^d} \tilde{g}(x',w)\tilde{f}(x',w)e^{i\xi\cdot(x-x')}dx'$$

where as before $\tilde{g}(x',w) = g(x' + ih(w))$ and $\tilde{f}(x',w) = f(x'+ih(w),w)$.

Note that $(\widehat{I}f)(z,w,\xi)$ is an entire function of $\xi \in \mathbb{C}^d$ (since g has compact x'-support). We can use Cauchy's theorem to change each ξ_j-integral in (8) to an integral over the contour $\{\hat{\xi}_j = \xi_j + i|\xi|(x_j - x_j');\ \xi_j \in \mathbb{R}\}$ in the complex plane $(1 \le j \le d)$. The integral over the side contours $\{|\xi_j| = R\}$ appearing in the change of contour process disappear as $R \to \infty$, because the measure of these side contours is $\mathcal{O}(R^d)$ whereas the integrand is $\mathcal{O}(R^{-(d+1)})$ by the estimate on $|\widehat{I}f|$ in Lemma 3. By replacing ξ by $\hat{\xi} = \xi + i|\xi|(x - x')$ in the exponential term in $\widehat{I}f$, we obtain

$$e^{i\hat{\xi}\cdot(x-x')} = e^{i\xi\cdot(x-x') - |\xi||x-x'|^2} \tag{9}$$

which is the exponential term in the definition of If. Moreover, we have

$$\begin{aligned}
d\hat{\xi} &= d\hat{\xi}_1 \wedge \ldots \wedge d\hat{\xi}_d \\
&= d_\xi(\xi_1 + i|\xi|(x_1 - x_1')) \wedge \ldots \wedge d_\xi(\xi_d + i|\xi|(x_d - x_d')) \\
&= \sum_{j=1}^{d} d\xi_1 \wedge \ldots \wedge d\xi_{j-1} \wedge (i d_\xi|\xi|(x_j - x_j')) \wedge d\xi_{j+1} \wedge \ldots \wedge d\xi_d \\
&\quad + d\xi_1 \wedge \ldots \wedge d\xi_d \\
&= \sum_{j=1}^{d} d\xi_1 \wedge \ldots \wedge d\xi_{j-1} \wedge \left(\frac{i\xi_j(x_j - x_j')}{|\xi|}\right) d\xi_j \wedge d\xi_{j+1} \wedge \ldots \wedge d\xi_d \\
&\quad + d\xi_1 \wedge \ldots \wedge d\xi_d \\
&= \Delta(x - x', \xi)d\xi.
\end{aligned} \tag{10}$$

From (8) and Cauchy's theorem, we obtain

$$\begin{aligned} f(z,w) &= \lim_{R\to\infty} \int_{-R}^{R} \cdots \int_{-R}^{R} \widehat{I}f(z,w,\xi)d\xi \\ &= \lim_{R\to\infty} \int_{-R}^{R} \cdots \int_{-R}^{R} \widehat{I}f(z,w,\hat{\xi})d\hat{\xi} \quad (\text{with } \hat{\xi} = \xi + i|\xi|(x-x')) \\ &= \lim_{R\to\infty} \int_{-R}^{R} \cdots \int_{-R}^{R} If(z,w,\xi)d\xi \quad (\text{by (9) and (10)}) \\ &= \int_{\xi\in\mathbb{R}^d} (If)(z,w,\xi)d\xi. \end{aligned}$$

This completes the proof of Theorem 1. ∎

16.2 The hypoanalytic wave front set

To summarize our progress so far, we have shown in the last section (Theorem 1) that

$$f(z,w) = \int_{\xi\in\mathbb{R}^d} (If)(z,w,\xi)d\xi \quad \text{for} \quad (z,w) \in M \text{ near the origin} \tag{1}$$

where

$$(If)(z,w,\xi) = (2\pi)^{-d} \int_{z'\in M_w} g(z')f(z',w)e^{i\xi\cdot(z-z')-|\xi|[z-z']^2}\Delta(z-z',\xi)dz'.$$

If $f\colon M \to \mathbb{C}$ is of class C^{d+1}, then we have shown (Lemma 3) that $|(If)(z,w,\xi)| \leq C(|\xi|+1)^{-(d+1)}$ for $\xi \in \mathbb{R}^d$ and $(z,w) \in M$ and hence the integral in (1) is well defined.

Note from its definition that $(If)(z,w,\xi)$ is analytic in $z \in \mathbb{C}^d$. If $(If)(z,w,\xi)$ is exponentially decreasing in $\xi \in \mathbb{R}^d$, then the right side of (1) also defines an analytic function of $z \in \mathbb{C}^d$. Later, we will see that if f is a CR function near the origin on M, then the right side of (1) also defines an analytic function of $w \in \mathbb{C}^{n-d}$ near the origin. In this case, (1) shows that f is the restriction of an ambiently defined holomorphic function. All of this is to serve as motivation for examining the set of vectors $\xi \in \mathbb{R}^d$ in which $(If)(z,w,\xi)$ is exponentially decreasing. Roughly speaking this is the complement of the hypoanalytic wave front set of the function f at the origin. More precisely, we make the following definition.

Fix a smooth function g: $\mathbb{C}^d \to \mathbb{R}$ with compact support which is identically one on a neighborhood of the origin.

DEFINITION 1 *Let $\mathcal{F}$ be a set of continuous functions on M. A vector $\xi \in \mathbb{R}^d$ is not in the hypoanalytic wave front set of $\mathcal{F}$ at the origin if there exist*

(a) *a cone $\widehat{\Gamma}$ in $\mathbb{R}^d$ containing ξ_0*

(b) *a neighborhood U of the origin in $\mathbb{C}^n$*

(c) *a constant $\epsilon > 0$*

such that if f belongs to $\mathcal{F}$, then there is a constant $C > 0$ such that

$$|(If)(z, w, \xi)| \leq Ce^{-\epsilon|\xi|} \quad \text{for} \quad \xi \in \widehat{\Gamma},\ (z, w) \in U.$$

In our application, $\mathcal{F}$ will be the class of CR functions of class C^{d+2} on an open neighborhood ω of the origin in M. However, the definition does not require the function f to be CR, which is the reason for the more general definition.

The order of the quantifiers is important. The cone $\widehat{\Gamma}$, the neighborhood U in $\mathbb{C}^n$, and the constant $\epsilon > 0$ are independent of the function $f \in \mathcal{F}$. The constant $C > 0$ is allowed to depend on f. In the literature, it is more common to see the concept of the hypoanalytic wave front set of a single function. However, we wish to holomorphically extend all CR functions defined on an open set ω to a fixed open set Ω in $\mathbb{C}^n$. For this reason, we have modified the more standard definition to the one given above.

We leave it as an exercise to show that the above definition is independent of the cutoff function g. This follows easily from examining the term $-|\xi|[z - z']^2$ in the exponent of the definition of $(If)(z, w, \xi)$.

We denote the hypoanalytic wave front set of $\mathcal{F}$ at the origin by $WF_0(\mathcal{F})$. The set $WF_0(\mathcal{F})$ is closed in $\mathbb{R}^d$. We identify this copy of $\mathbb{R}^d$ with the space of vectors that are normal to M at 0, denoted $N_0(M) = \{(0, y, 0); y \in \mathbb{R}^d\}$. In the literature, $WF_0(\mathcal{F})$ is often considered part of the totally real tangent (or cotangent) space of M at 0, $(X_0(M))$, which in our coordinates is the space $\{(x, 0, 0); x \in \mathbb{R}^d\}$. However, since we are extending CR functions in directions that are normal to M, it will be more convenient for us to think of $WF_0(\mathcal{F})$ as a subset of $N_0(M)$ rather than $X_0(M)$.

It is possible to define $WF_p(\mathcal{F})$ for any point $p \in M$, by first using a coordinate change so that $p = 0$ and so that M is graphed above its tangent space as we have done. A more invariant definition of $WF_p(\mathcal{F})$ is also available. However, we shall not use this definition. Instead, we refer the reader to [BRT] or [BR2].

For an open set $\omega \subset M$ containing the origin, let CR(ω) be the set of CR functions of class C^{d+2} on ω. If $WF_0(CR(\omega))$ is empty then $(If)(z, w, \xi)$ is exponentially decreasing for all $\xi \in \mathbb{R}^d$ and all $f \in CR(\omega)$. In this case, we will show that each $f \in CR(\omega)$ can be holomorphically extended to a neighborhood of the origin. We also wish to show that if $WF_0(CR(\omega))$ is contained in some cone, then elements in $CR(\omega)$ can be holomorphically extended to some open subset Ω which lies to one side of M, as in the conclusion of the CR extension theorem stated in Chapter 14. To state this criterion for holomorphic extension, we need some additional notation.

For a cone $\Gamma \subset \mathbb{R}^d$, define the *polar* of Γ, by

$$\Gamma^0 = \{v \in \mathbb{R}^d; v \cdot \xi \geq 0 \quad \text{for all} \quad \xi \in \Gamma\}.$$

Note that Γ^0 is a closed convex cone in $\mathbb{R}^d$. It is also easy to show that $\{\Gamma^0\}^0$ is the closure of the convex hull of Γ (this uses the fact that any point not in a convex set can be separated from it by a real hyperplane).

THEOREM 1

Let $\omega \subset M$ be an open neighborhood of the origin in M. Let $\Gamma \subset N_0(M)$ be a closed convex cone. If the interior of Γ^0 is nonempty, then the following are equivalent:

- ***(a)*** *$WF_0(CR(\omega))$ is contained in Γ.*
- ***(b)*** *For each open cone $\Gamma_1 < \Gamma^0$, there is a neighborhood ω_1 of the origin in M and there exists $\epsilon_1 > 0$ such that for each $f \in CR(\omega)$, there is a holomorphic function F defined on the set $\omega_1 + \{\Gamma_1 \cap B_{\epsilon_1}\}$ which is continuous up to ω_1 with $F|_{\omega_1} = f$.*

The proof of this theorem will show that even if f is not CR, then part (a) implies that f extends to a function that is holomorphic in $z \in \mathbb{C}^d$ (but not $w \in \mathbb{C}^{n-d}$) for $(z, w) \in \omega_1 + \{\Gamma_1 \cap B_{\epsilon_1}\}$. In the next section, we will relate the hypoanalytic wave front set to the convex hull of the image of the Levi form. This relationship together with the (a) $\Rightarrow$ (b) part of the above theorem will complete the proof of the CR extension theorem.

PROOF Let $\omega \subset M$ be a neighborhood of the origin. Let U_1, U_1', and U_2 be the open sets that satisfy the conclusions of Lemma 3 and Theorem 1 in the previous section. We also require $\{U_1 \times U_2\} \cap M \subset \omega$. Since the definition of $WF_0(\mathrm{CR}(\omega))$ is independent of the cutoff function g, we choose $g \in \mathcal{D}(U_1)$ with $g = 1$ on U_1'.

We first show that (a) implies (b). We start with the Fourier inversion formula

$$f(z, w) = \int_{\xi \in \mathbb{R}^d} (If)(z, w, \xi) d\xi \tag{2}$$

which holds for $(z, w) \in M \cap \{U_1' \times U_2\}$. We wish to show that if $WF_0(\mathrm{CR}(\omega))$ is contained in a convex cone $\Gamma \subset N_0(M)$, then the right side of (2) extends analytically to an open set of the form $\omega_1 + \{\Gamma_1 \cap B_{\epsilon_1}\}$ where Γ_1 is any given cone that is smaller than Γ^0 and where $\omega_1 \subset M$ and $\epsilon_1 > 0$ both depend on Γ_1. If $\Gamma_1 < \Gamma^0$ then $\overline{\Gamma}_1 \cap S$ is a compact subset of the interior of $\{\Gamma^0\}$ where S is the unit sphere in $\mathbb{R}^d$. Therefore, there is an $\alpha > 0$ and an open cone $\widetilde{\Gamma}$ which is slightly larger than Γ (i.e., $\Gamma < \widetilde{\Gamma}$) with

$$v \cdot \xi \geq \alpha |v||\xi| \quad \text{for} \quad v \in \Gamma_1, \xi \in \widetilde{\Gamma}. \tag{3}$$

The integral in (2) can be split into two integrals to obtain

$$\begin{aligned} f(z,w) &= \int\limits_{\xi \in \widetilde{\Gamma}} (If)(z,w,\xi)d\xi + \int\limits_{\xi \in \mathbb{R}^d - \widetilde{\Gamma}} (If)(z,w,\xi)d\xi \\ &= F_1(z,w) + F_2(z,w) \end{aligned} \tag{4}$$

for $(z,w) \in M \cap \{U_1' \times U_2\}$.

Now $\{\mathbb{R}^d - \widetilde{\Gamma}\} \cap S$ is a compact subset of the complement of the hypoanalytic wave front set. Using Definition 1 together with a finite open covering argument, we can shrink $U_1 \times U_2 \subset \mathbb{C}^d \times \mathbb{C}^{n-d}$ if necessary and find constants $C, \epsilon > 0$ such that

$$|(If)(z,w,\xi)| \leq Ce^{-\epsilon|\xi|} \quad \text{for} \quad (z,w) \in U_1 \times U_2, \ \xi \in \mathbb{R}^d - \widetilde{\Gamma}. \tag{5}$$

Since $(If)(z,w,\xi)$ is holomorphic in $z \in \mathbb{C}^d$, an application of Morea's theorem and Fubini's theorem shows that $F_2(z,w)$ is holomorphic in z for $(z,w) \in U_1 \times U_2$.

Next, we show that $F_1(z,w)$ is holomorphic in $z \in \mathbb{C}^d$ for $(z,w) \in \omega_1 + \{\Gamma_1 \cap B_{\epsilon_1}\}$ where ω_1 and $\epsilon_1 > 0$ have yet to be chosen.

Let $\omega_1 = \{U_1' \times U_2\} \cap M$. Points in the set $\omega_1 + \Gamma_1$ are the form

$$(z,w) = (x + i(h(w) + v), w)$$

where

$$(x + ih(w), w) \in \omega_1 \quad \text{and} \quad v \in \Gamma_1.$$

From part (b) of Lemma 3 from the previous section with $N = d + 2$ and $j = 1$, we have

$$|D\{(If)(z,w,\xi)\}| \leq C(|\xi| + 1)^{-(d+1)} e^{-v \cdot \xi + |v|^2 |\xi|}$$

where D is any first-order derivative in $z, \bar{z}, w, \bar{w}$ in $\mathbb{C}^n$. C is a uniform constant that is independent of $z = x + i(h(w) + v) \in U_1'$ with $v \in \Gamma_1$ and $w \in U_2$. Using (3), this estimate becomes

$$|D\{(If)(z,w,\xi)\}| \leq C(|\xi| + 1)^{-(d+1)} e^{-(\alpha - |v|)|v||\xi|}$$

for $\xi \in \widetilde{\Gamma}$ and $v \in \Gamma_1$.

Let $\epsilon_1 = \alpha/2 > 0$. For $(z, w) = (x + ih(w), w) + (iv, 0) \in \omega_1 + \{\Gamma_1 \cap B_{\epsilon_1}\}$, we have $|v| < \alpha/2$ and so

$$|D\{(If)(z, w, \xi)\}| \leq C(|\xi| + 1)^{-(d+1)} e^{-\alpha/2|\xi||v|} \quad \text{for} \quad \xi \in \widetilde{\Gamma}.$$

Due to the exponential decay in this estimate and the analyticity of $(If)(z, w, \xi)$ in $z \in \mathbb{C}^d$, the function

$$F_1(z, w) = \int\limits_{\xi \in \widetilde{\Gamma}} (If)(z, w, \xi) d\xi$$

is holomorphic in $z \in \mathbb{C}^d$ for $(z, w) \in \omega_1 + \{\Gamma_1 \cap B_{\epsilon_1}\}$. The above estimate on $|D\{If\}|$ also implies that F_1 is C^1 up to ω_1 because $(|\xi| + 1)^{-(d+1)}$ is integrable in $\xi \in \mathbb{R}^d$. Since we have already shown that $F_2(z, w)$ is holomorphic in $z \in \mathbb{C}^d$ for $(z, w) \in U_1' \times U_2$, (4) shows that f is the restriction on ω_1 of the function $F(z, w) = F_1(z, w) + F_2(z, w)$ which is holomorphic in $z \in \mathbb{C}^d$ for $(z, w) \in \omega_1 + \{\Gamma_1 \cap B_{\epsilon_1}\}$ and C^1 up to ω_1. So far, we have not used the fact that f is CR. The next lemma shows that if f is CR, then $F(z, w)$ is also holomorphic in $w \in \mathbb{C}^{n-d}$ for $(z, w) \in \omega_1 + \{\Gamma_1 \cap B_{\epsilon_1}\}$ and thus the proof of (a) $\Rightarrow$ (b) will be complete.

LEMMA 1

Suppose f is a CR function of class C^1 defined on the set $\omega_1 \subset M$. Suppose $F(z, w)$ is holomorphic in $z \in \mathbb{C}^d$ for $(z, w) \in \omega_1 + \{\Gamma_1 \cap B_{\epsilon_1}\}$ and that F is C^1 up to ω_1. If $F|_{\omega_1} = f$, then F is also holomorphic in $w \in \mathbb{C}^{n-d}$ and thus F is the holomorphic extension of f on $\omega_1 + \{\Gamma_1 \cap B_{\epsilon_1}\}$.

PROOF As shown in Theorem 3 in Section 7.2, a local basis $\{\overline{L}_1, \ldots, \overline{L}_{n-d}\}$ for $H^{0,1}(M)$ is given by

$$\overline{L}_j = \frac{\partial}{\partial \bar{w}_j} - 2i \sum_{k=1}^{d} \frac{\partial h_k}{\partial \bar{w}_j}(w) \frac{\partial}{\partial \bar{z}_k} \qquad 1 \leq j \leq n - d.$$

These vector fields are defined on the ambient $\mathbb{C}^n$ as well as on M. Since $F(z, w)$ is holomorphic in z

$$\frac{\partial F}{\partial \bar{w}_j}(z, w) = (\overline{L}_j F)(z, w) \quad \text{for} \quad (z, w) \in \omega_1 + \{\Gamma_1 \cap B_{\epsilon_1}\}.$$

Furthermore, since F is C^1 up to ω_1 and f is CR on ω_1, we have

$$\begin{aligned} \left.\frac{\partial F}{\partial \bar{w}_j}\right|_{\omega_1} &= \overline{L}_j F|_{\omega_1} = \overline{L}_j f \\ &= 0 \quad (\text{since } f \text{ is CR}). \end{aligned}$$

So for fixed w, $(\partial F / \partial \bar{w}_j)(z, w)$ is a holomorphic function of $z \in \mathbb{C}^d$ for $(z, w) \in \omega_1 + \{\Gamma_1 \cap B_{\epsilon_1}\}$, and $(\partial F / \partial \bar{w}_j)(z, w) = 0$ on ω_1. Now ω_1 is parameterized by $(x+ih(w),w)$ for $x \in \mathbb{R}^d, w \in \mathbb{C}^{n-d}$ near the origin. Therefore, $(\partial F/\partial \bar{w}_j)(x+i(h(w)+y),w)$ is holomorphic in $z = x+iy \in \mathbb{C}^d$ for $y \in \{\Gamma_1 \cap B_{\epsilon_1}\}$ and vanishes when $y = 0$. By the identity theorem for holomorphic functions, $\partial F/\partial \bar{w}_j$ must vanish identically on $\omega_1 + \{\Gamma_1 \cap B_{\epsilon_1}\}$ and so F is holomorphic on $\omega_1 + \{\Gamma_1 \cap B_{\epsilon_1}\}$, as desired.

Note that this version of the identity theorem follows from the elementary theory of analytic functions of one complex variable. To see this, fix any vector $v \in \Gamma_1 \subset N_0(M)$ with $|v| = 1$. Then Jv is a totally real tangent space vector. The function $G(x+iy) = (\partial F/\partial \bar{w}_j)(-x(Jv)+iyv, w)$ is a holomorphic function of $x+iy \in \mathbb{C}$ with $0 < y < \epsilon_1$, which vanishes for $y = 0$. Therefore $G(x+iy) = 0$ for all $0 \le y < \epsilon_1$ by the Schwarz reflection principle and identity theorem for analytic functions of one complex variable. Since $v \in \Gamma_1$ with $|v| = 1$ is chosen arbitrarily, we have $\partial F/\partial \bar{w}_j \equiv 0$ on $\omega_1 + \{\Gamma_1 \cap B_{\epsilon_1}\}$, as desired.

As mentioned just before Lemma 1, this lemma completes the proof of (a) $\Rightarrow$ (b).

For the converse, we assume (b) holds — that is, we assume there is a closed, convex cone $\Gamma \subset N_0(M)$ which satisfies statement (b). We must show that if $\xi_0 \notin \Gamma$ then $\xi_0 \notin WF_0(CR(\omega))$. Since Γ is closed and convex, we have $\{\Gamma^0\}^0 = \Gamma$. Therefore, if ξ_0 does not belong to Γ, then there is a vector v_0 belonging to the interior of Γ^0 with

$$|v_0| = 1 \quad \text{and} \quad v_0 \cdot \xi_0 < 0. \tag{6}$$

Choose an open cone $\Gamma_1 < \Gamma^0$ which contains v_0. From (b), we know that any $f \in CR(\omega)$ extends to a holomorphic function F defined on the set

$$\omega_1 + \{\Gamma_1 \cap B_{\epsilon_1}\} \subset \mathbb{C}^n$$

where $\epsilon_1 > 0$ and where ω_1 is some open neighborhood of the origin in M. We may assume ω_1 is of the form

$$\omega_1 = \{(x+ih(w),w); x \in \mathbb{R}^d, w \in \mathbb{C}^{n-d} \quad \text{with} \quad |x|, |w| < \delta\}$$

for some $\delta > 0$.

Now the idea is to deform the domain of integration in If using Cauchy's theorem and then estimate the resulting integrand. To carry out the details, we choose $g \in \mathcal{D}(\mathbb{C}^d)$ such that if $|y| \le \epsilon_1/2$, then $g(x+iy) = 1$ for $|x| \le 3\delta/4$ and $g(x+iy) = 0$ for $|x| \ge \delta$. We also choose $g_1 \in \mathcal{D}(\mathbb{C}^d)$ such that if $|y| \le \epsilon_1/2$, then $g_1(x+iy) = 1$ for $|x| \le \delta/2$ and $g_1(x+iy) = 0$ for $|x| \ge 3\delta/4$. For $0 < t < \epsilon_1/2$, let

$$M_{t,w} = \{z' = x' + i(h(w) + t\tilde{g}_1(x',w)v_0); \; x' \in \mathbb{R}^d\}$$

where $\tilde{g}_1(x',w) = g_1(x'+ih(w))$. Note that if $|h(w)| \le \epsilon_1/2$, then $M_{t,w} \subset M_w + \{\Gamma_1 \cap B_{\epsilon_1}\}$ and $M_{t,w} \cap \{|x'| \ge 3\delta/4\} \subset M_w$ (since $g_1(x',w) = 0$ if $|x'| \ge 3\delta/4$).

Since $F|_{\omega_1} = f$, we have

$$(If)(z,w,\xi) = (2\pi)^{-d} \int\limits_{z' \in M_w} F(z',w)g(z')e^{i\xi\cdot(z-z')-|\xi|[z-z']^2}\Delta(z-z',\xi)dz'.$$

For fixed w with $|h(w)| \leq \epsilon_1/4$, the integrand is holomorphic in z' for $z' = x' + i(h(w)+v)$ provided $|x'| \leq 3\delta/4$, $|v| \leq \epsilon_1/4$ and $v \in \Gamma_1$ (note $g(z') = 1$ here). Therefore by Cauchy's theorem, we can deform the domain of integration from M_w to $M_{t,w}$ to obtain

$$(If)(z,w,\xi) = (2\pi)^{-d} \int\limits_{z' \in M_{t,w}} F(z',w)g(z')e^{i\xi\cdot(z-z')-|\xi|[z-z']^2}\Delta(z-z',\xi)dz'.$$

provided $|h(w)| \leq \epsilon_1/4$ and $0 < t < \epsilon_1/4$.

Let

$$q(z,z',w)(\xi) = i\xi\cdot(z-z') - |\xi|[z-z']^2$$

which is the exponent appearing in the integrand. The variables $z \in \mathbb{C}^d$ and $w \in \mathbb{C}^{n-d}$ are independent variables. The variable z' depends on w, the independent variable x', and the parameter t through the equation $z' = x' + i(h(w) + t\tilde{g}_1(x',w)v_0) \in M_{t,w}$. To show that the vector ξ_0 does not belong to $WF_0(CR(w))$, we must choose t and show that there is an $\hat{\epsilon} > 0$, a conical neighborhood $\widehat{\Gamma}$ of ξ_0 in $\mathbb{R}^d$, and a neighborhood U of the origin in $\mathbb{C}^n$ such that

$$\mathrm{Re}\{q(z,z',w)(\xi)\} \leq -\hat{\epsilon}|\xi| \tag{7}$$

for $\xi \in \widehat{\Gamma}$, $(z,w) \in U$, and $z' \in M_{t,w} \cap \{|\mathrm{Re}\ z'| < \delta\}$. Note that since q does not involve the function f, the choice of $\hat{\epsilon}$, $\widehat{\Gamma}$, and U will be independent of f. It suffices to show that there exist $\hat{\epsilon} > 0$ and $0 < t < \epsilon_1/4$ such that

$$\mathrm{Re}\{q(z=0,z',w=0)(\xi_0)\} \leq -\hat{\epsilon} \tag{8}$$

for $z' \in \{M_{t,w=0}\} \cap \{|\mathrm{Re}\ z'| < \delta\}$. This is because q is homogeneous of degree 1 in ξ and continuous in (z,w) and x' and therefore the inequality (7) will hold with $\hat{\epsilon}$ replaced by $\hat{\epsilon}/2$ for (z,w) in a neighborhood of the origin in $\mathbb{C}^n$ and for ξ in a conical neighborhood of ξ_0.

A point $z' \in \{M_{t,w=0}\} \cap \{|\mathrm{Re}\ z'| < \delta\}$ is of the form

$$z' = x' + it\tilde{g}_1(x',0)v_0 \quad \text{with} \quad |x'| < \delta.$$

We obtain

$$\begin{aligned}\mathrm{Re}\{q(z=0,z',w=0)(\xi_0)\} &= \mathrm{Im}\{z'\}\cdot\xi_0 - |\xi_0|\ \mathrm{Re}\{[z']^2\} \\ &= t\tilde{g}_1(x',0)v_0\cdot\xi_0 \\ &\quad -|\xi_0|(|x'|^2 - t^2\tilde{g}_1(x',0)^2|v_0|^2).\end{aligned} \tag{9}$$

From (6), recall that $v_0 \cdot \xi_0 < 0$ and $|v_0| = 1$. If $|x'| \leq \delta/2$, then $\tilde{g}_1(x', 0) = 1$ and (9) becomes

$$\operatorname{Re}\{q(z = 0, z', w = 0)(\xi_0)\} \leq t(v_0 \cdot \xi_0) + t^2|\xi_0|. \tag{10a}$$

If $\delta/2 < |x'| \leq \delta$, then (9) becomes

$$\operatorname{Re}\{q(z = 0, z', w = 0)(\xi_0)\} \leq -|\xi_0|\frac{\delta^2}{4} + t^2|\xi_0|. \tag{10b}$$

Since $v_0 \cdot \xi_0 < 0$, we can choose t suitably small with $0 < t < \epsilon_1/4$ so that the right sides of (10a) and (10b) are both negative. This establishes (8) and by the discussion after (8), ξ_0 does not belong to $WF_0(CR(\omega))$, as desired. The proof of (b) $\Rightarrow$ (a) is now complete. ∎

16.3 The hypoanalytic wave front set and the Levi form

In this section, we relate the convex hull of the image of the Levi form and $WF_0(CR(\omega))$, thereby completing the Fourier integral approach to the proof of the CR extension theorem. Let Γ_0 be the convex hull of the image of the Levi form of M at 0. Let $\Gamma = \{\overline{\Gamma}_0\}^0$ (i.e., the polar of the closure of Γ_0). Note that $\Gamma^0 = \overline{\Gamma}_0$. According to Theorem 1 in the previous section, the CR extension theorem will follow once we show that $WF_0(CR(\omega))$ is contained in Γ. This is the content of the next theorem.

THEOREM 1
Suppose Γ is the polar of the closure of the convex hull of the image of the Levi form of $M = \{y = h(w)\}$ at the origin (i.e., $\Gamma = \{\overline{\Gamma}_0\}^0$). Then, for any neighborhood $\omega \subset M$ of the origin, $WF_0(CR(\omega))$ is contained in Γ.

PROOF Given $\xi_0 \notin \Gamma$, we must show there is a neighborhood U of the origin in $\mathbb{C}^n$, a conical neighborhood $\widehat{\Gamma}$ of ξ_0 in $\mathbb{R}^d$, and a number $\epsilon > 0$ such that if f is an element of $CR(\omega)$ then there is a constant $C > 0$ such that

$$|(If)(z, w, \xi)| \leq Ce^{-\epsilon|\xi|}$$

for all $(z, w) \in U$ and $\xi \in \widehat{\Gamma}$.

As usual, we identify $H_0^{1,0}(M)$ with $\{(0, w); w \in \mathbb{C}^{n-d}\}$ and $N_0(M)$ with $\{(0, y, 0); y \in \mathbb{R}^d\}$. Let $\tilde{\mathcal{L}}_0$: $\mathbb{C}^{n-d} \to \mathbb{R}^d$ be the extrinsic Levi form of M at 0. We have

$$\tilde{\mathcal{L}}_0(w) = \sum_{j,k=1}^{n-d} \frac{\partial^2 h(0)}{\partial w_j \partial \bar{w}_k} w_j \bar{w}_k \quad \text{for} \quad w = (w_1, \ldots, w_{n-d}) \in \mathbb{C}^{n-d}.$$

By definition, the cone Γ_0 is the convex hull (in $\mathbb{R}^d$) of the image of $\tilde{\mathcal{L}}_0$. The cone, Γ, is closed and $\Gamma^0 = \overline{\Gamma}_0$. Since $\xi_0 \notin \Gamma$, there must exist a vector $v_0 \in \Gamma_0$ with

$$v_0 \cdot \xi_0 < 0. \tag{1}$$

If $v \cdot \xi_0 \geq 0$ for all vectors v that lie in the image of $\tilde{\mathcal{L}}_0$, then the same inequality holds for all v in the convex hull of the image of $\tilde{\mathcal{L}}_0$. Therefore, we may assume (1) holds for some vector v_0 that lies in the image of $\tilde{\mathcal{L}}_0$. By a complex linear change of coordinates in the w-variables, we may assume

$$v_0 = \tilde{\mathcal{L}}_0(e_1)$$

where

$$e_1 = (1, 0, \ldots, 0) \in \mathbb{C}^{n-d}.$$

Now we examine the second-order Taylor expansion of h about the origin. We may assume there are no second-order pure terms in this expansion (Theorem 2 in Section 7.2). Therefore

$$h(w) = \tilde{\mathcal{L}}_0(w) + \mathcal{O}(3)$$

where $\mathcal{O}(3)$ involves terms that are third order in w and $\bar{w}$.

Let $w' = (w_2, \ldots, w_{n-d}) \in \mathbb{C}^{n-d-1}$. We have

$$\begin{aligned} h(w) &= \tilde{\mathcal{L}}_0(w_1 e_1) + \mathcal{O}(|w'||w|) + \mathcal{O}(3) \\ &= |w_1|^2 v_0 + \mathcal{O}(|w'||w|) + \mathcal{O}(3). \end{aligned}$$

Since $v_0 \cdot \xi_0 < 0$, there is a conical neighborhood $\widehat{\Gamma}$ of ξ_0 in $\mathbb{R}^d$ and a number $\alpha > 0$ such that

$$v_0 \cdot \xi \leq -\alpha|\xi| \quad \text{for} \quad \xi \in \widehat{\Gamma}.$$

We obtain

$$h(w) \cdot \xi \leq \left(\frac{-\alpha}{2} |w_1|^2 + C|w'||w| + C|w'|^3 \right) |\xi| \quad \text{for} \quad \xi \in \widehat{\Gamma} \tag{2}$$

where C is a positive constant that is independent of $\xi \in \hat{\Gamma}$ and $w \in \mathbb{C}^{n-d}$ near the origin.

Now we parameterize M_w by $z' = x' + ih(w)$ for $x' \in \mathbb{R}^d$. With $z = x + iy \in \mathbb{C}^d$, we obtain

$$\begin{aligned} &(If)(z, w, \xi) = \\ &\quad (2\pi)^{-d} \int\limits_{x' \in \mathbb{R}^d} \tilde{g}(x', w) \tilde{f}(x', w) e^{i\xi \cdot (x - x' + i(y - h(w))) - |\xi|[x - x' + i(y - h(w))]^2} \\ &\qquad \cdot \Delta(x - x' + i(y - h(w)), \xi) dx' \end{aligned}$$

where $\tilde{g}(x', w) = g(x' + ih(w))$ and $\tilde{f}(x', w) = f(x' + ih(w), w)$. Let

$$q(x, x', y, w)(\xi) = i\xi \cdot (x - x' + i(y - h(w))) - |\xi|[x - x' + i(y - h(w))]^2.$$

The estimate on $|If|$ requires an estimate on Re q. We have

$$\begin{aligned} \text{Re } q(x, x', y, w)(\xi) &= h(w) \cdot \xi - y \cdot \xi - |\xi||x - x'|^2 + |\xi||y - h(w)|^2 \\ &\leq \left(\frac{-\alpha}{2}|w_1|^2 + C|w'||w| + C|w'|^3 \right. \\ &\quad \left. + |y| + 2(|y|^2 + |h(w)|^2) \right) |\xi| \end{aligned}$$

where the last inequality follows from (2). We restrict $(z, w) \in \mathbb{C}^n$ so that $|z = x + iy| < \delta, |w_1| = r$ and $|w'| < r\delta$ where r and δ will be chosen later. We have $|h(w)| \leq C|w|^2 \leq C(r^2 + r^2\delta^2) \leq 2Cr^2$ provided $\delta < 1$. We obtain

$$\text{Re } q(x, x', y, w)(\xi) \leq \left(\frac{-\alpha}{2} r^2 + C\delta r^2 + C\delta^3 r^3 + \delta + 2\delta^2 + 4C^2 r^4 \right) |\xi|.$$

First choose $r > 0$ small enough so that $(-\alpha/2)r^2 + 4C^2r^4 < 0$. Then choose $\delta > 0$ small (depending on r) so that

$$\frac{-\alpha}{2} r^2 + 4C^2 r^4 + [C\delta r^2 + C\delta^3 r^3 + \delta + 2\delta^2] < 0.$$

Denote the number on the left by $-\epsilon$ (with $\epsilon > 0$). We obtain

$$\text{Re } q(x, x', y, w)(\xi) \leq -\epsilon|\xi| \quad \text{for} \quad \xi \in \widehat{\Gamma}.$$

It follows that

$$|(If)(z, w, \xi)| \leq Ce^{-\epsilon|\xi|} \tag{3}$$

for $\xi \in \widehat{\Gamma}$ and $(z, w) \in \mathbb{C}^n$ with $|z| < \delta$, $|w_1| = r$, $|w'| < \delta r$.

To prove that ξ_0 does not belong to $WF_0(CR)(\omega))$, we need to establish (3) for $|w_1| \leq r$ (rather than just for $|w_1| = r$ as we have done above). To accomplish this, we use the next lemma which allows us to approximate $(If)(z, w, \xi)$ by a function that is analytic in both z and w (whereas $(If)(z, w, \xi)$ is only analytic in z). Then the above estimate will hold for $|w_1| \leq r$ by the maximum principle.

LEMMA 1

There is a neighborhood U of the origin in $\mathbb{C}^n$ and a constant $\epsilon > 0$ such that for each $f \in CR(\omega)$ there is a function $G(f)$: $U \times \mathbb{R}^d \to \mathbb{C}$ which is analytic *in* $(z,w) \in U$ *and a constant* $C > 0$ *such that*

$$|(If)(z,w,\xi) - G(f)(z,w,\xi)| \le Ce^{-\epsilon|\xi|}$$

for all $(z,w) \in U$ *and* $\xi \in \mathbb{R}^d$.

Here again, the open set U and the constant ϵ are independent of $f \in CR(\omega)$. However, the constant C is allowed to depend on f.

Assuming the lemma for the moment, we complete the proof of Theorem 1 and hence the proof of the CR extension theorem. The estimate in (3) together with Lemma 1 implies that there is an $\epsilon > 0$ so that for each $f \in CR(\omega)$ there is a constant $C > 0$ such that

$$|G(f)(z,w,\xi)| \le Ce^{-\epsilon|\xi|}$$

for $\xi \in \widehat{\Gamma}$ and $(z,w) \in \mathbb{C}^n$ with $|z| < \delta$, $|w_1| = r$, and $|w'| < \delta r$ provided $r > 0$ and $\delta > 0$ are chosen suitably small as above. Since $G(f)(z,w,\xi)$ is analytic in (z,w), the maximum principle implies that the above inequality also holds for $|w_1| \le r$, and $|z| < \delta$, $|w'| < \delta r$. This together with another application of the lemma yields the estimate

$$|(If)(z,w,\xi)| \le Ce^{-\epsilon|\xi|}$$

for $\xi \in \widehat{\Gamma}$ and $(z,w) \in \mathbb{C}^n$ with $|z| < \delta$, $|w_1| \le r$, and $|w'| < \delta r$. By definition, the vector $\xi_0 \in \widehat{\Gamma}$ does not belong to $WF_0(CR(\omega))$, as desired.

PROOF OF LEMMA 1 Since $(If)(z,w,\xi)$ is already analytic in $z \in \mathbb{C}^d$, the idea is to solve the appropriate $\bar{\partial}$-problem in the w-variables to find the appropriate analytic $G(f)(z,w,\xi)$.

From the formula for $I(f)$, we may write

$$(If)(z,w,\xi) = \int\limits_{x' \in \mathbb{R}^d} E(z - x' - ih(w), \xi)\tilde{f}(x',w)\tilde{g}(x',w)dx'$$

where

$$E(\zeta,\xi) = e^{i\xi\cdot\zeta - |\xi|[\zeta]^2}\Delta(\zeta,\xi)$$

is a holomorphic function of $\zeta \in \mathbb{C}^d$. As usual, $\tilde{f}(x',w) = f(x' + ih(w), w)$ and $\tilde{g}(x',w) = g(x' + ih(w))$.

Differentiating If,we obtain

$$\frac{\partial}{\partial \bar{w}_j}(If)(z,w,\xi) = \int\limits_{x' \in \mathbb{R}^d} E(z - x' - ih(w), \xi)\frac{\partial}{\partial \bar{w}_j}\{(\tilde{f}\tilde{g})\}(x',w)dx'$$

$$- i\int\limits_{x' \in \mathbb{R}^d} \sum_{k=1}^{d} \frac{\partial h_k(w)}{\partial \bar{w}_j} \cdot \frac{\partial E}{\partial \zeta_k}(z - x' - ih(w), \xi)\tilde{f}(x',w)\tilde{g}(x',w)dx'. \quad (4)$$

The second term on the right can be rewritten as

$$i \int\limits_{x' \in \mathbb{R}^d} \sum_{k=1}^{d} \frac{\partial h_k(w)}{\partial \bar{w}_j} \cdot \frac{\partial}{\partial x'_k} \{E(z - x' - ih(w), \xi)\} \tilde{f}(x', w) \tilde{g}(x', w) dx'.$$

After integrating by parts, this term can be combined with the first term on the right side of (4) to yield

$$\frac{\partial}{\partial \bar{w}_j} (If)(z, w, \xi) = \int\limits_{x' \in \mathbb{R}^d} E(z - x' - ih(w), \xi) \overline{\tilde{L}}_j \{\tilde{f} \tilde{g}\}(x', w) dx' \tag{5}$$

where

$$\overline{\tilde{L}}_j = \frac{\partial}{\partial \bar{w}_j} - i \sum_{k=1}^{d} \frac{\partial h_k(w)}{\partial \bar{w}_j} \frac{\partial}{\partial x'_k}.$$

From Theorem 3 in Section 7.2, a basis for $H^{0,1}(M)$ is given by $\overline{L}_1, \dots, \overline{L}_{n-d}$ where

$$\overline{L}_j = \frac{\partial}{\partial \bar{w}_j} - 2i \sum_{k=1}^{d} \frac{\partial h_k(w)}{\partial w_j} \frac{\partial}{\partial \bar{z}'_k} \qquad 1 \leq j \leq n - d.$$

Let π: $\mathbb{C}^d \times \mathbb{C}^{n-d} \to \mathbb{R}^d \times \mathbb{C}^{n-d}$ be the projection $\pi(z' = x' + iy', w) = (x', w)$. From the definitions of $\tilde{f}$ and $\tilde{g}$, note that $f = \tilde{f} \circ \pi$ and $g = \tilde{g} \circ \pi$. Also note

$$\overline{\tilde{L}}_j = \pi_* \overline{L}_j \qquad 1 \leq j \leq n - d.$$

Since f is CR on ω (i.e., $\overline{L}_j f \equiv 0$ on ω) and since $g = 1$ on a neighborhood of the origin in M, we have

$$\begin{aligned} \overline{\tilde{L}}_j(\tilde{f}\tilde{g}) &= \pi_* \overline{L}_j \{\tilde{f}\tilde{g}\} \\ &= \overline{L}_j \{fg\} \\ &= 0 \qquad 1 \leq j \leq n - d \end{aligned}$$

near the origin in M. So there exists $\delta > 0$ such that if $|z|, |w| < \delta$, then the integrand in (5) vanishes for $|x'| < \delta$.

Therefore, to estimate (5), we only need to estimate the real part of the exponent, q, for $|x'| \geq \delta$, where

$$q(z, x', w)(\xi) = i\xi \cdot (z - x' - ih(w)) - |\xi|[z - x' - ih(w)]^2.$$

With $z = x + iy$, we have

$$\text{Re}\{q(z, x', w)(\xi)\} = (h(w) - y) \cdot \xi - |\xi|(|x - x'|^2 - |y - h(w)|^2).$$

If $z = x + iy = 0$, $w = 0$, and $|x'| \geq \delta$, then

$$\text{Re } q(z = 0, x', w = 0)(\xi) \leq -\delta^2|\xi|.$$

Since q is homogeneous of degree one in ξ and continuous in $z, x'w$, there is a neighborhood U of the origin in $\mathbb{C}^n$ such that if $(z, w) \in U$ and $|x'| \geq \delta$, then

$$\text{Re } q(z, x', w)(\xi) \leq -\frac{\delta^2}{2}|\xi|.$$

Note δ is independent of $f \in CR(\omega)$, $x' \in \mathbb{R}^d$ with $|x'| \geq \delta$, $(z, w) \in U$ and $\xi \in \mathbb{R}^d$. From this estimate and (5), we have

$$|\bar{\partial}_w(If)(z, w, \xi)| \leq C_1 e^{-\frac{\delta^2}{2}|\xi|} \tag{6}$$

for $(z, w) \in U$ and $\xi \in \mathbb{R}^d$. Here, C_1 is a constant that depends on f but it is independent of $(z, w) \in U$ and $\xi \in \mathbb{R}^d$. By shrinking U, we may assume that $U = U_1 \times U_2$ where U_1 is a neighborhood of the origin in $\mathbb{C}^d$ and U_2 is a strictly convex neighborhood (such as a ball) of the origin in $\mathbb{C}^{n-d}$. Since $\bar{\partial}_w(If)(z, w, \xi)$ is $\bar{\partial}_w$-closed in U_2, we can use the $\bar{\partial}_w$-theory on strictly convex domains to find a solution $K(f)(z, w, \xi)$ to the equation

$$\bar{\partial}_w\{K(f)(z, w, \xi)\} = \bar{\partial}_w(If)(z, w, \xi)$$

with the estimate

$$\sup_{w \in U_2} |K(f)(z, w, \xi)| \leq C \sup_{w \in U_2} |\bar{\partial}_w(If)(z, w, \xi)| \tag{7}$$

for $z \in U_1$, $\xi \in \mathbb{R}^d$. Here, C is a uniform constant that is independent of $z \in U_1$, $\xi \in \mathbb{R}^d$, and $f \in CR(\omega)$. For the reader who is not familiar with the solution to the $\bar{\partial}$-problem on a strictly convex domain with sup-norm estimates, an integral kernel solution to this problem is provided in Part IV (see Theorem 1 in Section 20.3).

Combining (6) and (7), we obtain

$$|K(f)(z, w, \xi)| \leq \widetilde{C} e^{-\frac{\delta^2}{2}|\xi|} \tag{8}$$

for $(z, w) \in U_1 \times U_2$ and $\xi \in \mathbb{R}^d$. Here, $\widetilde{C}$ depends on f but it is independent of $(z, w) \in U_1 \times U_2$ and $\xi \in \mathbb{R}^d$. From the integral kernel formula for $K(f)(z, w, \xi)$, this solution is holomorphic in $z \in U_1$ (since $(If)(z, w, \xi)$ is holomorphic in $z \in U_1$). Define

$$G(f)(z, w, \xi) = (If)(z, w, \xi) - K(f)(z, w, \xi).$$

Clearly, $G(f)(z, w, \xi)$ is analytic in both $z \in U_1$ and $w \in U_2$. From (8), we obtain the desired estimate for the lemma, with $\epsilon = \delta^2/2$.

This completes the proof of Lemma 1. As stated prior to the statement of Lemma 1, the proof of Theorem 1 is now complete. As stated at the beginning of this section, Theorem 1 together with Theorem 1 in the previous section completes the Fourier integral approach to the proof of the CR extension theorem. ∎

It is worthwhile to note the key role played by the term $|\xi|[z - z']^2$ in the exponent of the transform $(If)(z, w, \xi)$. Without this term, one would not be able to prove estimate (6) in the proof of the above theorem or estimate (10b) in the proof of Theorem 1 in Section 16.2. This is the reason for using the transform $(If)(z, w, \xi)$ rather than the simpler transform $(\widehat{If})(z, w, \xi)$.

17

Further Results

In this chapter, we discuss some extensions of the techniques developed earlier in Part III. We begin by outlining the modifications needed for the Fourier transform approach to the proof of the CR extension theorem for nonrigid CR manifolds. Next, we discuss the holomorphic extension of CR distributions. The chapter ends with a discussion of CR extension near points of higher type and analytic hypoellipticity.

17.1 The Fourier integral approach in the nonrigid case

In Chapter 16, we present the Fourier integral approach to the CR extension theorem in the rigid case $M = \{y = h(w)\}$. As usual, our coordinates for $\mathbb{C}^n$ are given by (z, w) where $z = x + iy \in \mathbb{C}^d$, $w \in \mathbb{C}^{n-d}$ with $d = \text{codim}_{\mathbb{R}}(M)$. In this section, we outline the major changes needed for the nonrigid case $M = \{y = h(x, w)\}$ and we leave some of the details to the reader (or see [BR2]).

As in Section 16.1, we start with the transform

$$\hat{I}f(z, w, \xi) = (2\pi)^{-d} \int\limits_{z' \in M_w} g(z')f(z', w)e^{i\xi\cdot[z-z']}\, dz'$$

where f is a continuous function on M and where g is a smooth function on M with compact support which is identically 1 near the origin. For $\epsilon > 0$, let

$$F_\epsilon(z, w) = (2\pi)^{-d} \int\limits_{\xi\in\mathbb{R}^d} \hat{I}f(z, w, \xi)e^{-\epsilon(\xi)^2}\, d\xi$$

where we have used the notation $(\xi)^2 = \xi_1^2 + \ldots + \xi_d^2$. Note that the map $\xi \mapsto (\xi)^2$, $\xi \in \mathbb{R}^d$ analytically continues to $\{\xi \in \mathbb{C}^d\}$. Lemmas 1 and 2 from

Section 16.1 hold without change to show

$$F_\epsilon(z,w) \mapsto f(z,w) \quad \text{as } \epsilon \mapsto 0$$

for $(z,w) \in M$ near the origin. For each $\epsilon > 0$, F_ϵ is an entire function of $z \in \mathbb{C}^d$.

The first change needed for the nonrigid case is to rewrite F_ϵ so that it involves an integral over a carefully chosen contour in $\mathbb{C}^d$. For $z = x + iy \in \mathbb{C}^d$ and $w \in \mathbb{C}^{n-d}$, let $\gamma_{x,w}$ be the contour parameterized by the map

$$\xi \mapsto \hat{\xi} = \left(\left[I + i\frac{\partial h(x,w)}{\partial x}\right]^{-1}\right)^t (\xi) \in \mathbb{C}^d.$$

Since $\hat{I}f(z,w,\xi)$ is an entire function of ξ, we can use Cauchy's theorem to show

$$F_\epsilon(z,w) = (2\pi)^{-d} \int\limits_{\hat{\xi}\in\gamma_{x,w}} \hat{I}f(z,w,\hat{\xi})e^{-\epsilon(\hat{\xi})^2}\, d\hat{\xi}.$$

Note that since $\partial h(0,0)/\partial x = 0$, the contour $\gamma_{x,w}$ is close to the copy of $\mathbb{R}^d$ given by $\{\mathrm{Im}\hat{\xi} = 0\}$. For the same reason, we have

$$|\hat{\xi}| \geq \frac{1}{2}|\xi|$$

for

$$\hat{\xi} = \left(\left[I + i\frac{\partial h(x,w)}{\partial x}\right]^{-1}\right)^t (\xi) \in \gamma_{x,w}$$

provided (x,w) is sufficiently close to the origin. The presence of the term $e^{-\epsilon(\hat{\xi})^2}$ means that the integrals over the side contours $\{|\hat{\xi}_j| = r\}$ appearing in the change of contour process disappear as $R \mapsto \infty$. The reason we need this change of contour will become clear at a later time. Note in the rigid case, nothing has changed (since $\partial h(x,w)/\partial x = 0$).

Now, we make the analogous change of contour as in the proof of Theorem 1 in Section 16.1. That is, we set

$$\eta(\hat{\xi}) = \hat{\xi} + i\langle\hat{\xi}\rangle(z - z')$$

where we have used the notation

$$\langle\hat{\xi}\rangle = \sqrt{\hat{\xi}_1^2 + \ldots + \hat{\xi}_d^2}.$$

The expression $\langle \hat{\xi} \rangle$ is well defined and holomorphic in $\hat{\xi} \in \mathbb{C}^d$ for provided $|\mathrm{Re}\hat{\xi}| > |\mathrm{Im}\hat{\xi}|$. After dropping the hat and changing the contour of integration, as done in the proof of Theorem 1 in Section 16.1, we obtain

$$F_\epsilon(z,w) = \int\limits_{\xi \in \gamma_{x,w}} If(z,w,\xi) e^{-\epsilon(\eta(\xi))^2}\, d\eta(\xi)$$

where

$$If(z,w,\xi) = (2\pi)^{-d} \int\limits_{z' \in M_w} g(z') f(z',w) e^{i\xi \cdot [z-z'] - \langle \xi \rangle [z-z']^2}\, \Delta(z-z',\xi) dz'.$$

Since $\eta = \xi + i\langle \xi \rangle (z - z')$, we can restrict z and z' so that $|\eta| \geq 1/2|\xi|$. The presence of the term $e^{-\epsilon(\eta)^2}$ means that the integrals over the side contours $\{|\eta_j| = R\}$ appearing in the change of contour process disappear as $R \mapsto \infty$. As mentioned earlier, $F_\epsilon(z,w)$ is holomorphic in $z \in \mathbb{C}^d$ and $F_\epsilon \mapsto f$ on M as $\epsilon \mapsto 0$.

The definition of the hypoanalytic wave front set given in Definition 1 in Section 16.2 is unchanged for the nonrigid case except that the cone $\hat{\Gamma}$ must be a conical neighborhood of ξ_0 in $\mathbb{C}^d$ rather than an $\mathbb{R}^d$-conical neighborhood as stated. The reason for this is that $If(z,w,\xi)$ involves an integral over $\gamma_{x,w}$ which is contained in $\mathbb{C}^d$ rather than an integral over $\mathbb{R}^d$ as in the rigid case.

Theorem 1 in Section 16.2 holds without change in the nonrigid case. The basic ideas of the proof are similar to the ideas given in the rigid case except one must use the new transform $If(z,w,\xi)$ defined above. For example, let us outline the key steps in the proof of the (a) implies (b) part of this theorem for the nonrigid case. Here, we are assuming that the hypoanalytic wave front set of a given CR function f is contained in a cone Γ and we are to show that f holomorphically extends to a set of the form $\omega + \{\Gamma_1 \cap B_{\epsilon_1}\}$ where Γ_1 is a smaller subcone of the polar of Γ. As in the proof for the rigid case, the key idea is to estimate the real part of the exponent $q(z,z',w)(\hat{\xi})$ appearing in the definition of $If(z,w,\hat{\xi})$ where $z' = x' + ih(x',w) \in M_w$, $z = x + i(h(x,w) + v)$ (where v belongs to Γ_1) and $\hat{\xi} = \left([I + i(\partial h(x,w)/\partial x)]^{-1}\right)^t (\xi) \in \gamma_{x,w}$. To carry out this estimate, we shall require that there are no second-order pure terms in the Taylor expansion of h. In particular, we need

$$\frac{\partial^2 h(0,0)}{\partial x_i \partial x_j} = 0 \quad \text{for } 1 \leq i,j \leq d.$$

We have

$$\begin{aligned} z - z' &= x - x' + i(h(x,w) - h(x',w) + v) \\ &= \left[I + i\frac{\partial h(x,w)}{\partial x} \right] (x - x') + \circ(|x - x'|^2) + iv. \end{aligned}$$

The notation $\circ(t^j)$ for $j \geq 0$ indicates terms that are small in absolute value when compared to $|t|^j$. Taking the Euclidean dot product of $z - z'$ with $\hat{\xi} = \left([I + i(\partial h(x,w)/\partial x)]^{-1}\right)^t(\xi) \in \gamma_{x,w}$, we obtain

$$i(z-z')\cdot\hat{\xi} - \langle\hat{\xi}\rangle[z-z']^2 = i(x-x')\cdot\xi - v\cdot\xi - |\xi||x-x'|^2 + 2iv\cdot(x-x')|\xi| + o(|x-x'|^2)|\xi| + \mathcal{O}(|x|+|w|)|v||\xi| + \mathcal{O}(|v|^2|\xi|).$$

Therefore

$$\begin{aligned}\operatorname{Re}\{i(z-z')\cdot\hat{\xi} - \langle\hat{\xi}\rangle[z-z']^2\} &\\ &\leq -v\cdot\xi - \frac{1}{2}|\xi|(|x-x'|^2 + C((|x|+|w|)|v| + |v|^2)|\xi| \\ &\leq -v\cdot\xi + C((|x|+|w|)|v| + |v|^2)|\xi| \qquad (1)\end{aligned}$$

provided z, z', and w are contained in a suitably small neighborhood of the origin. This is analogous to the estimate on the exponent given at the end of the proof of Lemma 3 in Section 16.1 for the rigid case. It is here that the reason for introducing the contour $\{\hat{\xi} \in \gamma_{x,w}\}$ becomes apparent. For if we compute $i\xi\cdot(z-z')$ for $\xi \in \mathbb{R}^d$ as we did for the rigid case, then we encounter the following additional error term:

$$\left[\frac{\partial h(x,w)}{\partial x}\right](x-x')\cdot\xi.$$

We cannot absorb this error term by the expression $|\xi||x-x'|^2$ in the way we absorbed the term $o(|x-x'|^2|\xi|)$ to obtain (1).

If f is a function of class C^{d+2} on M, then we can integrate by parts as done in the proof of Lemma 3 in Section 16.1 to show that $If(z,w,\hat{\xi})$ is dominated by an integrable function of $\hat{\xi}$ provided w and z are close to the origin and $z = x + iy$ with $y \in \Gamma_1$. After letting $\epsilon \mapsto 0$, we conclude that f is the boundary values of a function $(z,w) \mapsto F(z,w)$ which is holomorphic in z provided (z,w) belongs to a sufficiently small neighborhood of the origin in $M + \Gamma_1$. If f is CR, then F is holomorphic in w as well, as shown in Lemma 1 of Section 16.2.

The proof of the converse of this theorem and the proof of Theorem 1 in Section 16.3 are similar to the proofs given for the rigid case after making suitable modifications as above.

17.2 The holomorphic extension of CR distributions

First, we mention that the approximation theorem (Theorem 1 in Chapter 13) holds for other classes of CR functions. For example, if f is a CR function of class C^k then the approximation theorem produces a sequence of entire functions that converges to f in the topology of C^k. The analogous statement holds true for the class of $\mathcal{L}^p_{\text{loc}}$ CR functions, for $1 \leq p \leq \infty$. The proofs of these facts are easy modifications of the proof given for the case of the sup-norm.

The approximation theorem for the class of CR distributions is obtained by dualizing the proof of the approximation theorem for the class of smooth CR functions. To carry out the details (and the details for the remainder of the proofs of the theorems in Part III for CR distributions), it will be necessary to restrict the given CR distribution to totally real slices of the CR manifold M. Let $M = \{y = h(x, w)\}$. We identify a given function (or distribution) $f : M \mapsto \mathbb{C}$ with the function $f_1 : \mathbb{R}^d \times \mathbb{C}^{n-d} \mapsto \mathbb{C}$, where $f_1(x, w) = f(x + ih(x, w), w)$. It is possible to view a given CR distribution f_1 as a smooth (C^∞) map from the set $\{w \in \mathbb{C}^{n-d}\}$ to the space of distributions on $\mathbb{R}^d$ (in the x-variables). This means that for a given smooth, compactly supported function $\phi : \mathbb{R}^d \mapsto \mathbb{C}$, the map

$$w \mapsto \Phi(w) = \big(f_1(x, w), \phi(x)\big)_x$$

is a smooth function of w. To see this, we use Theorem 3 from Section 13.2 to write the projection of the tangential Cauchy–Riemann vector fields onto $\mathbb{R}^d \times \mathbb{C}^{n-d}$ as

$$\overline{L}_j = \frac{\partial}{\partial \overline{w}_j} - i \sum_{k=1}^{d} \frac{\partial h_k}{\partial \overline{w}_j} M_k$$

with

$$M_k = \sum_{l=1}^{d} \mu_{lk} \frac{\partial}{\partial x_l}$$

where μ_{lk} is the (l, k)th entry of

$$\left(I + i\frac{\partial h}{\partial x}\right)^{-1}.$$

If f_1 is a CR distribution, then $\overline{L}_j f_1 = 0$ and we can trade w-derivatives for x-derivatives. Together with a bootstrap argument, this shows that Φ is smooth.

The CR extension theorem (Theorem 1 in Section 14.2) can now be generalized to include the class of CR distributions. However, we need to explain what is meant by stating that a CR distribution is the boundary values of a holomorphic function F defined on a wedge $M + i\Gamma$. For the rigid case, $M = \{y = h(w)\}$, we define the *boundary values* of F (denoted bF) by

$$(bF(x, w), \phi(x, w))_{x,w} = \lim_{t \mapsto 0+} \int F(x+i(h(w)+tv), w)\phi(x, w)dxdv(w) \quad (1)$$

for $\phi \in \mathcal{D}(\mathbb{R}^d \times \mathbb{C}^{n-d})$ where $v \in \Gamma$, with $|v| = 1$. For the nonrigid case, $F(x + i(h(w) + tv), w)$ is replaced by $F(Z(x + itv, w), w)$ where for each w, $Z(\cdot, w) : \mathbb{C}^d \mapsto \mathbb{C}^d$ is a smooth function with $Z(x, w) = x + ih(x, w)$ for $x \in \mathbb{R}^d$ and where $Z(z, w)$ satisfies the Cauchy–Riemann equations on

Imz $= 0$ to infinite order. Suppose F has polynomial growth (this means $|F(Z(x+itv,w),w)| \leq Ct^{-N}$, for some positive constants C and N). It is shown in [BCT] that bF is a well-defined CR distribution on M which is independent of $v \in \Gamma$. For the proof, the main idea is to show that the integral on the right side of (1) and all of its t-derivatives are bounded by Ct^{-N}. The constant C is allowed to depend on the number of t-derivatives but N is not allowed to depend on the number of t-derivatives. This is accomplished by taking t-derivatives of the integral on the right side of (1); using the Cauchy–Riemann equations to turn them into x-derivatives, and then integrating by parts to put the x-derivatives onto the function ϕ. Repeated use of the Fundamental Theorem of Calculus shows that if a function of t, $t \geq 0$ and all of its t-derivatives are bounded by Ct^{-N}, then this function is continuous (in fact smooth) up to $t = 0$.

The converse of the above boundary value result is the CR extension theorem in the distributional category: under the same hypothesis on M and using the same notation as in Theorem 1 of Section 14.2, for a given CR distribution f on M, there is a holomorphic function F on $\omega_1 + i\{\Gamma_1 \cap B_{\epsilon_1}\}$ with polynomial growth such that $bF = f$ on ω_1. A short cut for the proof of this theorem can be obtained by using the following fact (see [BR2]): for a given CR distribution f_1 there is an positive integer N and a CR function $\hat{f}_1$ which is smooth (say C^4) such that

$$f_1 = \left(\sum_{k=1}^{d} M_k\right)^N \{\hat{f}_1\}$$

where M_k is the vector field involving $\partial/\partial x_1 \ldots \partial/\partial x_d$ defined above. In the rigid case, we have $M_k = \partial/\partial x_k$. To extend the given CR distribution f_1, we first extend the smooth CR function $\hat{f}_1$ to a holomorphic function $\hat{F}_1$ (which we already know how to do) and then note that the CR distribution f_1 is the boundary values of the holomorphic function $(\sum_{k=1}^{d} \partial/\partial z_k)^N \{\hat{F}_1\}$. For the nonrigid case, we refer the reader to [BR2].

In the Fourier transform proof of the CR extension theorem given in Chapter 16, we require the CR function f to be of class C^{d+2}. Under this smoothness assumption, we show (in the rigid case) that the transform $If(z,w,\xi)$ is an integrable function of $\xi \in \mathbb{R}^d$. This allows us to use the dominated convergence theorem to get rid of the annoying factor of $e^{-\epsilon|\xi|^2}$ appearing in Lemma 2 of Section 16.1 and thus simplify the proof of the CR extension theorem. This results in an apparent loss of derivatives. That is, if f is a CR function of class C^{d+2} then its holomorphic extension appears to only be of class C^1 up to M. However, by using a little potential theory, Baouendi, Jacobowitz, and Treves have shown in Lemma 2.4 in [BJT] that there is no loss of derivatives. In particular, the holomorphic extension of a continuous CR function is continuous up to M.

17.3 CR extension near points of higher type

If the convex hull of the image of the Levi form has empty interior, then CR extension to some open set in $\mathbb{C}^n$ may be impossible. For example, CR extension to a fixed open set from the submanifold $M = \{(z_1, z_2, w_1, w_2) \in \mathbb{C}^4;\ \text{Im}z_1 = 0,\ \text{Im}z_2 = |w_1|^2\}$ is impossible (because M is contained in the plane $\{\text{Im}z_1 = 0\}$). If the convex hull of the image of the Levi form has empty interior, then higher order conditions are required to ensure CR extension to an open set in $\mathbb{C}^n$. For a long time, the best theorem available in this context is a result of Baouendi and Rothschild [BR2] which handles the case of a point p of finite type in a semirigid CR submanifold, M (for the definitions of finite type and semirigid, see Chapter 12). Under these conditions, their theorem states that if ω is an open subset of M that contains p, then there is an open wedge W_Γ in $\mathbb{C}^n$ such that all CR functions on ω holomorphically extend to W_Γ. By definition, a wedge in $\mathbb{C}^n$ is an open set of the form $W_\Gamma = \omega' + \{\Gamma \cap B(0,\epsilon)\}$ where ω' is an open subset of M which contains p and where Γ is an open convex cone in the normal space of M at p. The simplest example that illustrates this theorem is the codimension two submanifold $M = \{(z_1, z_2, w) \in \mathbb{C}^3;\ \text{Im}z_1 = |w|^2,\ \text{Im}z_2 = |w|^2\text{Re}w\}$. This manifold is rigid (and hence semirigid). The image of the Levi form at the origin is the ray $\{\text{Im}z_1 \geq 0,\ \text{Im}z_2 = 0\}$ which has empty interior in the two-dimensional normal space of M at the origin. So the CR extension theorem in Section 14.2 does not apply. In fact, the hypothesis of the CR extension theorem in Section 14.2 is never satisfied by a codimension two submanifold in $\mathbb{C}^3$ (because the holomorphic tangent space of such a manifold is one-dimensional and so the image of the Levi form is always a one-dimensional ray in the two-dimensional normal space). In the above example, the origin is a point of type $(2,3)$ and so Baouendi and Rothschild's result states that CR functions near the origin in M holomorphically extend to a wedge in $\mathbb{C}^3$.

Recently, Tumanov [T] has significantly strengthened the above result. Not only has he shown that the above semirigid condition is unnecessary but he has also shown that finite type is unnecessary in certain situations. A CR structure $(M, \mathbb{L})$ is *minimal* at a point $p \in M$ if there does *not* exist a proper submanifold N of M which contains p such that $\mathbb{L}|_N \subset T^{\mathbb{C}}(N)$. If a point $p \in M$ is a point of finite type (i.e., all the Hörmander numbers are finite) then M is minimal at p. This is because the existence of a proper submanifold N with $\mathbb{L}|_N \subset T^{\mathbb{C}}(N)$ means that all the Lie brackets of all lengths of vector fields from $\mathbb{L} \oplus \overline{\mathbb{L}}$ at p must be contained in $T_p^{\mathbb{C}}(N)$ (since $T^{\mathbb{C}}(N)$ is involutive). So any vector in $T_p^{\mathbb{C}}(M)$ that is not in $T_p^{\mathbb{C}}(N)$ cannot be realized as a Lie bracket of any order at p of vector fields in $\mathbb{L} \oplus \overline{\mathbb{L}}$ and so p is not a point of finite type. If M is real analytic, then minimality is equivalent to finite type (see [BR3]). Tumanov has shown that if a generic smooth CR submanifold M of $\mathbb{C}^n$ is minimal at p, then CR functions on M near p holomorphically extend to some wedge in $\mathbb{C}^n$. In addition, Baouendi and Rothschild [BR3] have shown that the converse holds, that is, if CR functions on M near $p \in M$ holomorphically extend to a wedge in $\mathbb{C}^n$, then M must be minimal at p. We should mention that these necessary and sufficient conditions for the case of a real hypersurface had previously been discovered by Trepreau [Tr].

The reader should not get the impression that the above-mentioned results of Tumanov and Baouendi and Rothschild are the last word on CR extension. There are many interesting harmonic analysis problems remaining in the field of CR extension. For example, if a CR function is locally integrable, then what can be said about pointwise limits (on M) of its holomorphic extension? There is extensive literature on the approach regions allowed for pointwise limits of holomorphic functions for a domain bounded by a real hypersurface (see [St] or [BDN]). It would be interesting to obtain analogous results for wedges in $\mathbb{C}^n$ where the edge of the wedge is a higher codimension submanifold of the type we have been discussing (i.e., finite type or minimal). Some results along these lines for points where all the Hörmander numbers are 2 will be forthcoming in [BN].

In the case where each Hörmander number of a point $p \in M$ is 2, a precise description of the wedge W_Γ is given in the CR extension theorem in Section 14.2. Less can be said of the geometry and size of the wedge if the point p has higher type. However, a few observations can be noted. First, the wedge always points in the direction(s) in the normal space which can be generated by (J of) the Lie brackets of $H^{\mathbb{C}}(M)$ of lowest order. In the example above, $M = \{(z_1, z_2, w) \in \mathbb{C}^3;\ \text{Im} z_1 = |w|^2,\ \text{Im} z_2 = |w|^2 \text{Re} w\}$, the wedge is centered around the positive $\text{Im} z_1$ axis since this is the direction that is in the image of the Levi form at the origin. The second observation is that the size of the wedge is governed by the radius of the set $\omega \subset M$ on which the CR functions are defined and by the Hörmander numbers $(m_1, \ldots, m_d)$ which make up the type of the point p. If ω contains a Euclidean ball of radius $\delta > 0$ in M, then the dimensions of the normal cross section of the wedge at p are proportional to $\delta^{m_1}, \ldots, \delta^{m_d}$. The following figure illustrates the case where $(m_1, m_2) = (2, 3)$.

This result can be seen by the following scaling argument. Suppose the point $p \in M$ is the origin and the defining functions for M are put in the Bloom–Graham normal form (see Section 12.1)

$$
\begin{aligned}
y_1 &= p_{m_1}(w) + e_1(x, w) \\
y_2 &= p_{m_2}(x_1, w) + e_2(x, w) \\
&\;\; \vdots \\
y_d &= p_{m_d}(x_1, \ldots, x_{d-1}, w) + e_d(x, w)
\end{aligned}
$$

where p_{m_j} is a polynomial of weight m_j and where e_j is a smooth function of weight greater than m_j. Consider the biholomorphism $H_\delta : \mathbb{C}^n \mapsto \mathbb{C}^n$ given by $H_\delta(z, w) = (\delta^{m_1} z_1, \ldots, \delta^{m_d} z_d, \delta w)$. Let ω_δ be an open set in M that contains

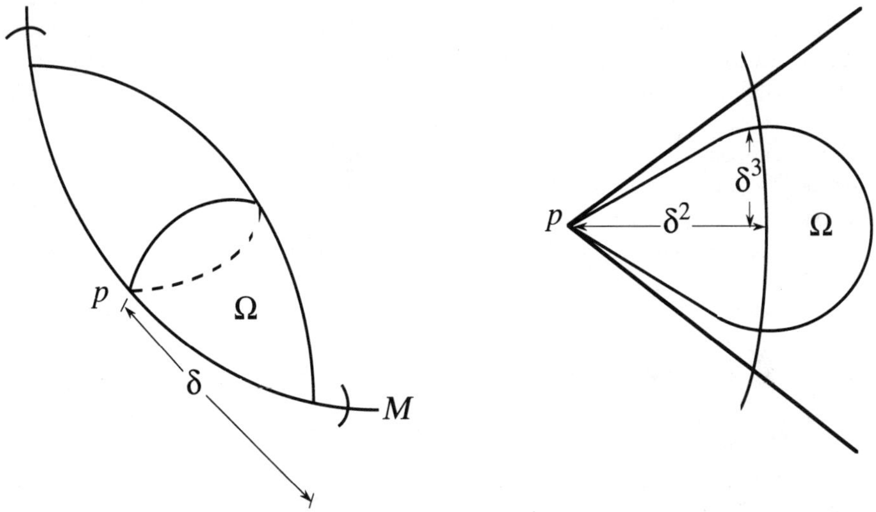

FIGURE 17.1

a Euclidean ball of radius δ in M about the origin. Let $\tilde{\omega}_\delta$ and M_δ be the image of ω_δ and M under the map H_δ^{-1}, respectively. Note that $\tilde{\omega}_\delta$ contains a ball of radius 1 in M_δ about the origin. The terms of lowest weight in the defining equations for M_δ are the same as the terms of lowest weight for the defining equations for M. Moreover, the terms of higher weight converge to zero as $\delta \mapsto 0$. Therefore, the above-mentioned result of Tumanov shows that CR functions on $\tilde{\omega}_\delta \subset M_\delta$ holomorphically extend to some wedge in $\mathbb{C}^n$ that is independent of δ. It follows that all CR functions on $\omega_\delta = H_\delta\{\tilde{\omega}_\delta\} \subset M$ holomorphically extend to a wedge with the dimensions described above. This argument works just as well if we only assume ω_δ contains a nonisotropic ball of radius δ of the type defined by Nagel, Stein, and Wainger (see [NSW]).

Up to proportionality constants, the above size estimate on the wedge is sharp. That is, we cannot expect to holomorphically extend CR functions to a larger wedge. For example, consider $M = \{(z_1, z_2, w) \in \mathbb{C}^3; y_1 = |w|^2,\ y_2 = |w|^2 \mathrm{Re} w\}$. For $\delta > 0$, let ω_δ be the intersection of M with the Euclidean ball of radius δ in $\mathbb{C}^3$ centered at the origin. Also let Γ_δ be the cone in $\mathbb{R}^2$ given by $\{0 \leq y_1,\ |y_2| \leq \delta y_1\}$. The convex hull of the set ω_δ is contained in the wedge $\omega_\delta + \{\Gamma_\delta \cap \{0 \leq y_1 \leq \delta^2\}\}$. This wedge has the dimensions described above for a point (in this case, the origin) of type $(2, 3)$. Since CR functions on a set cannot (in general) be holomorphically extended past its convex hull, the set $\omega_\delta + \{\Gamma_\delta \cap \{0 \leq y_1 \leq \delta^2\}\}$ contains the largest set to which all CR functions on ω_δ holomorphically extend.

Finally we mention that even if a point $p \in M$ does not satisfy the hypothesis of Tumanov's result mentioned above, CR functions on M may locally extend to CR functions on a CR manifold of higher dimension (rather than to holomorphic functions on an open set in $\mathbb{C}^n$). Suppose the type of the point p is $(m_1, \ldots, m_d)$ with $m_1 \leq \ldots \leq m_d$. If m_1 is finite (the other m_i are allowed to be infinite), then it can be shown that near p, M is the boundary of a manifold $\hat{M}$ with $\dim_{\mathbb{R}}(\hat{M}) = \dim_{\mathbb{R}}(M) + 1$ such that CR functions on M locally extend to CR functions on $\hat{M}$. If $m_1 = 2$, then the Levi form of M at p is not identically zero and this result is due to Hill and Taiani [HT1]. The case $m_1 \geq 2$ is handled in [BPi].

17.4 Analytic hypoellipticity

Suppose $L_1, \ldots, L_m$ are vector fields with real analytic, complex-valued coefficients defined on a real analytic manifold M. We say that the system $\{L_1, \ldots, L_m\}$ is *analytic hypoelliptic* near a point $p \in M$ if whenever $g_1, \ldots, g_m$ are real analytic functions near p and whenever u is a distributional solution to the equations

$$L_1 u = g_1, \ldots, L_m u = g_m \tag{1}$$

near p in M, then u is also real analytic near p in M. Let $\mathbb{L}$ be the subbundle of $T^{\mathbb{C}}(M)$ generated by $L_1, \ldots, L_m$. If $\mathbb{L}$ is involutive, then $(M, \mathbb{L})$ is a real analytic CR structure which (according to Theorem 1 in Chapter 11) we can imbed into $\mathbb{C}^n$. In this case, the real analytic hypoellipticity of the system $\overline{L}_1, \ldots, \overline{L}_m$ near a point $p \in M$ is equivalent to showing that the each CR distribution near p in M is real analytic near p. This equivalency is seen as follows. It is clear that if the system $\overline{L}_1, \cdots, \overline{L}_m$ is analytic hypoelliptic at p, then each CR distribution near p must be real analytic. For the converse, suppose (1) has a solution u. We can use the Cauchy–Kowalevsky theorem to locally solve the system of equations (1) with a real analytic solution u_1. The function $u - u_1$ is a CR function and u is real analytic if and only if $u - u_1$ is real analytic. The converse now follows.

If M is such that each CR distribution near a point $p \in M$ holomorphically extends to a neighborhood in $\mathbb{C}^n$ which contains p, then each CR distribution near p must be real analytic and thus by the discussion above, the system $\overline{\mathbb{L}} = H^{0,1}(M)$ is real analytic hypoelliptic at p. The converse also holds and this is shown in Section 5 of [BT1].

There are a number of situations where CR functions near a point p in M locally holomorphically extend to a neighborhood in $\mathbb{C}^n$ which contains a given point p. By the above discussion, this means that the resulting system of tangential Cauchy–Riemann equations is analytic hypoelliptic at p. The first situation is the one described in Theorems 2 and 3 in Section 14.2. Here, the convex hull of the image of the Levi form at the given point p is the entire normal space of M at p. This situation also occurs if the Levi form at p vanishes identically, but the commutators of length at most 3 at p generated from $H^{\mathbb{C}}(M)$ span $T_p(M)$ (over $\mathbb{R}$). This result is contained in [B]. For hypersurfaces, a result of Baouendi and Treves [BT3] states that if p is a point of odd type, then CR functions on M near p locally holomorphically extend to both sides of M.

Necessary and sufficient conditions real analytic hypoellipticity have been found by Baouendi and Treves in the tube like case. A submanifold $M = \{y = h(x, w)\}$ of $\mathbb{C}^n$ is said to be *tube like* if the graphing function h depends only on $u = \text{Re} w$. As usual, we are using (z, w) as coordinates for $\mathbb{C}^n$ where $z = x + iy \in \mathbb{C}^d$ and $w = u + iv \in \mathbb{C}^{n-d}$ ($d = \text{codim}_{\mathbb{R}}(M)$). In this case, Baouendi and Treves [BT2] have shown that the system of tangential Cauchy–Riemann equations for M is analytic hypoelliptic at the origin if and only if for each vector ξ in the normal space of M at 0 ($N_0(M) \approx \mathbb{R}^d$), the origin is not a local minimum of the map $u \mapsto h(u) \cdot \xi$, $u \in \mathbb{R}^{n-d}$.

There is also a considerable amount of literature on the related problem of deciding when a smooth CR map between two real analytic CR manifolds must be real analytic. The simplest case to consider is the case of two real analytic hypersurfaces M and M' in $\mathbb{C}^n$ and a smooth (say C^∞) CR diffeomorphism $\Phi : M \mapsto M'$. The theorem in this context is the following: if the Levi form of M at p is nondegenerate (i.e., the matrix representing the Levi form at p has maximal rank) then any smooth CR diffeomorphism from a neighborhood of p in M to M' must be real analytic near p. Of course, if the Levi form of M at p has eigenvalues of opposite sign, then every CR function near p holomorphically extends to both sides of M and therefore must be real analytic as discussed above. However, if M is strictly pseudoconvex at p, then CR functions only extend to one side of M and so some new ideas are needed to prove this theorem. One approach, taken in [L3] or [P], uses a reflection principle. Another approach, taken in [BJT], uses CR extension results. Roughly, the latter approach is the following. Since the Levi form of M at p is not identically zero the component functions of a CR map Φ (which are CR functions) must holomorphically extend to one side of M. Therefore, the conjugates of the component functions of Φ must holomorphically extend to the other side of M. Since the Levi form of M at p has maximal rank and since Φ is a CR diffeomorphism, the Levi form of M' at $\Phi(p)$ must also have maximal rank. This fact together with the implicit function theorem and the fact that the defining functions for M' are real analytic can be used to show that the component functions of Φ are holomorphic functions of the conjugates of the component functions of Φ and their derivatives. Since the conjugates of the component functions of Φ holomorphically extend to the other side of M, one obtains that the component functions of Φ holomorphically extend to the other side of M as well, and hence the component functions of Φ holomorphically extend to both sides of M. Therefore, Φ is real analytic on M near p, as desired.

This theorem and its proof generalize to higher codimension as shown in

[BJT]. Assume the convex hull of the image of the Levi form of M at $p \in M$ has nonempty interior as in the statement of our CR extension theorem. So the components of a CR map Φ holomorphically extend to a wedge of the form $W_\Gamma = \omega + i\Gamma$ as in the conclusion of our CR extension theorem. The conjugates of the component functions of Φ must therefore holomorphically extend to a wedge of the form $W_{-\Gamma} = \omega - i\Gamma$. A nondegeneracy condition on the Levi form given in Section 3 of [BJT] and the implicit function theorem can be used to show that the component functions of Φ are holomorphic functions of the conjugates of the component functions of Φ and their derivatives. Thus, Φ holomorphically extends to the union of the wedges $W_\Gamma \cup W_{-\Gamma}$. By the edge of the wedge theorem (see [Ro]), Φ holomorphically extends to a neighborhood of p in $\mathbb{C}^n$ and therefore Φ is real analytic near p, as desired.

Baouendi, Jacobowitz, and Treves have generalized these theorems to handle higher order situations (i.e., cases where the Levi forms vanish at a point). Results along these lines can be found in [BJT] and [BR1] (see also the references given in these papers).

We also define the notion of C^∞ hypoellipticity by replacing real analytic with C^∞ in the above definition of real analytic hypoellipticity. C^∞ hypoellipticity is a more difficult concept to analyze than real analytic hypoellipticity. For one thing, the Cauchy–Kowalevsky theorem does not hold for the C^∞ category (see Chapter 23, where Hans Lewy's example is discussed). Also, Maire [M] has found an example of a system of first-order vector fields with real analytic coefficients which is not C^∞ hypoelliptic. This example is tube like and satisfies the hypothesis of the above-mentioned result for tube like CR structures (see [BT2]) and so this system of vector fields is real analytic hypoelliptic. As far as positive results concerning C^∞ hypoellipticity, we mention a result of Shaw [Sh1]. Her result handles the case of a set of vector fields $\mathbb{L} = L_1, \ldots, L_m$ with smooth coefficients defined on an open subset of $\mathbb{R}^N$. Unlike the results mentioned above, she does not need to assume that $\mathbb{L}$ is involutive. She shows that if p is a point in $\mathbb{R}^N$ such that $(T_p^{\mathbb{C}}(M)/\mathbb{L} \oplus \overline{\mathbb{L}})$ is the convex hull of the image of the Levi form of $\mathbb{L}$ at p, the system $\mathbb{L}$ is C^∞ hypoelliptic at p.

Part IV

Solvability of the Tangential Cauchy–Riemann Complex

In Part II, we defined a complex to be a collection of vector spaces $\{\mathcal{A}_q, q \geq 0\}$ together with maps $d_q\colon \mathcal{A}_q \to \mathcal{A}_{q+1}$ with the property that $d_{q+1} \circ d_q = 0$. We have discussed three classes of complexes: the first is where $\mathcal{A}_q = \mathcal{E}^q(M)$ — the space of smooth q-forms on a manifold M — with $d_q = d_M$ — the exterior derivative on M; the second is where M is a complex manifold and $\mathcal{A}_q = \mathcal{E}^{p,q}(M)$ (p fixed) and $d_q = \bar{\partial}$ — the Cauchy–Riemann operator on M; the third is where M is a CR manifold and $\mathcal{A}_q = \mathcal{E}_M^{p,q}(M)$ and $d_q = \bar{\partial}_M$ — the tangential Cauchy–Riemann operator on M. For any complex $\{d_q\colon \mathcal{A}_q \to \mathcal{A}_{q+1};\ q \geq 0\}$, a natural question to ask is whether or not it is solvable. That is, if $f \in \mathcal{A}_q$ with $d_q f = 0$, does there exist $u \in \mathcal{A}_{q-1}$ with $d_{q-1} u = f$. For our three classes of complexes, this solvability question can be posed locally or globally. In the global problem, say for the tangential Cauchy–Riemann operator, we assume that $f \in \mathcal{E}_M^{p,q}(M)$ with $\bar{\partial}_M f = 0$ on all of M and then we ask whether or not there exists $u \in \mathcal{E}_M^{p,q-1}(M)$ with $\bar{\partial}_M u = f$ on all of M. In the local question, we ask whether or not there is a local neighborhood basis $\{\omega_\lambda;\ \lambda > 0\}$ about any given point $p \in M$ such that for each $f \in \mathcal{E}_M^{p,q}(\omega_\lambda)$ with $\bar{\partial}_M f = 0$ on ω_λ there is an element $u \in \mathcal{E}_M^{p,q-1}(\omega_\lambda)$ with $\bar{\partial}_M u = f$ on ω_λ. In all three classes of complexes, the answer to either the global or the local question does not automatically yield the answer to the other. For example, we cannot use the local solvability of $\bar{\partial}_M$ together with a partition of unity argument to obtain the global solvability of $\bar{\partial}_M$ because the product of a $\bar{\partial}_M$-closed form with a smooth cutoff function is typically no longer $\bar{\partial}_M$-closed. For the same reason, we cannot deduce local solvability from global solvability.

For the exterior derivative and the Cauchy–Riemann operators, the answer to the solvability question is well understood. Both complexes can always be solved locally. The global solvability of the exterior derivative depends on global topological conditions on the manifold (the cohomology groups). From the theory of several complex variables, the global solvability of the Cauchy–Riemann operator depends on the holomorphic convexity of the complex manifold.

Much less is known about the global and local solvability of the tangential Cauchy–Riemann operator on a CR manifold M. However, the theory is pretty well understood if M is a strictly pseudoconvex hypersurface in $\mathbb{C}^n$, and this is the subject of much of Part IV of this book. We show that the tangential Cauchy–Riemann complex is both locally and globally solvable except at top degree. For $1 \leq q < n-1$, the equation $\bar{\partial}_M u = f$ for $f \in \mathcal{E}^{p,q}(M)$ is overdetermined (more equations than unknown functions), and the condition $\bar{\partial}_M f = 0$ is the correct compatibility condition required for both local and global solvability. If $q = n-1$ (top degree), the equation $\bar{\partial}_M u = f$ is no longer overdetermined. In fact, if $u \in \mathcal{E}^{0,n-2}(M)$ and $f \in \mathcal{E}^{0,n-1}(M)$ then the equation $\bar{\partial}_M u = f$ consists of only one first-order partial differential equation. Based upon one's experience with d and $\bar{\partial}$, one would expect that the tangential Cauchy–Riemann complex should at least be locally solvable at top degree. In fact, if M and f are real analytic then local solvability follows from the Cauchy–Kowalevsky theorem. The surprise (provided by Hans Lewy's nonsolvability example [L2]) is that local solvability does not necessarily hold if f is only smooth. Adding the condition $\bar{\partial}_M f = 0$ does not help matters since this condition always holds if f has top degree. Instead, there are other criterion for both local and global solvability at top degree, which are discussed toward the end of Part IV.

We employ the integral kernel approach of Henkin [He1], [He2] for the solution of the tangential Cauchy–Riemann complex. We shall not discuss the $\mathcal{L}^2$ technique of Hörmander, Kohn et al. because there are ample references for this approach (see Folland and Kohn's book [FK]). Aside from the aesthetic appeal of exhibiting an explicit integral kernel formula for the solution to the tangential Cauchy–Riemann equations, the kernel approach makes estimating the solution rather easy (at least in the global case).

Part IV is organized as follows. In Chapter 18, we introduce the calculus of kernels and we define the concept of a fundamental solution for d, $\bar{\partial}$, and $\bar{\partial}_M$. Various fundamental solutions for d and $\bar{\partial}$ are discussed in Chapter 19. As an application we prove Bochner's global CR extension theorem [Boc] for the boundary of a smooth bounded domain in $\mathbb{C}^n$. In Chapter 20, we introduce Henkin's kernels, which along with the fundamental solution for $\bar{\partial}$ form the building blocks for the integral kernel solution to the tangential Cauchy–Riemann operator. Instead of using Henkin's notation, we use a more streamlined notation due to Harvey and Polking [HP]. Two global fundamental solutions to $\bar{\partial}_M$ are introduced in Chapter 21 along with a criterion for global solvability at top degree. In Chapter 22, one of these global fundamental solutions is modified to yield a solution to the local problem. Hans Lewy's local nonsolvability example is given in Chapter [illegible], along with a more general criterion of Henkin's [He2] for local solvability.

18

Kernel Calculus

Our presentation in this chapter closely follows Section 1 in [HP].

18.1 Definitions

To motivate our discussion of kernels, let K be a smooth form on $\mathbb{R}^{N'} \times \mathbb{R}^N$ of degree q $(0 \le q \le N + N')$. We can view K as an operator from $\mathcal{D}^p(\mathbb{R}^{N'})$ to $\mathcal{E}^{q+p-N'}(\mathbb{R}^N)$ by defining

$$(K\phi)(x) = \int\limits_{y \in \mathbb{R}^{N'}} K(y, x) \wedge \phi(y), \qquad x \in \mathbb{R}^N.$$

Here, it is understood that the only contributing term to this integral is the component of $K(x, y) \wedge \phi(y)$ containing $dy = dy_1 \wedge \ldots \wedge dy_{N'}$. The dx's appearing in this component are moved to the right (outside the integral) and then the y and dy are integrated. Since the degree of $K(y, x) \wedge \phi(y)$ is $p + q$, $(K\phi)(x)$ is a differential form in x of degree $q + p - N'$.

Recall (see Chapter 6) that as a current on $\mathbb{R}^N$, the form $K(\phi)$ acts on elements of $\mathcal{D}^s(\mathbb{R}^N)$ (with $s = N + N' - (q + p)$) by integration. So if $\psi \in \mathcal{D}^s(\mathbb{R}^N)$, then

$$\begin{aligned} \langle K\phi, \psi \rangle_{\mathbb{R}^N} &= \int\limits_{x \in \mathbb{R}^N} \left(\int\limits_{y \in \mathbb{R}^{N'}} K(y, x) \wedge \phi(y) \right) \wedge \psi(x) \\ &= \langle K, \phi \otimes \psi \rangle_{\mathbb{R}^{N'} \times \mathbb{R}^N} . \end{aligned}$$

The term $\langle K, \phi \otimes \psi \rangle_{\mathbb{R}^{N'} \times \mathbb{R}^N}$ is well defined for any current $K \in \mathcal{D}'^q(\mathbb{R}^{N'} \times \mathbb{R}^N)$ and any $\phi \in \mathcal{D}^p(\mathbb{R}^{N'})$, $\psi \in \mathcal{D}^s(\mathbb{R}^N)$. We can also replace $\mathbb{R}^N$ and $\mathbb{R}^{N'}$ with oriented manifolds X and Y of dimensions N and N', respectively. In this case, the manifold $Y \times X$ has an orientation induced on it by Y and X.

That is, if $\mathcal{Y} = (y_1, \ldots, y_{N'})\colon V \to \mathbb{R}^{N'}$ and $\chi = (x_1, \ldots, x_N)\colon U \to \mathbb{R}^N$ are orientation-preserving coordinate charts for X and Y, respectively, then $dy_1 \wedge \ldots \wedge dy_{N'} \wedge dx_1 \wedge \ldots \wedge dx_N$ determines an orientation for $Y \times X$ in the sense described in Section 2.5.

The above discussion motivates the following definition.

DEFINITION 1 *A current K in $\mathcal{D}'^q(Y \times X)$ is called a kernel of degree q on $Y \times X$. This kernel can be regarded as an operator $K\colon \mathcal{D}^p(Y) \to \mathcal{D}'^{q+p-N'}(X)$ by*

$$\langle K(\phi), \psi \rangle_X = \langle K, \phi \otimes \psi \rangle_{Y \times X}$$

for $\phi \in \mathcal{D}^p(Y)$ and $\psi \in \mathcal{D}^s(X)$ with $s = N + N' - (q + p)$.

Note that if $\psi_n \to \psi$ in $\mathcal{D}^*(X)$ then $\phi \otimes \psi_n \to \phi \otimes \psi$ in $\mathcal{D}^*(Y \times X)$; therefore, $\langle K(\phi), \psi_n \rangle_X \to \langle K(\phi), \psi \rangle_X$. This shows that $K(\phi)$ is a well-defined current on X. If $q = N' + r$, then we say that K is a *kernel of type* r. In this case, K is an operator from $\mathcal{D}^p(Y)$ to $\mathcal{D}'^{p+r}(X)$. For our applications r will usually be $0, -1$, or -2.

Example 1

As already mentioned, any smooth form K on $Y \times X$ defines a kernel. Many of the theorems about kernels are motivated by considering this class of kernels. Another closely related class of kernels is the space of forms on $Y \times X$ with locally integrable coefficients. □

Example 2

Let $X = Y$ be an oriented smooth manifold of dimension N. Let $\Delta = \{(x, x);\ x \in X\}$ be the diagonal of $X \times X$. The current $[\Delta]$ given by integration over the diagonal is a kernel of type 0 since the degree of $[\Delta]$ is N. As an operator, $[\Delta]\colon \mathcal{D}^p(X) \to \mathcal{D}'^p(X)$ represents the identity map. For if $\phi \in \mathcal{D}^p(X)$, then

$$\begin{aligned} \langle [\Delta](\phi), \psi \rangle_X &= \langle [\Delta], \phi \otimes \psi \rangle_{X \times X} \\ &= \int_\Delta \phi \otimes \psi. \end{aligned}$$

A parameterization for Δ is given by $x \mapsto (x, x)$, $x \in X$. We have

$$\begin{aligned} \langle [\Delta](\phi), \psi \rangle_X &= \int_{x \in X} \phi(x) \wedge \psi(x) \\ &= \langle \phi, \psi \rangle_X. \end{aligned}$$

So, $[\Delta](\phi) = \phi$, as claimed. This is analogous to the fact that the operation of convolution with the delta function is the identity operator (see Chapter 5). This analogy is made even clearer by writing the current $[\Delta]$ as a form with distribution coefficients

$$[\Delta] = \delta_0(x - y)d(x_1 - y_1) \wedge \ldots \wedge d(x_N - y_N). \qquad \square$$

Example 3

Let X and Y be oriented manifolds and suppose $f\colon X \to Y$ is a smooth map. Let $Gr\{f\}$ be the graph of f in $Y \times X$, i.e.,

$$Gr\{f\} = \{(y, x);\ y = f(x)\}.$$

The map $F\colon X \to Gr\{f\}$ given by $F(x) = (f(x), x)$ is a global parameterization for $Gr\{f\}$. Let us orient $Gr\{f\}$ by pushing the orientation on X to $Gr\{f\}$ via F. That is, the collection $\{\phi_1, \ldots, \phi_N\} \subset T^*\{Gr\{f\}\}$ is said to be positively oriented if and only if $\{F^*\phi_1, \ldots, F^*\phi_N\}$ is positively oriented on X. With this orientation on $Gr\{f\}$, the map F is orientation preserving.

The current (or kernel) $[Gr\{f\}]$ on $Y \times X$ has dimension N and therefore degree N' (where $N = \dim X$ and $N' = \dim Y$). As an operator from $\mathcal{D}^p(Y)$ to $\mathcal{D}'^p(X)$, we claim $[Gr\{f\}]$ represents the pull back map f^*. To see this, let $\phi \in \mathcal{D}^p(Y)$ and $\psi \in \mathcal{D}^{N-p}(X)$, then

$$\begin{aligned}\langle [Gr\{f\}](\phi), \psi\rangle_X &= \langle [Gr\{f\}], \phi \otimes \psi\rangle_{Y\times X} \\ &= \int_{Gr\{f\}} \phi \otimes \psi.\end{aligned}$$

Since $F\colon X \to Gr\{f\}$ is orientation preserving, we have

$$\begin{aligned}\langle [Gr\{f\}](\phi), \psi\rangle_X &= \int_X F^*(\phi \otimes \psi) \\ &= \int_X (f^*\phi) \wedge \psi \\ &= \langle f^*\phi, \psi\rangle_X.\end{aligned}$$

Therefore, $[Gr\{f\}](\phi) = f^*\phi$, as claimed. Note that if $X = Y$ and $f(x) = x$, $x \in X$, then this example reduces to Example 2 above. $\square$

Now let us discuss the *adjoint* of a kernel K denoted by K'. If K is an operator from $\mathcal{D}^*(Y)$ to $\mathcal{D}'^*(X)$, then K' is an operator from $\mathcal{D}^*(X)$ to $\mathcal{D}'^*(Y)$ and it is defined by

$$\langle K(\phi), \psi\rangle_X = \langle \phi, K'(\psi)\rangle_Y \qquad \phi \in \mathcal{D}^*(Y),\ \psi \in \mathcal{D}^*(X).$$

If K is of type r then K' is of type $N' - N + r$.

We wish to find a convenient way of computing K' from K. To motivate the calculation, we first suppose that K is a smooth form on $Y \times X$. If $\phi \in \mathcal{D}^*(Y)$ and $\psi \in \mathcal{D}^*(X)$, then

$$\begin{aligned}\langle K(\phi), \psi\rangle_X &= \langle K, \phi \otimes \psi\rangle_{Y\times X} \\ &= \int\limits_{x\in X}\int\limits_{y\in Y} K(y,x) \wedge \phi(y) \wedge \psi(x) \\ &= \pm \int\limits_{y\in Y} \phi(y) \wedge \left(\int\limits_{x\in X} K(y,x)\wedge\psi(x)\right) \\ &= \pm \left\langle \phi(y), \int\limits_{x\in X} K(y,x)\wedge\psi(x)\right\rangle_{y\in Y} .\end{aligned}$$

The sign in front depends on the degrees of the forms involved and it will be resolved later. From the above calculation, we have

$$K'(\psi)(y) = \pm \int\limits_{x\in X} K(y,x)\wedge\psi(x).$$

Note that the roles of x and y are reversed (recall that y is the variable of integration in $K(\phi)(x)$). If $X = Y$, then switching x with y shows that $K'(y,x) = \pm K(x,y)$. Switching x and y means that dx_j is also switched with dy_j.

To generalize this calculation for more general kernels, we introduce the switch map s: $X\times Y \to Y \times X$, $s(x,y) = (y,x)$. Formally, the pull back s^* is the identity map on forms, i.e., $s^*(\phi\otimes\psi)(x,y) = \phi(y)\wedge\psi(x)$ for $\phi \in \mathcal{D}^*(Y)$ and $\psi \in \mathcal{D}^*(X)$. However in local coordinates (x, y) for $X \times Y$, we have

$$\mathrm{Det}(Ds) = (-1)^{NN'}.$$

Hence, s^* changes the orientation by the factor $(-1)^{NN'}$. For $\Phi \in \mathcal{D}^{N+N'}(Y \times X)$, we obtain

$$\int\limits_{Y\times X} \Phi = (-1)^{NN'} \int\limits_{X\times Y} s^*\Phi. \tag{1}$$

This formula extends to currents on $Y \times X$. Since s is a diffeomorphism, s^*K is well defined for $K \in \mathcal{D}'^*(Y \times X)$ (see Definition 2 in Section 6.2). In addition, $s^*\Phi = (-1)^{NN'} s_*^{-1}\Phi$ for $\Phi \in \mathcal{D}^*(Y \times X)$ in view of the remark after

Lemma 2 in Section 6.2. We obtain

$$\begin{aligned}\langle K, \Phi\rangle_{Y\times X} &= \langle s^{-1^*} \circ s^* K, \Phi\rangle_{Y\times X} \\ &= \langle s^* K, s_*^{-1}\Phi\rangle_{X\times Y} \\ &= (-1)^{NN'}\langle s^* K, s^*\Phi\rangle_{X\times Y}. \qquad (2)\end{aligned}$$

This equation is the analogue of (1) for the pairing between currents and smooth forms.

Now we compute the kernel of K' and keep careful track of the minus signs. Suppose $K \in \mathcal{D}'^{N'+r}(Y \times X)$ is a kernel of type r. Let $\phi \in \mathcal{D}^p(Y)$ and $\psi \in \mathcal{D}^{N-p-r}(X)$. We have

$$\begin{aligned}\langle K(\phi), \psi\rangle_X &= \langle K, \phi\otimes\psi\rangle_{Y\times X} \\ &= (-1)^{NN'}\langle s^* K, s^*(\phi\otimes\psi)\rangle_{X\times Y} \quad \text{(by (2))}.\end{aligned}$$

Now, $s^*(\phi\otimes\psi)(x,y) = \phi(y)\wedge\psi(x)$. In addition, the degree of ϕ and ψ is p and $N-p-r$, respectively. Commuting $\phi(y)$ and $\psi(x)$ yields

$$\langle K(\phi), \psi\rangle_X = (-1)^{NN'+p(N-p-r)}\langle (s^*K)(\psi), \phi\rangle_Y.$$

The degree of $(s^*K)(\psi)$ is $N'-p$ and the degree of ϕ is p. Commuting $(s^*K)(\psi)$ and ϕ yields

$$\langle K(\phi), \psi\rangle_X = (-1)^{NN'+p(N+N'-r)}\langle \phi, (s^*K)(\psi)\rangle_Y.$$

So we have

$$K'(\psi) = (-1)^{NN'+p(N+N'-r)}(s^*K)(\psi).$$

To simplify the notation, we introduce the map c: $\mathcal{D}'^q(Y) \to \mathcal{D}'^q(Y)$

$$cT = (-1)^{N'-q}T \quad \text{for} \quad T \in \mathcal{D}'^q(Y).$$

Since the degree of $(s^*K)(\psi)$ is $N'-p$, we have $c((s^*K)(\psi)) = (-1)^p(s^*K)(\psi)$. We obtain

$$K'(\psi) = (-1)^{NN'}c^{N+N'-r}s^*K.$$

Define the *transpose* of K, $K^t \in \mathcal{D}'^{N'+r}(X \times Y)$, by

$$K^t = (-1)^{NN'}s^*K.$$

We have proved the following.

LEMMA 1

Suppose X and Y are smooth, oriented manifolds of dimensions N and N', respectively. Suppose K is a current of type r (i.e., $K \in \mathcal{D}'^{N'+r}(Y \times X)$). Then

$$K' = c^{N+N'-r}\{K^t\}.$$

Since K^t is easy to compute, this lemma gives a convenient formula for the adjoint, K'.

It may happen that a kernel sends smooth, compactly supported forms to smooth forms (rather than to currents with nonsmooth coefficients). We single out this special class of kernels.

DEFINITION 2 *Suppose $K \in \mathcal{D}'^{N'+r}(Y \times X)$.*

(a) *If K is a continuous map from $\mathcal{D}^p(Y)$ to $\mathcal{E}^{p+r}(X)$, then K is called a regular kernel.*

(b) *If K extends to a continuous operator from $\mathcal{E}'^p(Y)$ to $\mathcal{D}'^{p+r}(X)$, then K is called an extendable kernel.*

(c) *If both K and K' are regular kernels, then K is called biregular.*

LEMMA 2
If a kernel $K \in \mathcal{D}'^{N'+r}(Y \times X)$ is regular, then K' is extendable.

PROOF If $T \in \mathcal{E}'^*(X)$, then define

$$\langle \phi, K'(T)\rangle_Y = \langle K(\phi), T\rangle_X, \quad \text{for} \quad \phi \in \mathcal{D}^*(Y).$$

The right side is well defined because K is regular (so $K(\phi) \in \mathcal{E}^*(X)$). If $\phi_n \to \phi$ in $\mathcal{D}^*(Y)$, then by hypothesis $K(\phi_n) \to K(\phi)$ in $\mathcal{E}^*(X)$; therefore $\langle K(\phi_n), T\rangle_X \to \langle K(\phi), T\rangle_X$. This shows that $K'(T)$ is a well-defined $\mathcal{D}'$-current on Y. In a similar manner, the reader can show that if $T_n \to T$ in $\mathcal{E}'^*(X)$ then $K'(T_n) \to K'(T)$ in $\mathcal{D}'^*(Y)$. Thus, K' is extendable, as desired. ∎

The lemma implies that if K is biregular, then both K and K' are extendable.

An important class of biregular kernels is the class of kernels of convolution type, which we now describe. Let τ: $\mathbb{R}^N \times \mathbb{R}^N \to \mathbb{R}^N$ be defined by

$$\tau(y, x) = x - y.$$

The linear map τ has maximal rank and therefore the pull back τ^*: $\mathcal{D}'^*(\mathbb{R}^N) \to \mathcal{D}'^*(\mathbb{R}^N \times \mathbb{R}^N)$ is well defined (see Definition 2 and Lemma 5 in Section 6.2).

DEFINITION 3 *A kernel $K \in \mathcal{D}'^*(\mathbb{R}^N \times \mathbb{R}^N)$ is said to be of convolution type if there exists a current $k \in \mathcal{D}'^*(\mathbb{R}^N)$ with $K = \tau^* k$.*

So, for example, if $k = u dy^I$ with $u \in \mathcal{L}^1_{\text{loc}}(\mathbb{R}^N)$, then the kernel $K(y,x) = (\tau^* k)(y, x) = u(x-y) d(x-y)^I$ is a kernel of convolution type.

LEMMA 3
A kernel of convolution type is biregular.

PROOF Suppose $k \in \mathcal{D}'^{q}(\mathbb{R}^N)$, $\phi \in \mathcal{D}^p(\mathbb{R}^N)$, and $\psi \in \mathcal{D}^{2N-(q+p)}(\mathbb{R}^N)$. From the definitions, we have

$$\begin{aligned}\langle \tau^* k(\phi), \psi \rangle_{\mathbb{R}^N} &= \langle \tau^* k, \phi \otimes \psi \rangle_{\mathbb{R}^N \times \mathbb{R}^N} \\ &= \langle k, \tau_*(\phi \otimes \psi) \rangle_{\mathbb{R}^N} . \end{aligned} \tag{3}$$

Suppose $k = u dx^I$ with $u \in \mathcal{D}'(\mathbb{R}^N)$, and suppose $\phi = g dy^J, \psi = h dy^{J'}$ with $g, \psi \in \mathcal{D}(\mathbb{R}^N)$. From Lemma 1 in Section 6.2, we obtain

$$\tau_*(\phi \otimes \psi)(x) = \int\limits_{y \in \mathbb{R}^N} g(y) h(x+y) dy^J \wedge d(x+y)^{J'}$$

where the only nontrivial contribution to this integral comes from the terms involving $dy = dy_1 \wedge \ldots dy_N$. Depending on the multiindices J and J', the coefficient of the form $\tau_*(\phi \otimes \psi)$ is either 0 or the following function of x (up to a + or − sign)

$$\begin{aligned}\int\limits_{y \in \mathbb{R}^N} g(y) h(x+y) dy &= \int\limits_{y \in \mathbb{R}^N} g(y-x) h(y) dy \\ &= (\check{g} * h)(x)\end{aligned}$$

where $\check{g}(t) = g(-t)$. Together with (3), this means that $\langle \tau^* k(\phi), \psi \rangle_{\mathbb{R}^N}$ is either 0 or it is the term

$$(u, \check{g} * h)_{\mathbb{R}^N} = (u * g, h)_{\mathbb{R}^N}$$

up to a + or − sign. So the coefficient function of $(\tau^* k)(\phi)$ is either 0 or $u * g$, up to a + or − sign which depends only on the indices I, J, and J'. If $u \in \mathcal{D}'(\mathbb{R}^N)$, then the operator $g \mapsto u * g$ for $g \in \mathcal{D}(\mathbb{R}^N)$ is a continuous linear map from $\mathcal{D}(\mathbb{R}^N)$ to $\mathcal{E}(\mathbb{R}^N)$ (see Lemma 1 in Section 5.2). Therefore, $K = \tau^* k$ is regular.

If $K = \tau^* k$, then we leave it to the reader to show that $K^t = (-1)^N \tau^* \check{k}$ where $\check{k} = v^* k$ and v: $\mathbb{R}^N \to \mathbb{R}^N$ is defined by $v(x) = -x$. So K^t is also a kernel of convolution type. It follows that K^t is regular, and so K is biregular, as desired. ∎

The above definitions of kernels, type, regularity, etc., also apply to the case when X and Y are complex manifolds or CR manifolds. The only difference is that we must keep track of bidegrees rather than just degrees. For example, if X and Y are complex manifolds of complex dimension n and m respectively, a current of bidegree $(m+r, m+s)$ on $Y \times X$ is said to be a kernel of type (r, s) and K can be regarded as an operator from $\mathcal{D}^{p,q}(Y)$ to $\mathcal{D}'^{p+r,q+s}(X)$. Since a complex manifold has even real dimension, many of the minus signs in the above discussion disappear. For example, $K^t = s^* K$ in the complex manifold setting.

For CR manifolds, the bidegree counting is a little more complicated. Suppose M and M' are CR manifolds. Suppose $\dim_{\mathbb{R}} M = 2m + d$ and $\dim_{\mathbb{R}} M' = 2m' + d'$ where d and d' are the CR codimensions of M and M', respectively. A form of top degree on M' has bidegree $(m' + d', m')$ (see Section 8.1 or 8.2). Suppose K is a kernel of bidegree $(m' + d' + r, m' + s)$ on $M' \times M$. Then K is said to have type (r, s) and K can be regarded as an operator from $\mathcal{D}^{p,q}_{M'}$ to $\mathcal{D}'^{p+r,q+s}_{M}$. For most of our applications, $M = M'$ will be a hypersurface in $\mathbb{C}^n$, which means $m = m' = n - 1$ and $d = d' = 1$. In addition, r will usually be 0 and s will be 0 or -1.

18.2 A homotopy formula

In this section, we develop a homotopy formula for the exterior derivative, the Cauchy–Riemann operator and the tangential Cauchy–Riemann operator. Suppose X is a smooth manifold and let $K \in \mathcal{D}'^{N+r}(X \times X)$. Consider the current dK where d is the exterior derivative on $X \times X$. Viewed as a kernel operator, dK is a map from $\mathcal{D}^p(X)$ to $\mathcal{D}'^{p+r+1}(X)$. The homotopy formula given in the next theorem relates this operator with K and the exterior derivative on X. The $\bar{\partial}$ and $\bar{\partial}_M$ versions are also given for the case of a complex manifold and CR manifold, respectively.

THEOREM 1 HOMOTOPY FORMULA

(a) *Suppose X is a smooth manifold of real dimension N. If $K \in \mathcal{D}'^{N+r}(X \times X)$, then*

$$dK = (-1)^N (d \circ K + (-1)^{r+1} K \circ d)$$

as operators on $\mathcal{D}^*(X)$.

(b) *Suppose X is a complex manifold of dimension n. If $K \in \mathcal{D}'^{n+r,n+s}(X \times X)$, then*

$$\bar{\partial} K = \bar{\partial} \circ K + (-1)^{r+s+1} K \circ \bar{\partial}$$

as operators on $\mathcal{D}^*(X)$.

(c) *Suppose M is a CR manifold of real dimension $2m + d$ where d is the CR codimension of M. If $K \in \mathcal{D}'^{m+d+r,m+s}_{M \times M}$, then*

$$\bar{\partial}_{M \times M} K = (-1)^d (\bar{\partial}_M \circ K + (-1)^{r+s+1} K \circ \bar{\partial}_M)$$

as operators on $\mathcal{D}^*_M$.

On the left sides of the equations in parts (a) and (b), d and $\bar{\partial}$ refer to the d and $\bar{\partial}$ operators on the product space $X \times X$, whereas on the right sides, d and

$\bar{\partial}$ refer to the d and $\bar{\partial}$ operators on X. The notation in part (c) for the tangential Cauchy–Riemann operator is more clear in this respect.

PROOF We shall prove part (a). Part (b) follows from part (a) by taking the piece of bidegree $(n+r, n+s+1)$ of the equation in part (a). Likewise, part (c) follows by taking the piece of bidegree $(m+r+d, m+s+1)$ of the equation in part (a) and the intrinsic definition of the $\bar{\partial}_M$ operator (see Section 8.2).

If K is a form with C^1 coefficients, then (a) reads

$$\int\limits_{y\in X} (d_{yx}K)(y,x) \wedge \phi(y) = (-1)^N d_x \left\{ \int\limits_{y\in X} K(y,x) \wedge \phi(y) \right\} + (-1)^{N+r+1} \int\limits_{y\in X} K(y,x) \wedge d_y \phi(y)$$

where d_{yx} is the exterior derivative on $X \times X$. The above equation is established by first writing $d_{yx} = d_x + d_y$. The d_x can be taken outside the integral to give the first term on the right side of the above equation. The factor of $(-1)^N$ results from commuting d_X past N $dy's$ so that all the $dx's$ appear to the right of the y-integral as required by the convention set down at the beginning of this chapter. The second term on the right is obtained by an integration by parts with d_y and noting that $K(y,x)$ has degree $N+r$ (resulting in the factor of $(-1)^{N+r+1}$).

If K has distribution coefficients, then the argument is essentially the same. Suppose $\phi \in \mathcal{D}^p(X), \psi \in \mathcal{D}^{N-p-r-1}(X)$. From Definition 1 in the previous section, we obtain

$$\begin{aligned} \langle (d_{yx}K)(\phi), \psi \rangle_X &= \langle (d_{yx}K), \phi \otimes \psi \rangle_{X\times X} \\ &= (-1)^{N+r+1} \langle K, d\phi \otimes \psi + (-1)^p \phi \otimes d\psi \rangle_{X\times X}. \end{aligned} \tag{1}$$

The second equality follows from the definition of the exterior derivative of a current, and the product rule for d. We have

$$\langle K, d\phi \otimes \psi \rangle_{X\times X} = \langle K(d\phi), \psi \rangle_X$$

and

$$\begin{aligned} \langle K, \phi \otimes d\psi \rangle_{X\times X} &= \langle K(\phi), d\psi \rangle_X \\ &= (-1)^{p+r+1} \langle d(K(\phi)), \psi \rangle_X. \end{aligned}$$

Inserting these two equations into (1) yields

$$\langle (d_{yx}K)(\phi), \psi \rangle_X = (-1)^N \langle d(K(\phi)) + (-1)^{r+1} K(d\phi), \psi \rangle_X$$

and the homotopy formula follows. ∎

Of special significance is the current equation

$$dK = (-1)^N [\Delta] \tag{2}$$

where $[\Delta]$ is the current given by integration over the diagonal $\Delta = \{(x,x);\ x \in X\}$. This is because as an operator, $[\Delta]$ is the identity I, as shown in the previous section. Note that $[\Delta]$ has degree N and so K has degree $N-1$. If K satisfies (2), then in view of Theorem 1, we have

$$d \circ K + K \circ d = I$$

as operators on $\mathcal{D}^*(X)$. In particular, if $\phi \in \mathcal{D}^p(X)$ with $d\phi \equiv 0$ on X, then

$$d\{K(\phi)\} = \phi.$$

So the equation $du = \phi$ has a solution $u = K(\phi)$. Analogous statements hold for $\bar{\partial}$ on a complex manifold. This discussion motivates the following definition.

DEFINITION 1 *(a) Suppose X is a smooth manifold of real dimension N. A current $K \in \mathcal{D}'^{N-1}(X \times X)$ that satisfies*

$$dK = (-1)^N [\Delta] \quad \text{on} \quad X \times X$$

is called a fundamental solution for d.

(b) Suppose X is a complex manifold of complex dimension n. A current $K \in \mathcal{D}'^{n,n-1}(X \times X)$ that satisfies

$$\bar{\partial} K = [\Delta]$$

is called a fundamental solution for $\bar{\partial}$.

The reader should note the analogy with the concept of a fundamental solution T for a partial differential operator $P(D)$ with constant coefficients (i.e., a solution to $P(T) = \delta_0$). With the d or $\bar{\partial}$ operators, $[\Delta]$ takes the place of δ_0 and a solution to $du = \phi$ or $\bar{\partial} u = \phi$ is then obtained by setting $u = K(\phi)$. This is analogous to the solution to $P(D)u = \phi$ which is obtained by setting $u = T * \phi$ (see Section 5.4).

At this point, the reader may wonder why we have not defined the concept of a fundamental solution for $\bar{\partial}_M$ for a CR manifold M. The reason for this is that except for very special M (such as a foliation of complex manifolds) no solution to the equation $\bar{\partial}_{M \times M} K = [\Delta]$ exists. However, if M is a strictly pseudoconvex hypersurface in $\mathbb{C}^n$, we shall construct kernels, K, which solve this equation up to an error term which, as an operator, only acts nontrivially on forms of bottom and top degree. We shall then call such a kernel a fundamental solution for $\bar{\partial}_M$.

If K is an extendable current that is a fundamental solution to d, then the equation

$$d(K(T)) + K(dT) = T \tag{3}$$

holds for currents in $\mathcal{E}'^*(X)$. To see this, we approximate the given current $T \in \mathcal{E}'^*(X)$ by a sequence $\phi_n \in \mathcal{D}^*(X)$, $n = 1, 2, \ldots$ (see Theorem 1 in Section 22.2 and also Section 6.1). The exterior derivative is a continuous operator on $\mathcal{D}'^*(X)$ (with the weak topology). Therefore $d\phi_n \to dT$. Since K is extendable, $K(\phi_n) \to K(T)$ and $K(d\phi_n) \to K(dT)$. Equation (3) now follows. Similar remarks hold for $\bar{\partial}$. For future reference, we summarize this discussion in the following theorem.

THEOREM 2

(a) Suppose K is an extendable current that is a fundamental solution for d on a smooth oriented manifold X. If $T \in \mathcal{E}'^*(X)$, then

$$d(K(T)) + K(dT) = T.$$

(b) Suppose K is an extendable current that is a fundamental solution for $\bar{\partial}$ on a complex manifold X. If $T \in \mathcal{E}'^*(X)$, then

$$\bar{\partial}(K(T)) + K(\bar{\partial}T) = T.$$

Of particular interest is the current of degree 1 given by $T = [\partial D]f$ where $D \subset X$ is an open set with smooth boundary and $f \in \mathcal{D}^*(\partial D)$. If ∂D is given the usual boundary orientation (see Section 8.5), then $d\chi_D = -[\partial D]$ where χ_D is the characteristic function on D. If K is an extendable current, then $K([\partial D] \wedge f)$ is a well-defined current on X. An example of an extendable current is a differential form K with smooth coefficients on $X \times X$. In this case

$$\begin{aligned}
K([\partial D] \wedge f)(x) &= \langle K(y,x), [\partial D] \wedge (f)(y)\rangle_{y \in X} \\
&= (-1)^{\deg K} \langle [\partial D], K(y,x) \wedge f(y)\rangle_{y \in X} \\
&= (-1)^{\deg K} \int\limits_{y \in \partial D} K(y,x) \wedge f(y), \quad x \in X.
\end{aligned}$$

As in Part I, the notation $\langle\ ,\ \rangle_y$ indicates that y is the "variable of integration." This formula also holds for forms $K(y,x)$ whose coefficients are locally integrable in $y \in \partial D$.

Suppose $D \subset\subset X$ is a domain with smooth boundary and let $f \in \mathcal{E}^*(X)$. If K is a fundamental solution for d on X then by letting $T = \chi_D f$ in part (a) of Theorem 2, we obtain

$$d\{K(\chi_D f)\} + K(\chi_D df) - K([\partial D] \wedge f) = \chi_D f.$$

If $df = 0$ on D then the boundary integral term $K([\partial D] \wedge f)$ is the obstruction to solving the equation $du = f$ on D.

Similar remarks hold for $\bar{\partial}$ on a complex manifold X. Let $D \subset X$ be a domain in X with smooth boundary. The equation $d\chi_D = -[\partial D]$ splits into two equations: $\bar{\partial}\chi_D = -[\partial D]^{0,1}$ and $\partial\chi_D = -[\partial D]^{1,0}$. If $K(\zeta, z)$ is a form with locally integrable coefficients on ∂D, then for $f \in \mathcal{D}^*(X)$,

$$K([\partial D]^{0,1} \wedge f)(z) = (-1)^{\deg K}\langle [\partial D]^{0,1}, K(\zeta, z) \wedge f(\zeta)\rangle_{\zeta \in X}$$

$$= (-1)^{\deg K} \int\limits_{\zeta \in \partial D} [K(\zeta, z) \wedge f(\zeta)]^{n,n-1} \quad z \in X$$

where $[\ ]^{n,n-1}$ indicates the piece of bidegree $(n, n-1)$ in ζ.

If K is a fundamental solution for $\bar{\partial}$ on X, then by letting $T = \chi_D f$ in part (b) of Theorem 2, we obtain

$$\bar{\partial}\{K(\chi_D f)\} + K(\chi_D \bar{\partial} f) - K([\partial D]^{0,1} \wedge f) = \chi_D f. \tag{4}$$

If $\bar{\partial} f = 0$ on D, then the boundary integral $K([\partial D]^{0,1} \wedge f)$ is the obstruction to solving the equation $\bar{\partial} u = f$ on D. In the next chapter, fundamental solutions for d and $\bar{\partial}$ will be constructed. In Chapter 20, additional kernels will be presented which will allow us to solve the equation $\bar{\partial} u = K([\partial D]^{0,1} \wedge f)$ provided D is strictly convex. This will enable us to solve the equation $\bar{\partial} u = f$ on a strictly convex domain D.

The case where $f \in \mathcal{E}^{p,0}(X)$ deserves special note. If $K \in \mathcal{D}'^{n,n-1}(X \times X)$ then K is of type $(0, -1)$ and so $K(\chi_D f) = 0$. In simpler terms, $n - d\zeta's$ and $n - d\bar{\zeta}'s$ are required for integration on X, but K has at most $(n-1) - d\bar{\zeta}'s$ and f has none; thus $K(f) = 0$. If $\bar{\partial} f = 0$ on D, then (4) becomes

$$-K([\partial D]^{0,1} \wedge f) = \chi_D f.$$

In other words, we have

$$f(z) = \int\limits_{\zeta \in \partial D} [K(\zeta, z) \wedge f(\zeta)]^{n,n-1} \quad \text{for} \quad z \in D.$$

This means that any fundamental solution for $\bar{\partial}$ on X reproduces holomorphic functions on D. For future reference, we summarize this discussion in the following theorem.

THEOREM 3

Let X be a complex manifold and suppose $D \subset\subset X$ is an open set with smooth boundary in X. Suppose K is an extendable kernel that is a fundamental solution for $\bar{\partial}$ on X. Suppose f is holomorphic on D and continuous on $\overline{D}$. Then

$$\chi_D f = -K([\partial D]^{0,1} \wedge f).$$

19

Fundamental Solutions for the Exterior Derivative and Cauchy–Riemann Operators

In the previous chapter, we defined a fundamental solution for the exterior derivative operator d: $\mathcal{D}^q(\mathbb{R}^N) \to \mathcal{D}^{q+1}(\mathbb{R}^N)$ to be a kernel $K \in \mathcal{D}'^{N-1}(\mathbb{R}^N \times \mathbb{R}^N)$ which satisfies the equation

$$dK = (-1)^N[\Delta]$$

where d on the left is the exterior derivative on $\mathbb{R}^N \times \mathbb{R}^N$. For a fundamental solution for the Cauchy–Riemann operator $\bar{\partial}$: $\mathcal{D}^{p,q}(\mathbb{C}^n) \to \mathcal{D}^{p,q+1}(\mathbb{C}^n)$, the analogous equation is

$$\bar{\partial}K = [\Delta].$$

In this chapter, we construct the ray kernel and the spherical kernel which are fundamental solutions for d on $\mathbb{R}^N$. We then construct the Cauchy kernel on a slice and the Bochner–Martinelli kernel which are fundamental solutions for $\bar{\partial}$ on $\mathbb{C}^n$. All of these kernels are convolution kernels and therefore they are biregular (and hence extendable to currents).

Even though the fundamental solutions for the exterior derivative do not play a role in the construction of the solution to the tangential Cauchy–Riemann equations, we present them for three reasons. First, the fundamental solutions for d are easy to construct. Second, they are interesting in their own right. Third, it is interesting to see the parallels between the fundamental solutions for d and their counterparts for $\bar{\partial}$.

Our approach is to first solve the equations

$$dk = (-1)^N[0] \quad \text{on} \quad \mathbb{R}^N$$

and

$$\bar{\partial}k = [0] \quad \text{on} \quad \mathbb{C}^n.$$

Here, $[0]$ is the current of degree N on $\mathbb{R}^N$ (or degree $2n$ on $\mathbb{C}^n$) given by evaluation at the origin, i.e.,

$$\langle [0], f\rangle = f(0) \quad \text{for} \quad f \in \mathcal{D}^0(\mathbb{R}^N) \text{ (or } \mathcal{D}^0(\mathbb{C}^n)).$$

Our fundamental solution K for d (or $\bar{\partial}$) is then given by $K = \tau^* k$ where τ: $\mathbb{R}^N \times \mathbb{R}^N \to \mathbb{R}^N$ is defined by $\tau(y, x) = x - y$ (or $\tau(\zeta, z) = z - \zeta$ for $\zeta, z \in \mathbb{C}^n$). By definition, K is a kernel of convolution type. To see that K is a fundamental solution, we note that $\tau^*[0] = [\Delta]$ (see Lemma 6 in Section 6.2) and note that τ^* commutes with d (in the case of $\mathbb{R}^N$) and $\bar{\partial}$ (in the case of $\mathbb{C}^n$).

19.1 Fundamental solutions for d on $\mathbb{R}^N$

We present two fundamental solutions for the exterior derivative.

The kernel on a ray

The easiest solution to the equation

$$dk = (-1)^N [0] \quad \text{on} \quad \mathbb{R}^N$$

is obtained by letting k be the current given by integration over a ray emanating from the origin. Let σ be a unit vector in $\mathbb{R}^N$. Let

$$k_\sigma = \{t\sigma; t \geq 0\}.$$

Orient k_σ so that σ is positively oriented. The current $[k_\sigma]$ given by

$$\langle [k_\sigma], f\rangle = \int_{k_\sigma} f \qquad f \in \mathcal{D}^1(\mathbb{R}^N)$$

is a current of dimension 1 and therefore degree $N - 1$. By Theorem 1 in Section 6.2, $d[k_\sigma] = (-1)^N[\partial k_\sigma] = (-1)^N[0]$ (this is just Stokes' theorem). Pulling back this equation via τ^* where $\tau(y, x) = x - y$, we obtain

$$d\{\tau^*[k_\sigma]\} = (-1)^N[\Delta].$$

Therefore, $\tau^*[k_\sigma]$ is a fundamental solution to d. This kernel is called the *ray kernel.*

The spherical kernel

The kernel $[k_\sigma]$ defined above has support in the ray k_σ and therefore its distribution coefficients are not locally integrable functions. Sometimes it is desirable to have a fundamental solution with integrable coefficients. The spherical kernel is such a kernel.

To construct the spherical kernel, we start with a fundamental solution for Δ on $\mathbb{R}^N$. Let

$$T(x) = (2 - N)^{-1}(\omega_{N-1})^{-1}|x|^{2-N} \quad \text{for} \quad x \in \mathbb{R}^N$$

where $\omega_{N-1} = (2\pi^{N/2}/\Gamma(N/2))$ is the volume of the unit sphere in $\mathbb{R}^N$. T is a locally integrable function which satisfies

$$\Delta T = \delta_0$$

in the sense of distribution theory (see Theorem 3 in Section 5.4).

Now the Laplacian operator can be extended to differential forms by defining

$$\Delta\{f dx^I\} = (\Delta f) dx^I \quad \text{for} \quad f \in \mathcal{E}(\mathbb{R}^N).$$

The same formula (for $f \in \mathcal{D}'(\mathbb{R}^N)$) extends the Laplacian operator to currents.

LEMMA 1

For a current $F \in \mathcal{D}'^*(\mathbb{R}^N)$

$$-\Delta F = d(d^* F) + d^*(dF)$$

where d^* *is the* $\mathcal{L}^2$*-adjoint of d.*

PROOF This is a straightforward calculation using the formula for d

$$d\{f dx^I\} = \sum_{j=1}^{N} \frac{\partial f}{\partial x_j} dx_j \wedge dx^I$$

and the formula for d^* (see Lemma 2 in Section 1.5)

$$d^*\{f dx^I\} = \sum_{j=1}^{N} -\frac{\partial f}{\partial x_j} \frac{\partial}{\partial x_j} \lrcorner dx^I.$$

Details will be left to the reader. ∎

We apply Δ to the current $T(x)dx$ where $dx = dx_1 \wedge \ldots \wedge dx_N$. We obtain

$$\Delta\{T dx\} = \delta_0 dx = [0].$$

On the other hand, $d\{T dx\} = 0$ because $T dx$ is a current of top degree. Therefore, Lemma 1 yields

$$\begin{aligned} dd^*\{(-1)^{N-1} T dx\} &= (-1)^N \Delta\{T dx\} \\ &= (-1)^N [0]. \end{aligned}$$

Letting $k = (-1)^{N-1}d^*\{Tdx\}$, we obtain

$$dk = (-1)^N[0].$$

The resulting fundamental solution for d is

$$K = \tau^* k$$

where $\tau(y,x) = x - y$ for $x, y \in \mathbb{R}^N$. This kernel is called the *spherical kernel.*

Using the formula for d^*, we can write down an explicit formula for k and hence K. We have

$$\begin{aligned} k(x) &= (-1)^{N-1}d^*\{T(x)dx_1 \wedge \ldots \wedge dx_N\} \\ &= (-1)^N \sum_{J=1}^{N} \frac{\partial T(x)}{\partial x_j}\frac{\partial}{\partial x_j} \lrcorner (dx_1 \wedge \ldots \wedge dx_N) \\ &= \frac{(-1)^N}{\omega_{N-1}} \sum_{j=1}^{N} \frac{(-1)^{j-1}x_j}{|x|^N} dx_1 \wedge \ldots \wedge \widehat{dx_j} \wedge \ldots \wedge dx_N \end{aligned}$$

where $\widehat{dx_j}$ indicates that dx_j has been removed. The last equality uses the equation

$$\frac{\partial T(x)}{\partial x_j} = \frac{1}{\omega_{N-1}} \frac{x_j}{|x|^N}$$

which follows from a straightforward calculation for $x \neq 0$. Since $x_j|x|^{-N}$ is a locally integrable function of $x \in \mathbb{R}^N$, an easy safety disc argument or something equivalent shows that this equation holds in the sense of distribution theory across the origin as well.

The corresponding spherical kernel is given by

$$\begin{aligned} K(y,x) &= (\tau^* k)(y,x) \\ &= \frac{(-1)^N}{\omega_{N-1}} \sum_{j=1}^{N} \frac{(-1)^{j-1}(x_j - y_j)}{|x-y|^N} d(x_1 - y_1) \wedge \ldots \wedge d\widehat{(x_j - y_j)} \\ &\quad \wedge \ldots \wedge d(x_N - y_N). \end{aligned}$$

Note that K has locally integrable coefficients on $\mathbb{R}^N \times \mathbb{R}^N$, because

$$|K(y,x)| \leq C|x-y|^{1-N}$$

for some uniform constant C.

It is an interesting fact (which will not be used in the sequel) that the spherical kernel is equal to the average of the ray kernels k_σ with respect to surface measure $d\sigma$ on the unit sphere $\{|\sigma| = 1\}$ in $\mathbb{R}^N$.

19.2 Fundamental solutions for $\bar\partial$ on $\mathbb{C}^n$

We start with the construction of the Cauchy kernel on a slice which is analogous to the ray kernel on $\mathbb{R}^N$. Then we construct the Bochner–Martinelli kernel [Boc] which is analogous to the spherical kernel on $\mathbb{R}^N$.

Cauchy kernel on a slice

To construct the Cauchy kernel on a slice, we start with the fundamental solution for $\partial/\partial\bar z$ on $\mathbb{C}$ given by

$$T(z) = \frac{1}{\pi z}.$$

As shown in Section 22.4, we have

$$\frac{\partial T}{\partial \bar z} = \delta_0 \quad \text{on} \quad \mathbb{C}.$$

We give $\mathbb{C}^n$ the coordinates (z_1, z') where $z_1 \in \mathbb{C}$ and $z' = (z_2, \ldots, z_n) \in \mathbb{C}^{n-1}$. Define

$$[0'] = \delta_0(z')dv'$$

where

$$\begin{aligned} dv' &= dx_2 \wedge dy_2 \wedge \ldots \wedge dx_n \wedge dy_n \\ &= (2i)^{1-n} d\bar z_2 \wedge dz_2 \wedge \ldots \wedge d\bar z_n \wedge dz_n \end{aligned}$$

is the volume form on $\mathbb{C}^{n-1}$.

Let

$$c_1(z_1) = \frac{1}{2i} T(z_1) dz_1$$

and define

$$c = c_1 \otimes [0'].$$

Since T is locally integrable on $\mathbb{C}$, c is a well-defined current in $\mathcal{D}'^{n,n-1}(\mathbb{C}^n)$.

Now $\bar\partial[0'] = 0$ because $[0']$ is a current of top degree on $\mathbb{C}^{n-1}$. We obtain

$$\begin{aligned} \bar\partial c &= \frac{1}{2i}\left(\frac{\partial}{\partial \bar z_1} T\right) d\bar z_1 \wedge dz_1 \otimes [0'] \\ &= \frac{1}{2i}\delta_0(z_1) d\bar z_1 \wedge dz_1 \otimes [0'] \\ &= \delta_0(z_1) \otimes \delta_0(z') dx_1 \wedge dy_1 \wedge dv' \\ &= [0]. \end{aligned}$$

We define the *Cauchy kernel on a slice* by

$$C = \tau^* c$$

where τ: $\mathbb{C}^n \times \mathbb{C}^n \to \mathbb{C}^n$ is given by $\tau(\zeta, z) = z - \zeta$. Since τ is holomorphic, τ^* and $\bar{\partial}$ commute. Pulling back the equation $\bar{\partial}c = [0]$ yields

$$\begin{aligned} \bar{\partial}C &= \tau^*[0] \\ &= [\Delta] \quad \text{on} \quad \mathbb{C}^n \times \mathbb{C}^n. \end{aligned}$$

Therefore, the Cauchy kernel on a slice is a fundamental solution for $\bar{\partial}$. We summarize this discussion in the following theorem.

THEOREM 1
The Cauchy kernel on a slice defined by

$$C(\zeta, z) = \frac{1}{2\pi i} \frac{d(\zeta_1 - z_1)}{\zeta_1 - z_1} \otimes [\Delta']$$

is a biregular fundamental solution for $\bar{\partial}$. Here, Δ' is the diagonal in $\mathbb{C}^{n-1} \times \mathbb{C}^{n-1}$, i.e., $\Delta' = \{(z', z');\ z' \in \mathbb{C}^{n-1}\}$.

The Cauchy kernel on a slice is constructed as the tensor product of a fundamental solution for $\partial/\partial\bar{z}_1$ with the diagonal in the other variables. We could have used the fundamental solution to the Cauchy–Riemann operator along the complex line in $\mathbb{C}^n$ corresponding to an arbitrary point σ in projective space, $\mathbb{C}P^{n-1}$. In this way, we can construct a family of Cauchy kernels, C_σ, indexed by $\sigma \in \mathbb{C}P^{n-1}$, and each one is a fundamental solution for $\bar{\partial}$. The collection of kernels $\{C_\sigma;\ \sigma \in \mathbb{C}P^{n-1}\}$ is analogous to the collection of ray kernels discussed in the previous section.

Note that for the Cauchy kernel on the z_1-slice defined above,

$$\text{supp}\ C \subset \Delta' = \{\zeta' = z'\}.$$

This support property will be crucial in the proof of Bochner's global CR extension theorem presented in the next section. Since C is supported on such a "thin" set, clearly the coefficients of C are not locally integrable on $\mathbb{C}^n \times \mathbb{C}^n$. A locally integrable fundamental solution for $\bar{\partial}$ is given by the Bochner–Martinelli kernel, which is our next topic.

The Bochner–Martinelli kernel

As with the construction of the spherical kernel, we start with the fundamental solution for Δ on $\mathbb{C}^n$ given by

$$T(z) = \frac{-(n-2)!}{4\pi^n}|z|^{2-2n} \quad \text{for} \quad \mathbb{C}^n$$

(see Theorem 3 in Section 5.4).

The analogue of Lemma 1 in the previous section for the Laplacian on $\mathbb{C}^n$ is the following lemma.

LEMMA 1

For a current $F \in \mathcal{D}'^*(\mathbb{C}^n)$

$$-\Delta\{F\} = 4(\bar{\partial}\bar{\partial}^* F + \bar{\partial}^*\bar{\partial}F)$$

where $\bar{\partial}^*$ is the $\mathcal{L}^2$-adjoint of $\bar{\partial}$.

PROOF The proof of this lemma is a straightforward calculation using the formula for $\bar{\partial}$

$$\bar{\partial}\{f dz^I \wedge d\bar{z}^J\} = \sum_{j=1}^{n} \frac{\partial f}{\partial \bar{z}_j} d\bar{z}_j \wedge dz^I \wedge d\bar{z}^J$$

and the formula for $\bar{\partial}^*$

$$\bar{\partial}^*\{f dz^I \wedge d\bar{z}^J\} = \sum_{j=1}^{n} -\frac{\partial f}{\partial z_j}\frac{\partial}{\partial \bar{z}_j} \lrcorner (dz^I \wedge d\bar{z}^J)$$

(see Lemma 6 in Section 3.3). The factor of 4 comes from the fact that the Laplacian operator on functions is given by

$$\Delta = 4\sum_{j=1}^{n} \frac{\partial^2}{\partial z_j \partial \bar{z}_j}.$$

Details will be left to the reader. ∎

Let dv be the volume form for $\mathbb{C}^n$, i.e.,

$$dv = dx_1 \wedge dy_1 \wedge \ldots \wedge dx_n \wedge dy_n$$

where $z_j = x_j + iy_j$ for $1 \leq j \leq n$. By applying Δ to $T(z)dv$, we obtain

$$\begin{aligned}\Delta\{Tdv\} &= \delta_0 dv \\ &= [0].\end{aligned}$$

Since dv is a form of top degree, clearly $\bar{\partial}\{Tdv\} = 0$. Lemma 1 yields

$$\begin{aligned}-4\bar{\partial}\bar{\partial}^*\{Tdv\} &= \Delta\{T\}dv \\ &= \delta_0 dv \\ &= [0].\end{aligned}$$

We let

$$b = -4\bar{\partial}^*\{Tdv\}.$$

Using the previous equation, we obtain

$$\bar{\partial}b = [0].$$

Define the *Bochner–Martinelli kernel* by

$$B = \tau^* b$$

where $\tau(\zeta, z) = z - \zeta$ for $\zeta, z \in \mathbb{C}^n$. Since τ is holomorphic, τ^* commutes with $\bar{\partial}$. Therefore, we have

$$\begin{aligned}\bar{\partial}B &= \tau^*\bar{\partial}b \\ &= \tau^*[0] \\ &= [\Delta].\end{aligned}$$

So the Bochner–Martinelli kernel is a fundamental solution for $\bar{\partial}$.

Our next goal is to obtain a working formula for the Bochner–Martinelli kernel which will be useful in later chapters. To do this, we compute b. By using the formula for $\bar{\partial}^*$, we obtain

$$\begin{aligned}b &= -4\bar{\partial}^*\left\{\frac{-(n-2)!}{4\pi^n}|z|^{2-2n}dv\right\} \\ &= \frac{-(n-2)!}{\pi^n}\sum_{j=1}^{n}\frac{\partial}{\partial z_j}\{|z|^{2-2n}\}(2i)^{-n}\frac{\partial}{\partial \bar{z}_j}\lrcorner(d\bar{z}_1 \wedge dz_1 \wedge \ldots \wedge d\bar{z}_n \wedge dz_n) \\ &= \frac{(n-1)!}{(2\pi i)^n}\sum_{j=1}^{n}\frac{\bar{z}_j}{|z|^{2n}}d\bar{z}_1 \wedge dz_1 \wedge \ldots \wedge \widehat{d\bar{z}_j} \wedge dz_j \wedge \ldots d\bar{z}_n \wedge dz_n.\end{aligned}$$

The last equation follows from a straightforward calculation if $z \neq 0$. Since $\bar{z}_j|z|^{-2n}$ is locally integrable in $z \in \mathbb{C}^n$, this equation also holds in the sense of distribution theory across the origin.

Define

$$\bar{z} \cdot dz = \sum_{j=1}^{n} \bar{z}_j dz_j$$

$$d\bar{z} \cdot dz = \sum_{j=1}^{n} d\bar{z}_j \wedge dz_j.$$

We obtain

$$b = (2\pi i)^{-n} \frac{(\bar{z} \cdot dz) \wedge (d\bar{z} \cdot dz)^{n-1}}{|z|^{2n}}.$$

Pulling this back via τ^* yields our desired formula for the Bochner–Martinelli kernel. We summarize the above discussion in the following theorem.

THEOREM 2

The Bochner–Martinelli kernel

$$B(\zeta, z) = (2\pi i)^{-n} \frac{((\overline{\zeta - z}) \cdot d(\zeta - z)) \wedge (d(\overline{\zeta - z}) \cdot d(\zeta - z))^{n-1}}{|\zeta - z|^{2n}}$$

is a biregular fundamental solution for $\bar{\partial}$ on $\mathbb{C}^n$.

The Bochner–Martinelli kernel is analogous to the spherical kernel on $\mathbb{R}^N$. This analogy can be carried one step further. It can be shown that the Bochner–Martinelli kernel is the average of the Cauchy kernel slices C_σ with respect to the Fubini–Study volume form $d\lambda(\sigma)$ on $\mathbb{C}P^{n-1}$ (see [HP]).

If $n = 1$, then

$$B(\zeta, z) = (2\pi i)^{-1} \frac{d(\zeta - z)}{\zeta - z}.$$

So in one complex variable, the Bochner–Martinelli kernel reduces to the Cauchy kernel.

Suppose f is a smooth function defined on a simple closed contour γ in the complex plane. Define

$$\begin{aligned} F(z) &= B([\gamma]^{0,1} f)(z) \\ &= \frac{-1}{2\pi i} \int\limits_{\zeta \in \gamma} \frac{f(\zeta) d\zeta}{\zeta - z}. \end{aligned}$$

It is a classical fact that the boundary value jump of F across γ (from outside to inside) is precisely f. It is our goal to generalize this jump formula for the Bochner–Martinelli kernel. This jump formula will be used in the proof of Bochner's global CR extension theorem given in the next section. It also will be of crucial importance in the construction of the solution to the tangential Cauchy–Riemann equations in later chapters.

Let M be the smooth boundary of an open set D in $\mathbb{C}^n$. Let $f \in \mathcal{D}^*(M)$. Since B is biregular, $B([M]^{0,1} \wedge f)$ is a well-defined current on $\mathbb{C}^n$. In fact, since $|B(\zeta, z)| \leq C|\zeta - z|^{1-2n}$, $B([M]^{0,1} \wedge f)$ is a current on $\mathbb{C}^n$ with locally integrable coefficients. In addition, $B(\zeta, z)$ is smooth for $\zeta \neq z$. Therefore, $B([M]^{0,1} \wedge f)$ is a smooth form on $\mathbb{C}^n - M$. Our intention is to show that f is the boundary value jump of $B([M]^{0,1} \wedge f)$ across M. If M is noncompact, then we must require the compactness of supp f. If M is compact, then $\mathcal{D}^*(M) = \mathcal{E}^*(M)$ and so f can be any smooth form on M.

To state this boundary value result, we need some additional notation. For a form $f \in \mathcal{E}^{p,q}(D)$, we say that F has *continuous nontangential boundary values* on $M = \partial D$ from D if for each $z_0 \in M$

$$\lim_{\substack{z \to z_0 \\ z \in C_{z_0}}} F(z) \quad \text{exists}$$

where C_{z_0} is any nontangential cone in D with vertex at z_0. A cone C_{z_0} in D is *nontangential* if there exists a $\lambda > 0$ such that

$$|\zeta - z| \geq \lambda |\zeta - z_0| \quad \text{for} \quad \zeta \in M,\ z \in C_{z_0}.$$

We also remind the reader that for $f \in \Lambda^{p,q}T^*(\mathbb{C}^n)$, the tangential piece of $f|_M$ is denoted $f_{t_M} \in \Lambda^{p,q}T^*(M)$. Suppose $D = \{z \in \mathbb{C}^n;\ \rho(z) < 0\}$ where ρ: $\mathbb{C}^n \to \mathbb{R}$ is smooth, with $|d\rho| = 1$ on M. Let $\mathbb{N} = 4\sum(\partial\rho/\partial z_j)(\partial/\partial \bar{z}_j)$ be the dual vector to $\bar{\partial}\rho$. As shown in Lemma 2 in Section 8.1,

$$f_{t_M} = \mathbb{N} \lrcorner (\bar{\partial}\rho \wedge f).$$

Now $\mathbb{N}$, $\bar{\partial}\rho$, and f are defined on a neighborhood of M in $\mathbb{C}^n$. So f_{t_M} is also defined on a neighborhood of M in $\mathbb{C}^n$.

THEOREM 3

Suppose M is the smooth boundary of a domain D in $\mathbb{C}^n$. Assume M has the induced boundary orientation as the boundary of D. Let $D^- = D$ and $D^+ = \mathbb{C}^n - D$. Suppose f is a (p,q)-form $(0 \leq p, q \leq n)$ on $\mathbb{C}^n$ with C^1, compactly supported coefficients. Then

$$\{B([M]^{0,1} \wedge f)\}_{t_M}\big|_{D^+} \quad \textit{and} \quad \{B([M]^{0,1} \wedge f)\}_{t_M}\big|_{D^-}$$

have continuous nontangential boundary values on M from D^+ and D^-, respectively, denoted by B^+f and B^-f. Moreover,

$$B^+f - B^-f = f_{t_M} \quad \textit{on} \quad M.$$

More delicate boundary value results are mentioned in Chapter 24.

PROOF The idea is to reduce the proof of this theorem to the case where f is a function, where the proof is easy.

We need the following lemma (from [HP]).

LEMMA 2
There is a constant $C > 0$ such that for any multiindices I and J

$$B(\zeta, z) \wedge \bar{\partial}\rho(\zeta) \wedge d\bar{\zeta}^J \wedge d\zeta^I = B(\zeta, z) \wedge \bar{\partial}\rho(\zeta) \wedge d\bar{z}^J \wedge dz^I + A_1(\zeta, z) + \bar{\partial}\rho(z) \wedge A_2(\zeta, z)$$

where A_1 and A_2 are differential forms which are smooth for $\zeta \neq z$. Moreover, there is a constant $C > 0$ such that

$$|A_1(\zeta, z)| \leq C|\zeta - z|^{2-2n}$$

for $\zeta, z \in \mathbb{C}^n$ in some neighborhood of M.

Assuming Lemma 2 for the moment, we complete the proof of Theorem 3. Let

$$f(\zeta) = f_1(\zeta) d\bar{\zeta}^J \wedge d\zeta^I \quad |I| = p \quad |J| = q$$

be an arbitrary (p, q)-form on $\mathbb{C}^n$ whose coefficient function f_1 is C^1 with compact support. If $M = \{\rho = 0\}$ where $|d\rho| = 1$ on M then

$$[M]^{0,1} = \mu\bar{\partial}\rho$$

where μ denotes Hausdorff $(2n-1)$-dimensional measure on M (see the end of Section 6.1). From Lemma 2, we have

$$\begin{aligned} B([M]^{0,1} \wedge f)(z) &= \langle B(\zeta, z), \mu(\zeta) f_1(\zeta) \bar{\partial}\rho(\zeta) \wedge d\bar{\zeta}^J \wedge d\zeta^I \rangle_{\zeta \in \mathbb{C}^n} \\ &= B([M]^{0,1} f_1)(z) d\bar{z}^J \wedge dz^I \\ &\quad + \langle A_1(\zeta, z), \mu(\zeta) f_1(\zeta) \rangle_{\zeta \in \mathbb{C}^n} \\ &\quad + \bar{\partial}\rho(z) \wedge \langle A_2(\zeta, z), \mu(\zeta) f_1(\zeta) \rangle_{\zeta \in \mathbb{C}^n} . \end{aligned}$$

The tangential piece of the third term on the right vanishes due to the presence of $\bar{\partial}\rho$. We obtain

$$\{B([M]^{0,1} \wedge f)(z)\}_{t_M} = B([M]^{0,1} f_1)(z) \{d\bar{z}^J \wedge dz^I\}_{t_M} + \left\{ \int_{\zeta \in M} \tilde{A}_1(\zeta, z) f_1(\zeta) d\sigma(\zeta) \right\}_{t_M}$$

where $\tilde{A}_1(\zeta, z)$ is the coefficient of the piece of $A_1(\zeta, z)$ of bidegree (n, n) in ζ.

Now let $z_0 \in M$ and suppose C_{z_0} is a nontangential cone in either D^- or D^+. For some $\lambda > 0$ depending only on the aperture of C_{z_0}, we have

$$|\zeta - z| \geq \lambda |\zeta - z_0| \quad \text{for} \quad \zeta \in M, z \in C_{z_0} .$$

Combining this with the estimate on A_1 given in Lemma 2, we obtain

$$\sup_{z \in C_2} |\tilde{A}_1(\zeta, z)| \leq C|\zeta - z_0|^{2-2n} \quad \text{for} \quad \zeta \in M.$$

Since $|\zeta - z_0|^{2-2n}$ is integrable in $\zeta \in M$, the dominated convergence theorem implies that

$$\lim_{\substack{z \to z_0 \\ z \in C_{z_0}}} \int_{\zeta \in M} \tilde{A}_1(\zeta, z) f_1(\zeta) d\sigma(\zeta)$$

exists and that this limit is the same regardless of whether or not C_{z_0} is contained in D^+ or D^-. So the boundary value jump of $\int_{\zeta \in M} \tilde{A}_1(\zeta, z) f_1(\zeta) d\sigma(\zeta)$ across M at z_0 vanishes.

Therefore, it suffices to show that $B([M]^{0,1} f_1)$ has continuous nontangential boundary values from D^- and D^+ and that the boundary value jump across M from D^+ to D^- is f_1. In other words, we have reduced the proof of Theorem 3 to the case where $f = f_1$ is a C^1 function with compact support.

Since the Bochner–Martinelli kernel has only diagonal singularities, $B([M]^{0,1} f_1)$ is smooth (C^∞) on $\mathbb{C}^n - \{\text{supp } f_1\}$. In particular, the boundary value jump across $M - \{\text{supp } f_1\}$ is zero. Thus, it suffices to prove Theorem 3 at points in $\{\text{supp } f_1 \cap M\}$.

Fix $z_0 \in \{\text{supp } f_1\} \cap M$. Let $\phi \in \mathcal{D}(\mathbb{C}^n)$ with $\phi \equiv 1$ on a neighborhood of $\{\text{supp } f_1\}$. Define

$$g_1(\zeta) = \phi(\zeta) f_1(z_0)$$
$$h_1(\zeta) = (f_1(\zeta) - f_1(z_0))\phi(\zeta).$$

Note that $g_1 + h_1 = f_1$. Applying part (b) of Theorem 2 in Section 18.2 to the degree zero current $T = \chi_D g_1$, we obtain

$$\chi_D f_1(z_0)\phi = -B([M]^{0,1} g_1) + B(\chi_D \bar{\partial}\phi) f_1(z_0).$$

(Note that $B(T) = 0$ since T has degree zero). Since $\bar{\partial}\phi = 0$ near z_0, the second term on the right is smooth in a neighborhood of z_0 in $\mathbb{C}^n$ and so its boundary value jump across M is zero near z_0. From the above equation, $B([M]^{0,1} g_1)$ has a smooth extension to M from $D^- = D$ and $D^+ = \mathbb{C}^n - \overline{D}$ and the boundary value jump of $B([M]^{0,1} g_1)$ across M from D^+ to D^- is $f_1(z_0)\phi = f_1(z_0)$, near z_0, i.e.,

$$B^+(g_1) - B^-(g_1) = f_1(z_0) \tag{1}$$

on M near z_0.

Since f_1 is of class C^1, we have

$$|h_1(\zeta)| \leq C|\zeta - z_0|$$

for some uniform constant $C > 0$. As with $\tilde{A}_1$ above, this estimate yields

$$\sup_{z \in C_{z_0}} |B(\zeta, z) h_1(\zeta)| \leq C|\zeta - z_0|^{2-2n} \quad (\zeta \in M)$$

where C_{z_0} is any nontangential cone contained in either D^+ or D^-. By the dominated convergence theorem

$$\lim_{\substack{z \to z_0 \\ z \in C_{z_0}}} B([M]^{0,1} h_1)(z)$$

exists and this limit is the same regardless of whether C_{z_0} is contained in D^- or D^+. Therefore, its boundary value jump across M at z_0 vanishes, i.e.,

$$B^+(h_1)(z_0) - B^-(h_1)(z_0) = 0. \tag{2}$$

Since $f_1 = g_1 + h_1$, we see that $B([M]^{0,1} f_1)$ has continuous nontangential boundary values on M from D^- and D^+. Equations (1) and (2) imply

$$B^+(f_1)(z_0) - B^-(f_1)(z_0) = f_1(z_0)$$

as desired. This establishes Theorem 3 for functions f_1, and so the proof of Theorem 3 is complete. ∎

PROOF OF LEMMA 2 Let $\mathbb{N}$ be the dual vector to $\bar{\partial}\rho$. In particular

$$\mathbb{N} \lrcorner \bar{\partial}\rho = 1$$

near M in $\mathbb{C}^n$. Let

$$D(\zeta, z) = B(\zeta, z) \wedge \bar{\partial}\rho(\zeta) \wedge (d\bar{\zeta}^J \wedge d\zeta^I - d\bar{z}^J \wedge dz^I).$$

From the product rule for $\lrcorner$ (see Lemma 1 in Section 1.5), we have

$$D(\zeta, z) = A_1(\zeta, z) + \bar{\partial}\rho(z) \wedge A_2(\zeta, z)$$

where

$$A_1(\zeta, z) = \mathbb{N}_z \lrcorner (\bar{\partial}\rho(z) \wedge D(\zeta, z))$$
$$A_2(\zeta, z) = \mathbb{N}_z \lrcorner D(\zeta, z).$$

The term $\bar{\partial}\rho(z) \wedge A_2(\zeta, z)$ gives the third term on the right side of the equation stated in Lemma 2. So it suffices to show $A_1(\zeta, z)$ satisfies the estimate stated in the lemma. Actually, we shall show

$$|\bar{\partial}\rho(z) \wedge D(\zeta, z)| \leq C|\zeta - z|^{2-2n} \tag{3}$$

for a uniform constant C.

From the expression for $B(\zeta, z)$ given in Theorem 2, we may write

$$B(\zeta, z) = \alpha(\zeta, z) \wedge d(\zeta - z)$$

where α is a form of bidegree $(0, n-1)$ in ζ, z and where

$$d(\zeta - z) = d(\zeta_1 - z_1) \wedge \ldots \wedge d(\zeta_n - z_n).$$

Since $d(\zeta - z) \wedge (d\zeta_j - dz_j) = 0$, we have

$$B(\zeta, z) \wedge d\zeta^I = B(\zeta, z) \wedge dz^I.$$

From the definition of D, we therefore have

$$D(\zeta, z) = B(\zeta, z) \wedge \bar{\partial}\rho(\zeta) \wedge (d\bar{\zeta}^J - d\bar{z}^J) \wedge dz^I.$$

Since $\bar{\partial}\rho(\zeta) \wedge \bar{\partial}\rho(\zeta) = 0$, we obtain

$$\bar{\partial}\rho(z) \wedge D(\zeta, z) = (\bar{\partial}\rho(z) - \bar{\partial}\rho(\zeta)) \wedge B(\zeta, z) \wedge \bar{\partial}\rho(\zeta)(d\bar{\zeta}^J - d\bar{z}^J) \wedge dz^I. \quad (4)$$

We have

$$\bar{\partial}\rho(z) - \bar{\partial}\rho(\zeta) = \sum_{j=1}^{n} \left(\frac{\partial \rho}{\partial \bar{z}_j}(z) - \frac{\partial \rho}{\partial \bar{z}_j}(\zeta) \right) d\bar{z}_j + \sum_{j=1}^{n} \frac{\partial \rho}{\partial \bar{\zeta}_j}(\zeta) d(\bar{z}_j - \bar{\zeta}_j). \quad (5)$$

Since $|B(\zeta, z)| \leq C|\zeta - z|^{1-2n}$, we have

$$\left| \frac{\partial \rho}{\partial \bar{z}_j}(z) - \frac{\partial \rho}{\partial \bar{z}_j}(\zeta) \right| |B(\zeta, z)| \leq C|\zeta - z|^{2-2n} \quad (6)$$

for some uniform constant C. We may also write

$$d(\bar{z}_j - \bar{\zeta}_j) \wedge B(\zeta, z) = \tilde{\alpha}(\zeta, z) \wedge d(\overline{\zeta - z})$$

where $\tilde{\alpha}$ is a form of bidegree $(n, 0)$ in (ζ, z) and where

$$d(\overline{\zeta - z}) = d(\overline{\zeta_1 - z_1}) \wedge \ldots \wedge d(\overline{\zeta_n - z_n}).$$

Since $d(\overline{\zeta - z}) \wedge (d\bar{\zeta}^J - d\bar{z}^J) = 0$, we have

$$d(\bar{z}_j - \bar{\zeta}_j) \wedge B(\zeta, z) \wedge (d\bar{\zeta}^J - d\bar{z}^J) = 0.$$

This together with (4), (5), and (6) yields the estimate stated in (3). From the definition of A_1, the estimate in (3) implies

$$|A_1(\zeta, z)| \leq C|\zeta - z|^{2-2n}$$

for $\zeta, z \in \mathbb{C}^n$ near M where C is a uniform constant. The proof of Lemma 2 is now complete. ∎

19.3 Bochner's global CR extension theorem

The Cauchy kernel on a slice and the Bochner–Martinelli kernels can be used to give an easy proof of Bochner's global CR extension theorem, which roughly states that any CR function of class C^1 on the smooth boundary of a bounded domain D in $\mathbb{C}^n$ $(n \geq 2)$ extends to a holomorphic function on D.

We first recall that both the Bochner–Martinelli kernel (B) and the Cauchy kernel on a slice (C) are biregular kernels. Therefore, we can apply B and C to currents with compact support (Lemma 2 in Section 18.1). Ultimately, the current we have in mind is $T = [M]^{0,1} f$ where M is the smooth boundary of the domain D and f is our given CR function on M. This current is $\bar{\partial}$-closed since f is CR (Lemma 5 in Section 8.2). As the next lemma shows, the Bochner–Martinelli and Cauchy kernel on a slice both agree when applied to $\bar{\partial}$-closed compactly supported currents of bidegree $(0, 1)$. This together with the support property of the Cauchy kernel and the jump formula for the Bochner–Martinelli kernel will yield the proof of Bochner's theorem.

LEMMA 1
Suppose T is a $\bar{\partial}$-closed, compactly supported current of bidegree (0,1) in $\mathbb{C}^n$ $(n \geq 2)$. Then

(a) *$C(T) \equiv 0$ on the unbounded component of $\mathbb{C}^n$ – supp T.*

(b) *$C(T) \equiv B(T)$ as currents on $\mathbb{C}^n$.*

PROOF Give $\mathbb{C}^n$ the coordinates (z_1, z') with $z_1 \in \mathbb{C}$ and $z' \in \mathbb{C}^{n-1}$. From the formula

$$C(\zeta, z) = (2\pi i)^{-1} \frac{d(\zeta_1 - z_1)}{(\zeta_1 - z_1)} \otimes [\zeta' = z']$$

we see that supp $C \subset \{\zeta' = z'\}$. Therefore, if

$$\text{supp}\, T \subset \{(\zeta_1, \zeta');\ |\zeta'| < R\}$$

then

$$\text{supp}\, C(T) \subset \{(z_1, z');\ |z'| < R\}.$$

Intuitively, this is because if z' is fixed with $|z'| \geq R$ then the ζ-support of $C(\zeta, z)$ misses supp T. It is here that we have used the assumption $n \geq 2$.

Since $\bar{\partial} T = 0$, part (b) of Theorem 2 in Section 18.2 yields

$$\bar{\partial}\{C(T)\} = T.$$

So $C(T)$ is a holomorphic function on $\mathbb{C}^n$ – supp T. Since $C(T)$ vanishes for $\{|z'| \geq R\}$, part (a) follows from the identity theorem for holomorphic functions.

For part (b), first note that since both B and C are fundamental solutions for $\bar{\partial}$ and since $\bar{\partial}T \equiv 0$, we have

$$\bar{\partial}\{B(T) - C(T)\} = T - T = 0 \quad \text{on} \quad \mathbb{C}^n.$$

Therefore, $B(T) - C(T)$ is an entire function on $\mathbb{C}^n$.

We claim $B(T) - C(T)$ vanishes at ∞ and therefore it vanishes identically by Liouiville's theorem for entire functions. From part (a), we know $C(T)(z) = 0$ for $|z|$ large. So it suffices to show $B(T)$ vanishes at ∞.

Let Ω be a neighborhood of supp T. For fixed $z \notin \Omega$, the form $\zeta \mapsto B(\zeta, z)$ is smooth for $\zeta \in \Omega$. For $z \notin \Omega$, we have

$$\begin{aligned} B(T)(z) &= \langle B(\zeta, z), T(\zeta)\rangle_{\zeta\in\Omega} \\ &= (-1)^{\deg T}\langle T(\zeta), B(\zeta, z)\rangle_{\zeta\in\Omega} \end{aligned}$$

where, as before, the notation $\langle\ ,\ \rangle_{\zeta\in\Omega}$ indicates that ζ is the variable of "integration." Since T is a continuous linear functional on $\mathcal{E}^*(\Omega)$, there must exist a constant $C > 0$ and an integer $N > 0$ such that

$$|\langle T(\zeta), B(\zeta, z)\rangle_{\zeta\in\Omega}| \leq C \sup_{\substack{|\alpha|\leq N \\ \zeta\in \text{ supp } T}} |D^{\alpha}_{\zeta}\{B(\zeta, z)\}|$$

for $z \notin \Omega$. By examining the formula for $B(\zeta, z)$, we see that the right side converges to zero as $|z| \to \infty$. Thus, $B(T)(z) \to 0$ as $|z| \to \infty$, as desired. ∎

Part (b) of the lemma is useful since the Cauchy kernel and the Bochner–Martinelli kernel each has properties not a priori possessed by the other. The Cauchy kernel has a nice support property described in part (a) which is not apparent for the Bochner–Martinelli kernel. On the other hand, the Bochner–Martinelli kernel has nice regularity properties since $B(\zeta, z)$ has locally integrable coefficients. Part (b) states that if T is a $\bar{\partial}$-closed compactly supported (0,1)-current, then $C(T)$ and $B(T)$ both enjoy these properties since $B(T) = C(T)$.

We now state and prove Bochner's global CR extension theorem.

THEOREM 1

Suppose D is a bounded open set in $\mathbb{C}^n$ $(n \geq 2)$ with smooth boundary. Suppose f is a CR function on ∂D of class C^1. Then there is a holomorphic function F on D which has a continuous nontangential extension to ∂D from D such that $F|_{\partial D} = f$. Moreover, F is given by either of the following integral formulas:

$$\begin{aligned} F &= -B([\partial D]^{0,1} f) \\ &= -C([\partial D]^{0,1} f). \end{aligned}$$

Note there are no convexity assumptions on ∂D for this global theorem.

PROOF Let $M = \partial D$. If f is a CR on M, then from Lemma 5 in Section 8.1, $[M]^{0,1}f$ is a $\bar{\partial}$-closed, compactly supported current of bidegree (0,1) on $\mathbb{C}^n$. Let

$$F = -B([M]^{0,1}f).$$

By part (b) of Lemma 1, we have

$$F = -C([M]^{0,1}f).$$

From part (a) of Lemma 1, $F \equiv 0$ on $\mathbb{C}^n - \overline{D}$. From Theorem 3 in Section 19.2, F has a continuous nontangential extension to M from D. In addition, the boundary value jump of F across M (from D to $\mathbb{C}^n - D$) is equal to f. Since $F \equiv 0$ on $\mathbb{C}^n - \overline{D}$, f is the boundary values of F from D.

It remains to show that F is holomorphic on D. Since $\bar{\partial}\{[M]^{0,1}f\} \equiv 0$, Theorem 2 in Section 18.2 yields

$$\bar{\partial}F = -[M]^{0,1}f$$

which has support in M. Therefore, F is holomorphic on $\mathbb{C}^n - M$ (and in particular on D), as desired. ∎

Note that Bochner's theorem does not hold for domains in $\mathbb{C}^1$. This is because every function on a closed contour in $\mathbb{C}$ is a CR function (there are no tangential Cauchy–Riemann equations). The above proof breaks down in $n = 1$ because part (a) of Lemma 1 does not hold for $n = 1$.

For a simple closed contour γ in $\mathbb{C}$, the condition that replaces $\bar{\partial}_M f = 0$ for Bochner's theorem is the *moment condition*, which means

$$\int_{\zeta \in \gamma} f(\zeta)\zeta^n d\zeta = 0 \quad \text{for} \quad n = 0, 1, 2, \ldots.$$

Note by Cauchy's theorem, this condition is necessary for f to be the boundary values of a holomorphic function defined on the inside of γ. To see that this is a sufficient condition, let

$$\begin{aligned} F(z) &= -C([\gamma]^{0,1}f)(z) \\ &= \frac{1}{2\pi i} \int_{\zeta \in \gamma} \frac{f(\zeta)d\zeta}{\zeta - z}. \end{aligned}$$

Theorem 3 in Section 19.2 still holds (regardless of whether or not f satisfies the moment condition). In particular, the boundary value jump of F across γ (from the inside to the outside of γ) is equal to f. From a series expansion of $1/(\zeta - z)$ in powers of ζ together with the moment condition, we see that $F(z) = 0$ for z outside γ. It follows that the boundary values of F from the inside of γ agree with f.

20

The Kernels of Henkin

In Chapter 19, we constructed two fundamental solutions for $\bar{\partial}$ on $\mathbb{C}^n$. As mentioned in Chapter 18, if K is a fundamental solution for $\bar{\partial}$ and if $f \in \mathcal{D}^*(\mathbb{C}^n)$ with $\bar{\partial} f \equiv 0$, then the equation $\bar{\partial} u = f$ can be solved on $\mathbb{C}^n$ by setting $u = K(f)$. Now suppose D is a bounded domain in $\mathbb{C}^n$ and suppose $f \in \mathcal{E}^*(\overline{D})$ with $\bar{\partial} f = 0$ on D. In this case, we cannot directly apply K to f without first extending f and then multiplying by suitable cutoff function so that f has compact support. This process produces an extended f which is no longer $\bar{\partial}$-closed. Another way to cut off f is to multiply f by the characteristic function on D (denoted χ_D). If $\bar{\partial} f = 0$ on D, then

$$\bar{\partial}(\chi_D f) = -[\partial D]^{0,1} \wedge f.$$

As mentioned in Section 18.2 (see Theorem 2), we have

$$\bar{\partial}\{K(\chi_D f)\} - K([\partial D]^{0,1} \wedge f) = \chi_D f.$$

Therefore, the term $K([\partial D]^{0,1} \wedge f)$ is the obstruction to solving the equation $\bar{\partial} u = f$ on D. In this chapter, we define a general class of kernels due to Henkin which allows us to solve the equation $\bar{\partial} u_1 = K([\partial D]^{0,1} \wedge f)$ on a strictly convex domain D and so in this case, $u = K(\chi_D f) - u_1$ solves the equation $\bar{\partial} u = f$ on D. We should also mention that a slightly different kernel approach to the solution of the equation $\bar{\partial} u = f$ was discovered by Ramirez [Ra].

20.1 A general class of kernels

The kernels we are about to define are due to Henkin; however we shall employ the more streamlined notation of Harvey and Polking [HP].

Let V be an open subset of $\mathbb{C}^n \times \mathbb{C}^n$. For each $1 \leq j \leq N$, suppose $u^j\colon V \to \mathbb{C}^n$ is a smooth map. We write

$$u^j(\zeta, z) = (u_1^j(\zeta, z), \ldots, u_n^j(\zeta, z))$$

and we use the notation

$$u^j(\zeta, z) \cdot (\zeta - z) = \sum_{k=1}^{n} u_k^j(\zeta, z)(\zeta_k - z_k)$$

$$u^j(\zeta, z) \cdot d(\zeta - z) = \sum_{k=1}^{n} u_k^j(\zeta, z) d(\zeta_k - z_k)$$

$$\bar{\partial} u^j(\zeta, z) \cdot d(\zeta - z) = \sum_{k=1}^{n} \bar{\partial} u_k^j(\zeta, z) \wedge d(\zeta_k - z_k).$$

Here, $\bar{\partial}$ refers to the Cauchy–Riemann operator on $\mathbb{C}^n \times \mathbb{C}^n$ (i.e., in both ζ and z).

For $1 \leq j \leq N$, define the 1-form

$$\omega_j(\zeta, z) = \frac{u^j(\zeta, z) \cdot d(\zeta - z)}{u^j(\zeta, z) \cdot (\zeta - z)}.$$

The 1-form ω^j is smooth on the set $V - A_j$ where

$$A_j = \{(\zeta, z) \in V;\ u^j(\zeta, z) \cdot (\zeta - z) = 0\}.$$

Let $\mathbb{N}$ be the set of nonnegative integers. Let

$$\mathbb{N}^p = \{\alpha = (\alpha_1, \ldots, \alpha_p);\ \alpha_j \in \mathbb{N}\}.$$

For $\alpha \in \mathbb{N}^p$, set $|\alpha| = \alpha_1 + \cdots + \alpha_p$.

For an increasing multindex $I = \{i_1, \ldots, i_p\}$ with $1 \leq i_j \leq N$, let

$$\omega^I = \omega_{i_1} \wedge \ldots \wedge \omega_{i_p}.$$

For $\alpha \in \mathbb{N}^p$, let

$$(\bar{\partial}\omega^I)^\alpha = (\bar{\partial}\omega_{i_1})^{\alpha_1} \wedge \ldots \wedge (\bar{\partial}\omega_{i_p})^{\alpha_p}.$$

Here, $\bar{\partial}$ is the Cauchy–Riemann operator on $\mathbb{C}^n \times \mathbb{C}^n$. By definition, the form $(\bar{\partial}\omega_{i_j})^{\alpha_j}$ is the wedge product of $\bar{\partial}\omega_{i_j}$ with itself α_j times. Note that since $\bar{\partial}\omega_{i_j}$ a 2-form, $(\bar{\partial}\omega_{i_j})^{\alpha_j}$ is typically *not* zero. With this notation, we define the kernel

$$E_I = (2\pi i)^{-n} \omega^I \wedge \sum_{\substack{\alpha \in \mathbb{N}^p \\ |\alpha| = n-p}} (\bar{\partial}\omega^I)^\alpha.$$

We also define $E_I = 0$ if $I \in \mathbb{N}^0$. Note that $E_I = 0$ if $I \in \mathbb{N}^p$ with $p > n$. The form E_I has smooth coefficients on $V - \{\cup_{j=1}^{p} A_{i_j}\}$. Sometimes for emphasis, we shall write $E(u^{i_1}, \ldots, u^{i_p})$ for E_I.

The following lemma yields simpler formulas for the E_I.

LEMMA 1
For $k \geq 0$

$$\omega_j \wedge (\bar{\partial}\omega_j)^k = \left(\frac{u^j \cdot d(\zeta - z)}{u^j \cdot (\zeta - z)}\right) \wedge \left(\frac{\bar{\partial}u^j \cdot d(\zeta - z)}{u^j \cdot (\zeta - z)}\right)^k$$

for $(\zeta, z) \in V - A_j$.

PROOF From the formula

$$\omega_j = \frac{u^j \cdot d(\zeta - z)}{u^j \cdot (\zeta - z)}$$

we have

$$\bar{\partial}\omega_j = \frac{\bar{\partial}u^j \cdot d(\zeta - z)}{u^j \cdot (\zeta - z)} - \frac{(\bar{\partial}u^j \cdot (\zeta - z) \wedge (u^j \cdot d(\zeta - z))}{(u^j \cdot (\zeta - z))^2}.$$

The wedge product of ω_j with the second term on the right vanishes due to the repeated wedge product of the 1-form $u^j \cdot d(\zeta - z)$. The lemma now follows. ∎

Using this lemma, we single out some special cases of interest. With $p = 1$, we have

$$E_1 = E(u^1) = (2\pi i)^{-n} \left(\frac{u \cdot d(\zeta - z)}{u \cdot (\zeta - z)}\right) \wedge \left(\frac{\bar{\partial}u \cdot d(\zeta - z)}{u \cdot (\zeta - z)}\right)^{n-1}.$$

For example, from the expression for the Bochner–Martinelli kernel given in Theorem 2 of Section 19.2, we have

$$B = E(u)$$

where $u(\zeta, z) = \overline{\zeta - z}$.

With $p = 2$, we have

$$E(u^1, u^2) = (2\pi i)^{-n} \left(\frac{u^1 \cdot d(\zeta - z)}{u^1 \cdot (\zeta - z)}\right) \wedge \left(\frac{u^2 \cdot d(\zeta - z)}{u^2 \cdot (\zeta - z)}\right)$$
$$\wedge \sum_{\substack{j+k=n-2 \\ j,k \geq 0}} \left(\frac{\bar{\partial}u^1 \cdot d(\zeta - z)}{u^1 \cdot (\zeta - z)}\right)^j \wedge \left(\frac{\bar{\partial}u^2 \cdot d(\zeta - z)}{u^2 \cdot (\zeta - z)}\right)^k.$$

20.2 A formal identity

We now prove a formal identity which relates the above-defined kernels to the $\bar{\partial}$-operator.

Suppose $I = \{i_1, \ldots, i_p\}$ is an increasing multiindex. For $1 \le j \le p$, let

$$I_j = \{i_1, \ldots, \hat{i}_j, \ldots, i_p\}.$$

The notation $\hat{i}_j$ means that i_j has been omitted. So I_j is an increasing index of length $p - 1$.

THEOREM 1

Suppose $I = \{i_1, \ldots, i_p\}$ is an increasing index of length p.

(a) $\bar{\partial} E_I = \sum_{j=1}^{p} (-1)^j E_{I_j}$ *on* $V - \{\cup_{j=1}^{p} A_{i_j}\}$.

(b) *Suppose that M is a CR submanifold of $\mathbb{C}^n$, then* $\bar{\partial}_{M\times M} E_I = \sum_{j=1}^{p} (-1)^j \{E_{I_j}\}_{t_{M\times M}}$ *on* $\{M \times M\} \cap \{V - \cup_{j=1}^{p} A_{i_j}\}$.

The following special cases are important. If $p = 1$, then

$$\bar{\partial} E_1 = \bar{\partial} E(u^1) = 0 \quad \text{on} \quad V - A_1.$$

This generalizes a result we already know for the Bochner–Martinelli kernel (since $\bar{\partial} B = [\Delta]$, clearly $\bar{\partial} B \equiv 0$ on $\{(\zeta, z);\ \zeta \neq z\}$).

If $p = 2$, then

$$\bar{\partial} E_{12} = \bar{\partial} E(u^1, u^2) = E_1 - E_2 \quad \text{on} \quad V - \{A_1 \cup A_2\}.$$

If $p = 3$, then

$$\bar{\partial} E_{123} = \bar{\partial} E(u^1, u^2, u^3) = E_{13} - E_{12} - E_{23} \quad \text{on} \quad V - \{A_1 \cup A_2 \cup A_3\}.$$

The analogous identities hold for $\bar{\partial}_{M\times M}$.

PROOF From the definition of E_I and the product rule for $\bar{\partial}$, we have

$$\bar{\partial} E_I = \sum_{j=1}^{p} (-1)^{j-1} \omega_{i_1} \wedge \ldots \wedge \bar{\partial}\omega_{i_j} \wedge \ldots \wedge \omega_{i_p} \wedge \sum_{\substack{\alpha \in \mathbb{N}^p \\ |\alpha| = n-p}} (\bar{\partial}\omega^I)^\alpha$$

$$= \sum_{j=1}^{p} (-1)^{j-1} \left[\omega^{I_j} \wedge \sum_{\substack{\alpha = (\alpha_1, \ldots, \alpha_p) \\ |\alpha| = n-p+1 \\ \alpha_j \ge 1}} (\bar{\partial}\omega^I)^\alpha \right].$$

If $\alpha_j = 0$, then the expression within the brackets is exactly E_{I_j}. Therefore, we have

$$\bar{\partial} E_I = \sum_{j=1}^{p} (-1)^j E_{I_j} + \sum_{j=1}^{p} (-1)^{j-1} \omega^{I_j} \wedge \sum_{\substack{\alpha \in \mathbb{N}^p \\ |\alpha| = n-p+1}} (\bar{\partial} \omega^I)^\alpha .$$

The proof of part (a) will be complete provided we show the second term on the right vanishes. In fact, we will show

$$\sum_{j=1}^{p} (-1)^{j-1} \omega^{I_j} \wedge (\bar{\partial} \omega^I)^\alpha = 0 \tag{1}$$

for each multiindex $\alpha = (\alpha_1, \ldots, \alpha_p) \in \mathbb{N}^p$ with $|\alpha| = n - p + 1$. To establish (1), we use the vector field

$$\theta = (\zeta - z) \cdot \frac{\partial}{\partial \zeta} = \sum_{j=1}^{n} (\zeta_j - z_j) \frac{\partial}{\partial \zeta_j} .$$

From the formula for ω_j, we have

$$\theta \lrcorner \omega_j = 1 \qquad 1 \le j \le N.$$

From the product rule for $\lrcorner$, we obtain

$$\begin{aligned} \theta \lrcorner \bar{\partial} \omega_j &= \frac{-\bar{\partial} u^j \cdot (\zeta - z)}{u^j \cdot (\zeta - z)} + \frac{(\bar{\partial} u^j \cdot (\zeta - z))(u^j \cdot (\zeta - z))}{(u^j \cdot (\zeta - z))^2} \\ &= 0. \end{aligned}$$

We also note

$$\omega^I \wedge (\bar{\partial} \omega^I)^\alpha = 0$$

for $|\alpha| = n - p + 1$ because the form on the left involves a wedge product of degree $n+1$ generated by $\{d(\zeta_1 - z_1), \ldots, d(\zeta_n - z_n)\}$ and therefore one of the $d(\zeta_j - z_j)$ must be repeated. Using this together with $\theta \lrcorner \omega_j = 1$ and $\theta \lrcorner \bar{\partial} \omega_j = 0$, we obtain

$$\begin{aligned} 0 &= \theta \lrcorner (\omega^I \wedge (\bar{\partial} \omega^I)^\alpha) \\ &= \sum_{j=1}^{p} (-1)^{j-1} \omega_{i_1} \wedge \ldots \wedge (\theta \lrcorner \omega_{i_j}) \wedge \ldots \wedge \omega_{i_p} \wedge (\bar{\partial} \omega^I)^\alpha \\ &= \sum_{j=1}^{p} (-1)^{j-1} \omega^{I_j} \wedge (\bar{\partial} \omega^I)^\alpha . \end{aligned}$$

This proves (1) and thus completes the proof of part (a). Part (b) follows by taking the tangential piece of the equation in part (a) and by using the definition of the (extrinsic) $\bar{\partial}_{M \times M}$ complex (see Section 8.1). ∎

20.3 The solution to the Cauchy–Riemann equations on a convex domain

For a given convex domain D, we shall define an appropriate map u which when inserted into the kernel machinery of the last section will yield a solution to the $\overline{\partial}$-problem. The resulting kernels will also be used in subsequent chapters to solve the tangential Cauchy–Riemann equations on a strictly convex hypersurface. Although our focus will be on a strictly convex boundary, we will indicate in Chapter 24 how the kernels can be modified for strictly pseudoconvex and other geometries. Our eventual goal in Chapter 22 is the local solution for the tangential Cauchy–Riemann equations on a strictly pseudoconvex hypersurface. From Theorem 1 in Section 10.3, a strictly pseudoconvex hypersurface is locally biholomorphically equivalent to a strictly convex hypersurface. In addition, solvabilty of the tangential Cauchy–Riemann complex is invariant under a biholomorphic (or more generally, a CR diffeomorphic) change of variables (see Corollary 2 in Section 9.2). Therefore, the kernels that locally solve the tangential Cauchy–Riemann equations on a strictly convex hypersurface will also provide a local solution to the tangential Cauchy–Riemann equations on a strictly pseudoconvex hypersurface.

Let us suppose that D is a strictly convex domain in $\mathbb{C}^n$ with smooth boundary M. Let ρ be a smooth defining function for D, i.e.,

$$D = \{z \in \mathbb{C}^n; \rho(z) < 0\}.$$

Since D is strictly convex, the real hessian of ρ at a point on M is positive definite when restricted to the real tangent space of M. By replacing ρ with $\rho + C\rho^2$, for a suitable positive constant C, we may assume that the real hessian of ρ at a point on M is positive definite in all directions in $\mathbb{R}^{2n}$ — including the normal direction to M. This allows us to prove the following lemma.

LEMMA 1

Let $D = \{z \in \mathbb{C}^n; \rho(z) < 0\}$ be a strictly convex domain and suppose K is a compact set in $\mathbb{C}^n$.

(a) *There exists $\epsilon > 0$ such that if $\zeta, z \in K$ with $-\epsilon < \rho(\zeta) < \epsilon$, $z \neq \zeta$, and $(\partial\rho(\zeta)/\partial\zeta) \cdot (\zeta - z) = 0$, then $\rho(z) > \rho(\zeta)$.*

(b) *There exist constants $\epsilon > 0$ and $0 < C_1 < C_2 < \infty$ such that if ζ, $z \in K$ with $-\epsilon < \rho(\zeta) < \epsilon$ and $|\zeta - z| < \epsilon$, then*

$$\begin{aligned} C_1(|\zeta - z|^2) + \rho(\zeta) - \rho(z) &\leq 2Re\left\{\frac{\partial\rho}{\partial\zeta}(\zeta) \cdot (\zeta - z)\right\} \\ &\leq C_2(|\zeta - z|^2) + \rho(\zeta) - \rho(z). \end{aligned}$$

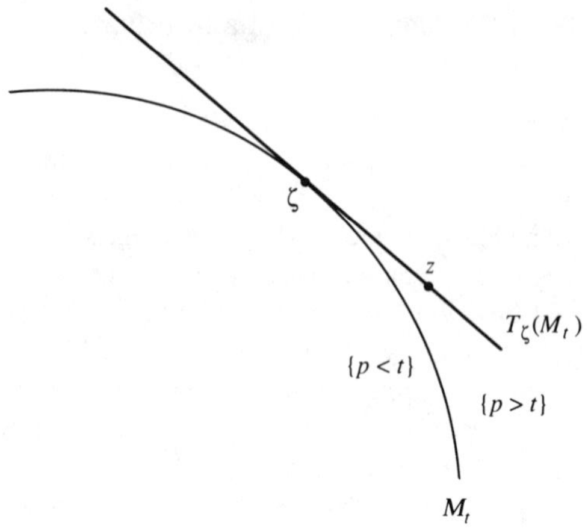

FIGURE 20.1

PROOF Since ρ is a strictly convex defining function for D, there exists $\epsilon > 0$ such that for each t with $-\epsilon < t < \epsilon$ the level set

$$M_t = \{\zeta \in K \subset \mathbb{C}^n ; \rho(\zeta) = t\}$$

is strictly convex. If $\zeta \in M_t$, then $\{w \in \mathbb{C}^n ; (\partial\rho(\zeta)/\partial\zeta) \cdot w = 0\}$ can be identified with the holomorphic tangent space, $H^{1,0}_\zeta(M_t)$, via the map $w = (w_1, \ldots, w_n) \mapsto \sum w_j(\partial/\partial\zeta_j)$. Therefore if $\rho(\zeta) = t$ and if $(\partial\rho(\zeta)/\partial\zeta) \cdot (\zeta - z) = 0$ then the vector $z - \zeta$ belongs to $H^{1,0}_\zeta(M_t) \subset T^{\mathbb{C}}_\zeta(M_t)$. The strict convexity of M_t implies that if $\zeta \neq z$ then $\rho(z) > t = \rho(\zeta)$, which completes the proof of part (a).

For part (b), we first note that for some $\epsilon > 0$ the real hessian of ρ at ζ is positive definite provided $-\epsilon < \rho(\zeta) < \epsilon$. Therefore, part (b) follows from a second-order Taylor expansion of ρ about the point ζ. ∎

Now we define the functions $u^j : \mathbb{C}^n \times \mathbb{C}^n \mapsto \mathbb{C}^n$ for $j = 1, 2, 3$, which will generate our desired kernels. Let

$$u^1(\zeta, z) = \frac{\partial\rho(\zeta)}{\partial\zeta} = \left(\frac{\partial\rho(\zeta)}{\partial\zeta_1}, \ldots, \frac{\partial\rho(\zeta)}{\partial\zeta_n}\right)$$

$$u^2(\zeta, z) = \frac{\partial\rho(z)}{\partial z} = \left(\frac{\partial\rho(z)}{\partial z_1}, \ldots, \frac{\partial\rho(z)}{\partial z_n}\right)$$

$$u^3(\zeta, z) = \overline{\zeta - z}.$$

Note that u^1 only depends on ζ and u^2 only depends on z, but we wish to think of both u^1 and u^2 as being defined on $\mathbb{C}^n \times \mathbb{C}^n$.

Using u^1, u^2, u^3, we define the kernels $E_1 = E(u^1)$, $E_2 = E(u^2)$, $E_3 = E(u^3)$, $E_{12} = E(u^1, u^2)$, $E_{13} = E(u^1, u^3)$, $E_{23} = E(u^2, u^3)$, and $E_{123} = E(u^1, u^2, u^3)$ as in Section 20.1. For historical reasons, we assign the following labels to these kernels:

$$L = E_1 \quad \text{(after Leray)}$$

$$H = E_{13} \quad \text{(after Henkin)}.$$

As already mentioned, the Bochner–Martinelli kernel is given by

$$B = E_3.$$

By definition, the transpose of a kernel is obtained by switching ζ with z. Since $u^3(z, \zeta) = -u^3(\zeta, z)$ and $u^2(z, \zeta) = u^1(\zeta, z)$, we obtain

$$H^t(\zeta, z) = E_{13}^t(z, \zeta) = E_{23}(\zeta, z)$$

$$L^t(\zeta, z) = E_1^t(z, \zeta) = E_2(\zeta, z)$$

and

$$B^t = B.$$

Recall that up to a sign, the transpose of a kernel is equal to its adjoint as an operator on forms (see Section 18.1).

The L and H kernels are smoothly defined on the set

$$\left\{(\zeta, z) \in \mathbb{C}^n \times \mathbb{C}^n; \frac{\partial \rho(\zeta)}{\partial \zeta} \cdot (\zeta - z) \neq 0\right\}.$$

In view of part (a) of Lemma 1, this set includes the set $\{(\zeta, z) \in M \times D\}$. Reversing the roles of ζ and z, we see that L^t and H^t are smoothly defined on the set $\{(\zeta, z); \zeta \in M \text{ and } 0 < \rho(z) < \epsilon\}$. From Theorem 1 in Section 20.2, we have

$$\bar{\partial} H = L - B \quad \text{on } M \times D.$$

$$\bar{\partial} H^t = L^t - B \quad \text{on } \{(\zeta, z); \zeta \in M \text{ and } 0 < \rho(z) < \epsilon\}.$$

LEMMA 2

Suppose K is a compact set in $\mathbb{C}^n$. For $f \in \mathcal{D}^*(K)$, the forms $L([M]^{0,1} \wedge f)$ and $H([M]^{0,1} \wedge f)$ are smoothly defined on D and the forms $L^t([M]^{0,1} \wedge f)$ and $H^t([M]^{0,1} \wedge f)$ are smoothly defined on the set $\{z \in \mathbb{C}^n; 0 < \rho(z) < \epsilon\}$.

Moreover, there is a constant $C > 0$ (independent of $f \in \mathcal{D}^*(K)$) such that

$$\sup_{z \in \overline{D} \cap K} |H([\partial D]^{0,1} \wedge f)(z)| \leq C \sup_{\zeta \in \partial D \cap K} |f(\zeta)|.$$

PROOF Since H and L have smooth coefficients on the set $\{(\zeta, z) \in M \times D\}$, the forms

$$H([M]^{0,1} \wedge f)(z) = \int_{\zeta \in M} [H(\zeta, z) \wedge f(\zeta)]^{n,n-1}$$

$$L([M]^{0,1} \wedge f)(z) = -\int_{\zeta \in M} [L(\zeta, z) \wedge f(\zeta)]^{n,n-1}$$

depend smoothly on $z \in D$. Similarly, the forms $H^t([M]^{0,1} \wedge f)$ and $L^t([M]^{0,1} \wedge f)$ are smooth on the set $\{z \in \mathbb{C}^n; 0 < \rho(z) < \epsilon\} \subset \mathbb{C}^n - \overline{D}$.

We shall defer the proof of the estimate given in the lemma until the next section where we shall prove a more general result. ∎

Another key fact is that L and L^t act nontrivially only on forms of highest and lowest degree, respectively, as the next lemma shows.

LEMMA 3

Suppose $f \in \mathcal{D}^{p,q}(\mathbb{C}^n)$. Then $L([M]^{0,1} \wedge f) = 0$ unless $q = 0$. In addition, $L^t([M]^{0,1} \wedge f) = 0$ unless $q = n - 1$.

PROOF Since $u_1(\zeta, z) = \partial\rho(\zeta)/\partial\zeta$ depends only on ζ, and examination of the formula for $L(\zeta, z)$ shows that the degree of L in $d\overline{\zeta}$ is $n - 1$. Therefore,

$$L([M]^{0,1} \wedge f)(z) = -\int_{\zeta \in M} [L(\zeta, z) \wedge f(\zeta)]^{n,n-1}$$
$$= 0$$

unless f has degree 0 in $d\overline{\zeta}$.

Similarly, the degree of $L^t(\zeta, z)$ in $d\overline{\zeta}$ is 0 and so $L^t([M]^{0,1} \wedge f) = 0$ unless the degree of f in $d\overline{\zeta}$ is $n - 1$. ∎

Now we present Henkin's integral kernel solution to the Cauchy–Riemann equations on a strictly convex domain. His solution has an L^∞ estimate which was one of the motivations for the construction of integral kernel solutions to the Cauchy–Riemann equations.

THEOREM 1

(See [He2].) Suppose D is a bounded strictly convex domain in $\mathbb{C}^n$ with smooth boundary M. Let $f \in \mathcal{E}^{p,q}(\overline{D})$ $1 \leq q \leq n$ with $\overline{\partial} f = 0$ on D. Then the form

$$u = B(\chi_D f) + H([M]^{0,1} \wedge f)$$

is a solution to the equation $\overline{\partial} u = f$ on D. Moreover, there is a constant $C > 0$ which is independent of f such that

$$\sup_D |u| \leq C \sup_D |f|.$$

PROOF We apply part (b) of Theorem 2 in Section 18.2 to the fundamental solution B with $T = \chi_D f$. Since $\overline{\partial}\chi_D = -[M]^{0,1}$ and since $\overline{\partial} f = 0$ on D, we obtain

$$\overline{\partial}\{B(\chi_D f)\} - B([M]^{0,1} \wedge f) = \chi_D f. \tag{1}$$

Using the equation $\overline{\partial} H = L - B$ on $M \times D$, we obtain

$$B([M]^{0,1} \wedge f) = -(\overline{\partial} H)([M]^{0,1} \wedge f) + L([M]^{0,1} \wedge f) \quad \text{on } D.$$

Since $f \in \mathcal{E}^{p,q}(\overline{D})$ and $q \geq 1$, the second term on the right vanishes by Lemma 2. Using the homotopy formula in part (b) of Theorem 1 in Section 18.2, the first term on the right equals

$$-\overline{\partial}\{H([M]^{0,1} \wedge f) + H(\overline{\partial}\{[M]^{0,1} \wedge f\})\}.$$

Since $\overline{\partial}\{[M]^{0,1} \wedge f\} = 0$, we obtain

$$B([M]^{0,1} \wedge f) = -\overline{\partial}\{H([M]^{0,1} \wedge f)\} \quad \text{on } D.$$

Inserting this into (1) yields our solution u.

The estimate on $|u|$ follows from the estimate given in Lemma 2 and from the fact that the Bochner–Martinelli kernel $B(\zeta, z)$ is uniformly integrable in $\zeta \in \overline{D}$ for $z \in \overline{D}$. ∎

20.4 Boundary value results for Henkin's kernels

In this section, we shall prove the estimate given in Lemma 2 of the previous section. We shall also examine the smoothness of these kernel operators up to the boundary of our given convex domain. The boundary values of these kernel operators are the key ingredients for the construction of one of the fundamental solutions for the tangential Cauchy–Riemann complex on the boundary.

Let $D^- = \{z \in \mathbb{C}^n;\ \rho(z) < 0\}$, let $D^+ = \{z \in \mathbb{C}^n;\ \rho(z) > 0\}$, and let M be the boundary of D^-. From Lemma 2 in the previous section, the forms

$H([M]^{0,1} \wedge f)$ and $L([M]^{0,1} \wedge f)$ are smooth on D^- and $H^t([M]^{0,1} \wedge f)$ and $L^t([M]^{0,1} \wedge f)$ are smooth on the set $D^+ \cap U$ where U is a neighborhood of M in $\mathbb{C}^n$. We shall examine the regularity of the boundary values (on M from D^-) of $H([M]^{0,1} \wedge f)$ and $L([M]^{0,1} \wedge f)$ along with their tangential derivatives. Likewise, we shall examine the regularity of the boundary values (on M from D^+) of $H^t([M]^{0,1} \wedge f)$ and $L^t([M]^{0,1} \wedge f)$ along with their tangential derivatives. Recall that a vector field X on $\mathbb{C}^n$ is considered tangential to M if $X\rho = 0$ near M where ρ is the defining function for M. Our main boundary value results for Henkin's kernels are contained in the following two theorems.

THEOREM 1

Suppose D^- is a strictly convex domain with smooth boundary M. There is a neighborhood U of M such that the following holds. Let N be any nonnegative integer and suppose $X_1, \ldots, X_N$ are tangential vector fields to M. If f is a compactly supported (p,q)-form with coefficients of class C^N on M, then $X_1 \ldots X_N\{H([M]^{0,1} \wedge f)\}|_{D^-}$ and $X_1 \ldots X_N\{H^t([M]^{0,1} \wedge f)\}|_{U\cap D^+}$ have continuous extensions to M. Moreover, for any compact set K in $\mathbb{C}^n$, there is a constant $C > 0$ which is independent of f such that

$$\sup_{D^-\cap K} |X_1 \ldots X_N\{H([M]^{0,1} \wedge f)\}| \leq C|f|_{C^N(M\cap K)}$$

$$\sup_{\{D^+\cap U\cap K\}} |X_1 \ldots X_N\{H^t([M]^{0,1} \wedge f)\}| \leq C|f|_{C^N(M\cap K)}$$

for any smooth form f on M with support in K.

Here, $|f|_{C^N(M)}$ is the usual C^N-norm of f on M.

THEOREM 2

Suppose D^- is a strictly convex domain with smooth boundary M. There is a neighborhood U of M such that the following holds. If f is any compactly supported (p,q)-form with coefficients that are of class C^{N+1} on M, then any derivative of order N of $L([M]^{0,1} \wedge f)|_{D^-}$ and of $L^t([M]^{0,1} \wedge f)|_{D^+\cap U}$ have continuous extensions to M.

If D^- is bounded, then the above two theorems also apply to forms without compact support. We remark that the case $N = 0$ (i.e., no derivatives) is given in [He2] (see also [He3]). Some refinements of these boundary value results due to Harvey and Polking are given in Chapter 24.

In Theorem 2, we are allowed to take normal derivatives of $L([M]^{0,1} \wedge f)$ and $L^t([M]^{0,1} \wedge f)$, whereas in Theorem 1, we are only allowed to take tangential derivatives of $H([M]^{0,1} \wedge f)$ and $H^t([M]^{0,1} \wedge f)$. There is no loss of differentiability in Theorem 1. That is, the form f is assumed to be of class C^N and we obtain a C^N-estimate. This contrasts with Theorem 2 where there is a loss of differentiability. As we shall see from the proofs, this is due to the fact that $L(\zeta, z)$ and $L^t(\zeta, z)$ are not integrable in $\zeta \in M$ for fixed $z \in M$, whereas both $H(\zeta, z)$ and $H^t(\zeta, z)$ are integrable in $\zeta \in M$ for fixed $z \in M$.

The rest of this section is devoted to the proofs of these theorems. The basic idea of the proofs is to localize and then make a change of variables that flattens out M. The strict convexity allows us to estimate the resulting kernels. We then show that tangential derivatives do not worsen these estimates.

We shall prove these theorems for H and L. The proofs for H^t and L^t are similar. By examining the L and H kernels, we see that we must analyze a term of the form

$$K(f)(z) = \int_{\zeta \in M} K(\zeta, z) f(\zeta)$$

where

$$K(\zeta, z) = \frac{k(\zeta, z) d\sigma(\zeta)}{\left(\frac{\partial \rho(\zeta)}{\partial \zeta} \cdot (\zeta - z)\right)^q |\zeta - z|^{2p}} \qquad p + q = n, \quad p,\ q \geq 0$$

where f is a function of class C^N on M and where $d\sigma(\zeta)$ denotes the volume form on M. For H, k is a smooth function satisfying

$$k(\zeta, z) = \mathcal{O}(|\zeta - z|)$$

and both p and q are at least 1. For L, k is simply a smooth function without any estimate and $q = n$, $p = 0$.

We fix a point $z_0 \in M$ and examine the regularity of $K(f)(z)$ as z approaches M from D^- near z_0. Fix $\zeta_0 \in M$. If $\zeta_0 \neq z_0$ then by Lemma 1, $((\partial \rho(\zeta)/\partial \zeta) \cdot (\zeta - z)) \neq 0$ for (ζ, z) in some neighborhood $U_1 \times U_2 \subset \mathbb{C}^n \times \mathbb{C}^n$ of (ζ_0, z_0). If ϕ is a smooth compactly supported function in U_1, then $K(\phi f)$ is smooth on U_2. In particular, $K(\phi f)$ is smooth up to $M \cap U_2$ from $U_2 \cap D^-$. By a partition of unity argument, it suffices to assume $\zeta_0 = z_0$. That is, we may assume that f has compact support in a set of the form $U \cap M$ where U is an open neighborhood of z_0 in $\mathbb{C}^n$ (to be chosen later). We must show that $K(f)$ has the desired regularity on $U \cap \overline{D^-}$.

We need the change of variables given in the following lemma.

LEMMA 1
For each $z_0 \in M$, there is a neighborhood U of z_0 in $\mathbb{C}^n$ and a smooth map $\Psi : U \times U \mapsto \mathbb{C}^n$ with the following properties:

(a) $\Psi(z, z) = (\rho(z), 0, \ldots, 0)$.

(b) *If we write $\Psi(\zeta, z) = (w_1(\zeta, z), \ldots, w_n(\zeta, z)) \in \mathbb{C}^n$ then*

$$w_1(\zeta, z) = \rho(\zeta) + i\, Im \left\{ \frac{\partial \rho(\zeta)}{\partial \zeta} \cdot (\zeta - z) \right\}.$$

(c) For each $z \in U$, the map $\Psi_z : U \mapsto \mathbb{C}^n$ given by $\Psi_z(\zeta) = \Psi(\zeta, z)$ is a diffeomorphism from U to $\Psi_z\{U\}$.

PROOF We let

$$w_1(\zeta, z) = \rho(\zeta) + i\text{Im}\left\{\frac{\partial\rho(\zeta)}{\partial\zeta} \cdot (\zeta - z)\right\}$$

as required by (b). We have

$$\begin{aligned} d_\zeta w_1(\zeta, z_0)|_{\zeta=z_0} &= d\rho(z_0) + i\text{Im}\{\partial\rho(z_0)\} \\ &= d\rho(z_0) - \frac{i}{2}J^* d\rho(z_0) \end{aligned}$$

where the last equation uses the fact that $\partial\rho = 1/2(d\rho - iJ^* d\rho)$ (see Lemma 5 in Section 3.3). The real vectors $d\rho(z_0)$ and $J^* d\rho(z_0)$ span (over $\mathbb{R}$) the 1-complex dimensional subspace generated by $\partial\rho(z_0)$. Since $\partial\rho \neq 0$ on M, we can find vectors $w_2, \ldots, w_n \in \mathbb{C}^n$ so that $\{(\partial\rho(z_0)/\partial z), w_2, \ldots, w_n\}$ form a basis for $\mathbb{C}^n$ over C. We let

$$\Psi(\zeta, z) = (w_1(\zeta, z), \ldots, w_n(\zeta, z))$$

where $w_1(\zeta, z)$ is defined above and

$$w_j(\zeta, z) = w_j \cdot (\zeta - z) \quad \text{for } 2 \leq j \leq n.$$

The real ζ-derivative of $\Psi(\zeta, z_0)$ at $\zeta = z_0$ is nonsingular. So property (c) follows from the inverse function theorem. Property (a) follows from the definition of Ψ. ∎

From now on, we require f to have support in $U \cap M$ where U is a neighborhood of z_0 which is small enough to satisfy Lemma 1. We shall also require U to be small enough so that the following estimate holds (from part (b) of Lemma 1 in Section 20.3)

$$\text{Re}\left\{\frac{\partial\rho(\zeta)}{\partial\zeta} \cdot (\zeta - z)\right\} \approx |\zeta - z|^2 + \rho(\zeta) - \rho(z) \quad \text{for } \zeta, z \in U.$$

Since $\text{Re}\{w_1(\zeta, z)\} = \rho(\zeta)$, $\Psi_z\{M \cap U\}$ contains a neighborhood of the origin in the copy of $\mathbb{R}^{2n-1}$ given by $\{w \in \mathbb{C}^n; \text{Re}w_1 = 0\}$ for each fixed $z \in M \cap U$. After pulling back the integral in $K(f)$ to this copy of $\mathbb{R}^{2n-1}$ via Ψ_z^{-1}, we obtain

$$K(f)(z) = \int_{\{\text{Re}w_1=0\}} K_1(w, z) f_1(w, z) dv(w)$$

where

$$K_1(w,z) = \frac{k_1(w,z)}{(p(w,z))^q |\Psi_z^{-1}(w) - z|^{2p}} \tag{1}$$

$$dv(w) = \text{volume form on } \{\text{Re} w_1 = 0\}$$

$$k_1(w,z)dv(w) = k(\Psi_z^{-1}(w), z)\Psi_z^{-1^*}(d\sigma)$$

$$f_1(w,z) = f(\Psi_z^{-1}(w))$$

$$p(w,z) = \frac{\partial \rho}{\partial \zeta}(\Psi_z^{-1}(w)) \cdot (\Psi_z^{-1}(w) - z).$$

Since f has compact support in U, there is an $\eta > 0$ such that the w-support of $f_1(w,z)$ is contained in $\{w \in \mathbb{C}^n; |w| < \eta\}$ for each z in U.

In view of property (b) of Lemma 1, we have

$$\text{Im} p(w,z) = y_1 \quad \text{where } y_1 = \text{Im} w_1. \tag{2}$$

We shall give $\{\text{Re} w_1 = 0\}$ the coordinates $w = (y_1, w')$ where $y_1 = \text{Im} w_1 \in \mathbb{R}$ and $w' \in \mathbb{C}^{n-1}$.

Certain components of the original kernel such as k (for the kernel H) and $(\partial\rho(\zeta)/\partial\zeta)\cdot(\zeta - z)$ and $|\zeta - z|^2$ vanish for $\zeta = z$. These terms can be estimated by some power of $|\zeta - z|$. We wish to transfer these estimates to the w-variables via the diffeomorphism $\zeta = \Psi_z^{-1}(w)$. We shall use the following notation. For a smooth function $g : \mathbb{C}^n \times \mathbb{C}^n \mapsto \mathbb{C}$ and a nonnegative integer j, we say

$$g = \mathcal{O}(\rho(z), y_1, w')^j$$

provided g is a homogeneous polynomial of at least degree j in the real coordinates of $(\rho(z), y_1, w')$ over the ring of smooth, complex-valued functions on $\mathbb{C}^n \times \mathbb{C}^n$. A typical term of such a g is of the form

$$a_{\alpha\beta\gamma\delta}(z,w)\ \rho(z)^\alpha y^\beta w^\gamma \overline{w}^\delta \tag{3}$$

$$|\alpha| + |\beta| + |\gamma| + |\delta| \geq j$$

where each $a_{\alpha\beta\gamma\delta}$ is a smooth, complex-valued function. We say

$$g \approx |(\rho(z), y_1, w')|^j$$

if $g = \mathcal{O}(\rho(z), y_1, w')^j$ and $g(w,z) \geq \epsilon |(\rho(z), y_1, w')|^j$ for some uniform $\epsilon > 0$.

With the above notation, we state and prove the following estimates.

LEMMA 2

(a) *If* $k(\zeta, z) = \mathcal{O}(|\zeta - z|^j)$ *for* $j \geq 0$, *then*

$$k(\Psi_z^{-1}(w), z) = \mathcal{O}(\rho(z), y_1, w')^j.$$

(b) *For* $w = (y_1, w')$

$$|\Psi_z^{-1}(w) - z|^2 \approx |(\rho(z), y_1, w')|^2 .$$

(c) *For* $w = (y_1, w')$

$$Re\{p(w, z)\} + \frac{1}{2}\rho(z) \approx |(\rho(z), y_1, w')|^2 .$$

PROOF Parts (a) and (b) follow from a Taylor expansion of $k(\Psi_z^{-1}(w), z)$ or $\Psi_z^{-1}(w) - z$ in w about the point $w = (\rho(z), 0, \ldots, 0)$ and the fact that $\Psi_z^{-1}(\rho(z), 0, \ldots, 0) = z$. Part (c) follows from the fact that $\Psi_z^{-1}\{\text{Re} w_1 = 0\} \subset M = \{\rho(\zeta) = 0\}$ and the estimate in part (b) of Lemma 1 in Section 20.3. ∎

Now suppose X_z is a vector field involving z or $\bar{z}$ derivatives. In general, if $f = \mathcal{O}(\rho(z), y_1, w')^j$ with $j \geq 1$, then $X_z f = \mathcal{O}(\rho(z), y_1, w')^{j-1}$ (note the exponent decreases by one). This is clear from (3) because $X_z\{\rho^\alpha\} = \alpha\rho^{\alpha-1}X_z\rho$. However, if X_z is a tangential vector field, then $X_z\rho = 0$. In this case, if $f = \mathcal{O}(\rho(z), y_1, w')^j$ then also $X_z f = \mathcal{O}(\rho(z), y_1, w')^j$. This is the key fact which we will use to show that the estimates on our kernels do not worsen when we differentiate them with a tangential vector field.

LEMMA 3

Suppose X_z is a tangential vector field to M.

(a) *If* $k(\zeta, z) = \mathcal{O}(|\zeta - z|^j)$ *then for* $w = (y_1, w')$

$$X_z\{k(\Psi_z^{-1}(w), z)\} = \mathcal{O}(\rho(z), y_1, w')^j.$$

(b) *For* $w = (y_1, w')$

$$X_z\{p(w, z)\} = \mathcal{O}(\rho(z), y_1, w')^2.$$

PROOF Part (a) follows immediately from the observations made preceding the statement of the lemma. For part (b), first note

$$p(w, z) = \text{Re}\{p(w, z)\} + iy_1.$$

Since $X_z\rho = X_z y_1 = X_z w' = 0$, part (b) follows from part (c) of Lemma 2. ∎

COMPLETION OF THE PROOF OF THEOREM 1 As already mentioned, for H, $k(\zeta, z) = \mathcal{O}(|\zeta - z|)$ and so

$$k_1(w, z) = \mathcal{O}(\rho(z), y_1, w').$$

Together with the other estimates in Lemma 2, we obtain for $w = (y_1, w')$

$$|K_1(w,z)| \le \frac{C(|\rho(z)| + |y_1| + |w'|)}{(\frac{-1}{2}\rho(z)+\rho(z)^2+|y_1|^2+|w'|^2+|y_1|)^q(|\rho(z)|^2+|y_1|^2+|w'|^2)^p}.$$

Since $\rho(z) \le 0$ for $z \in \overline{D^-}$, we obtain

$$|K_1(w,z)| \le \frac{C}{(|w'|^2 + |y_1|)^q |w'|^{2p-1}}$$

where C is some uniform positive constant.

Now suppose $X_z^1, \ldots, X_z^N$ are tangential vector fields to M. Differentiating $Kf(z)$ with $X_z^1, \ldots, X_z^N$ involves a sum of terms of the form

$$(X_z^1 \ldots X_z^m\{K_1(w,z)\}) \cdot (X_z^{m+1} \ldots X_z^N\{f_1(w,z)\})$$

with $0 \le m \le N$. The term involving the derivatives of f_1 is dominated above in absolute value by $|f|_{C^N}$. We now show that $|X_z^1 \ldots X_z^m\{K_1(w,z)\}|$ satisfies the same estimate that is satisfied by $|K_1(w,z)|$.

LEMMA 4

Suppose $X_z^1, \ldots, X_z^m$ are tangential vector fields to M. There is a uniform positive constant C such that

$$|X_z^1 \ldots X_z^m\{K_1(w,z)\}| \le \frac{C}{(|w'|^2 + |y_1|)^q |w'|^{2p-1}}$$

for $w = (y_1, w')$ and $z \in \overline{D^-}$.

PROOF This lemma follows easily from Lemma 3. For example, we have

$$X_z\{p(w,z)^{-q}\} = -q\frac{X_z p(w,z)}{(p(w,z))^{q+1}}$$

which by Lemmas 2 and 3 is dominated above in absolute value by

$$\frac{C(|\rho(z)^2| + |y_1|^2 + |w'|^2)}{(\frac{-1}{2}\rho(z) + \rho(z)^2 + |y_1|^2 + |w'|^2 + |y_1|)^{q+1}}.$$

Since $\rho(z) \le 0$ for $z \in \overline{D^-}$, this term is dominated by

$$\frac{C}{(|w'|^2 + |y_1|)^q}$$

which is the same term that dominates $|(p(w,z))^{-q}|$. Thus, differentiating $p(w,z)^{-q}$ with a tangential vector field does not worsen the estimate. ∎

Now we show that the function

$$\frac{1}{(|w'|^2 + |y_1|)^q |w'|^{2p-1}} \quad (p+q=n),\ q \ge 1$$

which dominates both $|K_1(w,z)|$ and $|X_z^1 \dots X_z^m\{K_1(w,z)\}|$ for $z \in \overline{D^-}$ is a locally integrable function of $w = (y_1, w') \in \mathbb{R}^{2n-1}$. For once this is done, the dominated convergence theorem will allow us to take limits of $Kf(z)$ or $X_z^1 \dots X_z^N\{Kf(z)\}$ as z approaches M from D^-. This will complete the proof of Theorem 1.

LEMMA 5
The function

$$g(w) = \frac{1}{(|w'|^2 + |y_1|)^q |w'|^{2p-1}} \quad p + q \le n, \quad q \ge 1$$

is a locally integrable function of $w = (y_1, w') \in \mathbb{R}^{2n-1}$.

PROOF First suppose $q > 1$. By integrating y_1, we obtain

$$\int_{|y_1| \le \eta,} \int_{|w'| \le \eta} \frac{dy_1 \, dv(w')}{(|w'|^2 + |y_1|)^q |w'|^{2p-1}} \le C \int_{|w'| \le \eta} \frac{dv(w')}{|w'|^{2n-3}}$$

where C is a uniform positive constant. Since this last expression is an integral in $w' \in \mathbb{C}^{n-1} \approx \mathbb{R}^{2n-2}$, it is easily shown to be finite by a standard polar coordinate calculation. If $q = 1$ then the above integral in y_1 results in a log term but otherwise the calculation is no different than the case $q > 1$. ∎

PROOF OF THEOREM 2 For the L kernel, we must analyze a term of the form

$$K_1(w,z) = \frac{k(w,z)}{(p(w,z))^n}$$

for $z \in \overline{D^-}$, where k is a smooth function. From part (c) of Lemma 2

$$|K_1(w,z)| \le \frac{C}{(|w'|^2 + |y_1|)^n}$$

for $\rho(z) \le 0$. The right side is not integrable in $w = (y_1, w') \in \mathbb{R}^{2n-1}$. Therefore, an integration by parts argument must be used to reduce the exponent in the expression for K_1. This explains the loss of one derivative in the statement of Theorem 2. Recall that $p(w,z) = \mathrm{Re}\{p(w,z)\} + iy_1$. We have

$$\frac{\partial p(w,z)}{\partial y_1} = \frac{\partial}{\partial y_1}\mathrm{Re}p(w,z) + i.$$

In particular, $(\partial/\partial y_1)p(w,z) \neq 0$.

Since $n \ge 2$, we have

$$p(w,z)^{-n} = \left((1-n)\frac{\partial}{\partial y_1}p(w,z)\right)^{-1} \frac{\partial}{\partial y_1}\left\{(p(w,z))^{1-n}\right\}.$$

Since $f_1(w,z) = f(\Psi_z^{-1}(w))$ has compact w-support, we may integrate by parts with $\partial/\partial y_1$ and obtain

$$Kf(z) = \int\int (p(w,z))^{-n} f_1(w,z) k(w,z) dy_1 dv(w')$$
$$= \frac{1}{n-1} \int\int (p(w,z))^{1-n}$$
$$\frac{\partial}{\partial y_1} \left\{ k(w,z) \left(\frac{\partial}{\partial y_1} \{p(w,z)\} \right)^{-1} f_1(w,z) \right\} dy_1 dv(w').$$

From part (c) of Lemma 2, we have

$$|p(w,z)|^{1-n} \le C \left(-\frac{\rho(z)}{2} + \rho(z)^2 + |y_1|^2 + |w'|^2 + |y_1| \right)^{1-n}$$
$$\le C(|w'|^2 + |y_1|)^{1-n} \tag{4}$$

for $z \in \overline{D^-}$ (because $\rho(z) \le 0$). The last expression on the right is a locally integrable function in $w = (y_1, w') \in \mathbb{R}^{2n-1}$ by Lemma 5 (with $p = 1/2$ and $q = n-1$). By the dominated convergence theorem, Kf is continuous up to M from D^-.

Derivatives are handled analogously. Suppose f is a function of class C^{N+1} with compact support on M. Let $X_z^1, \ldots, X_z^N$ be vector fields on $\mathbb{C}^n$ (not necessarily tangential to M). For $z \in D$, $X_z^1 \ldots X_z^N \{Kf(z)\}$ involves a sum of terms, typical of which is

$$\int\int (p(w,z)^{-(n+l)} D_z^q \{f_1(w,z)\} k_1(w,z) dy_1 dv(w')$$

where $l + q \le N$, where D_z^q is a differential operator in z or $\overline{z}$ of order q, and where k_1 is a smooth function obtained from various derivatives of k. Integrating by parts $l+1$ times with $\partial/\partial y_1$ yields another sum of terms, typical of which is

$$\int\int (p(w,z))^{1-n} D_{z,y_1}^{N+1} \{f_1(w,z)\} k_2(w,z) dy_1 dv(w')$$

where D_{z,y_1}^{N+1} is a differential operator in z, $\overline{z}$ and y_1 of order at most $N+1$ and where k_2 is a smooth function obtained from various derivatives of k_1. Since f_1 is of class C^{N+1}, the estimate in (4) and the dominated convergence theorem imply that $X_z^1 \ldots X_z^N \{Kf(z)\}$ is continuous up to M from D. This completes the proof of Theorem 2. ∎

21

Fundamental Solutions for the Tangential Cauchy–Riemann Complex on a Convex Hypersurface

As mentioned in Chapter 18, there does not exist a solution to the equation $\overline{\partial}_{M\times M}K = [\Delta]$, except for very specialized M. However, in this chapter, we shall construct a solution to the equation $\overline{\partial}_{M\times M}K = [\Delta]$ for a strictly convex M, modulo a kernel which as an operator acts nontrivially only on forms of bottom and top degree. We shall then abuse the notation and call K a fundamental solution for the tangential Cauchy–Riemann complex. We present two fundamental solutions for the tangential Cauchy–Riemann complex. The second solution will be derived from the first and it will play a key role in the local solution to the tangential Cauchy–Riemann equations in the same way that the Bochner–Martinelli kernel plays a role in the local solution to the Cauchy–Riemann equations.

For a form $f \in \mathcal{D}_M^{p,q}$, the first solution to the tangential Cauchy–Riemann equations $\overline{\partial}_M u = f$ will be constructed as the boundary values of terms of the form $K([M]^{0,1} \wedge f)$ where K is a kernel of the type introduced in Chapter 20. Since these kernels are ambiently defined, it will be convenient to use the extrinsic definition of the tangential Cauchy–Riemann complex presented in Section 8.1. The current $[M]^{0,1} \wedge f$ is well defined only if f is an ambiently defined form. However, as mentioned in Section 8.1, if F_1 and F_2 are two ambiently defined forms, with $\{F_1\}_{t_M} = f = \{F_2\}_{t_M}$ on M, then $[M]^{0,1} \wedge F_1 = [M]^{0,1} \wedge F_2$. Therefore, for $f \in \mathcal{D}_M^{p,q}$, we can unambiguously define $[M]^{0,1} \wedge f$ to be $[M]^{0,1} \wedge F$ where F is any ambiently defined (p,q)-form with $F_{t_M} = f$ on M.

21.1 The first fundamental solution for the tangential Cauchy–Riemann complex

Let us first summarize the results from Chapters 19 and 20. In Chapter 19, we constructed the Bochner–Martinelli kernel B, which is a fundamental solution for the Cauchy–Riemann complex. Suppose M is the smooth boundary of a domain D^- in $\mathbb{C}^n$ with the induced boundary orientation. For any element $f \in \mathcal{D}_M^{p,q}$, we showed that $\{B([M]^{0,1} \wedge f)\}_{t_M}$ has continuous nontangential extensions to M from D^- and D^+ $(= \mathbb{C}^n - \{\overline{D^-}\})$. These extensions are denoted $B^-(f)$ and $B^+(f)$, respectively. We also showed the following jump relation holds

$$\begin{aligned} B^+(f) - B^-(f) &= f_{t_M} \quad \text{on } M \\ &= f \quad (\text{since } f \in \mathcal{D}_M^{p,q}). \end{aligned}$$

In Chapter 20, we defined the H, H^t, L, and L^t kernels for a strictly convex domain D with smooth boundary M. We also showed that there is a neighborhood U of M such that

$$\begin{aligned} \overline{\partial} H &= L - B \quad \text{on } M \times D^- \\ \overline{\partial} H^t &= L^t - B \quad \text{on } M \times \{U \cap D^+\}. \end{aligned}$$

The kernels L and H have smooth coefficients on the set $M \times D^-$ and the kernels L^t and H^t have smooth coefficients on the set $M \times \{U \cap D^+\}$. Therefore, we may apply the above equations to the current $[M]^{0,1} \wedge f$. Using the equation $\overline{\partial}\{[M]^{0,1} \wedge f\} = -[M]^{0,1} \wedge \overline{\partial}_M f$ and the homotopy equation from part (b) of Theorem 1 in Section 18.2 (with $r = 0$ and $s = -2$), we obtain

$$\overline{\partial}\{H([M]^{0,1} \wedge f)\} + H([M]^{0,1} \wedge \overline{\partial}_M f) = L([M]^{0,1} \wedge f) - B([M]^{0,1} \wedge f) \quad (1a)$$

on D^- and

$$\overline{\partial}\{H^t([M]^{0,1} \wedge f)\} + H^t([M]^{0,1} \wedge \overline{\partial}_M f) = L^t([M]^{0,1} \wedge f) - B([M]^{0,1} \wedge f) \quad (1b)$$

on $U \cap D^+$. From Theorems 1 and 2 in Section 20.4, we can take boundary values of the terms appearing in the above equations. For $f \in \mathcal{D}_M^{p,q}$, we set

$$\begin{aligned} H(f) &= \text{extension of } \{H([M]^{0,1} \wedge f)\}_{t_M}|_{D^-} \text{ to } M \\ H^t(f) &= \text{extension of } \{H^t([M]^{0,1} \wedge f)\}_{t_M}|_{U \cap D^+} \text{ to } M \\ L^-(f) &= \text{extension of } \{L([M]^{0,1} \wedge f)\}_{t_M}|_{D^-} \text{ to } M \\ (L^t)^+(f) &= \text{extension of } \{L^t([M]^{0,1} \wedge f)\}_{t_M}|_{U \cap D^+} \text{ to } M. \end{aligned}$$

From the proof of Theorem 1 in Section 20.4, the kernels $H(\zeta, z)$ and $H^t(\zeta, z)$ are locally integrable in $\zeta \in M$ for each fixed $z \in M$. Therefore

$$H(f)(z) = \left\{ \int_{\zeta \in M} [H(\zeta, z) \wedge f(\zeta)]^{n,n-1} \right\}_{t_M} \quad \text{for } z \in M$$

$$H^t(f)(z) = \left\{ \int_{\zeta \in M} [H^t(\zeta, z) \wedge f(\zeta)]^{n,n-1} \right\}_{t_M} \quad \text{for } z \in M.$$

The kernels $L(\zeta, z)$ and $L^t(\zeta, z)$ are not integrable in ζ for each fixed $z \in M$. Therefore, these boundary values only exist in the sense described by Theorem 2 in Section 20.4. In particular, the above integral expressions with H replaced by L do not make sense for $z \in M$. For this reason we have used the $-$ and $+$ superscripts to denote the boundary values of $L([M]^{0,1} \wedge f)|_{D^-}$ and $L^t([M]^{0,1} \wedge f)|_{D^+}$, respectively.

The operator $\overline{\partial}_M = t_M \circ \overline{\partial}$ involves only vector fields that are tangential to M. Therefore

$$\{\overline{\partial}\{H([M]^{0,1} \wedge f)\}\}_{t_M}$$

extends from D^- to M as $\overline{\partial}_M\{H(f)\}$. Similarly

$$\{\overline{\partial}\{H^t([M]^{0,1} \wedge f)\}\}_{t_M}$$

extends from D^+ to M as $\overline{\partial}_M\{H^t(f)\}$. If we apply t_M to equations (1a) and (1b), subtract the result, and then use the jump relation $B^+(f) - B^-(f) = f$, we obtain

$$f = \overline{\partial}_M\{(H - H^t)(f)\} + (H - H^t)(\overline{\partial}_M f) + ((L^t)^+ - L^-)(f) \quad \text{on } M.$$

From Lemma 3 in Section 20.3, we have $L^-(f) = 0$ unless f is a $(p, 0)$-form and $(L^t)^+(f) = 0$ unless f is a $(p, n-1)$-form. Therefore, except for forms of bottom and top degree on M, the operator $K = H - H^t$ is a fundamental solution for $\overline{\partial}_M$. For this reason, we call K a fundamental solution for $\overline{\partial}_M$. From Theorem 1 in section 20.4, the operator K is regular (i.e., K sends compactly supported smooth forms to smooth forms). We also have $K^t = -K$ and so K is self-adjoint up to a sign. Therefore, K is a biregular fundamental solution. We summarize the above discussion in the following theorem.

THEOREM 1

Suppose M is a smooth, strictly convex hypersurface in $\mathbb{C}^n$. The operator $K = H - H^t$ is a biregular kernel that satisfies the following operator equation on $\mathcal{D}^*_M$:

$$I = (\overline{\partial}_M \circ K + K \circ \overline{\partial}_M) + ((L^t)^+ - L^-).$$

This operator equation has the following special cases:

(a) *For $f \in \mathcal{D}^{p,q}_M$ with $1 \leq q \leq n-2$*

$$f = \overline{\partial}_M\{K(f)\} + K(\overline{\partial}_M f).$$

(b) *For $f \in \mathcal{D}^{p,0}_M$*

$$f = K(\overline{\partial}_M f) - L^-(f).$$

(c) *For $f \in \mathcal{D}^{p,n-1}_M$*

$$f = \overline{\partial}_M\{K(f)\} + (L^t)^+(f).$$

Furthermore, given any positive integer N and any compact set K, there is a positive constant C such that

$$|K(f)|_{C^N(M\cap K)} \leq C|f|_{C^N(M\cap K)}$$

for all $f \in \mathcal{D}^*(M \cap K)$.

This theorem without the C^N-estimate appears in [He3].

The operator equation stated in the above theorem can also be stated in terms of currents as

$$[\Delta] = -(\overline{\partial}_{M\times M}K) + (L^t)^+ - L^-.$$

(See part (c) of Theorem 1 in Section 18.2 with $m = n-1$, $d = 1$, $r = 0$, and $s = -1$). We now state some easy corollaries to the above theorem. The first corollary (which follows from (a) above) provides a global solution to the equation $\overline{\partial}_M u = f$ except at the top degree.

COROLLARY 1

Suppose M is a smooth, strictly convex hypersurface in $\mathbb{C}^n$. If $f \in \mathcal{D}_M^{p,q}$ for $1 \leq q \leq n-2$ with $\overline{\partial}_M f = 0$, then $\overline{\partial}_M\{K(f)\} = f$ on M.

If M is compact, then the above corollary holds with $\mathcal{D}_M^{p,q}$ replaced by $\mathcal{E}_M^{p,q}$.

Now let us examine part (b) of Theorem 1. If $f \in \mathcal{D}_M^{p,0}$ with $\overline{\partial}_M f = 0$, then $f = -L^-(f)$. For $z \in D^-$, we have

$$\begin{aligned} -L([M]^{0,1} \wedge f)(z) &= \langle [M]^{0,1}, L(\zeta, z) \wedge f(\zeta)\rangle_\zeta \\ &= \int_{\zeta\in M} [L(\zeta, z) \wedge f(\zeta)]^{n,n-1}. \end{aligned}$$

Recall that $L = E(u)$ where $u(\zeta, z) = (\partial\rho(\zeta)/\partial\zeta)$. Since $u(\zeta, z)$ is independent of z, an examination of the formula for $L(\zeta, z)$ shows that $L(\zeta, z)$ is holomorphic in $z \in D$, for $\zeta \in M$. This yields the following corollary.

COROLLARY 2

Suppose $D = D^-$ is a strictly convex bounded domain with smooth boundary M in $\mathbb{C}^n$. If $f \in \mathcal{E}_M^{p,0}(M)$ with $\overline{\partial}_M f = 0$, then f has a holomorphic extension to D given by

$$\begin{aligned} F(z) &= -L([M]^{0,1} \wedge f)(z) \\ &= \int_{\zeta\in M} [L(\zeta, z) \wedge f(\zeta)]^{n,n-1}. \end{aligned}$$

The Bochner–Martinelli kernel also provides the holomorphic extension of a CR function as shown in Theorem 1 in Section 19.3. Furthermore, this theorem holds without any convexity assumption on D. The point of this corollary is that with the additional convexity assumption on D, we have a reproducing kernel for holomorphic functions (namely $L(\zeta, z)$) which is holomorphic in the variable z (this is not true for the Bochner–Martinelli kernel). We also note that on the unit sphere $\{\zeta \in \mathbb{C}^n; |\zeta| = 1\}$, the defining function for M is $\rho(\zeta) = |\zeta|^2 - 1$ and so $u(\zeta, z) = \overline{\zeta}$. In this case, $L = E(u)$ is the Szegö kernel which represents the L^2-projection onto the space of square integrable CR functions on M (see Krantz's book [Kr]).

Finally, we take a look at the case of top degree (part (c) in Theorem 1). Suppose M is the boundary of a bounded strictly convex domain $D = D^-$. Every form in $\mathcal{E}_M^{p,n-1}(M)$ is $\overline{\partial}_M$-closed. In this case, some other compatibility condition replaces $\overline{\partial}_M f = 0$ in order to solve the equation $\overline{\partial}_M u = f$. For motivation, let us suppose the equation $\overline{\partial}_M u = f \in \mathcal{E}^{p,n-1}(M)$ has a smooth solution u. Suppose $g \in \mathcal{E}_M^{n-p,0}(M)$ with $\overline{\partial}_M g = 0$. Then

$$\begin{aligned} \langle f, g\rangle_M &= \langle \overline{\partial}_M u, g\rangle_M \\ &= (-1)^{p+n-1}\langle u, \overline{\partial}_M g\rangle_M \\ &= 0. \end{aligned}$$

The second equation follows from the integration by parts formula given in Lemma 6 in Section 8.1. It follows that a necessary condition for solving the equation $\overline{\partial}_M u = f$ is the condition $\langle f, g\rangle_M = 0$ for all $g \in \mathcal{E}_M^{n-p,0}(M)$ with $\overline{\partial}_M g = 0$. Since CR functions on M extend holomorphically to D, we can replace this condition by requiring $\langle f, g\rangle_M = 0$ for all $g \in \mathcal{E}^{n-p,0}(\overline{D})$ with $\overline{\partial} g = 0$. This necessary condition is also a sufficient condition for the existence of a solution to the equation $\overline{\partial}_M u = f$ on M, as the next corollary shows.

COROLLARY 3

Suppose D is a bounded strictly convex domain in $\mathbb{C}^n$ with smooth boundary M. Suppose $f \in \mathcal{E}_M^{p,n-1}(M)$. The equation $\overline{\partial}_M u = f$ has a smooth solution $u \in \mathcal{E}_M^{p,n-2}(M)$ if and only if $\langle f, g\rangle_M = 0$ for all $g \in \mathcal{E}^{n-p,0}(\overline{D})$ with $\overline{\partial} g = 0$ on D.

PROOF The necessity has already been discussed prior to the statement of the corollary. For sufficiency, we note from Theorem 1, part (c),

$$f = \overline{\partial}_M\{K(f)\} + (L^t)^+(f). \tag{2}$$

Now $L^t = E(u^2)$ and $u^2(\zeta, z) = (\partial\rho(z)/\partial z)$ is independent of ζ. Therefore, the coefficients of $L^t(\zeta, z)$ are holomorphic in ζ near M for each fixed $z \in \{D^+ \cap U\}$. By Theorem 1 in Section 19.3 (or Corollary 2 above) $L^t(\zeta, z)$ extends to a form $G(\zeta, z)$ whose coefficients are holomorphic in $\zeta \in D$. Therefore, for each fixed $z \in \{D^+ \cap U\}$ we have

$$\begin{aligned} L^t([M]^{0,1} \wedge f)(z) &= -\int_{\zeta \in M} L^t(\zeta, z) \wedge f(\zeta) \\ &= -\int_{\zeta \in M} G(\zeta, z) \wedge f(\zeta) \\ &= -\langle G(\zeta, z), f(\zeta) \rangle_{\zeta \in M} \\ &= 0 \end{aligned}$$

where the last equation follows from the assumed compatibility condition on f. Therefore $(L^t)^+(f) = 0$, and from (2) we see that $u = K(f)$ is a solution to the equation $\overline{\partial}_M u = f$. ∎

21.2 A second fundamental solution to the tangential Cauchy–Riemann complex

In this section, we construct a simpler fundamental solution for the tangential Cauchy–Riemann complex on a strictly convex hypersurface. This fundamental solution involves only one kernel instead of the two kernels H and H^t that make up the fundamental solution described in the previous section.

Let us start by reviewing the notation. We suppose that our convex hypersurface M divides $\mathbb{C}^n$ into two open sets D^- and D^+, where D^- is convex. Let ρ be the defining function for D^-, i.e., $D^- = \{\zeta; \rho(\zeta) < 0\}$. As in Chapter 20, we set

$$\begin{aligned} u^1(\zeta, z) &= \frac{\partial \rho(\zeta)}{\partial \zeta} \\ u^2(\zeta, z) &= \frac{\partial \rho(z)}{\partial z} \\ u^3(\zeta, z) &= \overline{(\zeta - z)}. \end{aligned}$$

We have already defined the kernels $H = E(u^1, u^3)$, $L = E(u^1)$ and we have shown that their transposes are $H^t = E(u^2, u^3)$ and $L^t = E(u^2)$, respectively. As mentioned in Section 20.4, the H and H^t kernels are locally integrable on $M \times M$ and so in this section we shall denote their tangential pieces on $M \times M$ by the same symbols H and H^t. That is, on $M \times M$, we set

$$\begin{aligned} H &= \{E(u^1, u^3)\}_{tM \times M} \\ H^t &= \{E(u^2, u^3)\}_{tM \times M}. \end{aligned}$$

The L and L^t kernels are not locally integrable on $M \times M$. In Section 20.4, we defined the boundary value operators L^- and $(L^t)^+$.

We now define the kernel that forms our second fundamental solution for the tangential Cauchy–Riemann complex. This kernel is denoted R for Romanov and is given by

$$R = \{E(u^1, u^2)\}_{tM \times M}.$$

THEOREM 1

Suppose M is a smooth, strictly convex hypersurface in $\mathbb{C}^n$, $n \geq 3$. The R kernel is a biregular fundamental solution for the tangential Cauchy–Riemann complex on $M \times M$. As operators on $\mathcal{D}_M^*$

$$I = \overline{\partial}_M \circ R + R \circ \overline{\partial}_M + (L^t)^+ - L^-.$$

If N is a nonnegative integer and if K is a compact subset of M, then there is a positive constant C such that

$$|R(f)|_{C^N(K)} \leq C|f|_{C^N(K)}$$

for $f \in \mathcal{D}_M^*(K)$.

This theorem without the C^N-estimate appears in [He3].

The content of this theorem is that the kernel R can take the place of the kernel K in Theorem 1 in the previous section along with its corollaries. In terms of currents, the statement of the theorem reads

$$\overline{\partial}_{M \times M} R - (L^t)^+ + L^- = -[\Delta]. \tag{1}$$

(See part (c) of Theorem 1 in Section 18.2 with $m = n - 1$, $d = 1$, $r = 0$, $s = -1$).

PROOF In view of part (a) of Lemma 1 in Section 20.3, we have

$$\frac{\partial \rho(\zeta)}{\partial \zeta} \cdot (\zeta - z) \neq 0 \quad \text{for } \zeta, z \in M \text{ with } \zeta \neq z.$$

Reversing the roles of ζ and z yields

$$\frac{\partial \rho(z)}{\partial z} \cdot (\zeta - z) \neq 0 \quad \text{for } \zeta, z \in M \text{ with } \zeta \neq z.$$

Therefore, the kernels H, H^t, R, L, L^t and the kernel $E(u^1, u^2, u^3)$ are smoothly defined on the set $M \times M - \Delta$.

From part (b) of Theorem 1 in Section 20.2, we have

$$\overline{\partial}_{M \times M}\{E(u^1, u^2, u^3)\}_{tM \times M} = H - H^t - R \quad \text{on } M \times M - \Delta. \tag{2}$$

Our goal is to show that the singularity of $E(u^1, u^2, u^3)$ is mild enough so that this identity holds across Δ in the sense of currents. For once this is done, equation (1) follows by applying $\overline{\partial}_{M\times M}$ to both sides of equation (2) and then using the equation

$$\overline{\partial}_{M\times M}\{H - H^t\} = -[\Delta] + (L^t)^+ - L^-$$

which is the current equation associated to Theorem 1 in the previous section. To show that (2) holds across Δ, we need the following lemma.

LEMMA 1

(*a*) *The kernels R, H, H^t, and $E(u^1, u^2, u^3)$ are currents with locally integrable coefficients on $M \times M$.*

(*b*) *Given a compact set $K \subset M$, there is a positive constant C such that for each $\epsilon > 0$ and $z \in K$*

$$\int_{\substack{\zeta\in M\\ |\zeta - z|\leq\epsilon}} |E(u^1, u^2, u^3)(\zeta, z)| \leq C\epsilon^2.$$

Let us assume the lemma for the moment and show that equation (2) holds across Δ. This will complete the first part of Theorem 1.

For $\epsilon > 0$, let χ_ϵ be a smooth function on $M \times M$ such that

$$\chi_\epsilon(\zeta, z) = \begin{cases} 1 & \text{if } |\zeta - z| \geq \epsilon \\ 0 & \text{if } |\zeta - z| \leq \frac{\epsilon}{2}. \end{cases}$$

The function χ_ϵ can be chosen so that if D is any first-order derivative, then

$$|D\{\chi_\epsilon\}| \leq \frac{C}{\epsilon}$$

where C is a positive constant that is independent of ϵ.

Since χ_ϵ vanishes near Δ, we have from (2)

$$\begin{aligned}\overline{\partial}_{M\times M}\{\chi_\epsilon E(u^1, u^2, u^3)\} = {} & (\overline{\partial}_{M\times M}(\chi_\epsilon)) \wedge E(u^1, u^2, u^3) \\ & + \chi_\epsilon(H - H^t - R) \end{aligned} \tag{3}$$

on $M \times M$. From part (a) of Lemma 1, we have

$$\chi_\epsilon H \mapsto H, \quad \chi_\epsilon H^t \mapsto H^t$$

$$\chi_\epsilon R \mapsto R, \quad \chi_\epsilon E(u^1, u^2, u^3) \mapsto E(u^1, u^2, u^3)$$

as $\epsilon \mapsto 0$, where the convergence holds in the weak topology on the space of currents on $M \times M$. Since differentiation is a continuous operator in the space of currents with the weak topology, we have

$$\overline{\partial}_{M\times M}\{\chi_\epsilon E(u^1, u^2, u^3)\} \mapsto \overline{\partial}_{M\times M}\{E(u^1, u^2, u^3)\}$$

as $\epsilon \mapsto 0$. Moreover, the estimate

$$|\overline{\partial}_{M \times M}(\chi_\epsilon)| \leq \frac{C}{\epsilon}$$

together with the estimate given in part (b) of Lemma 1 implies

$$(\overline{\partial}_{M \times M}(\chi_\epsilon)) \wedge E(u^1, u^2, u^3) \mapsto 0$$

as $\epsilon \mapsto 0$, in the sense of currents. By taking limits of equation (3) as $\epsilon \mapsto 0$, we see that equation (2) holds on all of $M \times M$, as desired. This completes the proof of Theorem 1 except for the C^N-estimate on $R(f)$ which we will establish after giving the proof of Lemma 1. ∎

PROOF OF LEMMA 1 The proof of this lemma is much like the proof of Theorem 1 in Section 20.4. In fact, we have already shown in the proof of that theorem that the kernels H and H^t have locally integrable coefficients on $M \times M$. So we turn our attention to the kernels R and $E(u^1, u^2, u^3)$. An examination of these two kernels leads us to estimate a term of the form

$$K(\zeta, z) = \frac{k(\zeta, z)}{\left(\frac{\partial \rho(\zeta)}{\partial \zeta} \cdot (\zeta - z)\right)^q \left(\frac{\partial \rho(z)}{\partial z} \cdot (\zeta - z)\right)^r |\zeta - z|^{2s}}$$

where

$$q, r \geq 1, \quad s \geq 0$$

$$q + r + s = n.$$

For the R kernel, we have $s = 0$ and $q + r = n$. The function k involves coefficients of the differential form

$$\left(\frac{\partial \rho(\zeta)}{\partial \zeta} \cdot d(\zeta - z)\right) \wedge \left(\frac{\partial \rho(z)}{\partial z} \cdot d(\zeta - z)\right)$$

which vanishes when $\zeta = z$. Therefore, we have

$$k(\zeta, z) = \mathcal{O}(|\zeta - z|).$$

For the kernel $E(u^1, u^2, u^3)$, we have $s \geq 1$. In this case, k involves coefficients of the differential form

$$\left(\frac{\partial \rho(\zeta)}{\partial \zeta} \cdot d(\zeta - z)\right) \wedge \left(\frac{\partial \rho(z)}{\partial z} \cdot d(\zeta - z)\right) \wedge \left(\overline{(\zeta - z)} \cdot d(\zeta - z)\right)$$

which vanishes to second order on $\{\zeta = z\}$. Hence, we have

$$k(\zeta, z) = \mathcal{O}(|\zeta - z|^2).$$

Since the R and $E(u^1, u^2, u^3)$ kernels have coefficients that are smooth on $M \times M - \Delta$, we only need to estimate $|K(\zeta, z)|$ for ζ, z in a neighborhood in M of a fixed point $z_0 \in M$.

We use the change of variables $w = \Psi_z(\zeta)$ given in Lemma 1 in Section 20.4. This lemma implies that for a given $z_0 \in M$, there is a neighborhood U of z_0 in $\mathbb{C}^n$ such that for each $z \in U \cap M$, the map $\Psi_z : U \mapsto \mathbb{C}^n$ is a diffeomorphism onto its image; and that $\Psi_z\{M \cap U\}$ is an open neighborhood of the origin in the copy of $\mathbb{R}^{2n-1}$ given by $\{w = (w_1, \ldots, w_n); \mathrm{Re} w_1 = 0\}$. Letting $\zeta = \Psi_z^{-1}(w)$, we see that we must estimate

$$K_1(w, z) = \frac{k_1(w, z)}{(p(w, z)^q (p^*(w, z))^r |\Psi_z^{-1}(w) - z|^{2s}}$$

where

$$\begin{aligned} k_1(w, z) &= k(\Psi_z^{-1}(w), z) \\ p(w, z) &= \frac{\partial \rho(\Psi_z^{-1}(w))}{\partial \zeta} \cdot (\Psi_z^{-1}(w) - z) \\ p^*(w, z) &= \frac{\partial \rho(z)}{\partial z} \cdot (\Psi_z^{-1}(w) - z). \end{aligned}$$

As in Section 20.4, we give $\mathbb{R}^{2n-1}$ the coordinates $w = (y_1, w')$ where $y_1 \in \mathbb{R}$ and $w' \in \mathbb{C}^{n-1}$. Here, we use the following notation. Suppose $g : \mathbb{R}^{2n-1} \times \mathbb{C}^n \mapsto \mathbb{C}$ is smooth. We say that

$$g(w, z) = \mathcal{O}(w)^j$$

provided g is a polynomial of degree j in the real components of $w = (y_1, w') \in \mathbb{R}^{2n-1}$ over the ring of smooth functions defined on $\mathbb{C}^n$. If g is real valued and if $g = \mathcal{O}(w)^j$, then we say that

$$g \approx |w|^j$$

provided there is a uniform $\epsilon > 0$ such that $g(w, z) \geq \epsilon |w|^j$.

One of the differences between the proof of Lemma 1 and the proof of Theorems 1 and 2 in Section 20.4 is that here, the point z is always in M, whereas in Section 20.4, the point z lies to one side of M. In our present setting, this has the simplification that $\rho(z) = 0$ and so $\Psi_z(z) = 0$ and hence $\Psi_z^{-1}(0) = z$. For $z \in M$ and $w \in \mathbb{R}^{2n-1}$, we have

$$|\Psi_z^{-1}(w) - z| \approx |w|.$$

This is a key observation which will be used in the proof of the following analogue of Lemma 2 in Section 20.4.

LEMMA 2

There is a neighborhood U_1 of z_0 in M and there is a neighborhood U_2 of the origin in $\mathbb{R}^{2n-1}$ such that the following hold:

(a) *If $k(\zeta, z) = \mathcal{O}(\zeta - z)^j$ for $\zeta, z \in U_1$ then*

$$\begin{aligned} k_1(w, z) &= k(\Psi_z^{-1}(w), z) \\ &= \mathcal{O}(w)^j \end{aligned}$$

for $z \in U_1$ and $w \in U_2$.

(b) $|\Psi_z^{-1}(w) - z|^2 \approx |w|^2$ *for $z \in U_1$ and $w \in U_2$.*

(c) *$Rep(w, z) \approx |w|^2$ and $-Rep^*(w, z) \approx |w|^2$ for $z \in U_1$ and $w \in U_2$.*

(d) *$p(w, z) - p^*(w, z) = \mathcal{O}(w)^2$ for $z \in U_1$ and $w \in U_2$.*

(e) *There is a constant $\epsilon > 0$ such that*

$$|p(w, z)| \geq \epsilon(|w'|^2 + |y_1|)$$
$$|p^*(w, z)| \geq \epsilon(|w'|^2 + |y_1|)$$

for $z \in U_1$ and $w = (y_1, w') \in U_2$.

PROOF The proofs of parts (a)–(c) of this lemma are similar to the proofs of the corresponding parts of Lemma 2 in Section 20.4. As already mentioned, Ψ_z is a local diffeomorphism on an open set $U \subset \mathbb{C}^n$ and $\Psi_z\{M\}$ is the copy of $\mathbb{R}^{2n-1}$ given by $\{\text{Re}w_1 = 0\}$. Since $\Psi_z^{-1}(0) = z$ for $z \in M$, there is an open neighborhood U_1 of z_0 in M and an open neighborhood U_2 of the origin in $\mathbb{R}^{2n-1}$ with $\Psi_z^{-1}\{U_2\} \subset U \cap M$ for each $z \in U_1$. Parts (a) and (b) now follow from a Taylor expansion in $w \in \mathbb{R}^{2n-1}$ about the origin.

For part (c), we first note from part (b) of Lemma 1 in Section 20.3 that if ζ and z belong to $M = \{\rho = 0\}$ then

$$\text{Re}\left\{\frac{\partial\rho(\zeta)}{\partial\zeta} \cdot (\zeta - z)\right\} \approx |\zeta - z|^2.$$

With the roles of ζ and z reversed, we obtain

$$-\text{Re}\left\{\frac{\partial\rho(z)}{\partial z} \cdot (\zeta - z)\right\} \approx |\zeta - z|^2$$

for ζ and $z \in M$. Therefore, part (c) follows by letting $\zeta = \Psi_z^{-1}(w)$ and then using part (b).

Part (d) also follows from part (b) and the equation

$$\begin{aligned} \frac{\partial\rho(\zeta)}{\partial\zeta} \cdot (\zeta - z) - \frac{\partial\rho(z)}{\partial z} \cdot (\zeta - z) &= \left(\frac{\partial\rho(\zeta)}{\partial\zeta} - \frac{\partial\rho(z)}{\partial z}\right) \cdot (\zeta - z) \\ &= \mathcal{O}(\zeta - z)^2. \end{aligned}$$

The first estimate in part (e) follows from the estimate $\text{Re}p(w,z) \geq \epsilon|w|^2$ in part (c) together with the fact that $\text{Im}p(w,z) = y_1$. For the second estimate in part (e), we first note from part (c) that

$$|\text{Re}p^*(w,z)| \geq \epsilon|w'|^2$$

for $z \in U_1$ and $w \in U_2$. From part (d), we have

$$\begin{aligned}|\text{Im}p^*(z,w)| &\geq |\text{Im}p(w,z)| - |\text{Im}\{p(w,z) - p^*(w,z)\}| \\ &\geq |y_1| - C(|w'|^2 + |y_1|^2) \\ &\geq \frac{1}{2}|y_1| - C|w'|^2\end{aligned}$$

provided $|y_1|$ and $|w'|$ are suitably small, where C is some uniform positive constant. Therefore, we obtain

$$|p^*(w,z)| \geq \max\left\{\epsilon|w'|^2,\ \frac{1}{2}|y_1| - C|w'|^2\right\}. \tag{4}$$

Now we use an easily established inequality (also used in a similar context in [GL]): if $\alpha, \beta, \gamma > 0$ then

$$\max\{\alpha, \beta - \gamma\} \geq \left(2 + \frac{\gamma}{\alpha}\right)^{-1}(\alpha + \beta). \tag{5}$$

If $\alpha \geq \beta - \gamma$, then $2\alpha + \gamma \geq \alpha + \beta$ and this inequality follows by dividing through by $2 + (\gamma/\alpha)$. On the other hand if $\alpha \leq \beta - \gamma$, then

$$\begin{aligned}\frac{\alpha+\beta}{2+\frac{\gamma}{\alpha}} &\leq \frac{2\beta-\gamma}{2+\frac{\gamma}{\beta-\gamma}} \\ &= \beta - \gamma\end{aligned}$$

as desired.

If we use inequality (5) with $\alpha = \epsilon|w'|^2$, $\beta = (1/2)|y_1|$, and $\gamma = C|w'|^2$, then (4) becomes

$$|p^*(w,z)| \geq \left(2 + \frac{C}{\epsilon}\right)^{-1}\left(\epsilon|w'|^2 + \frac{1}{2}|y_1|\right)$$

and the second inequality in (e) is established. This completes the proof of Lemma 2. ∎

Now we return to the proof of Lemma 1. To show the R kernel has locally integrable coefficients, we must show that $K_1(w,z)$ is locally integrable in

$w = (y_1, w') \in \mathbb{R}^{2n-1}$. In this case

$$K_1(w,z) = \frac{k_1(w,z)}{p(w,z)^q p^*(w,z)^r}, \qquad q + r = n. \tag{6}$$

where $k_1(w,z) = \mathcal{O}(w)$ in view of part (a) of Lemma 2. Using part (e) of Lemma 2, we obtain

$$\begin{aligned} |K(w,z)| &\le \frac{C|w|}{(|w'|^2 + |y_1|)^n} \quad \text{for } z \in U_1 \quad w \in U_2 \\ &\le \frac{C}{(|w'|^2 + |y_1|)^{n-\frac{1}{2}}} \end{aligned}$$

where C is a uniform positive constant. The right side is a locally integrable function of $w = (y_1, w') \in \mathbb{R}^{2n-1}$ in view of Lemma 5 in Section 20.4 (with $q = n - (1/2)$ and $p = 1/2$). This completes the proof that the R kernel has locally integrable coefficients.

For $E(u^1, u^2, u^3)$, we must examine the term

$$K_1(w,z) = \frac{k_1(w,z)}{p(w,z)^q p^*(w,z)^r |\Psi_z^{-1}(w) - z|^{2s}}, \qquad q + r + s = n.$$

This time, $k_1(w,z) = \mathcal{O}(|w|^2)$ in view of the discussion at the beginning of the proof of Lemma 1. Lemma 2 yields the estimate

$$\begin{aligned} |K_1(w,z)| &\le \frac{C|w|^2}{(|w'|^2 + |y_1|)^{q+r}|w|^{2s}} \\ &\le \frac{C}{(|w'|^2 + |y_1|)^{q+r}|w'|^{2s-2}}. \end{aligned}$$

Since $s \ge 1$ and $q + r + s = n$, the right side is locally integrable in $w = (y_1, w') \in \mathbb{R}^{2n-1}$ by Lemma 5 in Section 20.4.

Part (b) of Lemma 1 also follows from the above estimate and by integrating the right side first with respect to y_1 and then with respect to w', i.e.,

$$\begin{aligned} \int\limits_{\substack{|y_1| \le \epsilon \\ |w'| \le \epsilon}} \int \frac{dy_1 \, dv(w')}{(|w'|^2 + |y_1|)^{q+r}|w'|^{2s-2}} &\le C \int\limits_{|w'| \le \epsilon} \frac{dv(w')}{|w'|^{2n-4}} \\ &\le C_1 \epsilon^2 \end{aligned}$$

where C_1 is a uniform positive constant. The last inequality follows from a standard polar coordinate integral calculation in $\mathbb{C}^{n-1} \simeq \mathbb{R}^{2n-2}$. This completes the proof of Lemma 1. ∎

As mentioned earlier, Lemma 1 shows that equation (2) holds across Δ from which the first part of Theorem 1 follows.

The only thing remaining in the proof of Theorem 1 is to establish the C^N-estimate on $R(f)$. Pulling back the integral appearing in $R(f)(z)$ to $\mathbb{R}^{2n-1}$ via Ψ_z^{-1}, we see that we must examine a term of the form

$$X_z^1 \dots X_z^N \left\{ \int_{w \in \mathbb{R}^{2n-1}} K_1(w,z) f_1(w,z) dv(w) \right\} \tag{7}$$

where K_1 is given in (6) and $f_1(w,z)$ is a coefficient of the pull back of the form f via the map $\zeta = \Psi_z^{-1}(w)$. Here, $X_z^1, \dots, X_z^N$ are tangential vector fields to M. The desired C^N-estimate can be established by differentiating under the integral sign which is valid provided we show

$$|X_z^1 \dots X_z^N \{K_1(w,z)\}|$$

is dominated by a locally integrable function in $w = (y_1, w') \in \mathbb{R}^{2n-1}$ uniformly in $z \in U_1 \subset M$. Since we already know that $K_1(w,z)$ is locally integrable in (y_1, w'), it suffices to show that differentiating $K_1(w,z)$ with a tangential vector field in z does not worsen the estimates. As with the proofs of Theorems 1 and 2 in Section 20.4, the key idea is to note that if g is a smooth, complex-valued function with $g(w,z) = \mathcal{O}(w)^j$ for $w \in \mathbb{R}^{2n-1}$ and $z \in M$ for some $j \geq 0$, then also $X_z\{g(w,z)\} = \mathcal{O}(w)^j$ provided X_z is a tangential vector field to M. From parts (a) (with $j = 1$), (c) and (d) of Lemma 2, we obtain

$$\begin{aligned} X_z\{k_1(w,z)\} &= \mathcal{O}(w) \\ X_z\{p(w,z)\} &= X_z\{\mathrm{Re}p(w,z)\} = \mathcal{O}(w)^2 \\ X_z\{p^*(w,z)\} &= X_z\{p^*(w,z) - p(w,z)\} + X_z\{p(w,z)\} \\ &= \mathcal{O}(w)^2. \end{aligned} \tag{8}$$

Note that Lemma 2 only holds for $z \in M$ and therefore the above estimates only hold for vector fields that are tangential to M.

Now we show that the estimates are no worse when we differentiate $K_1(w,z)$ with a tangential vector field. For example, we have

$$\begin{aligned} |X_z\{(p(w,z))^{-q}\}| &= q|X_z\{p(w,z)\}||p(w,z)|^{-q-1} \\ &\leq \frac{C|w|^2}{(|w'|^2 + |y_1|)^{q+1}} \quad \text{(from Lemma 2 and (8))} \\ &\leq \frac{C}{(|w'|^2 + |y_1|)^q} \end{aligned}$$

where C is some uniform positive constant that is independent of $z \in U_1 \subset M$ and $w = (y_1, w') \in U_2 \subset \mathbb{R}^{2n-1}$. This is the same estimate that is satisfied by $|p(w, z)^{-q}|$. Repeating the above arguments we can establish the following: if $X_z^1, \ldots, X_z^N$ are tangential vector fields to M, then there is a uniform positive constant C such that

$$\begin{aligned} |X_z^1 \ldots X_z^N \{K_1(w, z)\}| &\leq \frac{C|w|}{(|w'|^2 + |y_1|)^n} \\ &\leq \frac{C}{(|w'|^2 + |y_1|)^{n-\frac{1}{2}}} \end{aligned}$$

for $z \in U_1 \subset M$ and $w = (y_1, w') \in \mathbb{R}^{2n-1}$. Since the right side is locally integrable in $w \in \mathbb{R}^{2n-1}$ (by Lemma 5 in Section 20.4), we can differentiate under the integral sign in (7) and the proof of the desired C^N-estimate is complete.

22

A Local Solution to the Tangential Cauchy–Riemann Equations

In Chapter 21, we constructed the R kernel which is a biregular fundamental solution for the tangential Cauchy–Riemann complex on a strictly convex hypersurface. This kernel is analogous to the Bochner–Martinelli kernel which is a biregular fundamental solution for $\overline{\partial}$ on $\mathbb{C}^n$. In Theorem 1 in Section 20.3, we used the Bochner–Martinellí kernel together with a kernel of Henkin to construct a solution to the $\overline{\partial}$-equation on a strictly convex domain in $\mathbb{C}^n$. In this chapter, we shall use a similar procedure with the R kernel to construct a local solution to the tangential Cauchy–Riemann equations $\overline{\partial}_M u = f$ on a strictly pseudoconvex hypersurface. As with global solvability, there is an obstruction to the local solvability of the tangential Cauchy–Riemann equations at top degree. In Chapter 23, we will discuss necessary and sufficient conditions for local solvability at the top degree.

In this chapter, our goal is to prove the following theorem of Henkin's.

THEOREM 1

(See [He3].) Suppose M is a smooth real hypersurface in $\mathbb{C}^n$, $n \geq 3$, and let z_0 be a point in M. Suppose that M is strictly pseudoconvex at z_0. Then there is a local neighborhood basis C of open sets in M about z_0 with the following property. Suppose $\omega \in C$ and let f be a (p, q)-form, $1 \leq q \leq n-2$ which is C^1 on $\overline{\omega}$, with $\overline{\partial}_M f = 0$ on ω, then there exists a $(p, q-1)$-form u which is of class C^1 on ω with $\overline{\partial}_M u = f$ on ω.

The proof also exhibits a solution u by integral kernels.

By a local neighborhood basis about z_0, we mean a collection of open sets C in M about z_0 with the property

$$\cap_{\omega \in C} \omega = \{z_0\}.$$

Note that we are not shrinking the set on which we solve the tangential Cauchy–Riemann equations. That is, if f is $\overline{\partial}_M$-closed on $\omega \in C$, then there is a solution

u for the equation $\overline{\partial}_M u = f$ on all of ω. However, we are making no claims as to the regularity of the solution u at the boundary of ω. More will be said about boundary regularity in Chapter 24.

We should point out that not just any local neighborhood basis in M about z_0 will satisfy the above theorem. Such a neighborhood basis must be specially constructed as we shall do below. This is analogous to the situation with $\overline{\partial}$ on domains in $\mathbb{C}^n$. Using Theorem 1 in Section 20.3, we can find a local neighborhood basis consisting of balls in $\mathbb{C}^n$ on which we can solve the $\overline{\partial}$-equation. The reader familiar with the theory of several complex variables knows that the $\overline{\partial}$-equation cannot be solved on an arbitrary open neighborhood of a given point z_0 in $\mathbb{C}^n$. Such a set must be a domain of holomorphy. On a strictly convex hypersurface, it is unknown how much flexibility one has in constructing the neighborhood basis on which the tangential Cauchy–Riemann equations can be solved.

The rest of this chapter is devoted to the proof of Theorem 1. Since our analysis is local, we may choose holomorphic coordinates for $\mathbb{C}^n$ as in Theorem 1 in Section 10.3 so that M is a strictly convex hypersurface (near the origin) in $\mathbb{C}^n$. More precisely, we choose coordinates (z_1, z') for $\mathbb{C}^n$ so that the given point z_0 is the origin and so that a defining function for M is

$$\rho(z) = \text{Im} z_1 - h(\text{Re} z_1, z')$$

where $h : \mathbb{R} \times \mathbb{C}^{n-1} \mapsto \mathbb{R}$ is smooth with $h(0) = 0$, $Dh(0) = 0$ and we assume the real hessian of h at the origin is negative definite. We set $D^- = \{z \in \mathbb{C}^n;\ \rho(z) < 0\}$ and $D^+ = \{z \in \mathbb{C}^n;\ \rho(z) > 0\}$. Note that if U is a small enough ball centered at the origin then $U \cap D^-$ is convex.

As in Chapters 20 and 21, we let $u^1(\zeta, z) = (\partial\rho(\zeta)/\partial\zeta)$ and $u^2(\zeta, z) = (\partial\rho(z)/\partial z)$ and we form the kernels

$$\begin{aligned} L &= \{E(u^1)\}_{{}^tM \times M} \\ L^t &= \{E(u^2)\}_{{}^tM \times M} \\ R &= \{E(u^1, u^2)\}_{{}^tM \times M}. \end{aligned}$$

We restate the fundamental identity for these kernels

$$\overline{\partial}_M \circ R + R \circ \overline{\partial}_M + (L^t)^+ - L^- = I \tag{1}$$

(as operators on $\mathcal{D}^*_M(M \cap U)$).

Now we define our local neighborhood basis for Theorem 1. For $\lambda > 0$, let

$$\begin{aligned} \omega_\lambda &= \{z \in M;\ \text{Im} z_1 > -\lambda\} \\ &= \{z \in M;\ h(z) > -\lambda\}. \end{aligned}$$

Since $h(0) = 0$, $Dh(0) = 0$ and since the real hessian of h at the origin is negative definite, the diameter of ω_λ is proportional to $\sqrt{\lambda}$ provided λ is suitably small. So the collection

$$C = \{\omega_\lambda;\ \lambda > 0\}$$

is a local neighborhood basis for the origin in M.

We fix a small $\lambda > 0$ and show that the equation $\overline{\partial}_M u = f$ can be solved on ω_λ where f is a smooth (p,q)-form on $\overline{\omega_\lambda}$, $1 \le q \le n-2$, with $\overline{\partial}_M f = 0$ on ω_λ.

Define

$$u^3(\zeta, z) = (1, 0 \dots 0) \in \mathbb{C}^n .$$

We form the kernels

$$\begin{aligned} E_{123} &= \{E(u^1, u^2, u^3)\}_{t_{M\times M}} \\ E_{23} &= \{E(u^2, u^3)\}_{t_{M\times M}} \\ E_{13} &= \{E(u^1, u^3)\}_{t_{M\times M}} . \end{aligned}$$

Since $\overline{\partial}_{\zeta,z} u^3(\zeta,z) = 0$, we have $E(u^3) = 0$.

Note that E_{123} is self-adjoint up to a sign and $E_{23}^t = E_{13}$. The kernels E_{123}, E_{23}, and E_{13} are smoothly defined on the set

$$V = \{(\zeta, z) \in \{M \cap U\} \times \{M \cap U\};\ \zeta_1 \ne z_1\}.$$

This is because $u^3(\zeta,z)\cdot(\zeta - z) = \zeta_1 - z_1$ and because $u^1(\zeta,z)\cdot(\zeta-z)$ and $u^2(\zeta,z)\cdot(\zeta-z)$ are nonvanishing on $M \times M - \Delta$. Therefore, these kernels are biregular on the set V. This means that if U_1 and U_2 are open sets in M with $U_1 \times U_2 \subset V$, then these kernels represent continuous operators from $\mathcal{D}_M^*(U_1)$ to $\mathcal{E}_M^*(U_2)$. Moreover, these kernels extend to operators from $\mathcal{E}_M'^*(U_1)$ (compactly supported currents on U_1) to $\mathcal{D}_M'^*(U_2)$. From part (b) of Theorem 1 in Section 20.2, we have

$$\overline{\partial}_{M\times M} E_{123} = -R + E_{13} - E_{23} \quad \text{on } V. \tag{2}$$

In Lemma 3 in Section 20.3, we saw that the L kernel acts nontrivially only on forms of bidegree $(p,0)$ and L^t acts nontrivially only on forms of bidegree $(p, n-1)$. The following lemma describes the analogous behavior for the kernels E_{13} and E_{23}.

LEMMA 1

Suppose U_1 and U_2 are open sets in M with $U_1 \times U_2 \subset V$. Suppose g is a compactly supported current of bidegree (p,q) on U_1.

(a) *If $q \ne 1$, then $E_{13}(g) = 0$ on U_2.*

(b) *If $q \ne n-1$, then $E_{23}(g) = 0$ on U_2.*

PROOF As with the proof of Lemma 3 in Section 20.3, note that u^1 and u^3 are holomorphic in z and so the degree of $E_{13}(\zeta, z)$ in $d\overline{\zeta}$ is $n-2$ (see the formula for $E(u^1, u^3)$ given in Section 20.1). For $z \in U_2$, we have

$$E_{13}(g)(z) = \int_{\zeta \in U_1} \left[E_{13}(\zeta, z) \wedge g(\zeta)\right]^{n,n-1}$$
$$= 0$$

unless g is a form of bidegree $(p, 1)$. The above integral formula is well defined provided g is a form with continuous coefficients on U_1. If g is a more general current, then the above integral gets replaced by the pairing $\langle , \rangle_{\zeta \in M}$, keeping in mind the fact that E_{13} has smooth coefficients on $U_1 \times U_2$, and the same conclusion holds.

Part (b) follows by the same reasoning and by noting that the degree of $E_{23}(\zeta, z)$ in $d\overline{\zeta}$ is zero. ∎

Fix a smooth (p,q)-form f defined on $\overline{\omega_\lambda}$ with $\overline{\partial}_M f = 0$ on ω_λ. To solve the equation $\overline{\partial}_M u = f$ on ω_λ, we start by applying the R kernel to the current $\chi_\lambda f$ where χ_λ is the characteristic function on the set ω_λ. Since R is a bigregular kernel on $\{M \cap U\} \times \{M \cap U\}$, $R(\chi_\lambda f)$ is a well-defined current on $M \cap U$. In fact, since R has locally integrable coefficients (see Lemma 1 in Section 21.2) and since $\chi_\lambda f$ is bounded, $R(\chi_\lambda f)$ is a form with continuous coefficients on $U \cap M$. Also note that if $1 \le q \le n-2$, then $L^-(\chi_\lambda f) = 0$ and $(L^t)^+(\chi_\lambda f) = 0$ in view of Lemma 3 in Section 20.3. From (1), we obtain

$$\begin{aligned}\chi_\lambda f &= \overline{\partial}_M\{R(\chi_\lambda f)\} + R(\overline{\partial}_M\{\chi_\lambda f\}) \\ &= \overline{\partial}_M\{R(\chi_\lambda f)\} - R([\partial\omega_\lambda]^{0,1} \wedge f).\end{aligned}$$

The second equality uses the fact that $\overline{\partial}_M \chi_\lambda = -[\partial\omega_\lambda]^{0,1}$ as a current on M. This follows from the equation $\overline{\partial}_M \chi_\lambda = [d_M(\chi_\lambda)]^{0,1} = -[\partial\omega_\lambda]^{0,1}$ (by Stokes' theorem).

It is instructive to compare the above equation with the analogous $\overline{\partial}$-equation involving the Bochner–Martinelli kernel appearing in the proof of Theorem 1 in Section 20.3 In the proof of that theorem, we used the equation $\overline{\partial} H = L - B$ to rewrite the term $B([M]^{0,1} \wedge f)$ as $-\overline{\partial}\{H([M]^{0,1} \wedge f)\}$. Here, we want to use equation (2) to show that $R([\partial\omega_\lambda]^{0,1} \wedge f)$ belongs to the range of $\overline{\partial}_M$. With the help of part (c) of Theorem 1 in Section 18.2 and using $\overline{\partial}_M\{[\partial\omega_\lambda]^{0,1} \wedge f\} = -[\partial\omega_\lambda]^{0,1} \wedge \overline{\partial}_M f = 0$, we obtain

$$R([\partial\omega_\lambda]^{0,1}\wedge f) = \overline{\partial}_M\{E_{123}([\partial\omega_\lambda]^{0,1}\wedge f)\} + E_{13}([\partial\omega_\lambda]^{0,1}\wedge f) - E_{23}([\partial\omega_\lambda]^{0,1}\wedge f)$$

on ω_λ. If K is either E_{123}, E_{23}, or E_{13}, then $K([\partial\omega_\lambda]^{0,1} \wedge f)$ is smoothly defined on ω_λ. For if $z \in \omega_\lambda$ and $\zeta \in \partial\omega_\lambda$ then $\text{Im}z_1 > -\lambda$ and $\text{Im}\zeta_1 = -\lambda$ and so $\zeta_1 - z_1 \ne 0$. Hence, the denominators of $K(\zeta, z)$ are nonvanishing for $z \in \omega_\lambda$ and $\zeta \in \partial\omega_\lambda$.

If f is a (p,q)-form with $1 \le q \le n-3$, then $[\partial\omega_\lambda]^{0,1} \wedge f$ is a current of degree $(p,q+1)$ and $2 \le q+1 \le n-2$. So we have

$$E_{13}([\partial\omega_\lambda]^{0,1} \wedge f) = 0$$
$$E_{23}([\partial\omega_\lambda]^{0,1} \wedge f) = 0$$

in view of Lemma 1. Combining this with the previous two equations, we obtain a solution to the equation $\overline{\partial}_M u = f$ on ω_λ with

$$\begin{aligned} u(z) &= R(\chi_\lambda f)(z) - E_{123}([\partial\omega_\lambda]^{0,1} \wedge f)(z) \\ &= \int_{\zeta\in\omega_\lambda} R(\zeta,z) \wedge f(\zeta) - \int_{\zeta\in\partial\omega_\lambda} \left[E_{123}(\zeta,z) \wedge f(\zeta)\right]^{n,n-2}. \end{aligned} \tag{3}$$

(Compare this equation with the solution $u = B(\chi_D f) + H([\partial D]^{0,1} \wedge f)$ of the equation $\overline{\partial} u = f$ on a convex domain D in $\mathbb{C}^n$ given in Theorem 1 in Section 20.3.)

Theorem 1 also states that the equation $\overline{\partial}_M u = f$ can be solved when f is a form of bidegree $(p, n-2)$. In this case, $[\partial\omega_\lambda]^{0,1} \wedge f$ is a current of bidegree $(p, n-1)$ and therefore Lemma 1 cannot be used to show that the term $E_{23}([\partial\omega_\lambda]^{0,1} \wedge f)$ vanishes. However, the next lemma uses an approximation argument to show that $E_{23}([\partial\omega_\lambda]^{0,1} \wedge f) = 0$ and therefore the above form u solves the equation $\overline{\partial}_M u = f$ on ω_λ even when f is a form of bidegree $(p, n-2)$.

LEMMA 2

Suppose f is a smooth form of bidegree $(p, n-2)$ on $\overline{\omega_\lambda}$ with $\overline{\partial}_M f = 0$ on ω_λ. Then

$$E_{23}([\partial\omega_\lambda]^{0,1} \wedge f) = 0.$$

PROOF First note that the coefficients of $E_{23}(\zeta, z)$ are holomorphic in ζ provided the denominators of E_{23} are nonvanishing. We are tempted to write

$$[\partial\omega_\lambda]^{0,1} \wedge f = -\overline{\partial}_M\{\chi_\lambda f\} \quad (\text{since } \overline{\partial}_M f = 0)$$

and then integrate by parts with $\overline{\partial}_M$ to show

$$\begin{aligned} E_{23}(\overline{\partial}_M\{\chi_\lambda f\})(z) &= -\int_{\zeta\in\omega_\lambda} \overline{\partial}_\zeta\{E_{23}(\zeta,z)\} \wedge f(\zeta) \\ &= 0 \end{aligned}$$

for $z \in \omega_\lambda$. However, this argument is not valid because $E_{23}(\zeta, z)$ is not smoothly defined for $\zeta, z \in \omega_\lambda$ (when $\zeta_1 = z_1$) and so the integration by parts with $\overline{\partial}_M$ is not allowed.

Instead, we first approximate the coefficients of $E_{23}(\zeta, z)$ by entire functions of ζ and then we apply the above integration by parts argument. To carry this out, we fix $z \in \omega_\lambda$. Since f has bidegree $(p, n-2)$, we have

$$E_{23}([\partial\omega_\lambda]^{0,1} \wedge f)(z) = \int_{\zeta\in\partial\omega_\lambda} [E_z(\zeta) \wedge f(\zeta)]^{n,n-2}$$

where $E_z(\zeta)$ is a form of bidegree $(n-p, 0)$ in ζ. The coefficients of $E_z(\zeta)$ are holomorphic in ζ provided the denominators

$$\frac{\partial\rho(z)}{\partial z} \cdot (\zeta - z) \quad \text{and} \quad (\zeta_1 - z_1)$$

are nonvanishing. For fixed $z \in \omega_\lambda$, let

$$A = \{\zeta \in U \subset \mathbb{C}^n;\ \mathrm{Im}\zeta_1 = -\lambda \text{ and } \rho(\zeta) \le 0\}.$$

If $\zeta \in A$ (so $\mathrm{Im}\zeta_1 = -\lambda$) and if $z \in \omega_\lambda$ (so $\mathrm{Im} z_1 > -\lambda$) then $\zeta_1 - z_1 \neq 0$. Moreover, part (a) of Lemma 1 in Section 20.3 with the roles of z and ζ reversed implies that if $(\partial\rho(z)/\partial z) \cdot (\zeta - z) = 0$ then either $\zeta = z$ or $\rho(\zeta) > \rho(z)$. So if $\zeta \in A$ and $z \in \omega_\lambda \subset M$ (i.e., $\rho(\zeta) \le 0 = \rho(z)$ and $\zeta \neq z$) then $(\partial\rho(z)/\partial z) \cdot (\zeta - z) \neq 0$. Therefore if $z \in \omega_\lambda$, then the coefficients of $E_z(\zeta)$ are holomorphic in ζ for ζ in a neighborhood of A in $\mathbb{C}^n$.

Since A is a convex set, A is polynomially convex. Hence, a function which is holomorphic in a neighborhood of A can be uniformly approximated by a sequence of entire functions (see [Ho]). It follows that for fixed $z \in \omega_\lambda$, $E_z(\zeta)$ can be uniformly approximated for $\zeta \in A$ by a sequence of $(n-p, 0)$ forms $G_j(\zeta)$, $j = 1, 2, \ldots$, whose coefficients are entire functions of $\zeta \in \mathbb{C}^n$. Since $\partial\omega_\lambda \subset A$, we have

$$\begin{aligned}
\int_{\zeta\in\partial\omega_\lambda} E_z(\zeta) \wedge f(\zeta) &= \lim_{j\mapsto\infty} \int_{\zeta\in\partial\omega_\lambda} G_j(\zeta) \wedge f(\zeta) \\
&= \lim_{j\mapsto\infty} \int_{\zeta\in\omega_\lambda} d_\zeta\{G_j(\zeta) \wedge f(\zeta)\} \quad \text{(by Stokes' theorem)} \\
&= \lim_{j\mapsto\infty} \int_{\zeta\in\omega_\lambda} \overline{\partial}_M\{G_j(\zeta) \wedge f(\zeta)\}.
\end{aligned}$$

The last equation follows from the fact that $G_j \wedge f$ has bidegree $(n, n-2)$. Since G_j is entire and $\overline{\partial}_M f = 0$, the last limit vanishes and the proof of Lemma 2 is complete. ▮

As mentioned earlier, this lemma implies that the form u given in (3) is a solution to the equation $\overline{\partial}_M u = f$ on ω_λ when f is a $\overline{\partial}_M$-closed $(p, n-2)$-form. The proof of Theorem 1 is now complete. ▮

Let us examine the above analysis in the case when f is a smooth CR function defined on ω_λ. We shall see that the above formulas yield an integral kernel representation of the local holomorphic extension of f to the convex side of $\omega_\lambda \subset M$ given in Hans Lewy's CR extension theorem (Theorem 1 in Section 14.1).

First, note that if f is a CR function on ω_λ, then (1) yields

$$\chi_\lambda f = -R([\partial\omega_\lambda]^{0,1} \wedge f) - L^-(\chi_\lambda f).$$

Using (2), we obtain

$$R([\omega_\lambda]^{0,1} \wedge f) = E_{13}([\partial\omega_\lambda]^{0,1} \wedge f).$$

So f is the boundary values on ω_λ from D^- of the function

$$-E_{13}([\partial\omega_\lambda]^{0,1} \wedge f) - L(\chi_\lambda f) \quad \text{on } \omega_\lambda. \tag{4}$$

Both $E_{13}(\zeta, z)$ and $L(\zeta, z)$ are holomorphic in z provided $u(\zeta, z) \cdot (\zeta - z) \neq 0$ and $\zeta_1 \neq z_1$. Therefore, by part (a) of Lemma 1 in Section 20.3, $L(\chi_\lambda f)(z)$ is holomorphic in z for $z \in D^-$. By the same reasoning, $E_{13}([\partial\omega_\lambda]^{0,1} \wedge f)(z)$ is holomorphic in z for $z \in D^-$ with $\text{Im} z_1 > -\lambda$. Thus, the function in (4) is the holomorphic extension of f to the set $\{z \in \mathbb{C}^n;\ z \in D^- \text{ and } \text{Im} z_1 > -\lambda\}$.

23

Local Nonsolvability of the Tangential Cauchy–Riemann Complex

As was seen in Chapter 22, there is an obstruction to the local solvability of the tangential Cauchy–Riemann equations at the top degree. In this chapter, we discuss this obstruction in more detail. The system of tangential Cauchy–Riemann equations at the top degree is no longer an overdetermined system of partial differential equations. For example, if f is a form of bidegree $(n, n-1)$ on a real hypersurface in $\mathbb{C}^n$, then the equation $\overline{\partial}_M u = f$ consists of one partial differential equation and one unknown coefficient function. If M and f are real analytic, then by the Cauchy–Kowalevsky theorem, there is a local solution to the tangential Cauchy–Riemann equations. For some time, it was thought that "C^∞" could replace "real analytic" in the statement of the Cauchy–Kowalevsky theorem. However, in 1957, Hans Lewy [L2] found a counterexample which we present in Section 23.1. We then show that Hans Lewy's example can be recast in the language of the tangential Cauchy–Riemann complex of the Heisenberg group. In particular, Hans Lewy's example provides an example of the local nonsolvability of the tangential Cauchy–Riemann equations at the top degree. In Section 23.2, we consider a more general real analytic, strictly pseudoconvex hypersurface in $\mathbb{C}^n$ and we present Henkin's criterion on a smooth form f for the local solvability of the tangential Cauchy–Riemann equations $\overline{\partial}_M u = f$.

23.1 Hans Lewy's nonsolvability example

Give $\mathbb{R}^3$ the coordinates (x_1, x_2, y_2). Let $z_2 = x_2 + iy_2 \in \mathbb{C}$. Define the following differential operator on $\mathbb{R}^3$:

$$\overline{L} = \frac{\partial}{\partial \overline{z}_2} - iz_2 \frac{\partial}{\partial x_1}.$$

THEOREM 1

(See [L2].) Suppose f is a continuous real-valued function depending only on x_1. If there is a C^1 function u of (x_1, x_2, y_2) that satisfies $\overline{L}u = f$ in some neighborhood of the origin, then f is real analytic at $x_1 = 0$.

If f is a smooth function of x_1 that is not real analytic, then Hans Lewy's theorem implies that the equation $\overline{L}u = f$ has no locally defined C^1 solution. Therefore, the Cauchy–Kowalevsky theorem does not hold with "real analytic" replaced by "C^∞".

PROOF We follow the presentation given in [Fo]. Suppose the equation $\overline{L}u = f$ has a C^1 solution u defined on the set $\{(x_1, z_2); |x_1| < R \text{ and } |z_2| < R\}$. We write z_2 in polar coordinates as $z_2 = re^{i\phi}$. For $0 \leq r \leq R$ and $|x_1| < R$, define $V(x_1, r)$ by

$$V(x_1, r) = \int_{|z_2|=r} u(x_1, z_2)dz_2.$$

By Stokes' theorem, we have

$$\begin{aligned} V(x_1, r) &= \int_{|z_2|\leq r} d_{z_2}\{u(x_1, z_2)dz_2\} \\ &= \int_{|z_2|\leq r} \frac{\partial u}{\partial \overline{z}_2}(x_1, z_2)d\overline{z}_2 \wedge dz_2 \\ &= 2i \int_0^{2\pi} \int_0^r \frac{\partial u}{\partial \overline{z}_2}(x_1, \rho e^{i\phi})\rho d\rho d\phi. \end{aligned}$$

Therefore

$$\begin{aligned} \frac{\partial V(x_1, r)}{\partial r} &= 2i \int_0^{2\pi} \frac{\partial u}{\partial \overline{z}_2}(x_1, re^{i\phi})rd\phi \\ &= 2 \int_{|z_2|=r} \frac{\partial u}{\partial \overline{z}_2}(x_1, z_2)r\frac{dz_2}{z_2}. \end{aligned}$$

Now, we let $s = r^2$ and we think of V as a function of x_1 and s, which is C^1 on the region $\{|x_1| < R \text{ and } 0 < s < R^2\}$. We obtain

$$\begin{aligned} \frac{\partial V}{\partial s} &= \frac{\partial V}{\partial r}\frac{dr}{ds} \\ &= \frac{\partial V}{\partial r}\frac{1}{2r} \\ &= \int_{|z_2|=r} \frac{\partial u}{\partial \overline{z}_2}(x_1, z_2)\frac{dz_2}{z_2}. \end{aligned}$$

We use the equation $\overline{L}u = f$ to obtain

$$\frac{\partial V}{\partial s} = i \int\limits_{|z_2|=r} \frac{\partial u}{\partial x_1}(x_1, z_2)dz_2 + f(x_1) \int\limits_{|z_2|=r} \frac{dz_2}{z_2}$$

$$= i\frac{\partial V}{\partial x_1}(x_1, z_2) + 2\pi i f(x_1). \tag{1}$$

We set $F(x_1) = \int_0^{x_1} f(t)dt$ and we define

$$U(x_1, s) = V(x_1, s) + 2\pi F(x_1).$$

Equation (1) implies that U is holomorphic as a function of the complex variable $w = x_1 + is$ in the region $\{|x_1| < R \text{ and } 0 < s < R^2\}$. From the definition of V, we have $V(x_1, 0) = 0$. Therefore, U is continuous up to $s = 0$ (from $s > 0$) and $U(x_1, 0) = 2\pi F(x_1) \in \mathbb{R}$. By the Schwarz reflection principle, U holomorphically continues to the region $\{|x_1| < R \text{ and } |s| < R^2\}$. In particular, $U(x_1, 0) = 2\pi F(x_1)$ is a real analytic function of x_1 for $|x_1| < R$. So $F' = f$ is also a real analytic function of x_1 for $|x_1| < R$, as desired. ∎

The vector field $\overline{L}$ in Hans Lewy's counterexample also arises from the tangential Cauchy–Riemann complex of the Heisenberg group $M = \{(z_1, z_2) \in \mathbb{C}^2; \text{Im}z_1 = |z_2|^2\}$. From Theorem 3 in Section 7.2, a basis for $H^{0,1}(M)$ is given by $\tilde{\overline{L}} = (\partial/\partial\overline{z}_2) - 2iz_2(\partial/\partial\overline{z}_1)$. Define the map $\pi : \mathbb{C}^2 \mapsto \mathbb{R} \times \mathbb{C}$ by $\pi(z_1, z_2) = (x_1, z_2)$ where $x_1 = \text{Re}z_1$. We have $\pi_*\tilde{\overline{L}} = \overline{L}$. Therefore, the equation $\overline{L}u = f$ on $\mathbb{R} \times \mathbb{C}$ is equivalent to the equation $\tilde{\overline{L}}\tilde{u} = \tilde{f}$ on M where $\tilde{u}(x_1 + i|z_2|^2, z_2) = u(x_1, z_2)$ and $\tilde{f}(x_1 + i|z_2|^2, z_2) = f(x_1, z_2)$.

We claim the equation $\tilde{\overline{L}}\tilde{u} = \tilde{f}$ on M is equivalent to $\overline{\partial}_M\{\tilde{u}\} = \tilde{f}(d\overline{z}_2)_{t_M}$. To see this, we first note that $\rho(z_1, z_2) = |z_2|^2 - \text{Im}z_1$ is a defining function for M. In order to compute the tangential projection map t_M and the $\overline{\partial}_M$-operator, we need the following dual vector field $\mathbb{N}$ to $\overline{\partial}\rho$:

$$\mathbb{N} = (4|z_2|^2 + 1)^{-1}\left(4\overline{z_2}\frac{\partial}{\partial\overline{z}_2} + 2i\frac{\partial}{\partial\overline{z}_1}\right).$$

From Lemma 2 in Section 8.1, we have

$$\overline{\partial}_M\tilde{u} = \mathbb{N}\lrcorner(\overline{\partial}\rho \wedge \overline{\partial}\tilde{u})$$

$$= \left(z_2\frac{\partial\tilde{u}}{\partial\overline{z}_1} + \frac{i}{2}\frac{\partial\tilde{u}}{\partial\overline{z}_2}\right)\left(\frac{4\overline{z}_2 d\overline{z}_1 - 2id\overline{z}_2}{4|z_2|^2 + 1}\right)$$

and

$$\tilde{f}\{d\overline{z}_2\}_{t_M} = \tilde{f}\mathbb{N}\lrcorner(\overline{\partial}\rho \wedge d\overline{z}_2)$$

$$= \frac{i}{2}\tilde{f}\left(\frac{4\overline{z}_2 d\overline{z}_1 - 2id\overline{z}_2}{4|z_2|^2 + 1}\right).$$

Therefore, the equation $\overline{\partial}_M \tilde{u} = \tilde{f}(d\overline{z}_2)_{tM}$ is equivalent to the equation

$$\frac{\partial \tilde{u}}{\partial \overline{z}_2} - 2iz_2 \frac{\partial \tilde{u}}{\partial \overline{z}_1} = \tilde{f}$$

as claimed. So Theorem 1 implies the following local solvability (or nonsolvability) result for the tangential Cauchy–Riemann complex on the Heisenberg group.

COROLLARY 1
Let $M = \{(z_1, z_2) \in \mathbb{C}^2;\ Imz_1 = |z_2|^2\}$ and let f be a smooth, real-valued function of Rez_1. Suppose there exists a smooth function u defined on a neighborhood of the origin in M such that $\overline{\partial}_M u = f\{d\overline{z}_2\}_{tM}$, then f must be real analytic in a neighborhood of the origin.

The Heisenberg group $\{\mathrm{Im}z_1 = |z_2|^2\}$ is biholomorphic to the unit sphere via the Cayley transform: $\Phi : \mathbb{C}^2 \mapsto \mathbb{C}^2$, $\Phi(w_1, w_2) = (z_1, z_2)$ with

$$z_1 = i\left(\frac{1 - w_1}{1 + w_1}\right)$$

$$z_2 = \frac{w_2}{1 + w_1}.$$

In these coordinates, Φ takes the unit sphere in $\mathbb{C}^2$ biholomorphically to the Heisenberg group. Therefore, the above nonsolvability example on the Heisenberg group can be carried over to a nonsolvability example on the unit sphere. Another example of the local nonsolvability of the tangential Cauchy–Riemann equations on the unit sphere will be given at the end of the next section.

23.2 Henkin's criterion for local solvability at the top degree

In this section, we discuss local solvability of the tangential Cauchy–Riemann complex at the top degree for a general real analytic strictly pseudoconvex hypersurface M in $\mathbb{C}^n$, $n \geq 2$. As in Chapter 22, we may use a local biholomorphic change of coordinates and assume M is a strictly convex hypersurface in an open neighborhood U of a given point z_0 in $\mathbb{C}^n$. We assume M is defined by $\{z \in U; \rho(z) = 0\}$ and that M divides U into two open sets $U^+ = \{\rho > 0\}$ and $U^- = \{\rho < 0\}$ with U^- convex. As in Section 20, we construct the kernels L, L^t, H, and H^t. Suppose $f \in \mathcal{E}_M^{p,n-1}(M \cap U)$ and suppose $\phi \in \mathcal{D}(M \cap U)$ with $\phi = 1$ in a neighborhood of z_0. From part (c) of Theorem 1 in Section 21.1, we have

$$\phi f = \overline{\partial}_M\{(H - H^t)(\phi f)\} + (L^t)^+(\phi f). \tag{1}$$

The term $(L^t)^+(\phi f)$ is the obstruction to solving the the equation $\overline{\partial}_M u = f$ in a neighborhood of z_0 in M. This term is analyzed in the following theorem which is due to Henkin [He3].

THEOREM 1

Let M be a strictly convex real analytic hypersurface in an open set U in $\mathbb{C}^n$, $n \geq 2$ and let z_0 be a given point in $M \cap U$. Let $\phi \in \mathcal{D}(M \cap U)$ be identically one on a neighborhood of z_0 in M. If f is a smooth form of bidegree $(n, n-1)$ on $M \cap U$, then the equation $\overline{\partial}_M u = f$ has a locally defined smooth solution near z_0 on M if and only if $(L^t)^+(\phi f)$ is real analytic in a neighborhood of z_0 in M.

Before we prove this theorem, we show the statement "$(L^t)^+(\phi f)$ is real analytic near z_0" is independent of the cut-off function ϕ. For suppose ϕ_1 and ϕ_2 are two such cut-off functions. Then $v = (\phi_1 - \phi_2) f$ vanishes in a neighborhood of z_0 and has compact support in $M \cap U$. Since the defining equation for M (ρ) is real analytic, $L^t(\zeta, z)$ is real analytic in z provided $(\partial\rho(z)/\partial z) \cdot (\zeta - z) \neq 0$. So $(L^t)^+(v)$ is real analytic at a point $z \in M \cap U$ provided

$$\left\{\zeta \in M;\ \frac{\partial\rho(z)}{\partial z} \cdot (\zeta - z) = 0\right\} \cap \{\text{supp} v\} = \emptyset. \tag{2}$$

Since v vanishes in a neighborhood of z_0 and since M is strictly convex in U, part (a) of Lemma 1 in Section 20.3 (with the roles of z and ζ reversed) implies that (2) holds for z in a neighborhood of z_0 in M. So $(L^t)^+(vf)$ is real analytic in a neighborhood of z_0 in M. It follows that

$$(L^t)^+(\phi_1 f) = (L^t)^+(\phi_2 f) + (L^t)^+(vf)$$

is real analytic in a neighborhood of z_0 in M if and only if $(L^t)^+(\phi_2 f)$ is real analytic, as claimed.

PROOF Since the dimension of $\mathcal{E}_M^{n,n-1}(M \cap U)$ over the ring of smooth functions on M is one, the equation $\overline{\partial}_M u = f$ for $f \in \mathcal{E}_M^{n,n-1}(M \cap U)$ consists of one partial differential equation. If $(L^t)^+(\phi f)$ is real analytic in a neighborhood of z_0, then by the Cauchy–Kowalevsky theorem, there is a solution to the equation $\overline{\partial}_M v = (L^t)^+(\phi f)$ near z_0 in M. From (1), we obtain

$$\begin{aligned} f &= \phi f \quad (\text{near } z_0) \\ &= \overline{\partial}_M\{(H - H^t)(\phi f) + v\}. \end{aligned}$$

Conversely, let us suppose on some open neighborhood U_1 of z_0 in M, there is a solution to the equation $\overline{\partial}_M u = f$. Choose a smooth cut-off function ϕ with compact support in U_1 with $\phi = 1$ near z_0. We have

$$(L^t)^+(\phi f) = (L^t)^+(\phi \overline{\partial}_M u)$$

near z_0 on M. Recall that $(L^t)^+(\phi\overline{\partial}_M u)$ is the boundary values from U^+ of the form

$$-\int_{\{\zeta\in M\cap U\}} L^t(\zeta,z)\wedge(\overline{\partial}_M u(\zeta))\phi(\zeta).$$

Also recall that $L^t(\zeta,z)$ is holomorphic in $\zeta\in\overline{U^-}$ for $z\in U^+$. We can integrate by parts with $\overline{\partial}_M$ to obtain

$$(L^t)^+(\phi\overline{\partial}_M u) = -(L^t)^+(\overline{\partial}_M\phi\wedge u).$$

Since ρ is real analytic, $(L^t)^+(\overline{\partial}_M\phi\wedge u)(z)$ is real analytic at $z\in M$ provided

$$\left\{\zeta\in M;\ \frac{\partial\rho(z)}{\partial z}\cdot(\zeta-z)=0\right\}\cap\{\text{supp}\overline{\partial}_M\phi\}=\emptyset. \tag{3}$$

Since $\overline{\partial}_M\phi=0$ near z_0 in M, it follows from part (a) of Lemma 1 in Section 20.3 that (3) holds for z in a neighborhood of z_0 in M. Thus, $(L^t)^+(\phi f) = -(L^t)^+(\overline{\partial}_M\phi\wedge u)$ is real analytic near z_0 on M, as desired. ∎

As an application, we construct examples of local nonsolvability of the tangential Cauchy–Riemann equations on the unit sphere in $\mathbb{C}^n$. In this case, our hypersurface M is compact and so we set $\phi=1$. A defining function for M is $\rho(z)=|z|^2-1$. So $(\partial\rho(z)/\partial z)=\overline{z}$. The set U^- is the open unit ball $\{|z|<1\}$ and the set U^+ is the outside of the unit ball $\{|z|>1\}$. Suppose $f\in\mathcal{E}_M^{n,n-1}$. For $|z|>1$, we have

$$L^t([M]^{0,1}\wedge f)(z)$$
$$=-(2\pi i)^{-n}\int_{|\zeta|=1}\left[\frac{(\overline{z}\cdot d(\zeta-z))\wedge(d(\overline{\zeta-z})\cdot d(\zeta-z))^{n-1}}{(\overline{z}\cdot(\zeta-z))^n}\wedge f(\zeta)\right]^{n,n-1}.$$

Recall that the notation $[\psi(\zeta,z)]^{n,n-1}$ indicates the piece of ψ of bidegree $(n,n-1)$ in ζ. Since f has bidegree $(n,n-1)$, the term $L^t(\zeta,z)$ will only contribute $dz's$ and $d\overline{z}'s$ to the above integral. For $|z|>1$, we have

$$L^t([M]^{0,1}\wedge f)(z)=(-2\pi i)^{-n}\left(\int_{|\zeta|=1}\frac{f(\zeta)}{(1-\frac{\zeta\cdot\overline{z}}{|z|^2})^n}\right)\frac{(\overline{z}\cdot dz)\wedge(d\overline{z}\cdot dz)^{n-1}}{|z|^{2n}}.$$

For $f\in\mathcal{E}_M^{n,n-1}$, define the function If on U^- by

$$If(w)=(-2\pi i)^{-n}\int_{|\zeta|=1}\frac{f(\zeta)}{(1-\zeta\cdot w)^n}.$$

Note that If is holomorphic on the set $U^- = \{|w| < 1\}$. From the above expression for $L^t([M]^{0,1} \wedge f)$, we obtain

$$L^t([M]^{0,1} \wedge f)(z) = \left((If)\left(\frac{\overline{z}}{|z|^2}\right)\right) \frac{(\overline{z} \cdot dz) \wedge (d\overline{z} \cdot dz)^{n-1}}{|z|^{2n}}.$$

Since $L^t([M]^{0,1} \wedge f)$ extends from $U^+ = \{|z| > 1\}$ to a smooth form on $M = \{|z| = 1\}$ (see Theorem 2 in Section 20.4), the above equation shows that If extends from $\{|w| < 1\}$ to a smooth function on M. Since If is holomorphic on $\{|w| < 1\}$, the extension of If to M is a CR function. Also note that $(L^t)^+(f)$ has real analytic coefficients near $z_0 \in M$ if and only if the function $w \mapsto If(\overline{w})$ is real analytic in w near z_0. This in turn is equivalent to the real analyticity of the function $w \mapsto If(w)$ for w near $\overline{z}_0$. From Theorem 1 in Section 9.1, a real analytic CR function on M near a point $\overline{z_0}$ in M extends to a holomorphic function on a neighborhood of $\overline{z}_0$ in $\mathbb{C}^n$. Therefore from Theorem 1, we obtain the following corollary.

COROLLARY 1

The equation $\overline{\partial}_M u = f$ can be solved on $M = \{|z| = 1\}$ near $z_0 \in M$ if and only if the function If extends to a holomorphic function in a neighborhood of $\overline{z}_0$ in $\mathbb{C}^n$.

To construct examples of nonsolvability of the tangential Cauchy–Riemann equations on the sphere near a given point z_0, we first find a holomorphic function $\tilde{f}$ on $\{|z| < 1\}$ which extends smoothly to $M = \{|z| = 1\}$ but does not holomorphically continue past $\overline{z}_0$. For example, if $z_0 = (1, 0, \dots, 0)$ then we can let

$$\tilde{f}(z) = e^{\frac{-1}{\sqrt{1-z_1}}}$$

where we use the principal branch of the square root defined in the right half plane in $\mathbb{C}$. Next we define for $|\zeta| = 1$

$$f(\zeta) = \tilde{f}(\overline{\zeta})(\overline{\zeta} \cdot d\zeta) \wedge (d\overline{\zeta} \cdot d\zeta)^{n-1}.$$

The degree of f is $(n, n-1)$. We have $j^*\{d|\zeta|^2\} = 0$ on M where $j : M \mapsto \mathbb{C}^n$ is the inclusion map. It follows that $j^*(\overline{\zeta} \cdot d\zeta) = -j^*(\zeta \cdot d\overline{\zeta})$. Therefore

$$(j^* f)(\zeta) = -j^*\{(\tilde{f}(\overline{\zeta})(\zeta \cdot d\overline{\zeta})(d\overline{\zeta} \cdot d\zeta)^{n-1}\}$$

on M. We obtain

$$If(w) = (-2\pi i)^{-n} \int_{|\zeta|=1} \frac{-\tilde{f}(\overline{\zeta})(\zeta \cdot d\overline{\zeta}) \wedge (d\overline{\zeta} \cdot d\zeta)^{n-1}}{(1 - \zeta \cdot w)^n}.$$

After the change of variables $\zeta \mapsto \overline{\zeta}$ in the above integral, we obtain

$$\begin{aligned}
&If(w)\\
&= -(2\pi i)^{-n} \int_{|\zeta|=1} \frac{\tilde{f}(\zeta)(\overline{\zeta}\cdot d\zeta) \wedge (d\zeta \cdot d\overline{\zeta})^{n-1}}{(1-\overline{\zeta}\cdot w)^n}\\
&= -(2\pi i)^{-n} \int_{|\zeta|=1} \tilde{f}(\zeta) \wedge \left[\frac{\overline{\zeta}\cdot d(\zeta - w) \wedge (d(\zeta - w)\cdot d(\overline{\zeta - w}))^{n-1}}{(\overline{\zeta}\cdot(\zeta - w))^n}\right]^{n,n-1}.
\end{aligned}$$

The factor of $(-1)^n$ disappears due to the change in orientation of the map $\zeta \mapsto \overline{\zeta}$. From the formula for L (with z replaced by w), we obtain

$$If(w) = L([M]^{0,1} \wedge \tilde{f})(w).$$

Since $\tilde{f}$ is holomorphic on the set $\{|w| < 1\}$, $\tilde{f} = -L([M]^{0,1} \wedge \tilde{f})$ on $\{|w| < 1\}$ by Corollary 2 in Section 21.1. We obtain

$$If = -\tilde{f}$$

on $\{|w| < 1\}$. Since $\tilde{f}$ does not holomorphically continue past the point $\overline{z}_0 = z_0 = (1, 0, \ldots, 0)$, neither does If, and so the equation $\overline{\partial}_M u = f$ cannot be solved in any neighborhood of z_0 in M according to Corollary 1.

24

Further Results

24.1 More on the Bochner–Martinelli kernel

The boundary value result in Theorem 3 in Section 19.2 for the Bochner–Martinelli kernel can be strengthened. If we only assume that f is of class C^α for $0 < \alpha < 1$ on M, then $B([M]^{0,1} \wedge f)$ is of class C^α on $\overline{D^-}$ and $\overline{D^+}$. This can be established by adapting the proof of the corresponding result for the Cauchy kernel in Lemma 2 in Section 15.4. Cauchy's integral formula used in the proof of that lemma must be replaced by the equation

$$\begin{aligned} \int_{\zeta \in M} \{B(\zeta, z) f(\zeta)\}^{n,n-1} &= -B([M]^{0,1} f) \\ &= \chi_{D^-} f \end{aligned} \tag{1}$$

which holds for functions f which are holomorphic on D^- and continuous up to M (see Theorem 3 in Section 18.2). Here, χ_{D^-} is the characteristic function of D^- whose manifold boundary is M. The minus sign in the first equality in (1) results from commuting the current $[M]^{0,1}$ with the $(2n-1)$-form $B(\zeta, z)$. Adapting the computations in the proof of Lemma 2 in Section 9.4 for the Bochner–Martinelli kernel is somewhat tedious since the Bochner–Martinelli kernel is more complicated to unravel than the Cauchy kernel. However, the basic ideas are the same. This result for the Bochner–Martinelli kernel is due to Cirka [C], who also generalized Theorem 1 in Section 19.3 to the case where the ambient space is a Stein manifold (instead of merely $\mathbb{C}^n$).

It is interesting to compare the boundary values of the Bochner–Martinelli kernel with the principal value limits of the Bochner–Martinelli kernel. The latter is defined for $f \in \mathcal{D}_M^*$ by

$$B_b f(z) = \lim_{\epsilon \mapsto 0} \left\{ \int_{\substack{\zeta \in M \\ |\zeta - z| \geq \epsilon}} \{B(\zeta, z) \wedge f(\zeta)\}^{n,n-1} \right\}_{t_M} \qquad z \in M.$$

The following equations hold for $f \in \mathcal{D}_M^*$:

$$B^-(f) = B_b(f) - \frac{1}{2}f \tag{2}$$

$$B^+(f) = B_b(f) + \frac{1}{2}f. \tag{3}$$

Here, M is oriented by the equation $d\chi_{D^-} = -[M]$. These relationships are well known from residue theory for the Cauchy kernel. Note that subtracting these two equations gives the key boundary value jump formula $B^+(f) - B^-(f) = f$ which is crucial in Chapter 21 for the construction of the fundamental solution for $\overline{\partial}_M$. Equations (2) and (3) are due to Harvey and Lawson [HL] for the case where f is a function and to Harvey and Polking [HP] for the case of higher degree forms.

We shall sketch the ideas involved in the proof of (2). The proof of (3) is similar. The proof of (2) can be reduced to the case where f is a smooth function by using Lemma 2 as in the proof of Theorem 3 in Section 19.2. We assume that M is the boundary of a bounded domain D^-. The case where D^- is unbounded can be handled in a similar way by using cutoff functions as in the proof of Theorem 3 in Section 19.2. Fix $z_0 \in M$. If f is a smooth function that vanishes at z_0, then $|B(\zeta, z_0)f(\zeta)|$ is an integrable function of $\zeta \in M$. So $B_b f(z_0)$ exists and $B_b f(z_0) = B^+ f(z_0)$ by the dominated convergence theorem. This proves (2) at z_0 for the case of a smooth function f that vanishes at z_0. For more general f, we may write $f(\zeta) = (f(\zeta) - f(z_0)) + f(z_0)$. Since we know (2) holds at z_0 for the function $\zeta \mapsto f(\zeta) - f(z_0)$, it suffices to show (2) for the constant function $1(\zeta) = 1$.

For $z \in D^-$ and $\epsilon > 0$, we have

$$B([M]^{0,1}1)(z) = -\int\limits_{\substack{\zeta \in M \\ |\zeta - z_0| \geq \epsilon}} B(\zeta, z) - \int\limits_{M_\epsilon} B(\zeta, z) \tag{4}$$

where we have set $M_\epsilon = \{\zeta \in M;\ |\zeta - z_0| \leq \epsilon\}$. Also set $S_\epsilon^+ = \{\zeta \in D^+;\ |\zeta - z_0| = \epsilon\}$ and $D_\epsilon^+ = \{\zeta \in D^+;\ |\zeta - z_0| \leq \epsilon\}$ (see Figure 24.1).

Note that $[\partial D_\epsilon^+] = [S_\epsilon^+] - [M_\epsilon]$, where M_ϵ has the same orientation as M and S_ϵ^+ has the induced boundary orientation from D_ϵ^+. From (1) with D^+ instead of D^-, we have $B([\partial D_\epsilon^+]^{0,1})(z) = -\chi_{D_\epsilon^+}(z) = 0$ for $z \in D^-$. For $z \in D^-$, we obtain

$$\int\limits_{\zeta \in M_\epsilon} B(\zeta, z) = \int\limits_{\zeta \in S_\epsilon^+} B(\zeta, z) \mapsto \int\limits_{\zeta \in S_\epsilon^+} B(\zeta, z_0) \quad \text{as } z \mapsto z_0.$$

Equation (4) becomes

$$B^-(1)(z_0) = B_b(1)(z_0) - \lim_{\epsilon \mapsto 0} \int\limits_{\zeta \in S_\epsilon^+} B(\zeta, z_0).$$

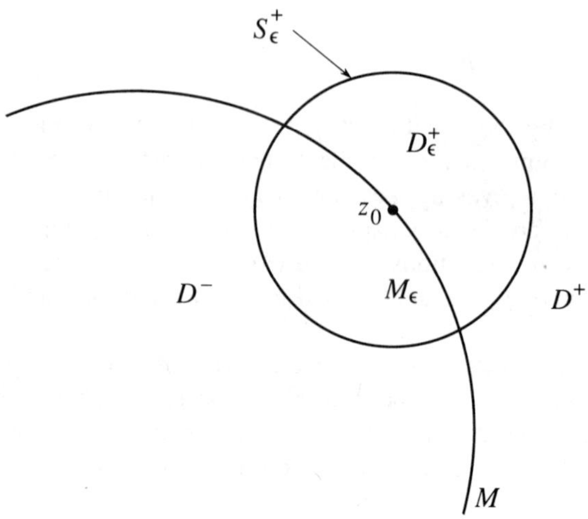

FIGURE 24.1

To prove (2), it therefore suffices to show

$$\lim_{\epsilon \mapsto 0} \int_{\zeta \in S_\epsilon^+} B(\zeta, z_0) = \frac{1}{2}.$$

By a translation, we may assume that z_0 is the origin. We may also pull back the resulting integral via the change of scale map $\zeta \mapsto \epsilon\zeta$ for $|\zeta| = 1$. The inverse image of the set S_ϵ^+ under this change of scale converges to half of a unit sphere in $\mathbb{C}^n$ as $\epsilon \mapsto 0$. Since the Bochner–Martinelli kernel is also invariant under the unitary group in $\mathbb{C}^n$, we may assume this half sphere is the set

$$S^+ = \{\zeta = (\zeta_1, \ldots, \zeta_n);\ |\zeta| = 1 \text{ and } \mathrm{Im}\zeta_1 > 0\}.$$

Therefore, we must show

$$\int_{\zeta \in S^+} b(\zeta) = \frac{1}{2} \tag{5}$$

where $b(\zeta) = B(\zeta, 0)$. To see (5), let S^- be the corresponding half of the unit sphere with $\mathrm{Im}\zeta_1 < 0$. In view of (1) with D^- as the unit ball in $\mathbb{C}^n$, we have

$$1 = \int_{\zeta \in S^+} b(\zeta) + \int_{\zeta \in S^-} b(\zeta).$$

Moreover, the conjugation map $\zeta \mapsto C(\zeta) = \overline{\zeta}$ is a diffeomorphism from S^- to S^+ which changes the orientation by a factor of $(-1)^n$. By an easy computation, $C^*b = (-1)^n b$ on the unit sphere. This uses the fact that on the sphere, $0 = d|\zeta|^2 = \zeta \cdot d\overline{\zeta} + \overline{\zeta} \cdot d\zeta$. We obtain

$$\begin{aligned} \int_{S^+} b &= (-1)^n \int_{S^-} C^* b \\ &= \int_{S^-} b. \end{aligned}$$

This together with the previous equation implies that (5) holds and so the proof of (2) is complete.

24.2 Kernels for strictly pseudoconvex boundaries

For a compact convex boundary $M = \partial D^-$, we use the complex gradient of a suitable defining function ρ as the generating function u for the kernels L, L^t, H, H^t, R, etc. The key properties possessed by u are the following.

(1) *The function $u(\zeta, z)$ is holomorphic in $z \in D^-$ for each fixed $\zeta \in M$.*

(2) *There are positive constants δ, ϵ, and C such that if ζ belongs to a δ-neighborhood of ∂D and if $|\zeta - z| < \epsilon$, then*

$$2Re\{u(\zeta, z) \cdot (\zeta - z)\} \geq \rho(\zeta) - \rho(z) + C(|\zeta - z|^2).$$

(3) *If z is in a δ-neighborhood of $\overline{D}$, ζ is in a δ-neighborhood of M, and if $|\zeta - z| \geq \epsilon$, then*

$$u(\zeta, z) \cdot (\zeta - z) \neq 0.$$

All the global results in Chapters 20 and 21 hold for any domain D with boundary M in which there is a generating function u that satisfies the above three properties.

In [He1], Henkin constructs a generating function u for a bounded, strictly pseudoconvex domain, such that u satisfies properties (1), a modified version of (2), and (3). In this section, we outline his construction. This construction is not necessary for the local solution for the tangential Cauchy–Riemann complex because a strictly pseudoconvex hypersurface is locally biholomorphic to a strictly convex hypersurface.

Henkin's construction of the kernel generating function requires the solution of the Cauchy–Riemann complex (with, say, Hörmander's $\mathcal{L}^2$-techniques [Ho]) on a bounded strictly pseudoconvex domain. This is somewhat unsatisfying since one of the applications of the generating function is the construction of integral kernel solutions to the Cauchy–Riemann complex on a bounded strictly pseudoconvex domain (i.e., the analogue of Theorem 1 in Section 20.3). Range has developed a self-contained (but more complicated) integral kernel approach to the problem of finding kernel generating functions and hence the solution to the Cauchy–Riemann complex and other results in several complex variables. We refer the reader to [Ran] for his approach.

Henkin's approach is to first locally construct the generating function u so that (1) and (2) hold (locally). He then modifies u by solving the appropriate $\overline{\partial}$-problem (globally) so that (1) and (3) hold. For the local construction, we start with a Taylor expansion of the defining function ρ for $D^- = \{\rho(z) < 0\}$. We obtain

$$\rho(z) = \rho(\zeta) - \mathrm{Re}\{\hat{Q}(\zeta, z)\} + \sum_{j,k=1}^{n} \frac{\partial^2 \rho(\zeta)}{\partial \zeta_j \partial \overline{\zeta}_k}(\zeta_j - z_j)(\overline{\zeta_k - z_k}) + \mathcal{O}(|\zeta - z|^3) \quad (4)$$

where

$$\hat{Q}(\zeta, z) = \frac{\partial \rho(\zeta)}{\partial \zeta} \cdot (\zeta - z) - \frac{1}{2} \sum_{j,k=1}^{n} \frac{\partial^2 \rho(\zeta)}{\partial \zeta_j \partial \zeta_k}(\zeta_j - z_j)(\zeta_k - z_k).$$

Define $\hat{u} = (\hat{u}_1, \ldots, \hat{u}_n) : \mathbb{C}^n \times \mathbb{C}^n \mapsto \mathbb{C}^n$ by

$$\hat{u}_j(\zeta, z) = \frac{\partial \rho(\zeta)}{\partial \zeta_j} - \frac{1}{2} \sum_{k=1}^{n} \frac{\partial^2 \rho(\zeta)}{\partial \zeta_j \partial \zeta_k}(\zeta_k - z_k).$$

Note that $\hat{u}(\zeta, z) \cdot (\zeta - z) = \hat{Q}(\zeta, z)$. Since D^- is strictly pseudoconvex, we may assume that ρ is strictly plurisubharmonic and so the complex hessian of ρ at each point ζ in a neighborhood of M is positive definite. Equation (4) now shows that property (2) holds for $\hat{u}$. Clearly, property (1) also holds for $\hat{u}$.

Before we modify $\hat{u}$, we first modify $\hat{Q}(\zeta, z)$ so that the resulting function is holomorphic in z and nonvanishing for ζ in a δ-neighborhood of M and for z in a δ-neighborhood of $\overline{D}$ with $|\zeta - z| \geq \epsilon$. We shall take the δ-neighborhood of M to be a set of the form $\{\zeta \in \mathbb{C}^n; |\rho(\zeta)| < \delta\}$ and the δ-neighborhood of $\overline{D}$ to be the set $D_\delta = \{z \in \mathbb{C}^n; \rho(z) < \delta\}$. If $|\rho(\zeta)| \leq \delta$ and if $z \in \overline{D_\delta}$ with $|\zeta - z| \geq \epsilon/2$, then the inequality in (2) implies

$$\mathrm{Re}\hat{Q}(\zeta, z) \geq -2\delta + C\left(\frac{\epsilon^2}{4}\right).$$

Choosing δ small relative to ϵ means that we can arrange $\text{Re}\hat{Q}(\zeta, z) > 0$ for $|\rho(\zeta)| \leq \delta$ and $z \in \overline{D_\delta}$ with $|\zeta - z| \geq \epsilon/2$. On this set, a branch of $\log \hat{Q}(\zeta, z)$ is well defined and holomorphic in z.

Choose a smooth cutoff function $\chi : \mathbb{C}^n \times \mathbb{C}^n \mapsto \mathbb{C}^n$ such that

$$\chi(\zeta, z) = \begin{cases} 1 & \text{if } |\zeta - z| \leq \frac{\epsilon}{2} \\ 0 & \text{if } |\zeta - z| \geq \frac{3\epsilon}{4}. \end{cases}$$

For fixed ζ with $|\rho(\zeta)| \leq \delta$, consider the 1-form

$$f_\zeta(z) = \begin{cases} \overline{\partial}_z \chi(\zeta, z) \log(\hat{Q}(\zeta, z)) & \text{for } z \in \overline{D_\delta} \text{ with } \frac{\epsilon}{2} \leq |\zeta - z| \leq \epsilon \\ 0 & \text{otherwise.} \end{cases}$$

The form $f_\zeta(\cdot)$ has smooth coefficients and it is $\overline{\partial}$-closed on $\overline{D_\delta}$. We can use the $\overline{\partial}$-theory for a bounded strictly pseudoconvex domain to find a solution to the equation $\overline{\partial} v_\zeta(\cdot) = f_\zeta(\cdot)$ on $\overline{D_\delta}$. The function $v_\zeta(z)$ is smooth for $z \in \overline{D_\delta}$ and $|\rho(\zeta)| \leq \delta$. Moreover, for each fixed ζ in this set, the function $\chi(\zeta, z) \log \hat{Q}(\zeta, z) - v_\zeta(z)$ is holomorphic in $z \in \overline{D_\delta}$ with $|\zeta - z| \geq \epsilon/2$.

For $|\rho(\zeta)| \leq \delta$, define

$$Q(\zeta, z) = \begin{cases} \hat{Q}(\zeta, z) e^{-v_\zeta(z)} & \text{if } z \in D_\delta \text{ with } |\zeta - z| \leq \frac{\epsilon}{2} \\ e^{\chi(\zeta, z) \log(\hat{Q}(\zeta, z)) - v_\zeta(z)} & \text{if } z \in D_\delta \text{ with } |\zeta - z| \geq \frac{\epsilon}{2}. \end{cases}$$

The function $Q(\zeta, z)$ is well defined and holomorphic in z for $z \in \overline{D_\delta}$ and $|\rho(\zeta)| \leq \delta$. Also, $Q(\zeta, z) \neq 0$ for $|\zeta - z| \geq \epsilon/2$.

Now the idea is to perform a division to find a function $u : \mathbb{C}^n \times \mathbb{C}^n \mapsto \mathbb{C}^n$ so that $u(\zeta, z)$ is holomorphic in $z \in \overline{D_\delta}$ (for $|\rho(\zeta)| \leq \delta$) and so that $u(\zeta, z) \cdot (\zeta - z) = Q(\zeta, z)$. The basic idea is the following. The function $Q(\zeta, z)$ vanishes on the diagonal $\{(z, z);\ z \in \overline{D_\delta}\}$. For fixed ζ, the functions $(\zeta_1 - z_1), \ldots, (\zeta_n - z_n)$ locally generate the sheaf of holomorphic functions (in z) which vanish at $z = \zeta$ over the ring of germs of holomorphic functions at ζ. Since D_δ is a domain of holomorphy, these functions also globally generate the space of global sections of this ideal (this follows from the theory of coherent analytic sheaves; see Chapter 24 in Hörmander's book [Ho]). So there exist functions $u_1(\zeta, z), \ldots, u_n(\zeta, z)$ which are holomorphic in $z \in D_\delta$ such that $\sum_{j=1}^n u_j(\zeta, z)(\zeta_j - z_j) = Q(\zeta, z)$. Care must be taken to ensure that the $u_j(\zeta, z)$ depend smoothly on ζ as well as z for $|\rho(\zeta)| < \delta$. We refer the reader to [He1] for details.

By the above construction, u satisfies properties (1) and (3). It is not clear that property (2) is satisfied. However for $|\zeta - z| \leq \epsilon/2$, we have

$$u(\zeta, z) \cdot (\zeta - z) = e^{-v_\zeta(z)} \hat{u}(\zeta, z) \cdot (\zeta - z)$$

and we know that $\hat{u}$ satisfies property (2). From the definitions of our kernels (see Section 20.1), it follows that if $c(\zeta, z)$ is a nonvanishing smooth function of ζ and z, then $E(c\hat{u}) = E(\hat{u})$ and $E(c\hat{u}_1, c\hat{u}_2) = E(\hat{u}_1, \hat{u}_2)$ and so forth. So for the parts of the proofs of the theorems in Chapters 20 and 21 for the strictly pseudoconvex case that require the estimate in (2), we can replace u by $\hat{u}$ in the appropriate kernels and then the proofs go through without further changes.

24.3 Further estimates on the solution to $\overline{\partial}_M$

Since $R(\zeta, z)$ is locally integrable in $\zeta \in M$ (locally uniformly in z), there is some gain in the regularity of the global solution for the tangential Cauchy–Riemann complex given by the R kernel. From the dominated convergence theorem, it follows that if f is a compactly supported form with $\mathcal{L}^\infty$ coefficients, then $R(f)$ is a form with continuous coefficients. More generally, Henkin [He3] has shown that if f is a compactly supported form with coefficients in $\mathcal{L}^p$ with $1 < p < 2n$, then $R(f)$ is a form with coefficients in $\mathcal{L}^r$ with $1/r = (1/p) - (1/2n)$. If $p = 1$, then Henkin obtains $r = (2n+2)/(2n+1) - \epsilon$, for any $\epsilon > 0$. In addition, $R(f)$ has $\mathcal{L}^\infty$-coefficients provided the coefficients of f are compactly supported and belong to $\mathcal{L}^{2n+\epsilon}$, for any $\epsilon > 0$.

Estimates on the local solution for the tangential Cauchy–Riemann complex are more difficult than for the global solution. Indeed, there does not yet appear to be a local kernel solution for the tangential Cauchy–Riemann complex that exhibits a gain in regularity (maybe none exists). The solution given in Chapter 22 does not even preserve regularity due to the presence of the term $E_{123}([\partial\omega_\lambda]^{0,1} \wedge f)$. The best result available in this context is a recent result of Shaw [Sh2] who shows that a modification of the solution operator given in Chapter 22 is continuous in $\mathcal{L}^p$ for $1 < p < \infty$.

The solution to the equation $\overline{\partial}_M u = f$ on ω_λ given in Chapter 22 does not necessarily have continuous boundary values on $\partial\omega_\lambda$ due again to the presence of the term $E_{123}([\partial\omega_\lambda]^{0,1} \wedge f)$. In Section 5 of [He3], Henkin modifies this solution to one that is continuous up to $\partial\omega_\lambda$ provided f is a continuous (p, q)-form with $1 \le q \le n-3$. This solution operator (different than Shaw's) is also continuous in the $\mathcal{L}^\infty$-norm on ω_λ.

24.4 Weakly convex boundaries

The full strength of properties (2) and (3) in Section 24.2 for the generating function u or $\hat{u}$ is not needed for some of the results in Chapters 20, 21, and 22 Many of these results only require property (1) in Section 24.2 (i.e., that $u(\zeta, z)$

is holomorphic in $z \in \overline{D_\delta}$ for ζ in M) and the following weaker version of properties (2) and (3).

> The function $u(\zeta, z)\cdot(\zeta - z)$ is nonvanishing for $(\zeta, z) \in V \cap \{\overline{D^+} \times D^-\}$, where V is a neighborhood of $M \times M$ in $\mathbb{C}^n \times \mathbb{C}^n$. Moreover, if $(\zeta_0, z_0) \in V$ with $u(\zeta_0, z_0)\cdot(\zeta_0 - z_0) = 0$, then the real derivatives of the maps $\zeta \mapsto u(\zeta, z_0)\cdot(\zeta - z_0)$ and $z \mapsto u(\zeta_0, z)\cdot(\zeta_0 - z)$ have maximal rank at the points $\zeta = \zeta_0$ and $z = z_0$, respectively.

For a bounded convex domain that is not strictly convex, a defining function can be found so that its complex gradient satisfies this weaker property. Harvey and Polking show in [HP] that integral kernels can be constructed for the solution of the $\overline{\partial}$-problem on domains that have generating functions that satisfy the above weaker condition. Their approach is somewhat different than Henkin's. Harvey and Polking take principal value limits across the singular sets of these kernels (i.e., the set $\{(\zeta, z) \in V; u(\zeta, z)\cdot(\zeta - z) = 0\}$). As a result, they obtain new integral kernel solutions for the equation $\overline{\partial} u = f$ on convex domains. These new solutions can be applied to more subtle problems. For example, if f is smooth on D^- (but not necessarily up to M), then they provide an integral kernel solution u (in certain bidegrees) which is smooth on D^-. If f has compact support then they also obtain a solution u with compact support (in certain bidegrees).

The boundary value results for the L, L^t, H, and H^t kernels given in Theorems 1 and 2 in Section 20.4 also only require that the generating function u satisfy the above weaker condition. So these theorems (and the result concerning a fundamental solution for $\overline{\partial}_M$ given in Theorem 1 of Section 21.1) also hold for boundaries of weakly convex domains. In addition, the boundary values of one normal derivative of $H([M]^{0,1} \wedge f)|_{D^-}$ and of $H^t([M]^{0,1} \wedge f)|_{D^+}$ exist on M (Theorems 1 and 2 in Section 20.4 only discuss the boundary values of tangential derivatives). These and related results can be found in Sections 8 and 9 of [HP].

24.5 Solvability of the tangential Cauchy–Riemann complex in other geometries

The goal of much of Part IV is the discussion of the local and global solvability of the tangential Cauchy–Riemann complex on a strictly pseudoconvex hypersurface in $\mathbb{C}^n$. In this section, we briefly discuss the local solvability of the tangential Cauchy–Riemann complex in other geometries. We fix a point z_0 in a smooth real hypersurface M in $\mathbb{C}^n$ and fix an integer q with $1 \leq q \leq n-1$. We ask the following question. What conditions on the Levi form of M at z_0 are needed to ensure that the equation $\overline{\partial}_M u = f$ can be solved near z_0 where f

is a given smooth $\overline{\partial}_M$-closed (p,q)-form on M near z_0? The answer is that M must satisfy the $Y(q)$ condition of J. J. Kohn at the point z_0. The hypersurface M is said to satisfy condition $Y(q)$ at the point z_0 if the Levi form of M at z_0 has either $\max\{n-q, q+1\}$ eigenvalues of the same sign or $\min\{n-q, q+1\}$ pairs of eigenvalues of opposite sign (i.e., $\min\{n-q, q+1\}$ positive eigenvalues and $\min\{n-q, q+1\}$ negative eigenvalues). If M is strictly pseudoconvex at z_0, then M satisfies condition Y(q) at z_0 for all $1 \leq q \leq n-2$. More generally, suppose the Levi form of M at z_0 has r-positive and s-negative eigenvalues with $r + s = n - 1$, then M satisfies condition Y(q) at z_0 for each $0 \leq q \leq n-1$ except $q = r$ and $q = s$. Folland and Kohn use $\mathcal{L}^2$ techniques in their book [FK] to show that if M satisfies condition Y(q) at every point z_0 in M, then the equation $\overline{\partial}_M u = f$ is globally solvable where f is any smooth $\overline{\partial}_M$-closed (p,q)-form. Local solvability under condition Y(q) was established by Andreotti and Hill [AnHi2]. We shall outline an integral kernel approach to Andreotti and Hill's result. For details, see [BS].

Since there are several cases to consider for the condition Y(q), we shall discuss the case where the Levi form of M at z_0 has $(q+1)$-positive eigenvalues with $q+1 \geq n-q$. The other cases of condition Y(q) are similar to this one.

We assume that z_0 is the origin and M is graphed over its tangent space at the origin as $M = \{z \in \mathbb{C}^n;\ \rho(z) = 0\}$ where

$$\rho(z) = \operatorname{Im} z_1 + \sum_{j=2}^{n} \mu_j |z_j|^2 + \mathcal{O}(|z|^3).$$

Here, we have diagonalized the Levi form of M at the origin and the numbers $\mu_2, \ldots, \mu_n$ are its eigenvalues. By replacing ρ with $\rho + C\rho^2$ as in the proof of Theorem 1 in Section 10.3, we may assume that the complex hessian of ρ at the origin is positive definite in the z_1-direction. Since the Levi form of M at the origin has $(q+1)$-positive eigenvalues, we may assume that $\mu_2, \ldots, \mu_{q+2}$ are positive. After a change in scale, we obtain

$$\rho(z) = \operatorname{Im} z_1 + \sum_{j=1}^{q+2} |z_j|^2 + \sum_{j=q+3}^{n} \mu_j |z_j|^2 + \mathcal{O}(|z|^3)$$

where

$$-1 \leq \mu_j \leq 0, \qquad \text{for } q+3 \leq j \leq n. \tag{1}$$

In the strictly pseudoconvex case, $q+2 = n$ and the function used to generate the kernels in Chapters 20 and 21 is given by the complex gradient of ρ. Here, we let $u^1 = (u_1^1, \ldots, u_n^1) : \mathbb{C}^n \times \mathbb{C}^n \mapsto \mathbb{C}^n$ be given by

$$u_j^1(\zeta, z) = \begin{cases} \dfrac{\partial \rho(\zeta)}{\partial \zeta_j} & \text{if } 1 \leq j \leq q+2 \\[2mm] \dfrac{\partial \rho(\zeta)}{\partial \zeta_j} + (\overline{\zeta}_j - \overline{z}_j) & \text{if } q+3 \leq j \leq n. \end{cases}$$

A Taylor expansion argument together with the estimate in (1) imply that there is an open set U in $\mathbb{C}^n$ containing the origin and a constant $C > 0$ such that

$$2\text{Re}\{u^1(\zeta, z) \cdot (\zeta - z)\} \geq \rho(\zeta) - \rho(z) + C(|\zeta - z|^2)$$

for $\zeta, z \in U$. Let $u^2(\zeta, z) = u^1(z, \zeta)$. Reversing the roles of ζ and z in the above inequality yields

$$2\text{Re}\{u^2(\zeta, z) \cdot (\zeta - z)\} \leq \rho(\zeta) - \rho(z) - C(|\zeta - z|^2)$$

for $\zeta, z \in U$.

We form the kernels $L = E(u^1)$, $L^t = E(u^2)$, and $R = E(u^1, u^2)$ as in Chapters 20 and 21. The above estimates imply that $L(\zeta, z)$ is smooth for $\zeta \in M \cap U$, $z \in U \cap D^-$ and $L^t(\zeta, z)$ is smooth for $\zeta \in M \cap U$ and $z \in U \cap D^+$, where $D^- = \{z \in \mathbb{C}^n;\ \rho(z) < 0\}$ and $D^+ = \{z \in \mathbb{C}^n;\ \rho(z) > 0\}$. Moreover, $R(\zeta, z)$ has only diagonal singularities on $M \times M$. The proof of Theorem 2 in Section 20.4 can be repeated using the above estimates to show that if $f \in \mathcal{D}^*_M(M \cap U)$ then $L(f)$ is smooth on the set $\overline{U \cap D^-}$ and $L^t(f)$ is smooth on the set $\overline{U \cap D^+}$. The proof of Theorem 1 in Section 21.2 is therefore valid in this context and we obtain

$$\overline{\partial}_M \circ R + R \circ \overline{\partial}_M + (L^t)^+ - L^- = I \tag{2}$$

as operators on $\mathcal{D}^*_M(M \cap U)$.

In the strictly pseudoconvex case, we show (Lemma 3 in Section 20.3) that the L kernel acts nontrivially only on $(p, 0)$-forms and the L^t kernel acts nontrivially only on $(p, n-1)$-forms. This together with (2) means that the R kernel is a fundamental solution for the tangential Cauchy–Riemann complex (acting on $\mathcal{D}^{p,q}_M(M \cap U)$) for $1 \leq q \leq n-2$.

In our setting where M satisfies condition Y(q) at the origin, we claim that $L(f)$ and $L^t(f)$ vanish if $f \in \mathcal{D}^{p,q}_M(M \cap U)$ and so the R kernel defined by our new support function u^1 is a fundamental solution to the tangential Cauchy–Riemann complex (acting on $\mathcal{D}^{p,q}_M(M \cap U)$). To see this, we examine the $(\overline{\partial} u^1 \cdot d(\zeta - z))^{n-1}$ part of the kernel $L(\zeta, z)$. Let $u(\zeta) = (\partial\rho(\zeta)/\partial\zeta)$. We have

$$\begin{aligned}(\overline{\partial}_{\zeta,z} u^1 \cdot d(\zeta - z))^{n-1} &= \left(\overline{\partial}_\zeta u \cdot d(\zeta - z) + \sum_{j=q+3}^{n} d(\overline{\zeta_j - z_j}) \wedge d(\zeta_j - z_j)\right)^{n-1} \\ &= \sum_{k=0}^{n-1} \binom{n-1}{k} \left(\overline{\partial}_\zeta u \cdot d(\zeta - z)\right)^{n-1-k} \\ &\quad \wedge \left(\sum_{j=q+3}^{n} d(\overline{\zeta_j - z_j}) \wedge d(\zeta_j - z_j)\right)^{k}.\end{aligned}$$

The only contributing terms to the sum on the right occur when $k \leq n-q-2$, for otherwise there will be a repeated wedge product of $d(\zeta_j - z_j)$ for some j with $q+3 \leq j \leq n$. Together with the fact that u only depends on ζ, we see that the degree of $(\bar{\partial}u^1 \cdot d(\zeta - z))^{n-1}$ in $d\bar{\zeta}$ is at least $n-1-(n-q-2) = q+1$. So if $f \in \mathcal{D}_M^{p,s}(M \cap U)$ with $s \geq n-q-1$, then $L(f) = 0$. Since $q \geq n-q-1$ (by assumption), we have $L(f) = 0$ for $f \in \mathcal{D}_M^{p,q}(M \cap U)$, as desired. In a similar manner, we can show $L^t(f) = 0$ for $f \in \mathcal{D}_M^{p,q}(M \cap U)$. It follows that R is a fundamental solution for $\bar{\partial}_M$ on $\mathcal{D}_M^{p,q}(M \cap U)$, i.e., $f = \bar{\partial}_M\{R(f)\} + R(\bar{\partial}_M f)$.

To modify the fundamental solution for the tangential Cauchy–Riemann complex to handle the local problem, we use a procedure similar to the one in Chapter 22. In that chapter, we consider a local neighborhood basis of open sets in M about the origin obtained by slicing M with the half space $\mathrm{Re} z_1 > -\lambda$ for $\lambda > 0$. Since M is strictly convex (in Chapter 22), these open sets shrink down to the origin as $\lambda \mapsto 0$.

In this section, M is only strictly convex in the $(z_1, \ldots, z_{q+2})$-directions. Therefore, we shall obtain a local neighborhood basis of open sets by intersecting M with a trough that is flat in the $(z_1, \ldots, z_{q+2})$-directions and that "bends up" in the $(z_{q+3}, \ldots, z_n)$-directions. More precisely, let

$$\omega_\lambda = \left\{ z \in M;\ \mathrm{Im} z_1 > -\lambda + 2 \sum_{j=q+3}^{n} |z_j|^2 \right\}.$$

The choice of the number 2 is motivated by the estimate (1) on the eigenvalues $\mu_{q+3}, \ldots, \mu_n$, which ensures that the above trough bends up faster than M in the directions $(z_{q+3}, \ldots, z_n)$. In fact, combining the defining equation for M and the defining equation for ω_λ, we obtain

$$\lambda \geq \sum_{j=1}^{q+2} |z_j|^2 + \sum_{j=q+3}^{n} (2 + \mu_j)|z_j|^2 + \mathcal{O}(|z|^3) \quad \text{for } z \in \omega_\lambda.$$

The estimate in (1) together with this inequality imply that the diameter of ω_λ is proportional to $\sqrt{\lambda}$. So ω_λ shrinks to the origin as $\lambda \mapsto 0$. From now on, we restrict λ so that ω_λ is contained in U.

Let

$$r(z) = \mathrm{Im} z_1 - 2 \sum_{j=q+3}^{n} |z_j|^2.$$

The defining equation for ω_λ is given by $\{z \in M;\ r(z) > -\lambda\}$. Define $u^3 : \mathbb{C}^n \times \mathbb{C}^n \mapsto \mathbb{C}^n$ by

$$u^3(\zeta, z) = \frac{\partial r(\zeta)}{\partial \zeta}.$$

As in Chapter 22, we form the kernels E_{13}, E_{23}, and E_{123}. As in the proof of part (a) of Lemma 1 in Section 20.3, the (weak) convexity of r implies that if

$z \in \omega_\lambda$ and $\zeta \in \partial\omega_\lambda$, then $u^3(\zeta, z) \cdot (\zeta - z) \neq 0$. So if $f \in \mathcal{E}_M^*(\overline{\omega_\lambda})$, then $E_{13}([\partial\omega_\lambda]^{0,1} \wedge f)$, $E_{23}([\partial\omega_\lambda]^{0,1} \wedge f)$, and $E_{123}([\partial\omega_\lambda]^{0,1} \wedge f)$ are smooth forms on ω_λ.

The same arguments used for the L kernel above allow us to show that the degree of $E_{13}(\zeta, z)$ in $d\overline{\zeta}$ is at least q. Therefore, E_{13} acts nontrivially only on currents of bidegree (p, s) with $s \leq n - q - 1$. Since $q \geq n - q - 1$, $E_{13}([\partial\omega_\lambda]^{0,1} \wedge f)$ vanishes for $f \in \mathcal{E}_M^{p,q}(\overline{\omega_\lambda})$.

As with the strictly pseudoconvex case, the term $E_{23}([\partial\omega_\lambda]^{0,1} \wedge f)$ does not vanish purely from bidegree considerations. However, if f is a $\overline{\partial}_M$-closed smooth (p, q)-form on ω_λ, then an approximation argument similar to that at the end of Chapter 22 can be carried out to show that $E_{23}([\partial\omega_\lambda]^{0,1} \wedge f)$ vanishes. In this case, the kernel $E_{23}(\zeta, z)$ is (formally) holomorphic only in the variables $\zeta_1, \ldots, \zeta_{q+2}$. So the approximation argument must be carried out in these variables with $\zeta_{q+3}, \ldots, \zeta_n$ treated as parameters. We refer the reader to the end of [BS] for details.

Since $E_{13}([\partial\omega_\lambda]^{0,1} \wedge f)$ and $E_{23}([\partial\omega_\lambda]^{0,1} \wedge f)$ both vanish if f is a smooth, $\overline{\partial}_M$-closed (p, q)-form on ω_λ, the procedure in Chapter 22 can be carried out in this context so that the solution to $\overline{\partial}_M u = f$ on ω_λ is given by (3) in Chapter 22.

In the strict pseudoconvex case, we showed in Chapter 23 (with Lewy's example) that the tangential Cauchy–Riemann complex is not locally solvable at top degree ($q = n - 1$). More generally, if the Levi form of a hypersurface at a point z_0 has p-positive eigenvalues and q-negative eigenvalues, then the tangential Cauchy–Riemann complex is not locally solvable in degrees p and q. This is a result of Andreotti, Fredricks, and Nacinovich [AFN].

Generalizations of the results in this section to manifolds of higher codimension have been obtained by Airapetyan and Henkin [AH1], [AH2].

Bibliography

[AH1] R. A. Airapetyan and G. M. Henkin, Integral representations of differential forms on Cauchy–Riemann manifolds and the theory of CR functions. *Russ. Math. Surveys* **39**, 41–118 (1984).

[AH2] R. A. Airapetyan and G. M. Henkin, Integral representations of differential forms on Cauchy–Riemann manifolds and the theory of CR functions, Part II. *Math USSR Sbornik* **55**, 91–111 (1986).

[AFN] A. Andreotti, G. Fredricks, and M. Nacinovich, *On the absence of a Poincaré lemma in tangential Cauchy–Riemann complexes, Ann. Scuola Norm. Sup. Pisa* **8**, 365–404 (1981).

[AnHi1] A. Andreotti and C. D. Hill, Complex characteristic coordinates and the tangential Cauchy–Riemann equations. *Ann. Scuola Norm. Sup. Pisa* **26**, 299–324 (1972).

[AnHi2] A. Andreotti and C. D. Hill, E. E. Levi convexity and the Hans Lewy problem, Parts I and II. *Ann. Scuola Norm. Sup. Pisa* **26**, 325–363 (1972) and **26**, 747–806 (1972).

[Bi] E. Bishop, Differentiable manifolds in complex Euclidean space. *Duke Math. J.* **32**, 1–22 (1965).

[BCT] M. S. Baouendi, C. H. Chang, and F. Treves, Microlocal hypo-analyticity and extension of CR functions. *J. Diff. Geom.* **18**, 331–391 (1983).

[BJT] M. S. Baouendi, H. Jacobowitz, and F. Treves, On the analyticity of CR mappings. *Ann. Math.* **122**, 365–400 (1985).

[BR1] M. S. Baouendi and L. P. Rothschild, Germs of CR maps between real analytic hypersurfaces. *Invent. Math.* **93**, 481–500 (1988).

[BR2] M. S. Baouendi and L. P. Rothschild, Normal forms for generic manifolds and holomorphic extension of CR functions. *J. Diff. Geom.* **25**, 431–467 (1987).

[BR3] M. S. Baouendi and L. P. Rothschild, Cauchy–Riemann functions on manifolds of higher codimension in complex space. Preprint.

[BRT] M. S. Baouendi, L. P. Rothschild, and F. Treves, CR structures with group action and extendability of CR functions. *Invent. Math.* **82**, 359–396 (1985).

[BT1] M. S. Baouendi and F. Treves, A property of the functions and distributions annihilated by a locally integrable system of complex vector fields. *Ann. Math.* **113**, 387–421 (1981).

[BT2] M. S. Baouendi and F. Treves, A microlocal version of Bochner's tube theorem. *Indiana University Math. J.* **31**, 885–895 (1982).

[BT3] M. S. Baouendi and F. Treves, About the holomorphic extension of CR functions on real hypersurfaces in complex Euclidean space. *Duke Math. J.* **51**, 77–107 (1984).

[BG] T. Bloom and I. Graham, On type conditions for generic real submanifolds of $\mathbb{C}^n$. *Invent. Math.* **40**, 217–243 (1977).

[Boc] S. Bochner, Analytic and meromorphic continuation by means of Green's formula. *Ann. Math.* **44**, 652–673 (1943).

[B] A. Boggess, CR extendibility near a point where the first Levi form vanishes. *Duke Math. J.* **48**, 665–684 (1981).

[BN] A. Boggess and A. Nagel, in preparation.

[BPi] A. Boggess and J. Pitts, CR extension near a point of higher type. *Duke Math. J.* **52**, 67–102 (1985).

[BP] A. Boggess and J. C. Polking, Holomorphic extension of CR functions. *Duke Math. J.* **49**, 757–784 (1982).

[BS] A. Boggess and M-C. Shaw, A kernel approach to the local solvability of the tangential Cauchy–Riemann equations. *Trans. Am. Math. Soc.* **289**, 643–658 (1985).

[C] E. M. Cirka, Analytic representation of CR functions. *Math. USSR* **27**, 526–553 (1975).

[D] J. P. D'Angelo, Real hypersurfaces, orders of contact, and applications. *Ann. Math.* **115**, 615–637 (1982).

[Fe] H. Federer, *Geometric Measure Theory*. Die Grundlehren der Math Wissenshaften, Band 153, Springer-Verlag, NY, 1969.

[Fo] G. B. Folland, *Introduction to Partial Differential Equations*. Princeton University Press, Princeton, NJ, 1976.

[FK] G. B. Folland and J. J. Kohn, *The Neumann Problem for the Cauchy–Riemann Complex*. Princeton University Press and University of Tokoyo Press, Princeton, NJ, 1972.

[Fr] M. Freeman, The Levi form and local complex foliations. *Trans. Am. Math. Soc.* **57**, 369–370 (1976).

[GL] H. Grauert and I. Lieb, Das Ramirezsche Integral und die Losung der Gleichung $\overline{\partial} f = \alpha$ in Bereich der Beschrankten Formen. Proc. Conf.

Rice University, 1969. *Rice University Studies* **56**, 29–50 (1970).

[Har] G. A. Harris, Higher order analogues to the tangential Cauchy–Riemann equations for real submanifolds of $\mathbb{C}^n$ with CR singularity. *Proc. Am. Math. Soc.* **74**, 79–86 (1979).

[HL] R. Harvey and B. Lawson, On boundaries of complex analytic varieties I. *Ann. Math.* **102**, 233–290 (1975).

[HP] R. Harvey and J. Polking, Fundamental solutions in complex analysis, Parts I and II. *Duke Math. J.* **46**, 253–300 and 301–340 (1979).

[He1] G. M. Henkin, Integral representations of functions holomorphic in strictly pseudoconvex domains and some applications. *Math. USSR* **7**, 597–616 (1969).

[He2] G. M. Henkin, Integral representations of functions in strictly pseudoconvex domains and applications to the $\overline{\partial}$-problem. *Mat. Sb.* **124**, 273–281 (1970).

[He3] G. M. Henkin, The Hans Lewy equation and analysis of pseudoconvex manifolds. *Uspehi Mat. Nauk* **32** (1977); English translation in *Math. USSR-Sbornik* **31**, 59–130 (1977).

[HT1] C. D. Hill and G. Taiani, Families of analytic discs in $\mathbb{C}^n$ with boundaries on a prescribed CR-submanifold. *Ann. Scuola Norm. Sup. Pisa* **4–5**, 327–380 (1978).

[HT2] C. D. Hill and G. Taiani, Real analytic approximation of locally embeddable CR manifolds. *Composito Mathematica* **44**, 113–131 (1981).

[Ho] L. Hörmander, *An Introduction to Complex Analysis in Several Variables*. Van Nostrand, Princeton, NJ, 1966.

[JT] H. Jacobowitz and F. Treves, Non-realizable CR structures. *Invent. Math.* **66**, 231–249 (1982).

[Jo] F. John, *Partial Differential Equations*. Springer-Verlag, 1971.

[KN] J. J. Kohn and L. Nirenberg, A pseudoconvex domain not admitting a holomorphic support function. *Math. Ann.* **201**, 265–268 (1973).

[KR] J. J. Kohn and H. Rossi, On the extension of holomorphic functions from the boundary of a complex manifold. *Ann. Math.* **81**, 451–472 (1965).

[Kr] S. Krantz, *Function Theory of Several Complex Variables*, John Wiley, New York, 1982.

[Ku] M. Kuranishi, Strongly pseudoconvex CR structures over small balls. Part I, An a priori estimate. *Ann. Math.* **115**, 451–500 (1982); Part II, A regularity theorem. *Ann. Math.* **116**, 1–64 (1982); Part III, An embedding theorem. *Ann. Math.* **116**, 249–330 (1982).

[L1] H. Lewy, On the local character of the solutions of an atypical linear differential equation in three variables and a related theorem for regular functions of two complex variables. *Ann. Math.* **64**, 514–522 (1956).

[L2] H. Lewy, An example of a smooth linear partial differential equation without solution. *Ann. Math.* **66**, 155–158 (1957).

[L3] H. Lewy, On the boundary behavior of holomorphic mappings. *Acad. Naz. Lincei* **35**, 1–8 (1977).

[M] H.-M. Maire, Hypoelliptic overdetermined systems of partial differential equations. *Comm. Partial Diff. Eq.* **5**, 331–380 (1980).

[NSW] A. Nagel, E. M. Stein, and S. Wainger, Balls and Metrics defined by vector fields, I: Basic properties. *Acta Math.* **155**, 103–147 (1985).

[NN] A. Newlander and L. Nirenberg, Complex analytic coordinates in almost complex manifolds. *Ann. Math.* **65**, 391–404 (1957).

[Nir] L. Nirenberg, On a question of Hans Lewy. *Russian Math. Surveys* **29**, 251–262 (1974).

[P] S. I. Pincuk, On the analytic continuation of holomorphic mappings. *Mat. Sb.* **98**, 416–435 (1975); or *Math. USSR Sb.* **27**, 375–392 (1975).

[PW] J. C. Polking and R. O. Wells Jr., Boundary values of Dolbeault cohomology classes and a generalized Bochner–Hartogs theorem. *Abhandlungen aus dem Mathematischen Seminar der Universitat Hamburg* **43**, 1–24 (1978).

[Ra] E. Ramirez de Arellano, Ein divisionproblem und randintegraldarstellungen in der komplexen analysis. *Math. Ann.* **184**, 172–187 (1970).

[Ro] J.-P. Rosay, A propos de "wedges" et d' "edges" et de prolongments holomorphes. *Trans. Am. Math. Soc.* **297**, 63–72 (1986).

[Ran] M. Range, *Holomorphic Functions and Integral Representations in Several Complex Variables*, Graduate Texts in Mathematics, Vol. 108. Springer-Verlag, NY, 1986.

[Sch] L. Schwartz, *Theorie des Distributions*, Vols. I and II. Hermann, Paris, 1950, 1951.

[Sh1] M-C. Shaw, Hypoellipticity of a system of complex vector fields. *Duke Math. J.* **50**, 713–728 (1983).

[Sh2] M-C. Shaw, $\mathcal{L}^p$ estimates for local solutions of $\overline{\partial}_b$ on strongly pseudoconvex CR manifolds. *Math. Ann.* **288**, 35–62 (1990).

[Sj] J. Sjöstrand, Singularities analytiques microlocales. *Soc. Math. Fr. Asterisque* **95**, 1–166 (1982).

[Sp] M. Spivak, *A Comprehensive Introduction to Differential Geometry*. Publish or Perish, Inc., Berkeley, CA, 1970.

[St] E. M. Stein, *Boundary Behavior of Holomorphic Functions of Several Complex Variables*. Princeton University Press, Princeton, NJ, 1972.

[Tai] G. Taiani, *Cauchy–Riemann Manifolds*. Monograph from Pace University Math. Dept., New York, 1989.

[Tom] G. Tomassini, Tracce delle functional olomorfe sulle sotto varieta analitiche reali d'una varieta complessa. *Ann. Scuola Norm. Sup. Pisa* **20**, 31–43 (1966).

[Tr] J. M. Trepreau, Sur le prolongement holomorphe des fonctions CR definis sur une hypersurface reelle de classe C^2 dans $\mathbb{C}^n$. *Invent. Math.* **83**, 583–592 (1986).

[T] A. E. Tumanov, Extending CR functions on manifolds of finite type to a wedge. *Mat. Sbornik* **136**, 128–139 (1988).

[W1] R. O. Wells Jr., Compact real submanifolds of a complex manifold with nondegenerate holomorphic tangent bundles. *Math. Ann.* **179**, 123–129 (1969).

[W2] R. O. Wells Jr. *Differential Analysis on Complex Manifolds*. Prentice-Hall, Englewood Cliffs, NJ, 1973.

[Y] K. Yosida, *Functional Analysis*. Springer-Verlag, NY, 1980.

Notation

The following is a partial list of notation. The page number refers to the page where the notation is first encountered or defined.

$\mathcal{E}(\Omega)$ the space of smooth functions on Ω — p. 2.
$\mathcal{D}(\Omega)$ the space of compactly supported smooth functions on Ω — p. 2.
$\mathcal{A}(\Omega)$ the space of real analytic functions on Ω — p. 4.
$\mathcal{O}(\Omega)$ the space of holomorphic functions on Ω — p. 6.
C^α the space of Hölder continuous functions with exponent α — p. 215.
$\|\cdot\|_\alpha$ the norm on the space C^α — p. 216.
$T(X)$ the tangent bundle to the manifold X — p. 24.
$T^{\mathbb{C}}(M)$ the complexified tangent bundle to M — p. 40.
$H(M)$ the complex tangent bundle to a CR manifold M — p. 97.
$H^{\mathbb{C}}(M)$ the complexification of $H(M)$ — p. 101.
$X(M)$ the totally real tangent bundle to a CR manifold M — p. 98.
$T^*(X)$ the cotangent bundle of X — p. 26.
$T^{*^{\mathbb{C}}}(M)$ the complexified cotangent bundle of M — p. 40.
$\Lambda^r T^*(M)$ the bundle of r-forms on M — p. 27.
$\Lambda^r T^{*^{\mathbb{C}}}(X)$ the bundle of complexified r-forms on X — p. 40.
J a complex structure map — pp. 7, 41.
J^* the induced complex structure map on forms — p. 42.
$V^{1,0}$ (resp. $V^{0,1}$) the $+i$ (resp. $-i$) eigenspace of J on the vector space V — p. 43.
$\Lambda^{p,q}T^*(M)$ the bundle of (p,q)-forms on M — p. 46.
Λ^s_M see p. 123.
$\mathcal{E}^r(M)$ the space of smooth r-forms on M — p. 27.
$\mathcal{D}^r(M)$ the space of smooth r-forms with compact support on M — p. 79.
$\mathcal{E}^{p,q}(M)$ the space of smooth (p,q)-forms on a complex manifold M — p. 46.
$\mathcal{D}^{p,q}(M)$ the space of elements of $\mathcal{E}^{p,q}(M)$ with compact support — p. 49.
$\mathcal{E}^{p,q}_M$ the space of smooth (p,q)-forms on a CR manifold M — p. 124.
$\mathcal{D}^{p,q}_M$ the space of elements of $\mathcal{E}^{p,q}_M$ with compact support — p. 124.
$\mathcal{E}^s_M$ the space of smooth sections of Λ^s_M — p. 124.

$\mathcal{D}^s_M$ the space of smooth sections of Λ^s_M with compact support — pp. 124, 128.
$\mathcal{D}'(\Omega)$ the space of distributions on Ω — p. 62.
$\mathcal{E}'(\Omega)$ the space of distributions with compact support on Ω — p. 62.
$\mathcal{D}'^r$ the space of currents of degree r on Ω — p. 81.
$\mathcal{E}'^r(\Omega)$ the space of currents of degree r with compact support on Ω — p. 81.
$\mathcal{D}'^{p,q}(\Omega)$ the space of currents of bidegree (p,q) on Ω — p. 83.
$\mathcal{D}'^{p,q}_M$ the space of currents of bidegree (p,q) on a CR manifold M — p. 129.
F_* the push forward map via F defined on vectors (or currents) — p. 8.
F^* the pull back map via F defined on forms (or currents) — p. 11.
DF the matrix which represents the derivative of F — p. 7.
d or d_M the exterior derivative operator — p. 28.
d^* the $\mathcal{L}^2$-adjoint of the exterior derivative on $\mathbb{R}^N$ — p. 16.
$\bar{\partial}$ the Cauchy–Riemann operator — p. 47.
$\bar{\partial}^*$ the $\mathcal{L}^2$-adjoint of $\bar{\partial}$ on $\mathbb{C}^N$ — p. 49.
$\bar{\partial}_M$ the tangential Cauchy–Riemann operator (either extrinsic or intrinsic) — p. 124.
f_{t_M} the tangential piece of the form f along M — p. 124.
$d\sigma$ a volume form — p. 36.
$d\mu_M$ surface measure on M — p. 63.
$\langle\ ,\ \rangle_p$ pairing between forms and vectors at the point p — p. 9.
$(\ ,\ )_\Omega$ pairing between distributions and functions on Ω — p. 62.
$\langle\ ,\rangle_\Omega$ pairing between currents and forms on Ω — p. 80.
$\mathcal{L}_p$ the intrinsic Levi form at p — p. 156.
$\tilde{\mathcal{L}}_p$ the extrinsic Levi form at p — p. 160.
Γ_p the convex hull of the image of $\tilde{\mathcal{L}}_p$ — p. 200.
$\Gamma_1 < \Gamma_2$ the cone Γ_1 is smaller than the cone Γ_2 — p. 200.
Γ^o the polar of the cone Γ — p. 239.
T the Hilbert transform in Part III — p. 214.
$\hat{I}f$ the Fourier transform of f — p. 231.
If the modified Fourier transform of f — p. 232.
WF the hypoanalytic wave front set — p. 238.
$[M]$ the current given by integration over the manifold M — p. 81.
$[\Delta]$ the current given by integration over the diaognal — p. 83.
K' the adjoint of the kernel K — p. 267.
K^t the transpose of the kernel K — p. 269.
E_I or $E(u^{i_1},\dots,u^{i_p})$ a particular type of kernel defined on p. 295.
B the Bochner–Martinelli kernel — p. 284.
L the Leray kernel — p. 301.
H the Henkin kernel — p. 301.
R the Romanov kernel — p. 318.
$\mathcal{O}(\rho, y_1, w')^j$ see p. 307.
$\approx |(\rho, y_1, w')|^j$ see p. 307.

Index

Airapetyan, A., 353
almost complex structure, 58
analytic disc, 206–228
  boundary of, 206
  theorem, 207, 221–228
Andreotti, A., 169, 350, 353
approximation
  by entire functions, 191–197
  by polynomials, 5

Baeouendi, M. S., 110, 145, 183, 184, 191, 229, 256–258, 260–262
bidegree
  of a current, 83
  of a differential form, 46
bidimension (of a current), 83, 134
Bishop, E., 214
Bishop's equation, 215, 220
Bloom, T., 105, 181
Bochner, S., 264, 284, 292
Bochner–Martinelli kernel, 284, 285, 300
  boundary values, 286
  as a fundamental solution for $\bar{\partial}$, 285
  jump formula, 286
  principal value limits of, 342
Bochner's extension theorem, 292
Boggess, A., 200, 258, 260, 261, 350
boundary point, 23, 32

Caley transform, 337
canonical local basis, 109
Cauchy kernel, 216, 217
  on a slice, 282
Cauchy–Riemann operator, 47
Chang, C. H., 229, 256
change of variables formula for integrals, 30, 33
Cirka, E. M., 342
complex, 134, 135, 263
complex structure
  on a more general vector space, 41
  on $\mathbb{C}^n$, 7
complex tangent space, 97
complexification (of a vector space), 39
complexified tangent bundle, 40
conjugation, 39
contraction (of a form by a vector), 15, 49
convolution
  of distributions, 66
  of functions, 3
coordinate
  chart, 17
  patch, 17
CR codimension, 98, 121
  distribution, 140, 255, 256
  equivalence, 154
  function, 140
  manifold, 121
  map, 149–155
  structure (same as CR manifold), 121
  submanifold, 99
current

definition of, 80
operations with, 84–94
cutoff function, 3

D'Angelo, J. P., 180
degree
of a current, 81
of a form, 12
delta function, 62, 68
diagonal, 83, 266
differentiable function (on a manifold), 18
differential form
on a manifold, 27, 46
on $\mathbb{R}^N$, 9, 12
dimension (of a current), 80, 81
distribution
definition, 62
operations with, 65–70
support of, 63
divergence theorem, 36
Dwilewicz, R., 258

extension theorems for CR functions, 141, 147, 198–202, 239–250, 292, 315, 333
exterior derivative of a current, 85
exterior derivative of a form
on a manifold, 28, 41, 47
on $\mathbb{R}^N$, 9, 12

Federer, H., 1
finite type, 181
Folland, G. B., 1, 57, 264, 335, 350
formal identity (for Henkin's kernels), 297
Fourier
inversion formula, 235, 251–254
transform (modified), 231, 232, 253
Fréchet space, 62
Fredricks, G., 353
Freeman, M., 186
Frobenius theorem
analytic version, 56–58
smooth version, 52, 159, 186
fundamental solution
for $\bar{\partial}$, 274, 281–285
for a partial differential operator, 75
for the Cauchy–Riemann operator on $\mathbb{C}$, 75
for the exterior derivative, 274, 278–280
for the Laplacian, 76
for the tangential Cauchy–Riemann operator, 314, 318

generic, 102, 187
Graham, I., 105, 181
graph of a function, 20, 267
Grauert, H., 323
Green's formula, 37

Harris, G. A., 188
Harvey, R., 264, 286, 343, 349
Heisenberg group, 113, 336
Henkin, G. M., 264, 294, 304, 315, 318, 327. 338, 345, 348, 353
Henkin's kernels, 295
as a fundamental solution for $\bar{\partial}_M$, 314
boundary values of, 304
definition of, 301
Hilbert transform, 214, 217
Hill, C. D., 169, 187, 260, 350
Hölder continuous, 215
holomorphic
function, 6, 19
map, 7
tangent space (same as complex tangent space), 97
homotopy formula, 272
Hörmander, L., 1, 59, 181, 210, 332, 345, 347
Hörmander number, 181, 257, 258
hypersurface, 20
hypoanalytic wave front set, 238, 253
hypoellipticity, 260

identity theorem, 6, 142, 148
imbeddability of CR manifolds, 121, 169–178
inner product, 44
Euclidean, 9
$\mathcal{L}^2$, 16, 49
integral curve, 54
integration of differential forms, 30–32, 41
involutive, 52, 101

Jacobowitz, H., 145, 172, 256, 261, 262
John, F., 1 , 57

kernel
  adjoint of, 267
  definition of, 266
  of convolution type, 270
  on a ray, 278
  transpose of, 269
Krantz, S., 1, 316
Kuranishi, M., 187

Lawson, B., 343
Leray kernel, 301
Levi flat, 158
Levi form
  coordinate formula for, 160, 162, 198
  definition of, 156, 160
Levi null set, 185
Lewy, H., 198, 261, 264, 334
Lie algebra
  bracket, 14, 179
  group, 112, 183
  subalgebra, 183
Lieb, I., 323
local defining system, 20

Maire, H.-M., 262
manifold
  complex, 18
  CR, 121
  point, 23, 32
  real analytic, 17
  smooth, 17
minimal, 257
moment condition, 293

Nacinovich, M., 353
Nagel, A., 258, 259
Newlander, A., 59, 121
Newlander–Nirenberg theorem, 59, 121, 159, 186
  abstract version, 59
  imbedded version, 59
Nirenberg, L., 59, 121, 165, 169, 172
Nirenberg's nonimbeddable example, 172–178
normal form, 105, 109
  Bloom–Graham, 181
  for codimension 2 quadrics, 114

orientation, 29
  induced, 30
  preserving map, 29
oriented, 29
  coordinate chart, 30
  manifold, 30

partition of unity, 3
Pincuk, S. I., 261
Pitts, J., 260
Poisson's integral formula, 216
polar (of a cone), 239
Polking, J. C., 130, 264, 286, 343, 349
polynomial growth, 256
projective space, 18
pseudoconvex, 165
  strictly, 164, 187, 346
pull back
  of a current, 91, 92
  of a form, 11, 27, 48
pure term, 106,
push forward
  of a current, 87
  of a vector, 8, 25

quadratic form, 111
quadric submanifold, 111, 210

Ramirez de Arellano, E., 294
Range, M., 346
real analytic
  function, 4, 19
  manifold, 17
rigid, 110, 229
Romanov kernel, 318
Rothschild, L. P., 110, 183, 184, 229, 256–258, 262
Rosay, J.-P., 262
Rossi, H., i

Schwartz, L., 1
second fundamental form, 166
semirigid, 184
Shaw, M.-C., 262, 348, 350
Sjöstrand, J., 229
solvability
  of $\bar{\partial}$ (globally), 315, 316, 318
  of $\bar{\partial}$ (locally), 327, 338
  of $\bar{\partial}$ on a strictly convex domain, 303
spherical kernel, 278–280

Spivak, M., 1
Stein, E. M., 258, 259
Stokes' theorem, 33
subbundle, 24, 40, 57
submanifold, 19
  with boundary, 22

Taiani, G., 187, 260
tangent
  bundle, 24
  space, 24
tangential Cauchy–Riemann complex, 122–139
  equivalence of extrinsic and intrinsic definitions, 135
  extrinsic definition, 124
  intrinsic definition, 131
tangential vector field, 304
Tomassini, G., 141
totally real, 98, 99
transversal Lie Group action, 183
Trepreau, J. M., 258
Treves, F., 110, 145, 172, 183, 191, 229, 256, 260–262
tube like, 261
Tumanov, A. E., 257
type
  of a kernel, 266, 271
  of a point, 179, 257

uniqueness theorems, 141, 142, 198–199, 200–201, 202

vector, 7, 23
  bundle, 25
  field, 8, 24, 40
volume form, 36

Wainger, S., 259
weak topology, 64, 68
wedge, 196
wedge product
  of currents, 84
  of forms, 10
Weierstrass theorem, 5, 191, 192
weight, 180
Wells, R. O. Jr., 25, 130, 188
Whitney extension theorem, 71, 72

Yosida, K., 1
Y(q) (condition), 350